Surface Analysis – The Principal Techniques
2nd Edition

Surface Analysis – The Principal Techniques

2nd Edition

Editors

JOHN C. VICKERMAN

Manchester Interdisciplinary Biocentre,
University of Manchester, UK

IAN S. GILMORE

National Physical Laboratory,
Teddington, UK

A John Wiley and Sons, Ltd., Publication

Library of Congress Cataloging-in-Publication Data

Surface analysis : the principal techniques /c [edited by] John Vickerman, Ian Gilmore. –2nd ed. p. cm.
  Includes bibliographical references and index.
  ISBN 978-0-470-01763-0
  1. Surfaces (Technology)–Analysis. 2.   Spectrum analysis. I. Vickerman, J. C. II. Gilmore, Ian.
  TA418.7.S726 2009
  620′.44–dc22

                          2008040278

A catalogue record for this book is available from the British Library

HB 978-0470-017630,   PB 978-0470-017647

Typeset in 10.5/12.5 Palatino by Laserwords Private Limited, Chennai, India

Contents

Contributors

David G. Castner *National ESCA & Surface Analysis Center for Biomedical Problems, Departments of Chemical Engineering and Bioengineering, University of Washington, Seattle, WA 98195-1750, USA*

Mark Dowsett *Department of Physics, University of Warwick, Coventry, CV4 7AL, UK*

Peter Gardner *Manchester Interdisciplinary Biocentre, School of Chemical Engineering and Analytical Science, The University of Manchester, Manchester, M1 7DN, UK*

Ian S. Gilmore *Surface and Nanoanalysis, Quality of Life Division, National Physical Laboratory, Teddington, Middlesex, TW11 0LW, UK*

Joanna L. S. Lee *Surface and Nanoanalysis, Quality of Life Division, National Physical Laboratory, Teddington, Middlesex, TW11 0LW, UK*

Graham J. Leggett *Department of Chemistry, University of Sheffield, Brook Hill, Sheffield S3 7HF, UK*

Christopher A. Lucas *Department of Physics, University of Liverpool, Liverpool, L69 3BX, UK*

Hans Jörg Mathieu *EPFL, Départment des Matériaux, CH-1015, Lausanne, Switzerland*

David McPhail *Department of Materials, Imperial College, Prince Consort Road, London, SW7 2AZ*

Martyn E. Pemble *Tyndall National Institute, 'Lee Maltings', Prospect Row, Cork, Ireland*

Buddy D. Ratner *Department of Bioengineering, University of Washington, Seattle, WA 98195-1720, USA*

Edmund Taglauer *Max-Planck-Institut für Plasmaphysik, EURATOM Association, D-8046, Garching bei München, Germany*

John C. Vickerman *Surface Analysis Research Centre, Manchester Interdisciplinary Biocentre, School of Chemical Engineering and Analytical Science, The University of Manchester, Manchester, M1 7DN, UK*

Preface

In today's world there are vast areas of high innovation technologies which benefit strongly from the application of surface analysis techniques in research, manufacture and quality control. Examples cover the gamut of industry sectors with strong growth in the use of surface analysis for nanotechnologies, biotechnologies, nanoparticle characterization, lightweight materials, energy efficient systems and energy storage. Over the years an enormous number of techniques have been developed to probe different aspects of the physics, chemistry and biology of surfaces. Some of these techniques have found wide application in basic surface science and applied surface analysis and have become very powerful and popular techniques. This book seeks to introduce the reader to the principal techniques used in these fields together with the computational methods used to interpret the increasingly complex data generated by them. Each chapter has been written by experts in the field. The coverage includes the basic theory and practice of each technique together with practical examples of its use and application and most chapters are followed by some review questions to enable the reader to develop and test their understanding. The aim has been to give a thorough grounding without being too detailed.

Chapter 1 introduces the concept of 'the surface' and the challenges implicit in distinguishing the composition of the surface of materials from the rest of the material. In *Chapter 2* Professor Hans Jörg Mathieu from Ecole Polytechnique, Lausanne, introduces perhaps the oldest widely used technique of surface analysis – Auger Electron Spectroscopy (AES). This technique has been exploited extensively and extremely effectively in Lausanne for metal and alloy analysis.

Electron Spectroscopy for Surface Analysis (ESCA) or X-ray Photoelectron Spectroscopy (XPS) is probably the most widely used surface analysis technique. It has been extremely effective for the solution of an enormous number of problems in both basic surface science and in applied analysis. Professors Buddy Ratner and Dave Castner from Washington State University have exploited the technique very successfully for polymer and biomaterials analysis and they introduce this technique in *Chapter 3*.

Secondary ion mass spectrometry (SIMS), introduced in *Chapter 4* by Professor John Vickerman, is a very powerful technique because of the mass spectral nature of the data. The group in Manchester have contributed particularly to the development of SIMS for molecular surface analysis and in addition to its application to inorganic materials analysis they have shown that it can be exploited very effectively to investigate the complexities of biological systems.

SIMS has also been very effectively and widely used in its so-called dynamic form to characterize the elemental composition of electronic materials. Professor Mark Dowsett from the University of Warwick and Dr David McPhail of Imperial College, London, provide an insight into the challenges and capabilities of the technique in *Chapter 5*.

Low energy ion scattering (LEIS) and Rutherford backscattering (RBS) are powerful for probing the elemental composition and structure of surfaces. Professor Edmund Taglauer from the Max Planck Institute in Garching is a widely recognized authority on these elegant techniques which are introduced in *Chapter 6*.

Vibrational spectroscopy is very widely used in chemistry for compound identification and analysis. There are now many variants which can be applied to the study of surfaces and particularly of molecules on surfaces. Professor Martyn Pemble of the Tyndall National Institute, Cork and Dr Peter Gardner of The University of Manchester, have been involved in the development of several of the techniques and they exploit them in research associated with the growth of electronic materials and in understanding biological processes. They discuss a number of these variants in *Chapter 7*.

In *Chapter 8* Dr Chris Lucas of the Department of Physics, The University of Liverpool, introduces techniques which use diffraction and other interference based methods for the analysis of surface structure. Low energy electron diffraction (LEED) has been an important technique in basic surface science for many years; however, more recently extended X-ray absorption fine structure (EXAFS) and the related techniques which probe local short range surface structure have become extremely valuable and are used extensively in many areas of materials characterization.

Surface studies have been significantly advanced by the scanning probe techniques – scanning tunnelling microscopy (STM) and atomic force microscopy (AFM). The impressive images with atomic resolution of metal surfaces have excited many surface analysts. The extension of the capabilities to bio-organic materials has resulted in considerable insights into the surface behaviour of these materials. Professor Graham Leggett, who is exploiting these techniques to study bio-organic surfaces at the University of Sheffield describes the theory and practice of these techniques in *Chapter 9*.

As the capabilities of the analytical techniques have advanced and the materials to be characterized have become ever more complex the need for computational methods to help interpret the multivariate character of the data has become a vital component of the analytical process for many of the techniques. In *Chapter 10* Joanna Lee and Dr Ian Gilmore of the National Physical Laboratory, introduce the main methods of multivariate data analysis as applied to surface analysis.

Two appendices have been provided. Since most, though not all, surface analysis techniques are carried out in vacuum based equipment, *Appendix 1* provided by Dr Rod Wilson, briefly describes the main features of the vacuum technology used in surface analysis. *Appendix 2* provides a listing of the main units, constants and conversions that require to be used in surface analysis.

Most surface problems, be they in basic surface science or applied surface analysis, require careful selection of the most appropriate technique to answer the questions posed. Frequently more than one technique will be required. It is anticipated that readers of this book will be equipped to make the judgements required. Thus the book should be of value to those who need to have a wide overview of the techniques in education or in industrial quality control or R&D laboratories. For those who wish to further develop their knowledge and practice of particular techniques, it should also give a good basic understanding from which to build.

John C. Vickerman

Manchester, UK

Ian S. Gilmore

Teddington, UK

1 Introduction

JOHN C. VICKERMAN

Manchester Interdisciplinary Biocentre, School of Chemical Engineering and Analytical Science, The University of Manchester, Manchester, UK

The surface behaviour of materials is crucial to our lives. The obvious problems of *corrosion* are overcome by special surface treatments. The *optical behaviour* of glass can be modified by *surface coatings* or by changing the surface composition. The surface chemistry of polymers can be tuned so that they cling for packaging, are non-stick for cooking or can be implanted into our bodies to feed in drugs or replace body components. The *auto-exhaust catalyst* which removes some of the worst output of the combustion engine is a masterpiece of surface chemistry as are the industrial catalysts which are vital for about 90 % of the output of the chemical industry. Thus whether one considers a car body shell, a biological cell, tissue or implant, a catalyst, a solid state electronic device or a moving component in an engine, it is the surface which interfaces with its environment. The surface reactivity will determine how well the material behaves in its intended function. It is therefore vital that the surface properties and behaviour of materials used in our modern world are thoroughly understood. Techniques are required which enable us to analyse the surface chemical and physical state and clearly distinguish it from that of the underlying solid.

1.1 How do we Define the Surface?

It is obvious that the surface properties of solids are influenced to a large extent by the solid state properties of the material. The question arises as

Surface Analysis – The Principal Techniques 2nd Edition Edited by John Vickerman and Ian Gilmore
© 2009 John Wiley & Sons, Ltd

to how we define the surface. Since the top layer of surface atoms are those that are the immediate interface with the other phases (gas, liquid or solid) impinging on it, this could be regarded as the surface. However, the structure and chemistry of that top layer of atoms or molecules will be significantly determined by the atoms or molecules immediately below. In a very real sense therefore, the surface could be said to be the top 2–10 atomic or molecular layers (say, 0.5–3 nm). However, many technologies apply surface films to devices and components, to protect, lubricate, change the surface optical properties, etc. These films are in the range 10–100 nm or sometimes even thicker, but the surface may be thought of in this depth range. However, beyond 100 nm it is more appropriate to begin to describe such a layer in terms of its bulk solid state properties. Thus we can consider the surface in terms of three regimes: the top surface monolayer, the first ten or so layers and the surface film, no greater than 100 nm. To understand fully the surface of a solid material, we need techniques that not only distinguish the surface from the bulk of the solid, but also ones that distinguish the properties of these three regimes.

1.2 How Many Atoms in a Surface?

It will be appreciated that it is not straightforward to probe a surface layer of atoms or molecules analytically and distinguish their structure and properties from that of the rest of the solid. One has only to consider the relatively small number of atoms involved in the surface layer(s) of an atomic solid to see that high sensitivity is required. How many atoms are we dealing with at the surface and in the bulk of a solid? We can consider a 1 cm cube of metal. One of the 1 cm^2 surfaces has roughly 10^{15} atoms in the surface layer. Thus the total number of atoms in the cube will be $\approx 10^{23}$. Therefore the percentage of surface to bulk atoms will be:

$$s/b \approx 10^{-8} \times 100 \approx 10^{-6} \%$$

Typically, a surface analysis technique may be able to probe in the region of 1 mm^2. Thus in the top monolayer there will be about 10^{13} atoms. In the top ten layers there will be 10^{14} atoms or 10^{-10} mol. Clearly in comparison with conventional chemical analysis we are considering very low concentrations. Things become more demanding when we remember that frequently the chemical species which play an important role in influencing surface reactivity may be present in very low concentration, so the requirement will be to analyse an additive or contaminant at the 10^{-3} or even 10^{-6} (ppm) atomic level, i.e. 10^{10} or 10^7 atoms or 10^{-14} or 10^{-17} mole levels respectively, perhaps even less.

Similar demands arise if the analysis has to be carried out with high spatial resolution. The requirement to map variations in chemistry across a surface can arise in a wide variety of technologies. There may be a need to monitor the homogeneity of an optical or a protective coating or the distribution of catalyst components across a support, a contaminant on an electronic device or a drug in a cell or tissue, etc. It is not unusual for $1\,\mu m$ spatial resolution to be demanded, frequently even less would be beneficial. If we continue the discussion above in an area of $1\,\mu m^2$ ($10^{-12}\,m^2$ or 10^{-8} cm^2) there are only $\approx 10^7$ atoms, so if we want to analyse to the 10^{-3} atom fraction level, there are only 10^4 atoms. The nano particles that are part of many technologies these days present far fewer atoms for analysis, making the surface analysis task even more demanding.

Thus surface analysis is demanding in terms of its surface resolution and sensitivity requirements. However, there are in fact many surface analysis techniques, all characterized by distinguishing acronyms – LEED, XPS, AES, SIMS, STM, etc. Most were developed in the course of fundamental studies of surface phenomena on single crystal surface planes. Such studies which comprise the research field known as *surface science* seek to provide an understanding of surface processes at the atomic and molecular level. Thus for example, in the area of catalysis, there has been an enormous research effort directed towards understanding the role of surface atomic structure, composition, electronic state, etc. on the adsorption and surface reactivity of reactant molecules at the surface of the catalyst. To simplify and systematically control the variables involved, much of the research has focused on single crystal surfaces of catalytically important metals and more recently inorganic oxides. The surface analysis techniques developed in the course of these and related research are, in the main, based on bombarding the surface to be studied with electrons, photons or ions and detecting the emitted electrons, photons or ions.

1.3 Information Required

To understand the properties and reactivity of a surface, the following information is required: the physical topography, the chemical composition, the chemical structure, the atomic structure, the electronic state and a detailed description of bonding of molecules at the surface. No one technique can provide all these different pieces of information. A full investigation of a surface phenomenon will always require several techniques. To solve particular problems it is seldom necessary to have all these different aspects covered; however, it is almost always true that understanding is greatly advanced by applying more than one technique to a surface study. This book does not attempt to cover all the techniques in existence. A recent count identified over 50! The techniques introduced here are those (excluding electron microscopy which is not covered but for which there are numerous introductions) that have made the most significant impact in both fundamental *and* applied surface analysis. They are tabulated (via their acronyms) in Table 1.1 according to the principal information they provide

Table 1.1   Surface analysis techniques and the information they can provide

Radiation IN	photon	photon	electron	ion	neutron
Radiation DETECTED	electron	photon	electron	ion	neutron

SURFACE INFORMATION					
Physical topography			SEM STM (9)		
Chemical composition	ESCA/XPS (3)		AES (2)	SIMS (5) ISS (6)	
Chemical structure	ESCA/XPS (3)	EXAFS (8) IR & SFG (7)	EELS (7)	SIMS (4)	INS (7)
Atomic structure		EXAFS (8)	LEED RHEED (8)	ISS (6)	
Adsorbate bonding		EXAFS (8) IR (7)	EELS (7)	SIMS (4)	INS (7)

ESCA/XPS – Electron analysis for chemical analysis/X-ray photoelectron spectroscopy. X-ray photons of precisely defined energy bombard the surface, electrons are emitted from the orbitals of the component atoms, electron kinetic energies are measured and their electron binding energies can be determined enabling the component atoms to be determined.

AES – Auger electron spectroscopy. Basically very similar to the above except that a keV electron beam may be used to bombard the surface.

SIMS – Secondary ion mass spectrometry. There are two forms, i.e. dynamic and molecular SIMS. In both a beam of high energy (keV) *primary* ions bombard the surface while secondary atomic and cluster ions are emitted and analysed with a mass spectrometer.

ISS – Ion scattering spectrometry. An ion beam bombards the surface and is scattered from the atoms in the surface. The scattering angles and energies are measured and used to compute the composition and surface structure of the sample target.

IR – Infrared (spectroscopy). Various variants on the classical methods – irradiate with infrared photons which excite vibrational frequencies in the surface layers; photon energy losses are detected to generate spectra.

EELS – Electron energy loss spectroscopy. Low energy (few eV) electrons bombard the surface and excite vibrations – the resultant energy loss is detected and related to the vibrations excited.

INS – Inelastic neutron scattering. Bombard a surface with neutrons – energy loss occurs due to the excitation of vibrations. It is most efficient in bonds containing hydrogen.

SFG – Sum frequency generation. Two photons irradiate and interact with an interface (solid/gas or solid liquid) such that a single photon merges resulting in electronic or vibrational information about the interface region.

LEED – Low energy electron diffraction. A beam of low energy (tens of eV) electrons bombard a surface; the electrons are diffracted by the surface structure enabling the structure to be deduced.

RHEED – Reflection high energy electron diffraction. A high energy beam (keV) of electrons is directed at a surface at glancing incidence. The angles of electron scattering can be related to the surface atomic structure.

EXAFS – Extended X-ray absorption fine structure. The fine structure of the absorption spectrum resulting from X-ray irradiation of the sample is analysed to obtain information on local chemical and electronic structure.

STM – Scanning tunnelling microscopy. A sharp tip is scanned over a conducting surface at a very small distance above the surface. The electron current flowing between the surface and the tip is monitored; physical and electron density maps of the surface can be generated with high spatial resolution.

AFM – Atomic force microscopy (not included in table). Similar to STM but applicable to non-conducting surfaces. The forces developed between the surface and the tip are monitored. A topographical map of the surface is generated.

and the probe/detection system they use. The number after each technique indicates the chapter in which it is described.

It is a characteristic of most techniques of surface analysis that they are carried out in vacuum. This is because electrons and ions are scattered by molecules in the gas phase. While photon based techniques can in principle operate in the ambient, sometimes gas phase absorption of photons can occur and as a consequence these may also require vacuum operation. This imposes a restriction on some of the surface processes that can be studied. For example, to study the surface gas or liquid interface it will usually be necessary to use a photon based technique, or one of the scanning probe techniques. Developments since the turn of the century are enabling the analysis of surfaces under ambient atmospheres using mass spectral methods analogous to SIMS (Chapter 4).

However, the vacuum based methods allow one to control the influence of the ambient on the surface under study. To analyse a surface uncontaminated by any adsorbate it is necessary to operate in ultra-high vacuum ($<10^{-9}$ mmHg) since at 10^{-6} mmHg a surface can be covered by one monolayer of adsorbed species within 1 s if the sticking coefficient (probability for adsorption) is 1. Controlled exposure of the surface to adsorbates or other surface treatments can then be carried out to monitor effects in a controlled manner. Appendix 1 on 'Vacuum Technology' will enable the reader to become familiar with the concepts and equipment requirements in the generation of vacua.

1.4 Surface Sensitivity

To generate the information, we require that a surface analysis technique should derive its data as near exclusively as possible from within the depth range discussed in Section 1.2. The extent to which a technique does this is a measure of its surface sensitivity. Ion scattering spectrometry (ISS) derives almost all its information from the top monolayer. It is very surface sensitive. Electron Spectroscopy for Chemical Analysis (ESCA) or X-ray Photoelectron Spectroscopy (XPS) samples the top ten or so layers of the surface, while infrared (IR) spectroscopy is not very surface sensitive and will sample deep into the solid, unless it is used as a reflection mode.

In general the surface sensitivity of an analytical method is dependent on the radiation *detected*. As already indicated, most of the methods of surface analysis involve bombarding the surface with a form of radiation – electrons, photons, ions, neutrons – and then collecting the resulting emitted radiation – electrons, photons, ions, neutrons. The scanning probe methods are a little different, although one could say that scanning tunnelling microscopy (STM) detects electrons. (Atomic force microscopy monitors the forces between the surface and a sharp tip, see

Chapter 9.) The surface sensitivity depends on the depth of origin of the detected species. Thus in XPS while the X-ray photons which bombard the surface can penetrate deep into the solid, the resultant emitted electrons which can be detected without loss of energy can only arise from within 1–4 or 8 nm of the surface. Electrons generated deeper in the solid may escape, but on the way out they will have collided with other atoms and lost energy. They are no use for analysis. Thus the surface sensitivity of ESCA is a consequence of the short distance electrons can travel in solids without being scattered (known as the *inelastic mean free path*). Similarly, in secondary ion mass spectrometry (SIMS) the surface is bombarded by high energy ions. They deposit their energy down to 30 or 40 nm. However, 95 % of the secondary ions that are knocked out (sputtered) of the solid arise from the top two layers.

There are techniques like infrared (IR) spectroscopy which, although they are not intrinsically very surface sensitive, can be made so by the methods used to apply them. Thus with IR a reflection approach can be used in which the incoming radiation is brought in at a glancing incidence. This enables vibrational spectra to be generated from adsorbates on single crystal surfaces. The technique is very surface sensitive. Surface sensitivity can be significantly increased even in surface sensitive methods like ESCA by irradiating the surface at glancing incidence – see Chapter 3.

Various terms are used to define surface sensitivity. With all the techniques described in this book the total signal detected will originate over a range of depths from the surface. An *information depth* may be specified which is usually defined as the average distance (in nm) normal to the surface from which a specified percentage (frequently 90, 95 or 99 %) of the detected signal originates. Sometimes, as in ESCA, a *sampling depth*, is defined. This is three times the *inelastic mean free path*, and turns out to be the information depth where the percentage is 95 %. Obviously a very small proportion of the detected signal does arise from deeper in the solid, but the vast majority of the useful analytical information arises from within the sampling depth region.

In molecular SIMS the information depth is the depth from which 95 % of the secondary ions originate. For most materials this is believed to be about two atomic layers, about 0.6 nm. However, it is sometimes difficult to be sure what a *layer* is. For example, there are surface layers used to generate new optical properties that are composed of long organic chains bonded to metal or oxide surfaces. The organic layer is much less dense than the substrate underneath. SIMS studies of these materials suggest that the analytical process may remove the whole molecular chain which can easily be >20 nm long. Surface sensitivity in this case is a very different concept from that which would apply to the surface of a metal or inorganic compound.

1.5 Radiation Effects – Surface Damage

To obtain the surface information required entails 'interfering' with the surface state in some way! Most of the techniques require the surface to be bombarded with photons, electrons or ions. They will affect the chemical and physical state of the surface being analysed. Thus in the course of analysing the surface, the surface may be changed. It is important to understand the extent to which this may happen, otherwise the information being generated from the surface may not be characteristic of the surface before analysis; rather it may reflect a surface damaged by the incident radiation.

Table 1.2 shows the penetration depth and influence of the 1000 eV particles. It can be seen that most of the energy is deposited in the near surface under ion and electron bombardment, so in general terms it would be expected that the extent of surface damage would vary as photons < electrons < ions. Consequently, it is sometimes carelessly suggested that ESCA/XPS is a low damage technique. However, the power input to the surface in the course of an experiment is considerably less in the ion bombardment method of SIMS compared to photon bombardment in ESCA (Table 1.3). SIMS is very obviously a phenomenon that depends on damage – ions bombard to knock out other ions! Without damage there is no information, but as will be seen in Chapter 4 it can be operated in a low damage mode to generate significant surface information. The X-ray photons which bombard the surface in XPS penetrate deep into the solid. However, if the material is delicate, e.g. a polymer, and if the power input is too high or the time under the beam too long, the sample can be literally 'fried'. The same effect is even more obvious for the methods involving

Table 1.2 Penetration depths of particles

Particle	Energy (eV)	Depths (Å)
Photon	1000	10 000
Electron	1000	20
Ions	1000	10

Table 1.3 Comparison of typical primary particle flux densities and energies and the resulting power dissipated in SSIMS, LEED and X-ray photoelectron experiments

	Primary flux (cm^{-2})	Primary energy	Power (W cm^{-2})
SIMS	10^{10} ions s^{-1}	3 keV	3×10^{-6}
LEED	10^{15} electrons s^{-1}	50 eV	5×10^{-3}
XPS	10^{14} photons s^{-1}	1.4 keV	2×10^{-2}

electron irradiation. It is consequently very difficult to analyse the surfaces of organic materials using any technique which relies on electron bombardment.

1.6 Complexity of the Data

The advance in the capability of surface analysis techniques has been enormous since the publication of the first edition of this book. The information content of many of them has escalated. The complexity of materials that they are now expected to characterize has also increased. As a consequence it is sometimes difficult to understand the data using the simple analysis routines employed when the techniques were in their infancy. SIMS is a good case in point. The spectra in molecular SIMS can be so complex that it is impossible to discern by a 'stare and compare' approach the important chemical differences between, say, what is supposed to be a 'good' sample and a 'bad' sample. This type of problem has become particularly acute as surface analysis techniques have begun to be applied to biological systems. Multiple factors may influence the spectral differences. To deal with this type of problem many analysts and researchers have turned to computational methods of multivariate analysis (MVA) that seek to isolate the crucial differences between the spectra of differing materials or treatments. MVA methods are introduced and discussed in a new Chapter 10.

Surface analysis techniques have been enormously successful in developing our understanding of surface phenomena. There are vast numbers of areas of technology which would benefit from the application of surface analysis techniques in both research and development and in quality control. Frequently these techniques are not being applied because of a lack of knowledge and understanding of how they can help. Hopefully, this book will help to develop increased awareness such that surface analysis will be increasingly applied to further our understanding of the surface states at both the fundamental and applied levels.

None of the techniques are analytical 'black boxes' delivering answers to problems at the push of a button. Two general rules should be remembered in surface analysis: (a) in every case it is important to understand the capabilities and limitations of the technique being used with regard to the material being studied and the information required; (b) no one technique gives the whole story.

2 Auger Electron Spectroscopy

HANS JÖRG MATHIEU
EPFL, Lausanne, Switzerland

2.1 Introduction

Auger Electron Spectroscopy (AES) represents today the most important chemical surface analysis tool for conducting samples. The method is based on the excitation of so-called 'Auger electrons'. Already in 1923 Pierre Auger [1] had described the β emission of electrons due to ionization of a gas under bombardment by X-rays. This ionization process can be provoked either by electrons – commonly known as the Auger process – or by photons as used by P. Auger. In the latter case we will call this method photon induced Auger Electron Spectroscopy (see also the chapter on ESCA/XPS). Today's AES is based on the use of primary electrons with typical energies between 3 and 30 keV and the possibility to focus and scan the primary electron beam in the nanometer and micrometer range analyzing the top-most atomic layers of matter. The emitted Auger electrons are part of the secondary electron spectrum obtained under electron bombardment with a characteristic energy allowing one to identify the emitting elements. The experimental setup is very similar to that of a Scanning Electron Microscope – with the difference that the electrons are not only used for imaging but also for chemical identification of the surface atoms.

Auger electrons render information essentially on the elemental composition of the first 2–10 atomic layers. Figure 2.1 shows schematically the distribution of electrons, i.e. primary, backscattered and Auger electrons together with the emitted characteristic X-rays under electron bombardment. We notice that under typical experimental conditions the

Surface Analysis – The Principal Techniques 2nd Edition Edited by John Vickerman and Ian Gilmore
© 2009 John Wiley & Sons, Ltd

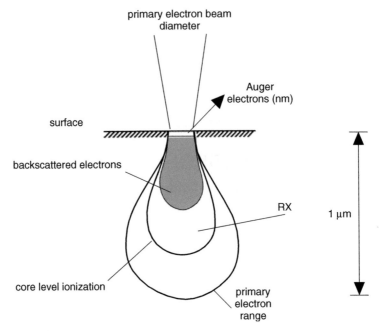

Figure 2.1 Distribution (schematic) of primary, backscattered and Auger electrons together with X-rays. Note the schematic here is for a broad electron beam focus of approximately a micrometer. Here, this is approximately the same diameter of the area of emitted backscattered electrons. Often, the electron beam focus is very much higher with a diameter in the nm range. See Figure 5.31 of Briggs and Seah [2] for more details

latter have a larger escape depth due to a much smaller ionization cross section with matter, i.e. a higher probability to escape matter. Auger electrons with energies up to 2000 eV, however, have a high probability to escape only from the first few monolayers because of their short attenuation length. Consequently, they are much better suited for surface analysis. A second important detail is shown in Figure 2.1 revealing that the diameter of the analyzed zone can be larger than the diameter of the primary beam due to scattering of electrons.

2.2 Principle of the AUGER Process

Before determining the kinetic energies of Auger electrons let us have a quick look at quantum numbers and nomenclature. A given energy state is characterized by four quantum numbers, i.e. n (principal quantum number), l (orbital), s (spin) and j (spin-orbit coupling with $j = l + s$). The latter can only have values with J always positive. The energy E (nlj) of a given electronic state can therefore be characterized by these three numbers as indicated in Table 2.1 for certain elements.

Table 2.1 Nomenclature of AES and XPS peaks

n	l	j	Index	AES notation	XPS notation
1	0	1/2	1	K	$1s_{1/2}$
2	0	1/2	1	L_1	$2s_{1/2}$
2	1	1/2	2	L_2	$2p_{1/2}$
2	1	3/2	3	L_3	$2p_{3/2}$
3	0	1/2	1	M_1	$3s_{1/2}$
3	1	1/2	2	M_2	$3p_{1/2}$
3	1	3/2	3	M_3	$3p_{3/2}$
3	2	3/2	4	M_4	$3d_{3/2}$
3	2	5/2	5	M_5	$3d_{5/2}$
etc.			etc.	etc.	etc.

2.2.1 KINETIC ENERGIES OF AUGER PEAKS

Figure 2.2 shows schematically the Auger process. The primary beam energy has to be sufficiently high to ionize a core level W (i.e. K, L, . . .) with energy E_W. The empty electron position will be filled by an electron from a level E_X closer to the Fermi level. The transition of the electron between levels W

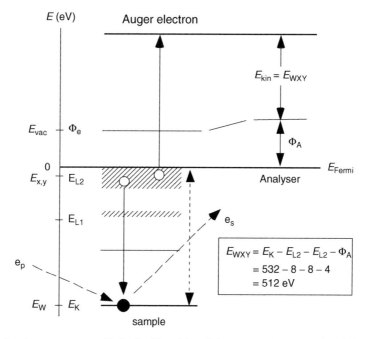

Figure 2.2 Auger process: E_F is the Fermi level (zero atomic energy level for binding energies of electrons) while Φ_e and Φ_A are the work functions of the sample (e) and analyzer (A), respectively

and X liberates an energy corresponding to $\Delta E = E_W - E_X$ which in turn is transferred to a third electron of the same atom at level E_Y. The kinetic energy of this third electron corresponds therefore to the difference of energy between the three electronic levels involved minus the sample work function, Φ_e. If the analyzer is in good contact with the sample holder (i.e. Fermi levels of sample and instrument are identical) we can determine the kinetic energy of an element with atomic number Z and an Auger transition between level W, X and Y as follows:

$$E_{WXY} = E_W(Z) - E_X(Z + \Delta) - E_Y(Z + \Delta) - \Phi_A \qquad (2.1)$$

where Φ_A represents the work function of the analyzer and W, X and Y the three energy levels of the Auger process involved (i.e. KLL, LMM, MNN – omitting to note the sub-levels). The term Δ (between 0 and 1) denotes the displacement of an electronic level towards higher binding energies after the ionization of the atom by the primary electron. $\Delta = 0.5$ represents a fair approximation for an estimate of the kinetic energy. The work function of the analyzer detector is typically 4 eV. Taking such values (see Table 2.2) we obtain for the transition of oxygen O_{KLL} an energy $E_{KLL} = 512$ eV, as indicated by Figure 2.2.

Figure 2.3 shows schematically an Auger spectrum in which the number of emitted electrons N is given as a function of the kinetic energy E.

We observe that the Auger peaks are superimposed on the spectrum of the secondary electrons. The elastic peak E_p represents the primary energy applied. We further notice on the tail of the elastic peak characteristic loss peaks from the ionization levels (E_W, E_X, etc.) and on the low kinetic energy side of the Auger peaks tails which are due to characteristic energy losses. The characteristic losses are used for quantification in scanning electron microscopy. The method is called Electron Energy Loss Spectroscopy

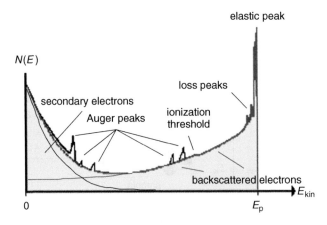

Figure 2.3   Schematic representation of an Auger spectrum

Table 2.2 Binding energies of some elements

Z	El	$1s_{1/2}$ K	$2s_{1/2}$ L$_1$	$2p_{1/2}$ L$_2$	$2p_{3/2}$ L$_3$	$3s_{1/2}$ M$_1$	$3p_{1/2}$ M$_2$	$3p_{1/2}$ M$_3$	$3d_{3/2}$ M$_4$	$3d_{5/2}$ M$_5$
1	H	14								
2	He	25								
3	Li	55								
4	Be	111								
5	B	188			5					
6	C	284			6					
7	N	399			8					
8	O	532	24		8					
9	F	686	31		9					
10	Ne	867	45		18					
11	Na	1072	63		31	1				
12	Mg	1305	89		52	2				
13	Ml	1560	118	74	73	1				
14	Si	1839	149	100	99	8				
15	P	2149	189	136	135	16		10		
16	S	2472	229	165	164	16		8		
17	Cl	2823	270	202	200	18		7		
18	Ar	3202	320	247	245	25		12		
19	K	3608	377	297	294	34		18		
20	Ca	4038	438	350	347	44		26	5	
21	Sc	4493	500	407	402	54		32	7	
22	Ti	4965	564	461	455	59		34	3	
23	V	5465	628	520	513	66		38	2	
24	Cr	5989	695	584	757	74		43	2	
25	Mn	6539	769	652	641	84		49	4	
26	Fe	7114	846	723	710	95		56	6	
27	Co	7709	926	794	779	101		60	3	
28	Ni	8333	1008	872	855	112		68	4	
29	Cu	8979	1096	951	932	120		74	2	
30	Zn	9659	1194	1044	1021	137		90	9	
31	Ga	10367	1299	1144	1117	160		106	20	
42	Mo	20000	2866	2625	2520	505	410	393	208	205
46	Pd	24350	36304	3330	3173	670	559	531	340	335
48	Ag	25514	3806	3523	3351	718	602	571	373	367
73	Ta[a]	67416	11681	11136	11544	566[a]	464[a]	403[a]	24[a]	22[a]
79	AuTa[a]	80724	14352	13733	14208	763[a]	643[a]	547[a]	88[a]	84[a]

[a]4s, 4p and 4f levels indicated, respectively.

(EELS). Auger transitions have been calculated and can be found in the literature . Figure 2.4 gives the principal Auger transitions of all elements starting from Li. Since for an Auger transition a minimum of three electrons is required, only elements with $Z \geq 3$ can be analyzed. Table 2.3 gives numerical values of the principal transitions together with other useful parameters in AES.

Table 2.3 AES transitions and their relative sensitivity factors

Legend:

```
                    Atomic Number
                 Z A  ─── Atomic Volume A/ρ* [ x 10⁻⁶ m³ / mol ]
                 S(5)
                 S(10)    ─── Element
                 LMM KE   ─── Kinetic Energy of AES transition (eV)

AES rel. sensitivity at 5 keV**
AES rel. sensitivity at 10 keV**
AES transition
```

* atomic weight A – mass density ρ
** valid for CMA only

Z	Element	Atomic Volume	S(5)	S(10)	Transition	KE
1	H	14.1				
2	He	31.8				
3	Li	13.1	0.160		KLL	43
4	Be	5.0	0.10	0.045	KLL	104
5	B	4.6	0.120	0.055	KLL	179
6	C	5.3	0.14	0.08	KLL	272
7	N	17.3	0.230	0.160	KLL	379
8	O	14	0.400	0.350	KLL	508
9	F	17.1	0.48	0.45	KLL	647
10	Ne	16.8			KLL	805
11	Na	23.7	0.25	0.23	KLL	990
12	Mg	14.0	0.13	0.13	KLL	1186
13	Al	10.0	0.19	0.15	LMM	68
14	Si	12.1	0.28	0.15	LMM	92
15	P	17.0	0.47	0.30	LMM	120
16	S	15.5	0.75	0.57	LMM	152
17	Cl	18.7	1.05	0.69	LMM	181
18	Ar	24.2			LMM	215
19	K	45.3	0.90	0.37	KLL	252
20	Ca	29.9	0.40	0.22	LMM	291
21	Sc	15.0	0.28	0.20	LMM	340
22	Ti	10.6	0.34	0.23	LMM	418
23	V	8.35	0.38	0.29	LMM	473
24	Cr	7.23	0.31	0.28	LMM	529
25	Mn	7.39	0.193	0.160	LMM	589
26	Fe	7.1	0.22	0.15	LMM	703
27	Co	6.7	0.23	0.19	LMM	775
28	Ni	6.6	0.27	0.22	LMM	848
29	Cu	7.1	0.23	0.20	LMM	920
30	Zn	9.2	0.19	0.18	LMM	994
31	Ga	11.8	0.16	0.14	LMM	1070
32	Ge	13.6	0.130	0.125	LMM	1147
33	As	13.1	0.12	0.11	LMM	1228
34	Se	16.5	0.092	0.088	LMM	1315
35	Br	23.5	0.075	0.074	LMM	1376
36	Kr	32.2			MNN	53
37	Rb	55.9	0.052	0.053	LMM	1565
38	Sr	33.7	0.043	0.045	LMM	1649
39	Y	19.8	0.11	0.01	MNN	127
40	Zr	14.1	0.16	0.15	MNN	147
41	Nb	10.8	0.21	0.18	MNN	167
42	Mo	9.4	0.28	0.28	MNN	186
43	Tc					
44	Ru	8.3	0.50	0.37	MNN	273
45	Rh	8.3	0.68	0.47	MNN	302
46	Pd	8.9	0.89	0.60	MNN	330
47	Ag	10.3	0.97	0.67	MNN	356
48	Cd	13.1	0.99	0.68	MNN	376
49	In	15.7	0.97	0.65	MNN	404
50	Sn	16.3	0.90	0.53	MNN	430
51	Sb	18.4	0.65	0.40	MNN	454
52	Te	20.5	0.47	0.28	MNN	483
53	I	25.7	0.34	0.21	MNN	511
54	Xe	42.9	0.24	0.15	MNN	532
55	Cs	70	0.17	0.12	MNN	563
56	Ba	39	0.12	0.08	MNN	584
57	La	22.5	0.88	0.60	MNN	625
58	Ce	21.0	0.068	0.045	MNN	661
59	Pr	20.8	0.055	0.038	MNN	699
60	Nd	20.6	0.047	0.032	MNN	730
61	Pm					
62	Sm	19.9	0.033	0.026	MNN	814
63	Eu	28.9	0.029	0.025	MNN	858
64	Gd	19.9	0.027	0.024	MNN	895
65	Tb	19.2	0.026	0.025	MNN	1073
66	Dy	19.0	0.027	0.027	MNN	1126
67	Ho	18.7	0.030	0.030	MNN	1175
68	Er	18.4	0.036	0.035	MNN	1393
69	Tm	18.1	0.042	0.040	MNN	1449
70	Yb	24.8	0.051	0.048	MNN	1514
71	Lu	17.8	0.062	0.058	MNN	1573
72	Hf	13.6	0.141		NNN	185
73	Ta	10.9	0.136	0.093	NNN	179
74	W	9.53	0.115	0.079	NNN	179
75	Re	8.85	0.096		NNN	176
76	Os	8.43	0.046		NOO	176
77	Ir	8.54	0.046		NOO	54
78	Pt	9.10	0.28		NOO	64
79	Au	10.2	0.34	0.21	NOO	69
80	Hg	14.8	0.030		NOO	76
81	Tl	17.2	0.42		NOO	84
82	Pb	18.3	0.40		NOO	94
83	Bi	21.3	0.37		NOO	101
84	Po	22.7				
85	At					
86	Rn					
87	Fr					
88	Ra	45				
89	Ac					
90	Th	19.9	0.286		OPP	65
91	Pa	15.0				
92	U	12.5	0.437 (3 keV)		OPP	72

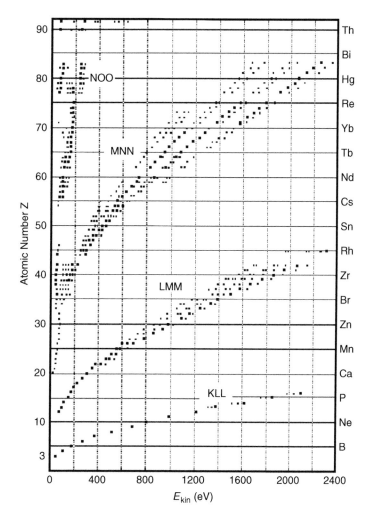

Figure 2.4 Principal transitions of AES. From Physical Electronics, Minnesota, USA

2.2.2 IONIZATION CROSS-SECTION

The probability of an Auger transition is determined by the probability of the ionization of the core level W and its de-excitation process involving the emission of an Auger electron or a photon. Primary electrons with a given energy E arriving at the surface will ionize the atoms starting at the surface of the sample. The cross-section, $\sigma_W(E)$, calculated by quantum mechanics for the Auger process at an energy core level W, can be estimated by:

$$\sigma_W = \text{const} \times \frac{C(E_p/E_W)}{E_W^2} \tag{2.2}$$

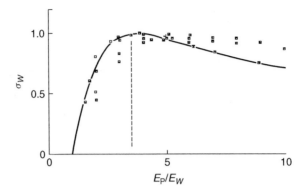

Figure 2.5 Variation of the ionization cross-section with the ratio of primary electron beam energy E_P and core level energy E_W

where the constant depends on the core level $W (= K, L, M)$; σ_W is a function of the primary energy E_P and the core level E_W. Figure 2.5 shows experimental results together with the calculated σ_W according to Equation (2.2) as a function of the ratio E_P/E_W. One observes that the ionization cross-section passes through a maximum at approximatively $E_P/E_W = 3$. Typical absolute values for σ_W are 10^{-3}–10^{-4}. This means that the probability of an ionization followed by an Auger de-excitation is 1 in 10^4. Thus one finds experimentally Auger electron transitions superimposed on a high secondary electron spectrum, as indicated in Figure 2.3.

2.2.3 COMPARISON OF AUGER AND PHOTON EMISSION

Figure 2.2 indicated schematically the Auger process. We have already learned that after ionization of the core level W the de-excitation takes place by an electron filling the place at level W. The liberated energy difference $\Delta E = E_W - E_X$ can either be transferred to an electron of the same atom or

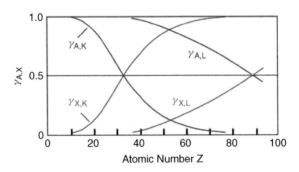

Figure 2.6 Emission probability of an Auger electron (Λ) or photon (X)

a photon with same energy $\Delta E = h\nu$. Again, whether an Auger electron or a photon is emitted is determined by quantum mechanical selection rules. The emission probability varies with the atomic number Z and the type of atomic level involved (K, L, M, etc.) leading to cross-sections γ_{AK} and γ_{XK} or γ_{AL} and γ_{XL} for detection via emission of an Auger electron (A) or a photon (X-ray (X)), respectively, as indicated by Figure 2.6. Probability of excitation via an Auger process is very high for light elements and transition of the KLL type, γ_{AK}. However, even for heavy elements one observes a relatively high probability for elements of type LMM (γ_{AL}) or MNN (γ_{AM} – not shown).

2.2.4 ELECTRON BACKSCATTERING

In Auger electron spectroscopy, primary electrons arrive at the sample surface with an energy of 3–30 keV. Monte Carlo calculations indicate that such electrons can penetrate up to a depth of several microns (compare with Figure 2.1). During their trajectory, these electrons lose a certain amount of energy, change their direction and are also backscattered. They may create secondary electrons, Auger electrons and photons. Some of the backscattered electrons can in turn produce themselves Auger electrons if they have sufficient energy. This way the backscattered electrons contribute to the total Auger current. Since the number of Auger electrons is proportional to the total Auger current one obtains:

$$I_{total} = I_o + I_M = I_o(1 + r_M) \qquad (2.3a)$$

Figure 2.7 shows the backscattering factor, r_M, calculated for various matrices with atomic number Z. One notices that r_M becomes larger for increasing Z, i.e. elements with more free electrons like gold ($Z = 79$), produce more backscattered electrons.

The backscattering factor can be estimated by the following equation:

$$1 + r_M = 1 + 2.8\left[1 - 0.9\frac{E_W}{E_p}\right]\eta(Z) \qquad (2.3b)$$

where E_W is the ionization energy of the core level and E_p the primary beam energy with:

$$\eta(Z) = -0.0254 + 0.16Z - 0.00186Z^2 + 8.3 \times 10^{-7}Z^3 \qquad (2.3c)$$

Inspection of Figure 2.7 indicates the importance of the variation of r_M for Auger analysis, especially for very thin films on a substrate that produces a large number of backscattered electrons.

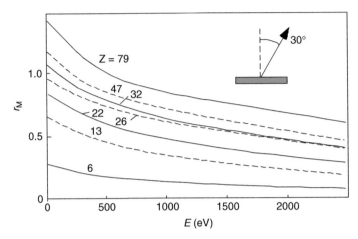

Figure 2.7   Electron backscattering factor r_M as a function of kinetic energy energy for a primary electron energy of 5 keV and an angle of incidence of $\theta = 30°$ (reprinted from [2], with permission from John Wiley & Sons, Ltd.)

2.2.5 ESCAPE DEPTH

The attenuation length of Auger electrons λ with a kinetic energy E_{kin} determines the escape depth Λ according to:

$$\Lambda = \lambda \cos \theta \qquad (2.4a)$$

where θ is the emission angle of the Auger electrons with respect to surface normal. The probability for an electron to travel over a distance x without any collision is proportional to $\exp(-x/\Lambda)$ with 95 % of the Auger intensity coming from within 3Λ of the surface. A rough estimate of λ is obtained for the elements by [2]:

$$\lambda = 0.41a^{1.5}E_{kin}^{0.5} \qquad (2.4b)$$

where a (in nm) is the monolayer thickness of a cubic crystal calculated by:

$$\rho N a^3 = A \qquad (2.4c)$$

with ρ the density (in kg/m^3), N the Avogadro constant ($N = 6.023 \times 10^{23}/$ mol), a (in m) and A the molecular mass (kg/mol) of the matrix in which the Auger electron is created. The ratio A/ρ (atomic volume) is given in Table 2.3. Figure 2.8 shows λ as a function of the kinetic energy. It reveals that λ varies from 2 to 20 monolayers for typical kinetic energies up to 2000 eV. The thickness of a monolayer is approximatively 0.2–0.25 nm for metals. Since the kinetic energy determines the escape depth, a measurement of two peaks of the same element but of different energy can be used as a measure for the variation of composition with depth. The effective attenuation

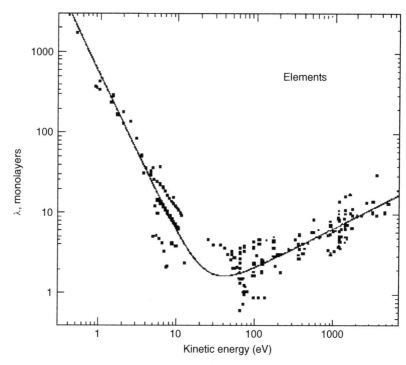

Figure 2.8 Dependence of attenuation length, λ, on kinetic energy (reprinted from [2], with permission from John Wiley & Sons, Ltd.)

length (EAL) for applications in Auger electron spectroscopy and X-ray photoelectron spectroscopy is defined for a measurement of an overlayer film thickness and the measurement of the depth of a thin marker layer by Auger electron spectroscopy and X-ray photoelectron spectroscopy. Results can be found in the literature, e.g. Powell and Jablonski [6].

2.2.6 CHEMICAL SHIFTS

A change of the oxidation state of an element results in a shift of the binding energy of the valence band level. Therefore, in principle, each time a change of a binding energy occurs, one observes also a 'chemical shift' for Auger transitions. The same phenomenon is found in ESCA. However, since three energy levels are involved in an Auger transition such shifts cannot always easily be correlated to a shift of one particular level. A fine structure of Auger peaks of certain elements is well known (i.e. C, Si, Al, etc.) allowing the experimentalist to distinguish between different states of oxidation, as indicated in Figure 2.9. Figure 2.10 illustrates an example of the variation of the levels of the different peaks of aluminum and indicates schematically differences of the density of electrons $\rho(E)$ of the M-level. The discrete levels

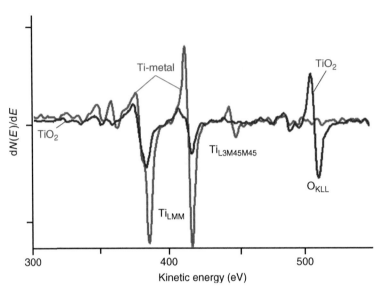

Figure 2.9   Examples of differentiated AES spectra of Ti metal and TiO$_2$ versus kinetic energy

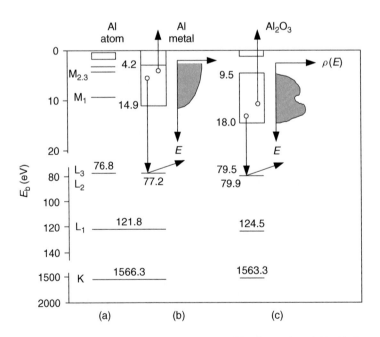

Figure 2.10   Energy levels of (a) Al atom, (b) Al metal and (c) Al$_2$O$_3$

of the Al atom are found as bands in the metal or in the oxide. One notices a shift in E_b from 14.9 eV to 18.0 eV when going from Al metal to Al_2O_3. Some examples are given in David [3].

2.3 Instrumentation

The main parts of an Auger spectrometer are the electron gun and the electrostatic energy analyzer. Both are placed in an ultra-high vacuum chamber with base pressures between 10^{-9} and 10^{-8} mbar. Such low pressures are necessary to guarantee a contamination-free surface to keep adsorption of residual gases below 10^{-3} monolayers/second. This is achieved for pressures of 10^{-9} mbar or below. Essential accessories of spectrometers are vacuum gauges for total pressure reading, a partial pressure analyzer controlling the rest gas, a fast introduction lock and a differentially pumped ion gun for sample cleaning or in-depth thin film analysis, together with a secondary electron collector for imaging. Figure 2.11 shows an example of a simple Auger spectrometer using a cylindrical mirror analyzer (CMA) with a variable potential applied between an inner and outer cylinder and resulting in a signal which is proportional to the number of detected electrons N at kinetic energy E. The other type of analyzer used in AES, i.e. an hemispherical analyzer (HPA), is shown in Figure 2.12, is often used in XPS

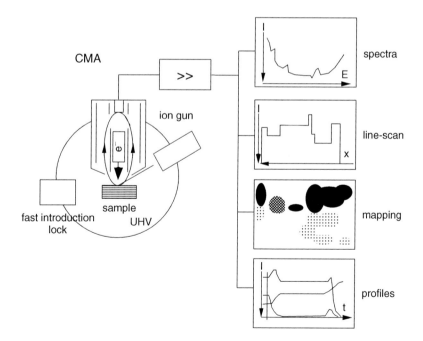

Figure 2.11   Cylindrical mirror analyzer

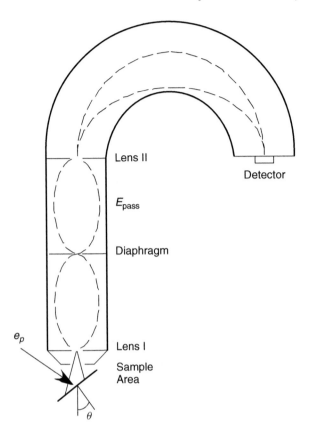

Figure 2.12   Hemispherical analyzer

analysis. In general, HPAs give a better energy resolution. The electron beam can be static (static AES) or scanned (Scanning Auger Microprobe (SAM)). The lateral resolution depends on the electron optics applied (electrostatic or electromagnetic lenses). Achievable lateral resolution of spectrometers in the Auger detection mode of 10 nm can be achieved.

2.3.1 ELECTRON SOURCES

Today's scanning Auger systems use three types of electron source with decreasing lateral resolution: (a) tungsten filament, (b) LaB$_6$ crystal or (c) a field emission gun (FEG). The classical W filament reaches a minimum beam diameter of 3–5 µm. Only LaB$_6$ or FEG sources give beam diameters ≤20 nm and their primary electron beam energy has to be increased to 20–30 keV. Lowest beam diameters for a given primary beam current are obtained by a field emission gun (see Figure 2.13) which in turn is more delicate and demands a better control of the vacuum.

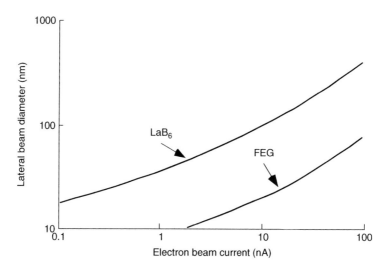

Figure 2.13 Comparison of electron sources (LaB_6 and field emission gun (FEG))

The two types of electron sources, thermionic [a,b] and field emission [c] are based on rather different physical principles. The former more common ones apply a certain thermal energy to remove an electron from the source. This energy is called the work function, which represents the barrier at the material surface necessary to free the electron. Typical work function energies are around 4–5 eV. For thermionic sources the material is heated by passing a certain current to obtain a sufficiently high temperature to allow the electrons to reach the vacuum. Field emission is based on the 'tunnelling' process of electrons which is probable, if a sufficiently high electrical field between the emitter and an extraction electrode is applied. Sharp needle-like points of typically 20–50 nm radii and short distances between emitter and extraction electrode (nm) are needed. In the end the ultimate limit of the lateral resolution is determined by the focussing lenses. Purely electrostatic electron guns allow focussing to 0.2 μm, whereas electromagnetic focussing allows one to decrease spot sizes down to 0.02 μm for LaB_6 or tungsten field emitters, respectively. Such field emitters are used in scanning electron microscopes as well. However, focussing of the electron beam may lead to beam damage, particularly for sample areas of low conductivity. To avoid beam damage, beam current densities above 1 mA cm^{-2} corresponding to 1 nA into a spot of 10 μm should be applied. Unfortunately, such limits cannot always be met, particularly in high lateral resolution work, leading in certain cases to local sample decomposition.

For beam currents ≥10 nA the LaB_6 source is superior to both thermionic and field emitting tungsten in terms of spot size obtainable and signal-to-noise ratio. However, for a better lateral resolution at beam currents below 1 nA, the field emitter is preferred.

2.3.2 SPECTROMETERS

As mentioned already, two types of analyzers are used in AES, either a CMA or an HPA. The CMA (Figure 2.11) has a larger electron transmission than the HPA (Figure 2.12). The transmission is defined as the ratio of emitted to detected Auger electrons. A scanning electron gun is built coaxially into the CMA avoiding, in many cases, shadowing effects since the analyzer and electron gun axis are identical. The CMA derives its name from the fact that the electron emitting spot on the sample surface is imaged by the CMA at the detector surface. Primary electrons of known energy which are reflected from the sample surface are used to optimize the signal intensity to find the analyzed spot and calibrate the analyzer.

In HPAs, the primary electron is off-axis allowing a simpler geometry and a better definition of the angle of emitted electrons (compare Figure 2.12). The working distance between sample and analyzer is generally larger for HPAs (approx. 10 mm). At the entrance of the analyzer a system of electrostatic lenses is placed to define the accepted analysis area. In the cylindrical part of the analyzer a diaphragm limits the analyzed area and a second electrostatic lens controls the pass energy of the electrons. A potential is applied to this second lens to reduce the kinetic energy of the Auger electrons allowing one to operate the analyzer at constant pass energy. The hemispherical part of the analyzer focuses the electrons in the plane of a detector which is an arrangement of different channeltron or channelplate electron multipliers. The detecting system measures directly the number of electrons at a certain kinetic energy $N(E)$. Such an analyzer can be used in two detection modes:

1. $\Delta E = $ constant (FAT = fixed Analyzer Transmission mode) applying a constant pass energy by controlling lens II (compare with Figure 2.12).

2. $\Delta E/E = $ constant (FRR Fixed Relative Resolution) applying a constant energy ratio where ΔE is the FWHM (Full Width at Half Maximum) of a given peak and E its kinetic energy.

Electron detectors lead to a signal-to-noise ratio (800:1 for Cu LMM line) and allow one to decrease the detection limit (1 % of a monolayer) at a given spatial resolution (0.1 µm) and fixed primary beam current (10 nA). For better radiation shielding each analyzer is made out of stainless steel and/or completely of mμ-metal.

2.3.3 MODES OF ACQUISITION

There are four modes of operation in Auger electron spectroscopy:

1. Point analysis.
2. Line scan.

3. Mapping.

4. Profiling.

   Figure 2.14 shows a typical survey spectrum of tungsten as a result of a point analysis indicating the number of detected electrons $N(E)/E$ as a function of kinetic energy (compare also Figure 2.3). One observes a large number of secondary electrons on which the Auger electrons are superimposed. Transitions of oxygen, carbon and tungsten have been labelled. Carbon is found as a surface contamination, often observed on samples introduced into the UHV. Peaks can be represented in their differentiated form after background subtraction, i.e. $dN(E)/dE$. Figure 2.15 illustrates different chemical states of titanium. However, as indicated already above, identification of oxidation states is generally easier in ESCA because of the involvement of three energy levels in the Auger process. The advantage of AES is the small spot size and the shorter acquisition time of the measurement for conducting samples.

   As we noticed already above, the escape depth of Auger electrons is limited to a few nm. Many practical problems require determination of the variation of an element with depth. Modern AES systems are equipped to perform different types of depth analysis as illustrated by Figure 2.16. It is evident that the principles shown in Figure 2.16 apply also to other methods like XPS or SIMS. For layers of thicknesses of a few nm one measures the detected intensity as a function of the angle θ making use of Equation (2.4a) and illustrated by Figure 2.17. Such angular resolved analysis (AREAS) is limited to a very shallow depth because λ is typically only a few nm as discussed above (Section 2.2.5)

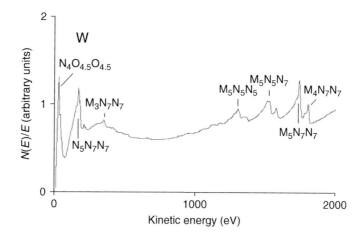

Figure 2.14   AES survey spectrum of W as a result of a point analysis

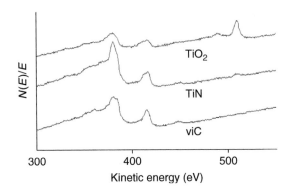

Figure 2.15   Normalized AES spectra of TiO_2, TiN and TiC versus kinetic energy

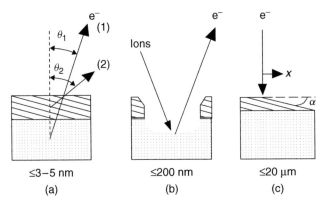

Figure 2.16   Principle of different types of in-depth measurements: (a) non-destructive measurement for layers ≤ 3–$5\,nm$ by variation of the angle of emission; (b) for layers $\leq 200\,nm$ by combining AES analysis with destructive sputter erosion; (c) line scan over a creater edge produced by ball cratering or taper sectioning under a small angle applied for layers $\leq 20\mu m$

Composition of thicker layers up to 0.2–1 µm can be determined by combining Auger analysis with Ar^+ (or Kr^+) sputtering by observation of the Auger signal at the bottom of the sputtered crater, either simultaneously or alternately. Sputter depth profiling will be discussed in Section 2.5 below in more detail.

Layers of even larger thickness (i.e. a few µm) should be analyzed by other methods, i.e. electron microprobe analysis or – if light elements are to be detected – by scanning the electron beam across a mechanically prepared ball crater or a tapered section. Figure 2.18 gives an example of the line scan mode: a section of stainless steel is covered with a layer of TiN as shown in Figure 2.19. The crater has been prepared by mechanical abrasion of the TiN layer by a stainless steel sphere – for more details see the ISO technical

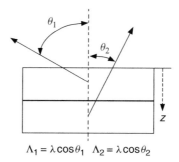

$$\Lambda_1 = \lambda \cos\theta_1 \quad \Lambda_2 = \lambda \cos\theta_2$$

Figure 2.17 Variation of escape depth with angle of emission

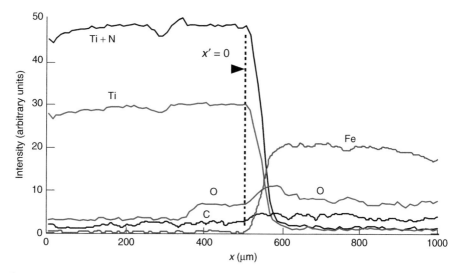

Figure 2.18 Example of a line scan over the crater edge produced by ball cratering showing the atomic concentration as a function of the displacement of the electron beam. The crater edge is located at approximately $x = 500\ \mu m$

report ISO/TR 15969 (2001). The electron beam is scanned from left to right over the TiN layer over the crater edge before reaching the substrate. The displacement x can be correlated to the thickness z of the layer by Equation (2.5), R is the radius of the sphere used during polishing and D the diameter of the crater produced at the surface with $R \gg D$ according to ISO/TR 15969 (2001).

$$z = \frac{D^2/4 - \left(D/2 - x'\right)^2}{2R} \tag{2.5}$$

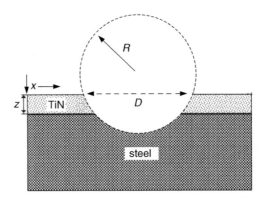

Figure 2.19 Section of a stainless steel sample covered with a layer of TiN of thickness z. The vertical arrows indicate the limits of displacement of the electron beam where R is the radius of the sphere used during polishing and D the diameter of the crater produced at the surface

Figure 2.20 AES mapping – scanning Auger micrographs of an Sn–Nb superconductor multi-wire: (a,b) SEM, (c) Sn and (d) Nb elemental distributions

An application of scanning Auger analysis is illustrated by Figure 2.20 which shows an Auger map of an Sn–Nb multi-wire alloy used as a superconductor in magnets: (a,b) SEM, (c) Sn and (d) Nb elemental distributions. It illustrates that SEM micrographs as well as scanning Auger micrographs can be obtained with high lateral resolution.

Figure 2.20 shows the elemental distribution of Sn and Nb. In this mode the electron beam is scanned over a selected area of the sample. The Auger intensity is measured at each point of the area by keeping the analyzer pass energy constant at the peak maximum and minimum of an elemental peak, respectively. The image displayed shows the peak intensity (maximum minus minimum) of each pixel.

2.3.4 DETECTION LIMITS

Identification of Auger peaks is often easier for light elements than for heavier elements because of the interference of peaks of heavier elements with a larger number of transitions. Peaks with higher kinetic energy have a larger width (FWHM is typically 3–10 eV) and therefore peak overlap is more likely. The sensitivity of elements varies only by one order of magnitude, where silver is the most sensitive and yttrium one of the least sensitive elements. The detection limits are set by the signal to noise ratio. Typical limits are:

concentration:	0.1–1 % of a monolayer
mass (volume of $1\,\mu m \times 1\,\mu m$ of $1\,nm$ thickness):	$10^{-16}-10^{-15}$ g
atoms:	$10^{12}-10^{13}$ atoms/cm^2

Scanning Auger analysis allows one to decrease the area of detection below the micron level. However, a finely focussed electron beam may provoke a change in composition if the power dissipated into a small area is too large. One should avoid exceeding a limit of 10^4 W/cm^2. In addition, the detection limit is drastically lower in the mapping mode as illustrated by Figure 2.21 for a pure Cu sample. Inspection reveals that a static measurement with a lateral resolution of 50 nm gives a detection limit between 0.1 and 0.01 of a monolayer depending on the kind of electron source used. As already mentioned earlier, the field emission gun has a higher brightness and therefore a better detection limit. However, for mapping, the detection limit deteriorates by approximatively a factor of 100 compared to the point analysis, because of the shorter acquisition time per pixel.

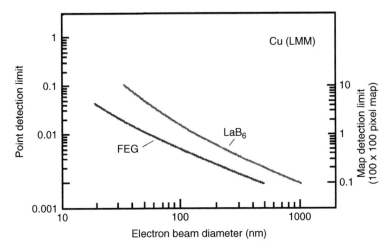

Figure 2.21 Detection limit (monolayer) as a function of lateral resolution for point analysis (left ordinate) and mapping (right ordinate) for the most common electron sources used in AES, i.e. LaB_6 crystal sources or field emission (FEG)

2.3.5 INSTRUMENT CALIBRATION

For meaningful measurements it is important to ensure the instrument is calibrated. For example, to identify chemical constituents from the peak energy correctly, the energy scale needs to be calibrated. To provide quantitative information, use sensitivity factors and compare with other instruments, while the intensity scale needs to be linear and also corrected for the intensity response function. The intensity response function (IRF) includes the angular acceptance of the analyzer to electrons, the transmission efficiency of electrons and the detection efficiency. Fortunately, the underpinning metrology for AES is highly developed and procedures for calibration have been developed under the ISO (International Standards Organization). Calibration of the IRF is provided in some manufacturers' software or from the NPL (http://www.npl.co.uk/server.php?show = ConWebDoc.606). The most relevant ISO standards are listed below:

- ISO 17973 – Medium resolution AES – calibration of energy scales for elemental analysis.

- ISO 17974 – High resolution AES – calibration of energy scales for elemental and chemical state analysis.

- ISO 21270 – XPS and AES – linearity of intensity scale.

- ISO 24236 – AES – repeatability and constancy of intensity scale.

For interested readers, an accessible and detailed overview of instrument calibration for AES (and XPS) is given in BCR-261T [7].

2.4 Quantitative Analysis

The Auger peak intensity of an element A can be correlated to its atomic concentration $c_A(z)$. Supposing the signal comes from a layer of thickness dz and depth z analyzed at an emission angle θ with respect to surface normal, one obtains the intensity I_A of an Auger peak by:

$$I_A = g \int_0^\infty c_A(Z) \exp\left(-\frac{z}{\lambda \cos\theta}\right) dz \qquad (2.6)$$

where the attenuation length λ (defined by the ISO Standard 18115:2001 on Surface Chemical Analysis) is calculated by Equations (2.4a)–(2.4c). Further information can be obtained in Seah [8] The parameter g is given by:

$$g = T(E)D(E)I_o\sigma\gamma(1 + r_M) \qquad (2.7)$$

neglecting the influence of the roughness R, where:

$c_A(z)$ – concentration of element A which has a function of depth z.

λ – attenuation length of the Auger electron.

θ – emission angle with respect to surface normal.

$T(E)$ – transmission factor which is a function of the kinetic energy E of the Auger electron.

$D(E)$ – detection efficiency of the electron multiplier, a factor which may vary with time.

I_o – primary current.

σ – cross-section of the Auger process.

γ – probability of an Auger transition (to be compared to the emission of a photon during the de-excitation process).

r_M – electron backscatter factor which is matrix (M) dependent (see Section 2.2.4).

Assuming that we have a flat surface and a homogeneous depth distribution of element A in a matrix M, integration of Equation (2.6) gives:

$$I_{A,M} = D(E)T(E)I_o\sigma_A\gamma_A(1 + r_{A,M})\lambda_{A,M}c_{A,M} \qquad (2.8)$$

Applying Equation (2.8) to a binary alloy one obtains (take as an example, Figure 2.20 with A = Sn and B = Nb):

$$\frac{I_{A,AB}}{I_{B,AB}} = \frac{\sigma_{A,A}}{\sigma_{B,B}} \bullet \frac{x_A}{x_B} \qquad (2.9)$$

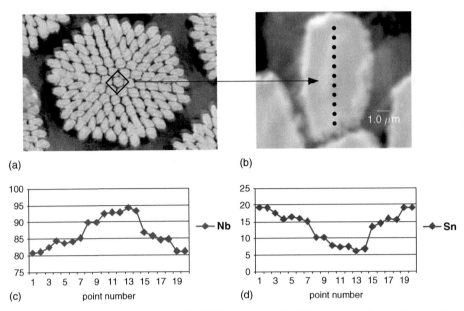

(a)   (b)

(c)   point number   (d)   point number

Figure 2.22 Sn–Nb multi-wire: (a) SEM micrograph; (b) point analysis of one wire giving the atomic concentrations in (c,d)

In general the elemental cross-sections are replaced by relative elemental sensitivity factors, s. The result illustrated in Figure 2.22 shows the Sn–Nb multi-wire and the point analysis on one wire giving the atomic concentrations in Figure 2.22(c,d). Table 2.4 shows the corresponding numerical values of the atomic concentrations of the line scan.

The composition of a sample of n elements can be calculated semi-quantitatively by the following:

$$x_A = \frac{I_A/s_A}{\sum\limits_i^n I_i/s_i} \qquad (2.10)$$

where I_i ($i = A, B, \ldots n$) are the intensities of the elemental peaks and s_i are the respective sensitivity factors. The elemental cross-sections have been converted into standardized elemental sensitivity factors, which are corrected for the transmission function of the analyzer used. In case spectra are measured in the $N(E)$ direct mode, area sensitivity factors are applied, whereas after differentiation to $dN(E)/dE$ peak-to peak intensities together with the corresponding elemental peak-to-peak sensitivity factors are used in Equation (2.10).

Figure 2.23 illustrates spectra of a non-homogeneous oxidized Fe–Cr–Nb alloy in (a) direct $N(E)$ form and (b) after differentiation, $dN(E)/dE$. This

Table 2.4 Numerical AES data of atomic concentrations (at%) of a Nb–Sb wire

Point	Sn (at%)	Nb (at%)
1	19.3	80.7
2	19.0	81.0
3	17.6	82.4
4	15.7	84.3
5	16.3	83.7
6	15.9	84.1
7	14.9	85.1
8	10.1	89.9
9	10.1	89.9
10	7.6	92.4
11	7.2	92.8
12	7.3	92.7
13	5.9	94.1
14	4.7	95.3
15	13.3	86.7
16	14.3	85.7
17	15.6	84.4
18	15.3	84.7
19	18.9	81.1
20	18.9	81.1

alloy shows different compositions, in particular the Nb content for different grains of the alloy. Inspection of this figure illustrates the differences of the Auger spectra of two different grains with different compositions. Table 2.4 illustrates the corresponding numerical data of the AES line scan. For the interested reader, an overview is provided in Seah [9] giving a more detailed approach to quantification including matrix effects and sensitivity factors.

2.5 Depth Profile Analysis

The fourth mode of data acquisition combines AES with ion beam sputtering yielding in-depth information beyond the escape depth limit of a few nm of the Auger electrons as discussed above. Sputtering is done either by simultaneous or alternating ion bombardment of a raster scanned noble ion beam of known beam energy and current over the sample surface. The ion beam has to be well aligned with the electron beam in order to avoid crater edge effects. Auger analysis should be performed in the center of the ion crater. Preferably the sputtered area should exceed the analyzed area by a factor of 3–10.

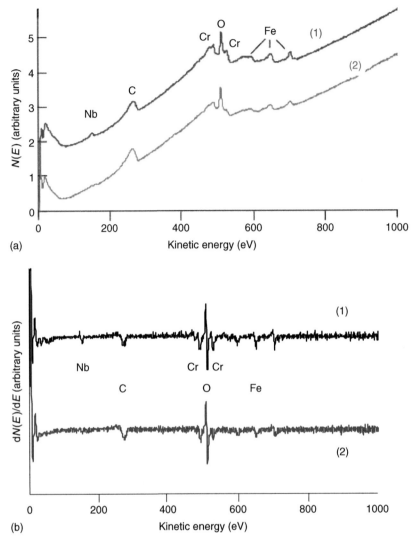

Figure 2.23 Oxidized Fe–Cr–Nb alloy: (a) direct $N(E)$ form; (b) after differentiation, $dN(E)/dE$

2.5.1 THIN FILM CALIBRATION STANDARD

The aim of depth profile analysis is to convert the sputter time into depth and the measured intensities into elemental concentrations. The latter can be performed as outlined in Section 2.4. Depth calibration is often performed by use of a thin film standard which consists of an anodic Ta_2O_5 film on metallic Ta [7]. A typical profile of Ta_2O_5 on Ta is shown in Figure 2.24 reporting in (a) the peak-to-peak amplitude of the oxygen and tantalum

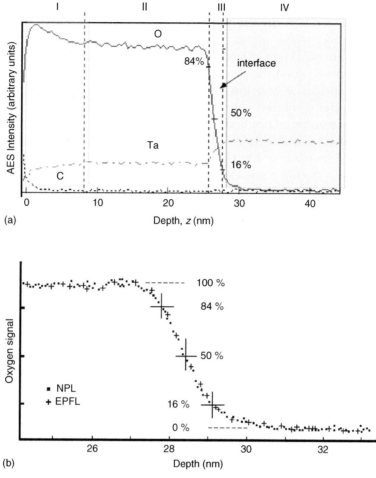

Figure 2.24 (a) Auger profile of a standard 30 nm Ta_2O_5/Ta film showing the variation of the Auger amplitude of oxygen and the metal as a function of sputter time. The thickness of zone III is $\Delta l = v \Delta t_I$ where v is the sputter rate of the oxide and t_I the time necessary to reach the center of the interface defined at 50 % of the amplitude in the steady state region (II). (b) Interlaboratory comparison of the interface of a profile of a 30 nm Ta_2O_5/Ta standard indicating the precision which which an interface can be measured under favourable conditions. Measurements were performed at two laboratories (NPL, Teddington, UK and EPF, Lausanne, Switzerland. (b) reproduced from [10], with permission from Elsevier.)

peak as a function of Ar^+ sputter time. The thickness of the anodic film has been calibrated separately by Nuclear Reaction Analysis (NRA). Such a depth profile is typical for thin films with an internal film/substrate interface. Four zones can be identified: in zone I the sample is cleaned from surface contamination (adsorbed species like C, CO, C_xH_y). Interaction of the ion with the oxide leads to preferential sputtering, i.e. a change of composition. In the present case oxygen is preferentially removed with respect to the metal. Actually, a chemical reduction of the oxide due to sputtering takes place. A steady state (zone II) is reached after a certain time depending on the sputtered material, crystallinity, ion beam energy, angle of incidence and ion beam current. The composition of zone II depends on the sputtered sample and on the experimental conditions of ion and electron beam applied. In zone III the interface between the oxide film and metal substrate is reached, characterized by the decrease of the oxygen and increase of the tantalum signal. In practice one measures the time to reach the interface, determines the steady state amplitudes in the film and converts the sputter time into a depth provided the film thickness is known from an independent measurement. Zone IV represents the substrate. Due to an interaction of the sputter ions with the surface atoms at the crater bottom atoms are not only removed from the surface but also knocked into the sample. This leads to atomic mixing and broadening of the interface. More details are discussed elsewhere [9]. For very thin films ($<10\,nm$) zones I and II are often not separated since no steady state is reached. In such cases it is difficult and almost impossible to determine the composition and stoichiometry within the film together with the width of the interface. The physical limit of depth resolution of approx. $1-2\,nm$ is determined by the ion beam energy and doses as well as by the escape depth of the electrons. Figure 2.24(b) shows an example of the interface between an oxide film and the metal substrate. It further underlines the repeatability obtainable in AES thin film analysis even in different laboratories.

2.5.2 DEPTH RESOLUTION

The resolution Δz of the interface is defined by the width which corresponds to the time necessary to reach 84 % and 16 % of the steady state value taken as 100 % of the oxygen amplitude as shown in Figure 2.24(b). For convenience, the 50 % point is taken as measure for the position of the interface. Figure 2.25 illustrates the dependence of Δz with thickness for various films sputtered. In particular, the resolution for amorphous Ta_2O_5 films formed anodically on Ta is given. For many samples, in particular for crystalline samples, the depth resolution Δz degrades proportionally with the square root of the film thickness z. Amorphous films such as Ta_2O_5/Ta (+) or SiO_2/Si exhibit a much better depth resolution because of a smaller sputter induced roughness at a given angle of incidence of the ion beam. By rotating the sample during analysis one can reduce the effect of induced roughness since the angle of incidence of the primary ions varies, thus avoiding cone formation during sputtering (for further details see Briggs and Seah [2]).

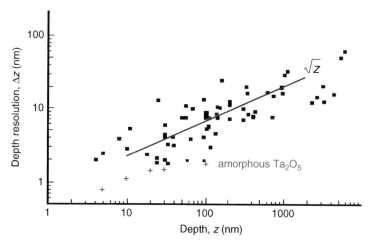

Figure 2.25 Depth resolution Δz as a function of thickness for different crystalline films and Ta_2O_5 films on Ta (+) (reproduced from [11], with permission from John Wiley & Sons, Ltd.)

Other important factors which influence the depth resolution are the original roughness and the purity of the the rest gas of the analysis chamber.

2.5.3 SPUTTER RATES

The objective of a depth profile is to determine the elemental concentration as a function of depth. As discussed above, Equation (2.10) can be used as a first-order approximation to calculate the atomic concentration of the elements measured. To convert the time axis into depth we apply the general relation between depth z and sputter time t given by:

$$z(t) = \int_0^\infty v dt \qquad (2.11)$$

where the sputter rate v of the elements is obtained by:

$$v = \frac{JYA}{\rho e Nn} \qquad (2.12)$$

with:

J ion current density (A/m^2)
Y sputter yield (atoms/primary ion)
A molecular weight (kg/mol)
ρ mass density (kg/m^3)
e electron charge $(= 1.602 \times 10^{-19}$ A s)
N Avogadro's constant $(= 6.023 \times 10^{23}/mol)$
n number of molecules in a molecule (i.e. 7 in Ta_2O_5).

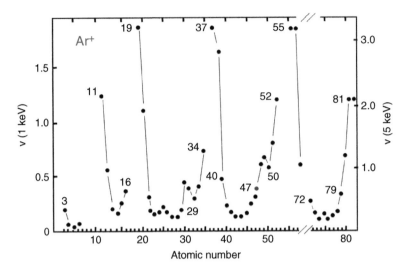

Figure 2.26 Normalized Ar^+ sputter rate in nm/s per $\mu A/mm^2$ for elements with Z between 3 and 82 for ion beam energies of 1 and 5 keV

The ratio A/ρ (atomic volume) and the sputter yield Y for 1 keV Ar^+ ions can be found in Tables 2.3 and Table 2.5, respectively. The elemental sputter rate (in nm/s) is calculated from the elemental yield after normalizing by the ion current density (in $\mu A/mm^2$) according to:

$$\frac{v}{J} = \frac{YA}{100\rho} \left(\frac{nm/s}{\mu A/mm^2} \right) \tag{2.13}$$

Application of Equation (2.13) results in the sputter rates shown in Figure 2.26 for 1 keV and 5 keV Ar^+ ions.

The sputter yield, Y, of elements and components depends on several parameters such as ion beam energy E_{ion} and angle of incidence, θ. This dependence is shown as an example in the next two figures for Ta_2O_5 on Ta. Inspection of Figure 2.27 reveals that Y varies approximatively with log E_{ion} for low ion beam energies. Therefore, knowledge of a sputter yield at a certain ion beam energy E_1 allows one to estimate the sputter yield at E_2. Figure 2.28 indicates that Y varies approximatively with $1/\cos \theta$ for angles between $\theta = 0°$ and 45°. Consequently, one concludes that the ion sputter rate v becomes independent of the angle θ below 45° because the primary ion current varies with $\cos \theta$ as experimentally shown (Figure 2.29). As a rule of thumb one retains that a sputter ion current density of a few $\mu A/cm^2$ leads to a sputter rate of a few Å/min for 1 keV Ar^+ sputtering for many elements and compounds.

Table 2.5 Sputter parameters of elements

Legend:

Box position	Parameter
Z	Atomic Number
A	Element
P	Ionisation Potential (eV)
S	Sputter yield for Ar$^+$ 1 keV
EA	Electron Affinity (eV)

Key cell layout:

Z	A
P	S
EA	

Periodic table (each cell lists: P (IP) / S / EA):

1	2	3	4	5	6	7	8	9	10	11	12	13	14	15	16	17	18
1 H 13.6 / — / 0.75																	**2 He** 24.59 / — / 0
3 Li 5.39 / 1.47 / 0.62	**4 Be** 9.32 / 0.9 / <0.5											**5 B** 8.30 / 0.58 / 0.28	**6 C** 11.26 / 0.5 / 1.26	**7 N** 14.53 / 0.83 / 0.75	**8 O** 13.62 / 1.87 / 1.46	**9 F** 17.42 / 3.4 / —	**10 Ne** 21.56 / 0 / —
11 Na 5.14 / 4.9 / 0.55	**12 Mg** 7.65 / 3.8 / 0											**13 Al** 5.99 / 1.84 / 0.44	**14 Si** 8.15 / — / 1.39	**15 P** 10.49 / — / 0.75	**16 S** 10.36 / — / 2.08	**17 Cl** 12.97 / — / 3.62	**18 Ar** 15.76 / — / 0
19 K 4.34 / 8.2 / 0.5	**20 Ca** 6.11 / 4.13 / <0.5	**21 Sc** 6.54 / 2.05 / 0.19	**22 Ti** 6.82 / 1.67 / 0.08	**23 V** 6.74 / 1.55 / 0.53	**24 Cr** 6.77 / 2.05 / 0.67	**25 Mn** 7.44 / 2.88 / 0	**26 Fe** 7.87 / 2.0 / 0.16	**27 Co** 7.86 / 1.96 / 0.66	**28 Ni** 7.64 / 2.03 / 1.16	**29 Cu** 7.73 / 2.52 / 1.23	**30 Zn** 9.39 / 6.7 / 0	**31 Ga** 6.0 / 3.43 / 0.3	**32 Ge** 7.9 / — / 1.2	**33 As** 9.81 / 3.1 / 0.81	**34 Se** 9.75 / 4.48 / 2.02	**35 Br** 11.81 / — / 3.36	**36 Kr** 14.00 / — / 0
37 Rb 4.18 / 12.2 / 0.49	**38 Sr** 5.70 / 2.4 / <0.5	**39 Y** 6.38 / 2.4 / 0.31	**40 Zr** 6.84 / 1.7 / 0.43	**41 Nb** 6.88 / 1.45 / 0.89	**42 Mo** 7.10 / 1.45 / 0.75	**43 Tc** 7.28 / 1.55 / 0.55	**44 Ru** 7.37 / — / 1.14	**45 Rh** 7.46 / 1.9 / 1.14	**46 Pd** 8.34 / 2.75 / 0.56	**47 Ag** 7.58 / 3.7 / 1.30	**48 Cd** 8.99 / 9.6 / 0	**49 In** 5.79 / 4.4 / 0.30	**50 Sn** 7.34 / — / 1.15	**51 Sb** 8.64 / — / 1.05	**52 Te** 9.01 / 4.55 / 1.97	**53 I** 10.45 / — / 3.06	**54 Xe** 12.13 / — / 0
55 Cs 3.89 / 15.3 / 0.47	**56 Ba** 5.21 / — / <0.5	**57 La** 5.58 / 2.63 / 0.5	**72 Hf** 7.0 / 2.05 / 0	**73 Ta** 7.89 / 1.65 / 0.32	**74 W** 7.98 / — / 0.82	**75 Re** 7.88 / — / 0.12	**76 Os** 8.7 / — / 1.12	**77 Ir** 9.1 / — / 1.57	**78 Pt** 9.0 / — / 2.13	**79 Au** 9.23 / — / 2.31	**80 Hg** 10.44 / — / 0	**81 Tl** 6.11 / 5.78 / 0.30	**82 Pb** 7.42 / — / 0.37	**83 Bi** 7.29 / — / 0.95	**84 Po** 8.42 / — / 1.9	**85 At** 9.5 / — / 2.8	**86 Rn** 10.75 / — / 0
87 Fr — / — / —	**88 Ra** 5.28 / — / —	**89 Ac** — / 6.9 / —															

Lanthanides:

58 Ce	59 Pr	60 Nd	61 Pm	62 Sm	63 Eu	64 Gd	65 Tb	66 Dy	67 Ho	68 Er	69 Tm	70 Yb	71 Lu
5.47 / 2.72 / <0.5	5.42 / 3.25 / <0.5	5.49 / 3.55 / <0.5	5.55 / — / <0.5	5.63 / — / <0.5	5.67 / — / <0.5	6.14 / 2.95 / <0.5	5.85 / 3.03 / <0.5	5.93 / — / <0.5	6.02 / 3.9 / <0.5	6.10 / 3.75 / <0.5	6.18 / — / <0.5	6.25 / 8.1 / <0.5	5.43 / 2.9 / <0.5

Actinides:

90 Th	91 Pa	92 U
7.0 / — / 2.12	5.89 / — / 2.12	6.08 / — / 2.4

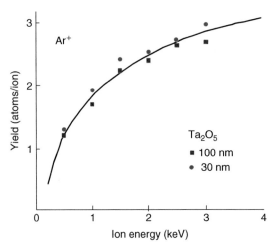

Figure 2.27 Ion sputter yield of Ta_2O_5 for typical sputter energies applied in AES profiles (reproduced from [11], with permission from John Wiley & Sons, Ltd.)

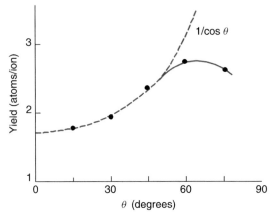

Figure 2.28 Variation of sputter yield of Ta_2O_5 with the angle of incidence θ defined with respect to surface normal (reproduced from [11], with permission from John Wiley & Sons, Ltd.)

2.5.4 PREFERENTIAL SPUTTERING

In the following chapter we will see that the sputter rate and yield varies from element to element. Consequently, while sputtering a multi-element sample one observes a change of composition because of a difference in the elemental sputter yields. This phenomenon is called *preferential sputtering*. Continued sputtering leads to a steady state as illustrated by Figure 2.24(a) for Ta_2O_5 on Ta. For a binary alloy we obtain by application of

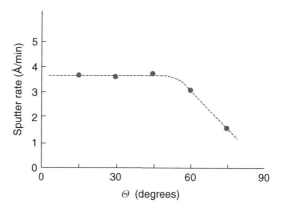

Figure 2.29 Ion sputter rate v as a function of angle of incidence for 1 keV Ar^+ and an ion current density of 4.2 $\mu A/cm^2$ determined for the Ta_2O_5/Ta standard (reproduced from [11], with permission from John Wiley & Sons, Ltd.)

Equation (2.9), use of elemental relative sensitivity factors, S, and introduction of elemental sputter yields, κ:

$$\frac{I_{A,AB}}{I_{B,AB}} = \frac{S_A}{S_B} \bullet \frac{x_A}{x_B} \bullet \frac{\kappa_B}{\kappa_A} \tag{2.14}$$

where κ_A/κ_B is the ratio of the sputter yields of A and B, Y_A and Y_B, divided by their atomic density $n_M = \rho/M$ (M, atomic weight of A or B):

$$\frac{\kappa_B}{\kappa_A} = \frac{Y_B/n_B}{Y_A/n_A} \tag{2.15}$$

2.5.5 λ-CORRECTION

The influence of the attenuation length of Auger electrons on the signal intensity is very important for thin films $\leq 3\Lambda$ with $\Lambda = \lambda\cos\theta$ where Λ is of the same magnitude as the film thickness. To correct for its influence, the measured intensity $I_A(z)$ given by Equation (2.6) can be replaced by $F_A(z)$, applying [2]:

$$F_A(z) = I_A(z) - \lambda\cos\theta\frac{dI_A(z)}{dz} \tag{2.16}$$

The transformation $F_A(z)$ is called the λ-correction. Figure 2.30 gives an example of the effectiveness of such a transformation showing for a binary alloy Fe–Cr the enrichment of chromium underneath the surface within the oxide film which without λ-correction would have passed almost unnoticed. In addition, the distribution of iron is totally different.

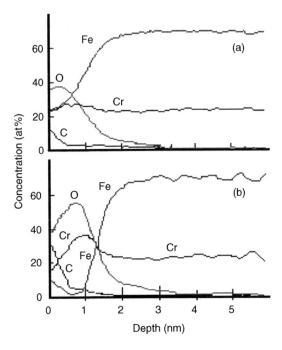

Figure 2.30   λ-correction applied to an oxide layer on an Fe–Cr alloy

2.5.6 CHEMICAL SHIFTS IN AES PROFILES

A chemical shift of the peak of element A is defined as the displacement of the kinetic or binding energy of that peak following a change in the state of oxidation (compare Section 2.2.6). Such displacements are also observed in AES, but they are less evident than in ESCA, because of the interaction of different electron levels of the Auger process. In AES profiling, such shifts are observed for either a variation of the state of oxidation with depth or during sputtering, in particular for oxides. However, elemental intensity plots – often used in AES profiling – do not directly allow one to identify such peak shifts because only the total intensity of a peak is reported, which results from the interaction of three electronic levels (compare Figure 2.2). As an illustration for a chemical shift in AES the evolution of the Al_{LMM} peak is shown in Figure 2.31 indicating the differentiated peaks of the natural oxide Al_2O_3 (Al-ox) and the Al metal (Al-met). One observes a shift from the metal to the oxide of approximately 15 eV. In addition, the shape and intensity of the peaks changes as a consequence of the electron density of the band changes illustrated in Figure 2.10 when comparing metal and oxide. As a consequence for AES depth profiling data where peak intensities are plotted as a function of depth, such chemical shifts as for the Al_{LMM} peak must be considered and corrected to avoid misinterpretation

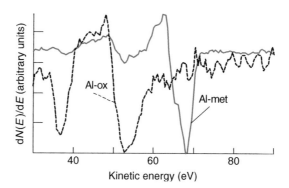

Figure 2.31 Example of the chemical shift of the Auger Al_{LMM} peak between Al_2O_3 (Al-ox) and Al metal (Al-met). The differential spectra of the oxide and metal show the peak intensity in arbitrary units

of the data. Needless to say in cases where a change of shape of a given peak during profiling is observed, standard elemental sensitivity factors become invalid. Peak separation for different oxidation states, re-definition of energy windows and deconvolution routines are necessary to avoid errors up to 50 % in quantification. Most software routines of recent AES equipments offer such procedures, which unfortunately however, exclude automatic quantification of the raw data.

2.6 Summary

Qualitative AES is an elemental analysis technique detecting all elements except H and He by measuring a characteristic kinetic electron energy specific for a given elemental Auger electron transition on conducting samples. The number of characteristic Auger electrons allows one to quantify data by application of experimentally determined elemental sensitivity factors yielding in general a precision of 10–50 % due to the influence of matrix factors and influence of the state of oxidation on the peak shape. The detection limit of point analysis is about 0.1 to 1 % of a monolayer corresponding to $10^{12}-10^{13}$ particles/cm^2. This limit may be increased for line scans or elemental mapping depending on the data acquisition time. Spatial Auger resolution is determined by the focus of the primary electron beam and the backscattering of the electrons within the analyzed matrix. Specifications of modern scanning Auger microprobes are as good as 10 nm.

Escape depth and electron attenuation lengths of most elements is 1–3 nm for kinetic energies below 2 keV, typical for Auger transitions. State of the art instruments use primary electron beams with energies up to 30 keV focussed down to 20 nm secondary electron spatial resolution.

In-depth information to a depth of 0.1–1 μm and resolution of 1–20 nm is obtained by combining AES with ion beam sputtering. Despite the fact that ion beam sputtering may change the composition of the layer analyzed, sputter depth profiling is one of the most important applications of AES, because it gives access to internal interfaces and allows one to identify relative composition changes with depth. The quantification of depth profiles is often limited by the absence of either precise sensitivity factors adapted for the measured oxidation state or well defined sputter yield ratios. A Ta_2O_5 film standard on Ta of known thickness serves as a welcome tool to calibrate sputter conditions.

Applications of AES cover all fields of materials science, physics and chemistry, namely nanotechnology, nanoparticle chemical analysis, thin film preparation of surfaces and thin films in micro-electronics, semi- and superconductors, corrosion and electrochemistry and catalysis, as well as metallurgy and tribology.

📖 References

1. P. Auger, *J. Phys. Radium* **6**, 205 (1925).

2. D. Briggs and M. P. Seah, *Practical Surface Analysis by Auger and X-ray Photoelectron Spectroscopy*, 2nd Edition, John Wiley & Sons, Ltd, Chichester, UK (1990).

3. D. David, *Méthodes usuelles de caractérisation des surfaces*, Eyrolles, Paris (1992).

4. J. P. Eberhart, *Analyse structurale et chimique des matériaux*, Dunod, Paris (1989).

5. D. Briggs and J. T. Grant, *Surface Analysis by Auger and X-Ray Photoelectron Spectroscopy*, Cromwell Press, Trowbridge, UK (2003).

6. C. J. Powell and A. Jablonski, *Surf. Interface Anal.* **33**, 211 (2002); *J. Phys. Chem. Ref. Data* **28**, 19 (1999).

7. BCR Reference Material, BCR-261T, c/o IRMM, Retieseweg, B-2440 Geel, Belgium.

8. M. P. Seah in *Surface Analysis by Auger and X-Ray Photoelectron Spectroscopy* (D. Briggs and J. T. Grant (Eds)), IM Publications and Surface Spectra Ltd, Chichester, UK, p. 167 (2003).

9. M. P. Seah in *Surface Analysis by Auger and X-Ray Photoelectron Spectroscopy* (D. Briggs and J. T. Grant (Eds)), IM Publications and Surface Spectra Ltd, Chichester, UK, p. 345 (2003).

10. C. P. Hunt, H. J. Mathieu and M. P. Seah, *Surf. Sci* **139**, 549 (1984).

11. C. P. Hunt and M. P. Seah, *Surf. Interface Anal.* **5**, 199 (1983).

Problems

1. Which elements can be detected by AES? Why it is not possible to detect hydrogen?

2. Can you imagine using an X-ray source to excite Auger electrons?

3. Is there any chemical information available in Auger spectra?

4. Insulating thin film oxides like SiO_2 or Ta_2O_5 can be detected by AES. How is it possible not to have charging problems?

5. What is the sensitivity limit for AES (in percentage of monolayers or kg/m^2)?

6. Determine the AES escape depth of Ta for its NNN peak at 179 eV as a function of the emission angle at $\theta = 90°$, using Equations (2.4a) – (2.4c).

7. Could one combine AES with electron microscopy? Discuss the benefits and disadvantages.

8. Explain why the work function of a specimen analyzed does not have any influence on the kinetic energy of the detected peaks.

9. Discuss the benefits and disadvantages of the use of $N(E)$ or $dN(E)/dE$ AES data.

10. Can you imagine performing angular sensitive measurements (variation of the take-off angle) with a CMA analyser?

3 Electron Spectroscopy for Chemical Analysis

BUDDY D. RATNER and DAVID G. CASTNER

**Departments of Chemical Engineering and Bioengineering,
University of Washington, Seattle, USA**

Of all the contemporary surface characterization methods, electron spectroscopy for chemical analysis (ESCA) is the most widely used. ESCA is also called X-ray photoelectron spectroscopy (XPS), and the two acronyms can be used interchangeably. The popularity of ESCA as a surface analysis technique is attributed to its high information content, its flexibility in addressing a wide variety of samples, and its sound theoretical basis. This chapter will introduce the ESCA method and describe its theory, instrumentation, spectral interpretation and application. The intent of this introduction is to provide a broad overview. Many general reviews on this subject exist and further reading about ESCA theory and applications is encouraged [1–15]. This review is aimed so that readers who have had little or no formal introduction to the ESCA method can profit from it – it should provide an appreciation of the power and limitations of the contemporary surface analytical method. The jargon associated with ESCA will also be introduced and discussed, thereby assisting the reader in digesting the specialist literature.

3.1 Overview

ESCA falls in the category of analytical methods referred to as electron spectroscopies, so called because electrons are measured. Other

Surface Analysis – The Principal Techniques 2nd Edition Edited by John Vickerman and Ian Gilmore
© 2009 John Wiley & Sons, Ltd

prominent electron spectroscopies include Auger electron spectroscopy (AES, Chapter 2) and high-resolution electron energy loss spectroscopy (HREELS, Chapter 7).

3.1.1 THE BASIC ESCA EXPERIMENT

The basic ESCA experiment is illustrated in Figure 3.1. The surface to be analyzed is placed in a vacuum environment and then irradiated with photons. For ESCA, the photon source is in the X-ray energy range. The irradiated atoms emit electrons (photoelectrons) after direct transfer of energy from the photon to core-level electrons. Photoelectrons emitted from atoms near the surface can escape into the vacuum chamber and be separated according to energy and counted. The energy of the photoelectrons is related to the atomic and molecular environment from which they originated. The number of electrons emitted is related to the concentration of the emitting atom in the sample.

3.1.2 A HISTORY OF THE PHOTOELECTRIC EFFECT AND ESCA

The discovery of the photoelectric effect, its explanation, and the development of the ESCA method are entwined with the revolution in physics

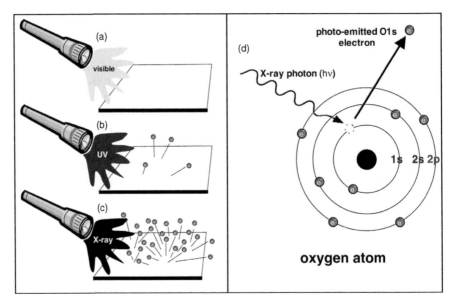

Figure 3.1 (a–c) A surface irradiated by a photon source of sufficiently high energy will emit electrons. If the light source is in the X-ray energy range, this is the ESCA experiment. (d) The X-ray photon transfers its energy to a core-level electron imparting enough energy for the electron to leave the atom

that took place in the early years of the twentieth century. This revolution led from classical physics based upon observational mechanics to quantum physics, whose impact is most clearly appreciated at the atomic scale. Some of the developments that took place from the 1880s up to the second half of the twentieth century that are fundamental to the development of ESCA will be briefly reviewed [16–18].

Hertz, in the 1880s, noticed that metal contacts in electrical systems, when exposed to light, exhibit an enhanced ability to spark. Hallwachs, in 1888, observed that a negatively charged zinc plate lost its charge when exposed to ultraviolet (UV) light, but that positively charged zinc plates were not affected. In 1899, J.J. Thompson found that subatomic particles (electrons) were emitted from the zinc plate exposed to light. Finally, in 1905, Einstein, using Planck's 1900 quantization of energy concept, correctly explained all these observations – photons of light directly transferred their energy to electrons within an atom, resulting in the emission of the electrons without energy loss. This process will be clarified in Section 3.2 of this chapter. Planck received the Nobel Prize for his contribution of the concept of the quantization of energy in 1918. Einstein received the Nobel Prize for explaining the photoelectric effect in 1921. To put the revolutionary aspects of these developments in perspective, a quotation from Max Planck, in nominating Einstein for the Prussian Academy in 1913, is illuminating. Planck said of Einstein, 'That he may sometimes have missed the target in his speculations, as for example, in his hypothesis of light quanta, cannot really be held against him'. Of course, history continues to support both the ideas of Planck and Einstein, and these ideas form the foundation for the theoretical understanding of ESCA.

As an analytical method, a more straightforward history can be presented. In 1914, Robinson and Rawlinson studied photoemission from X-ray irradiated gold and, using photographic detection, observed the energy distribution of electrons produced. Although they were hampered by poor vacuum systems and inhomogeneous X-ray sources they were still able to publish a recognizable gold photoemission spectrum. In 1951, Steinhardt and Serfass first applied photoemission as an analytical tool. Throughout the 1950s and 1960s, Kai Siegbahn (son of the 1924 Nobel Prize winner, Manne Siegbahn) developed the instrumentation and theory of ESCA to give us the method we use today. Siegbahn also coined the term 'electron spectroscopy for chemical analysis', later modified by his group to 'electron spectroscopy for chemical applications.' In 1981, Kai Siegbahn was rewarded for his contributions with the Nobel Prize in Physics.

3.1.3 INFORMATION PROVIDED BY ESCA

ESCA is an information-rich method (Table 3.1). The most basic ESCA analysis of a surface will provide qualitative and quantitative information on

Table 3.1 Information derived from an ESCA experiment

In the outermost 10 nm of a surface, ESCA can provide the following:
- Identification of all elements (except H and He) present at concentrations >0.1 atomic %.
- Semiquantitative determination of the approximate elemental surface composition (error < ±10 %).
- Information about the molecular environment (oxidation state, covalently bonded atoms, etc.).
- Information about aromatic or unsaturated structures or paramagnetic species from shake-up ($\pi^* \rightarrow \pi$) transitions.
- Identification of organic groups using derivatization reactions.
- Non-destructive elemental depth profiles 10 nm into the sample and surface heterogeneity assessment using (1) angular-dependent ESCA studies and (2) photo-electrons with differing escape depths.
- Destructive elemental depth profiles several hundred nanometers into the sample using ion etching.
- Lateral variations in surface composition (spatial resolutions down to 5 μm for laboratory instruments and spatial resolutions down to 40 nm for sychrotron-based instruments).
- 'Fingerprinting' of materials using valence band spectra and identification of bonding orbitals.
- Studies on hydrated (frozen) surfaces.

all the elements present (except H and He). More sophisticated application of the method yields a wealth of detailed information about the chemistry, electronic structure, organization, and morphology of a surface. Thus, ESCA can be considered one of the most powerful analytical tools available. The capabilities of ESCA highlighted in Table 3.1 will be elaborated upon throughout this article.

3.2 X-ray Interaction with Matter, the Photoelectron Effect and Photoemission from Solids

An understanding of the photoelectric effect and photoemission is essential to appreciate the surface analytical method, ESCA. When a photon impinges upon an atom, one of three events may occur: (1) the photon can pass through with no interaction, (2) the photon can be scattered by an atomic orbital electron leading to partial energy loss, and (3) the photon may interact with an atomic orbital electron with total transfer of the photon energy to the electron, leading to electron emission from the atom. In the first case, no interaction occurs and it is, therefore, not pertinent to this discussion. The second possibility is referred to as 'Compton scattering' and can be important in high-energy processes. The third process accurately describes the photoemission process that is the basis of ESCA. Total transfer of the photon energy to the electron is the essential element of photoemission.

Let us examine four observations associated with this photoelectric effect in more detail. First, no electrons will be ejected from an atom regardless of the illumination intensity unless the frequency of excitation is greater than or equal to a threshold level characteristic for each element. Thus, if the frequency (energy) of the excitation photon is too low, no photoemission will be observed. As the energy of this photon is gradually increased, at some value, we will begin to observe the photoemission of electrons from the atom (see Figure 3.1). Second, once the threshold frequency is exceeded, the number of electrons emitted will be proportional to the intensity of the illumination (i.e., once we've irradiated the sample with photons of sufficient energy to stimulate electron emission, the more photons we irradiate the sample with, the more photoelectrons will be produced). Third, the kinetic energy of the emitted electrons is linearly proportional to the frequency of the exciting photons – if we use photons of higher energy than our threshold value, the excess photon energy above the threshold value will be transferred to the emitted electrons. Finally, the photoemission process from excitation to emission is extremely rapid (10^{-16} s). The basic physics of this process can be described by the Einstein equation, simply stated as:

$$E_B = h\nu - KE \qquad (3.1)$$

where E_B is the binding energy of the electron in the atom (a function of the type of atom and its environment), $h\nu$ is the energy of the X-ray source (a known value), and KE is the kinetic energy of the emitted electron that is measured in the ESCA spectrometer. Thus, E_B, the quantity that provides us with valuable information about the photoemitting atom, is easily obtained from $h\nu$ (known) and KE (measured). Binding energies are frequently expressed in electron volts (eV; 1 eV $= 1.6 \times 10^{-19}$ joules). More rigorous descriptions of the photoemission process can be found elsewhere [19, 20].

The concept of the binding energy of an electron in an atom requires elaboration. A negatively charged electron will be bound to the atom by the positively charged nucleus. The closer the electron is to the nucleus, the more tightly we can expect it to be bound. Binding energy will vary with the type of atom (i.e., a change in nuclear charge) and the addition of other atoms bound to that atom (bound atoms will alter the electron distribution on the atom of interest). Different isotopes of a given element have different numbers of neutrons in the nucleus, but the same nuclear charge. Changing the isotope will not appreciably affect the binding energy. Weak interactions between atoms such as those associated with crystallization or hydrogen bonding will not alter the electron distribution sufficiently to change the measured binding energy. Therefore, the variations we see in the binding

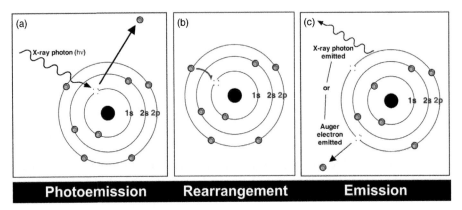

Figure 3.2 (a) The X-ray photon transfers its energy to a core-level electron leading to photoemission from the n-electron initial state. (b) The atom, now in an $(n-1)$-electron state, can reorganize by dropping an electron from a higher energy level to the vacant core hole. (c) Since the electron in (b) dropped to a lower energy state, the atom can rid itself of excess energy by ejecting an electron from a higher energy level. This ejected electron is referred to as an Auger electron. The atom can also shed energy by emitting an X-ray photon, a process called X-ray fluorescence

energy that provide us with the chemical information content of ESCA are associated with covalent or ionic bonds between atoms. These changes in binding energy are called binding energy shifts or chemical shifts and will be elaborated upon in Section 3.3.

For gases, the binding energy of an electron in a given orbital is identical to the ionization energy or first ionization potential of that electron. In solids, the influence of the surface is felt, and additional energy must be accounted for to remove an electron from the surface. This extra energy is called the work function and will be discussed in Section 3.3.

Irradiation of a solid by X-rays can also result in emission of Auger electrons (Figure 3.2). Auger electrons, discussed in detail in Chapter 2, differ in a number of respects from the photoelectrons that this chapter primarily deals with. A characteristic of Auger electrons is that their energy is independent of irradiation energy. Photoelectron energy is directly proportional to irradiation energy, according to Equation (3.1).

Much of the basic physics of photoemission, and the background material for other surface analysis methods as well, is couched in the jargon of solid state physics. An excellent 'translation' of this jargon is available [21].

3.3 Binding Energy and the Chemical Shift

The general concept of the electron binding energy and its relationship to the energy of the incident X-ray and the emitted photoelectron was introduced in the previous section. This section will develop this relationship in more detail, with particular emphasis placed on describing the quantities that affect the E_B magnitude and how measurement of the E_B can be used to characterize materials.

3.3.1 KOOPMANS' THEOREM

The E_B of an emitted photoelectron is simply the energy difference between the $(n-1)$-electron final state and the n-electron initial state (see Figure 3.2). This is written as:

$$E_B = E_f(n-1) - E_i(n) \qquad (3.2)$$

where $E_f(n-1)$ is the final state energy and $E_i(n)$ is the initial state energy. If no rearrangement of other electrons in the atom or material occurred during the photoemission process, then the observed E_B would be just the negative orbital energy, $-\varepsilon_k$ for the ejected photoelectron. This approximation comes from Koopmans' theorem [22] and is written as:

$$E_B \approx -\varepsilon_k \qquad (3.3)$$

The values of ε_k can be calculated using the Hartree–Fock method. These values are typically within 10–30 eV of the actual E_B values. The disagreement between E_B and $-\varepsilon_k$ is because Koopmans' theorem and the Hartree–Fock calculation method do not provide a complete accounting of the quantities that contribute to E_B. In particular, the assumption that other electrons remain 'frozen' during the photoemission process is not valid. During emission of the photoelectron, other electrons in the sample will respond to the creation of the core hole by rearranging to shield, or minimize, the energy of the ionized atom. The energy reduction caused by this rearrangement of electrons is called the 'relaxation energy'. Relaxation occurs for both electrons on the atom containing the core hole (atomic relaxation) and on surrounding atoms (extra-atomic relaxation). Relaxation is a final state effect and will be described in more detail later in this section. In addition to relaxation, quantities such as electron correlation and relativistic effects are neglected by the Koopmans/Hartree–Fock scheme. Thus, a more complete description of E_B is given by:

$$E_B = -\varepsilon_k - E_r(k) - \delta\varepsilon_{corr} - \delta\varepsilon_{rel} \qquad (3.4)$$

where $E_r(k)$ is the relaxation energy and $\delta\varepsilon_{corr}$ and $\delta\varepsilon_{rel}$ are corrections for the differential correlation and relativistic energies. Both the correlation and relativistic terms are typically small and usually can be neglected.

3.3.2 INITIAL STATE EFFECTS

As shown in Equation (3.2), both initial and final state effects contribute to the observed E_B. The initial state is just the ground state of the atom prior to the photoemission process. If the energy of the atom's initial state is changed, for example, by formation of chemical bonds with other atoms, then the E_B of electrons in that atom will change. The change in E_B, ΔE_B, is called the chemical shift. To a first approximation all core-level E_Bs for an element will undergo the same chemical shift (for example, if a silicon atom is bound to a chlorine atom, i.e. a Si—Cl bond, the chemical shift for the Si_{2p} and Si_{2s} Si—Cl peaks relative to the position of the Si^0 state for each peak will be similar).

It is usually assumed that initial state effects are responsible for the observed chemical shifts, so that, as the formal oxidation state of an element increases, the E_B of photoelectrons ejected from that element will increase. This assumes that final state effects such as relaxation have similar magnitudes for different oxidation states. For most samples, the interpretation of ΔE_B solely in terms of initial state effects (Equation (3.5)) is usually adequate.

$$\Delta E_B = -\Delta \varepsilon_k \tag{3.5}$$

Several examples showing a correlation between the initial state of an element and its ΔE_B are given in the classical work of Siegbahn and colleagues [8]. For example, Figure 3.3(a) shows their observation that E_B for the S_{1s} orbital increased by nearly 8 eV as the formal oxidation state of sulfur increased from -2 (Na_2S) to $+6$ (Na_2SO_4). Tables 3.2 and 3.3 list typical C_{1s} and O_{1s} E_B values for functional groups present in polymers. A more complete listing of E_B values for polymeric functional groups is given by Beamson and Briggs [23]. The C_{1s} E_B is observed to increase monotonically as the number of oxygen atoms bonded to carbon increases (C—C < C—O < C=O < O—C=O < O—(C=O)O—) since oxygen is more electronegative than carbon and will draw electrons away from carbon. This is consistent with an initial state effect, since, as the number of oxygen atoms bonded to a carbon increase, the carbon should become more positively charged, resulting in an increase in the C_{1s} E_B.

A caution must be made against solely using initial state effects for interpreting chemical shifts. There are examples where final state effects can significantly alter the relationship between the formal oxidation state and ΔE_B. Also, it is the changes in the distribution and density of electrons of an atom resulting from changes in its chemical environment that contribute to ΔE_B. These quantities do not necessarily have a straightforward relationship to the formal oxidation state. For example, the full charge implied by the formal oxidation state is only attained when the chemical bonding is completely ionic (no covalent character). The degree of ionic/covalent character can vary with the chemical environment. Thus, it is best to correlate ΔE_B with the charge on the atom, not its formal oxidation state. Siegbahn and colleagues have shown this yields consistent correlations for both inorganic and organic sulfur species [8]. Figure 3.3(b) shows the linear

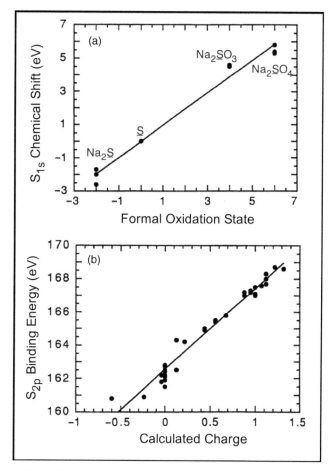

Figure 3.3 (a) The sulfur 1s chemical shifts versus formal oxidation state for several inorganic sulfur species. (b) The sulfur 2p binding energy versus calculated charge for several inorganic and organic sulfur species. Data taken from the results of Siegbahn *et al.*[8]

relationship they observed for the S_{2p} E_B versus calculated charge on the sulfur atom.

One approach developed to provide a physical basis for chemical shifts is the charge potential model [8]. This model relates the observed E_B to a reference energy E_B°, the charge q_i on atom i, and the charge q_j on the surrounding atoms j at distances r_{ij}, as follows:

$$E_B = E_B^\circ + kq_i + \sum_{j \neq i}(q_j/r_{ij}) \tag{3.6}$$

with the constant k. Generally, the reference state is considered to be E_B for the neutral atom. It is then apparent that, as the positive charge on the atom

Table 3.2   Typical C_{1s} binding energies for organic samples[a]

Functional group		Binding energy (eV)
Hydrocarbon	C—H,$\underline{C}$—C	285.0
Amine	$\underline{C}$—N	286.0
Alcohol, ether	$\underline{C}$—O—H, $\underline{C}$—O—C	286.5
Cl bound to carbon	$\underline{C}$—Cl	286.5
F bound to carbon	$\underline{C}$—F	287.8
Carbonyl	$\underline{C}$=O	288.0
Amide	N—$\underline{C}$=O	288.2
Acid, ester	O—$\underline{C}$=O	289.0
Urea	N—$\underline{C}$(=O)—N	289.0
Carbamate (urethane)	O—$\underline{C}$(=O)—N	289.6
Carbonate	O—$\underline{C}$(=O)—O	290.3
2F bound to carbon	—CH_2CF_2—	290.6
Carbon in PTFE	—CF_2CF_2—	292.0
3F bound to carbon	—CF_3	293–294

[a]The observed binding energies will depend on the specific environment where the functional groups are located. Most ranges are ±0.2 eV, but some (e.g., fluorocarbon samples) can be larger.

Table 3.3   Typical O_{1s} binding energies for organic samples[a]

Functional group		Binding energy (eV)
Carbonyl	C=$\underline{O}$, O—C=$\underline{O}$	532.2
Alcohol, ether	C—$\underline{O}$—H, C—$\underline{O}$—C	532.8
Ester	C—$\underline{O}$—C=O	533.7

[a]The observed binding energies will depend on the specific environment where the functional groups are located. Most ranges are ±0.2 eV.

increases by formation of chemical bonds, E_B will increase. The last term on the right-hand side of Equation (3.6) is often called the Madelung potential because of its similarity to the lattice potential of a crystal, $V_i = \Sigma q_j / r_{ij}$). This term represents the fact that the charge q_i removed or added by formation of a chemical bond is not displaced to infinity, but rather to the surrounding atoms. Thus, the second and third terms on the right-hand side of Equation (3.6) are of opposite sign. Using Equation (3.6) the chemical shift between states 1 and 2 can now be written as:

$$\Delta E_B = k[q_i(2) - q_i(1)] + V_i(2) - V_i(1) \qquad (3.7)$$
$$\Delta E_B = k\Delta q_i + \Delta V_i$$

where ΔV_i represents the potential change in the surrounding atoms.

3.3.3 FINAL STATE EFFECTS

As noted in Section 3.3.1, relaxation effects can have a significant impact on the measured E_B. In all cases the electron rearrangements that occur during photoemission result in the lowering of E_B. If the magnitude of the relaxation energy varies significantly as the chemical environment of an atom is changed, the E_B ranking that would be expected based on initial state considerations, can be altered. For example, the ranking of the Co $2p_{3/2}E_B$ values is Co^0 (778.2 eV) < Co^{+3} (779.6 eV) < Co^{2+} (780.5 eV) [24]. Also, both Cu^0 and Cu^{+1} have $2p_{3/2}E_B$ values of 932.5 eV ($\Delta E_B = 0$) [25]. Thus, for the Co and Cu systems, final state effects cause deviations in the E_B versus oxidation state ranking expected from initial state considerations. ESCA Cu 2p spectra in Figure 3.4(a) show the similar E_Bs of metallic Cu (Cu^0) and Cu_2O (Cu^{+1}).

Contributions to the relaxation energy arise from both the atom containing the core hole (atomic relaxation) and its surrounding atoms (extra-atomic relaxation). Most of the atomic relaxation component results from rearrangement of outer shell electrons, which have a smaller E_B than the emitted photoelectron. In contrast, the inner shell electrons (E_B larger than the emitted photoelectron) make only a small contribution to the atomic relaxation energy and can usually be neglected. The form of extra-atomic relaxation depends on the material being examined. For electrically conducting samples such as metals, valence band electrons can move from one atom to the next to screen the core hole. For ionically bonded solids such as the alkali halides, electrons are not free to move from one atom to the next. The

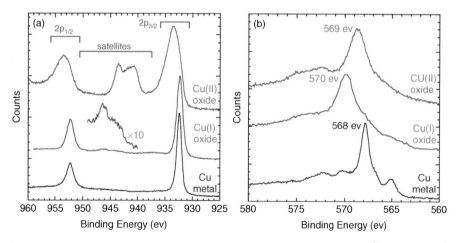

Figure 3.4 (a) The copper 2p photoemission spectra for metallic Cu (Cu^0), Cu_2O (Cu^{+1}) and CuO (Cu^{+2}). (b) The X-ray induced copper Auger LVV spectra for metallic Cu (Cu^0), Cu_2O (Cu^{+1}) and CuO (Cu^{+2})

electrons in these materials, however, can be polarized by the presence of a core hole. The magnitude of reduction in E_B produced by extra-atomic relaxation in ionic materials is smaller than the extra-atomic relaxation in metallic samples.

Other types of final state effects such as multiplet splitting and shake-up satellites can contribute to E_B. Multiplet splitting arises from interaction of the core hole with unpaired electrons in the outer shell orbitals. Shake-up satellites arise from the outgoing photoelectron losing part of its kinetic energy to excite a valence electron into an unoccupied orbital (e.g. $\pi \rightarrow \pi^*$ transition). These features and their presence in ESCA spectra are described in Section 3.6. Detailed discussion of final state effects is given elsewhere [26].

3.3.4 BINDING ENERGY REFERENCING

As is apparent from the preceding sections, accurate measurement of E_B can provide information about the electronic structure of a sample. As discussed in Section 3.2, E_B is determined by measuring the KE of the emitted photoelectron. To do this properly, a calibrated and suitably referenced ESCA spectrometer is required. The following paragraphs will describe how to set up an ESCA spectrometer to measure accurately the photoelectron KE (and therefore E_B) for different types of samples.

Conducting samples such as metals are placed in electrical contact with the spectrometer, typically by grounding both the sample and the spectrometer. This puts the Fermi level (E_F), the highest occupied energy level, of both the sample and spectrometer, at the same energy level. Then the photoelectron KE can be measured as shown in Figure 3.5. As can be seen in this figure, the sum of the KE and E_B does not exactly equal the X-ray energy, as implied in the Einstein equation. The difference is the work function of the spectrometer (ϕ_{sp}). The work function, ϕ, is related to the E_F and vacuum level (E_{vac}) by:

$$\phi = E_F - E_{vac} \tag{3.8}$$

Thus, ϕ is the minimum energy required to eject an electron from the highest occupied level into vacuum. The Einstein equation now becomes:

$$E_B^F = h\nu - KE - \phi_{sp} \tag{3.9}$$

Therefore, both KE and ϕ_{sp} must be measured to determine E_B^F. The superscript F on E_B means that E_B is referenced to E_F. For conducting samples it is the work function of the spectrometer (ϕ_{sp}) that is important (see Figure 3.5). This can be calibrated by placing a clean Au standard in the spectrometer and adjusting the instrumental settings such that the known E_B values for Au are obtained (e.g. $E_F = 0\,eV$, $4f_{7/2} = 83.96\,eV$). The linearity of the E_B

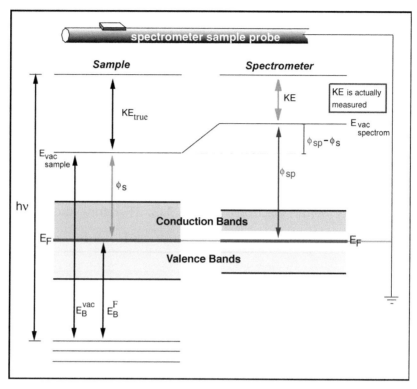

Figure 3.5 The energy level diagram for an electrically conducting sample that is grounded to the spectrometer. The Fermi levels of the sample and spectrometer are aligned ($E_f^s = E_f^{sp}$) so that E_B is referenced with respect to E_F. The measurement of E_B is independent of the sample work function, ϕ_s, but is dependent on the spectrometer work function, ϕ_{sp}

scale is then calibrated by adjusting the energy difference between two widely spaced lines of a sample (e.g. the 3s and $2p_{3/2}$ peaks of clean Cu) to their known values. The operator continues to iterate between the two calibration procedures until they converge to the accepted values. Further details of the calibration procedure have been described elsewhere [27–32]. A detailed procedure is given in ISO Standard 15472:2001. Once the spectrometer energy scale has been calibrated, it is assumed to remain constant. This is valid as long as the spectrometer is maintained in an UHV environment. If the pressure of the spectrometer is raised above the UHV range, particularly when exposed to a reactive gas, different species can adsorb to components in the analyzer. This will change the ϕ_{sp} and necessitate recalibration. It is always good practice to regularly (i.e., daily to weekly) check the instrument calibration.

3.3.5 CHARGE COMPENSATION IN INSULATORS

For measuring E_B, the procedure described in the previous section is the one of choice when the electrical conductivity of the sample is higher than the emitted current of photoelectrons. However, some materials do not have sufficient electrical conductivity or cannot be mounted in electrical contact with the ESCA spectrometer. These samples require an additional source of electrons to compensate for the positive charge built up by emission of the photoelectrons. Ideally, this is accomplished by flooding the sample with a monoenergetic source of low-energy (<20 eV) electrons. When the only source of compensating electrons is monoenergetic low-energy electrons, the vacuum level of the sample will be in electrical equilibrium with the energy of the electrons (Figure 3.6). Therefore, the measured E_B of an

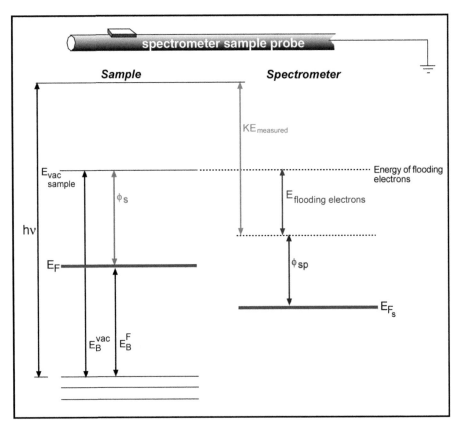

Figure 3.6 The energy level diagram for a sample electrically insulated from the spectrometer. The vacuum level of the sample (E_{vac}^s) is aligned with the energy of the charge neutralization electrons (ϕ_s) so that E_B is referenced with respect to ϕ_e. The measurement of E_B is dependent on the sample work function, ϕ_s

insulated sample depends on its work function (ϕ_s) and the energy of the flooding electrons, ϕ_e, as shown in Equation (3.10) and Figure 3.6 [33].

$$E_B^{\text{vac}} = E_B^F + \phi_s = h\nu - \text{KE} + \phi_e \tag{3.10}$$

Thus, for insulators E_B is referenced to E_{vac} and ϕ_e. This makes it difficult or impossible to measure absolute E_B values for samples not in electrical contact with the spectrometer. Under these conditions it is best to use an internal reference. For polymer and organic samples, the hydrocarbon component (C—C/C—H) of the C_{1s} peak is typically set to 285.0 eV. For supported catalysts, a major peak of the oxide support (Si_{2p}, Al_{2p}, etc.) is typically used. Internal referencing of the E_B scale allows the accurate measurement of other E_B values in the sample.

Usually the energy of the flooding electrons is varied to obtain the narrowest width of the photoemission peak. It is important to have the entire sample either electrically grounded or fully isolated. A sample in partial electrical contact with the spectrometer can lead to differential charging, which will produce distorted peak shapes and, under extreme conditions, new peaks. These experimental artifacts must be avoided to obtain a proper analysis of the sample. The analyst must be aware of the electrical properties of the sample and how they can affect the ESCA experiment. For example, a conducting metal substrate with a thin ($\sim$5 nm or less) insulating overlayer can usually be analyzed with the sample grounded. However, if the insulating overlayer becomes too thick ($\sim$10 nm or more), differential charging can occur. Then the entire sample must be electrically isolated from the spectrometer for proper analysis. For samples with electrical properties that vary with location on the sample, carefully designed experiments can be used to gain further information about the electrical and spatial properties of a sample [34–37].

3.3.6 PEAK WIDTHS

The observed width of a given photoelectron peak is determined by the lifetime of the core hole, instrumental resolution, and satellite features. The peak width due to the core hole lifetime can be calculated from the Heisenberg uncertainty relationship:

$$\Gamma = h/\tau \tag{3.11}$$

where Γ is the intrinsic peak width in eV, h is the Planck constant in eV-seconds, and τ is the core hole lifetime in seconds. For the C_{1s} orbital, Γ is $\sim$0.1 eV. For a given element, the value of Γ is typically larger for inner shell orbitals versus outer shell orbitals. This is because an inner shell core hole can be filled by electrons from the outer shells. Thus, the deeper the orbital, the shorter the core hole lifetime and the larger the intrinsic peak

width. For example, the intrinsic peak widths of Au increase in the order
4f < 4d < 4p < 4s, the order of increasing E_B. Similarly, the value of Γ
for a given orbital (e.g. 1s) increases as the atomic number of the element
increases, since the valence electron density, and therefore the probability of
filling the core hole, increases with increasing atomic number. The lineshape
due to the core hole lifetime is Lorenzian.

Instrumental effects that can broaden a photoemission peak width include
the energy spread of the incident X-rays and the resolution of the analyzer.
For insulating materials, additional peak broadening can occur from the
energy spread of the flooding electrons and the resulting energy spread
in the surface potential [38]. Typically it is assumed that instrumental
contributions to the photoemission peak have a Gaussian lineshape. The
contributions that the intrinsic and instrumental effects make to the peak-
width are given, to a first approximation, by:

$$\text{FWHM}_{tot} = (\text{FWHM}_n^2 + \text{FWHM}_x^2 + \text{FWHM}_a^2 + \text{FWHM}_{ch}^2 + \cdots)^{1/2} \quad (3.12)$$

where FWHM is the full-width at half-maximum of the observed peak
(tot), core hole lifetime (n), X-ray source (x), analyzer (a), and charging
contribution (ch).

The third contribution to peak widths is satellite features. These can arise
from several sources such as vibrational broadening, multiplet splitting, and
shake-up satellites. These features typically have asymmetric lineshapes
and, depending on their E_B, may or may not be resolvable from the main
photoemission peak. For example, metallic samples have a continuous band
of unfilled electron levels above E_F (the conduction band). Upon leaving the
sample, a photoelectron can transfer a portion of its KE to excite a valence
band electron into the conduction band. Because of the continuous ranges
of energies available to this process, an asymmetric tail on the high E_B (low
KE) side of the photoemission peak is observed for metallic samples. The
degree of peak asymmetry depends on the density of states near E_F [28].
Additional discussion of satellite features will be given in Section 3.6. Also,
further details have been published elsewhere [39].

3.3.7 PEAK FITTING

To maximize the information extracted from ESCA spectra, the area and E_B
of each subpeak for a given orbital (e.g. C_{1s}) must be determined. Typically,
the spacing between subpeaks is similar to observed peak widths (~1 eV).
Thus, it is rare when individual subpeaks are completely separated in an
experimental spectrum. This requires the use of a peak-fitting procedure to
resolve the desired peak parameters. Parameters used in such procedures
include the background, peak shape (Gaussian, Lorenzian, asymmetric, or
mixtures thereof), peak position, peak height, and peak width. The most
common method used to model the background (inelastic scattering) was

developed by Shirley [40]. After the background has been set, initial guesses, based on reasonable values obtained from previous experiments, are made for each peak parameter and then a least-squares fitting routine is used to iterate to the final values [41]. Caution must be exercised when performing the peak fit since many of the quantities are correlated. This can cause instabilities in the fitting algorithm or generation of non-unique results. For spectra containing severely overlapping peaks, the results obtained from peak fitting may depend on the starting parameters chosen (the algorithm converges to a local minimum instead of a global minimum). The experimenter must ensure that the results obtained from the peak fitting procedure are consistent with all other available information, as described in Section 3.6. It is best to start with accurate initial peak parameters. Also, additional independent information can be used to constrain peak parameters such as position, width, and area during the initial curve fitting iterations. Modern ESCA instruments that can produce spectra with narrow peak widths will permit more precise and detailed peak fits, as well as providing faster convergence of the peak fitting algorithm. Once the algorithm is close to convergence, the peak fitting constraints can be removed or relaxed. For example, appropriate purified standard specimens can be used to set peak positions. If the instrumental resolution dominates the peak width, then a 100 % Gaussian peak shape can be used. Under these conditions all peaks in a given spectrum should have similar widths. As the instrumental resolution is improved, mixtures of Gaussian–Lorenzian-asymmetric tailing must be used. For non-conducting polymer samples, when the C_{1s} peak width is <1 eV, Gaussian–Lorenzian mixtures are required. For narrow peaks of metallic samples, some asymmetric tailing should also be included. The peak fitting results obtained for the polyurethane C_{1s} spectrum are discussed in Section 3.6. This example shows how the concentration and E_B values for different functional groups determined by peak fitting correlate with information from the survey scan and other high-resolution scans.

Careful peak fitting allows detailed information to be extracted from ESCA spectra [42–45]. For example, where a C_{1s} spectrum of poly(2-chloroethyl methacrylate) was traditionally resolved into three peak components, using high-resolution spectrometers it becomes apparent that five peaks can be accurately fitted to this peak envelope [45]. High resolution instruments may even reveal vibrational components that broaden the ESCA peaks [28, 46].

3.4 Inelastic Mean Free Path and Sampling Depth

As illustrated schematically in Figure 3.7, while X-rays can readily travel through solids, electrons exhibit significantly less ability to do so. In fact, for X-rays of 1 keV (a typical order of magnitude for an ESCA excitation source), the X-rays will penetrate 1000 nm or more into matter while electrons of this

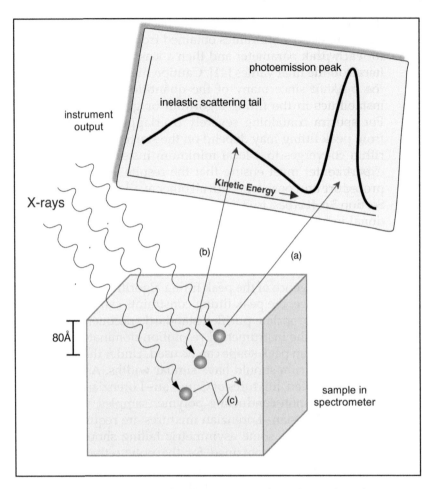

Figure 3.7   X-rays will penetrate deeply into a sample, and stimulate electron emisson throughout the specimen. Only those electrons emitted from the surface zone that have suffered no energy loss will contribute to the photoemission peak (a). Electrons emitted from the surface zone that have lost some energy due to inelastic interactions will contribute to the scattering background (b). Electrons emitted deep within a sample will lose all their kinetic energy to inelastic collisions and will not be emitted (c)

energy will only penetrate approximately 10 nm. Because of this difference, ESCA, in which only emitted electrons are measured, is surface sensitive. Electrons emitted from X-ray excitation below the uppermost surface zone cannot penetrate far enough to escape from the sample and reach the detector. Let us examine these relationships more quantitatively.

   In the ESCA experiment, we are concerned with only the intensity of the emitted photoelectrons (i.e., the total number emitted) that have not lost any energy. If an electron suffers energy loss, but still has sufficient energy to escape from the surface, it will contribute to the background signal, but not to the photoemission peak (Figure 3.7). Therefore, the ESCA sampling depth refers to a characteristic, average length in a solid that the electron can travel with no loss of energy. The decrease in the number of photoemitted electrons that have suffered no energy loss travelling through matter, where each unit thickness of matter the electrons travel through will absorb the same fraction of the energy, is described by Beer's law (Figure 3.8(a)). The inelastic mean free path (IMFP) term, λ, in this equation is that thickness of matter through which 63 % of the traversing electrons will lose energy. Table 3.4 presents a series of definitions of other terms commonly used in ESCA to describe this decrease in elastic electron intensity associated with transport through matter.

The Beer's law equation, as formulated in Figure 3.8(a), applies to transmission of electrons through a specimen of thickness, d. In ESCA, we

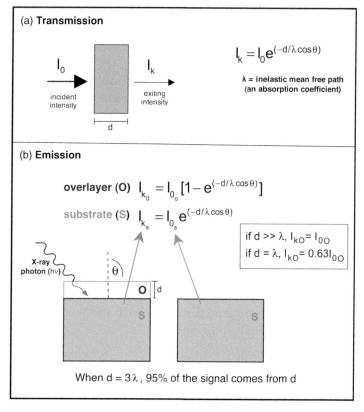

Figure 3.8 (a) For electrons transmitted through a sample, Beer's law of molecular absorption explains the total intensity loss for electrons that lose no energy in traversing the sample. (b) For electron emission from a thick sample, modifications of Beer's law can explain the photoemission intensity from an overlayer or from the substrate covered by an overlayer

Table 3.4 Definitions for electron transport in materials[a]

IMFP (λ) – Inelastic Mean Free Path. The average distance that an electron with a given energy travels between successive inelastic collisions.

ED – Mean Escape Depth. The average depth normal to the surface from which the photoelectrons escape.

AT – Attenuation Length. The quantity l in the expression $\Delta x/l$ for the fraction of a parallel beam of X-rays or electrons that are removed by passing through a thin layer Δx of a substance in the limit as Δx approaches zero (Δx is in the direction of the beam).

ID – Information Depth. The maximum depth, normal to the surface, from which useful information is obtained.

Sampling Depth $= 3\lambda$ (ID where percentage of detected electrons is 95 %).

[a]Definitions adapted from ISO 18115:2001 and Powell [47].

generally detect electrons escaping from a solid that is many times thicker than the escape depth of the electrons (Figure 3.7). Since, over the thickness range through which the electron flux is appreciably attenuated, the X-ray flux suffers essentially no diminution, the X-rays can be viewed as stimulating photoemission throughout the bulk of the sample. View the sample, then, as a source of electrons (I_0). We may now ask, how will the electron flux from this source (the sample) be attenuated, if we cap this electron source with a thin overlayer? This situation, and the equations that describe it, are presented in Figure 3.8(b). The equations in Figure 3.8(b) are useful for qualitatively and quantitatively describing the photoelectron emission intensity for many commonly encountered sample types. These equations will be applied to depth profiling in Section 3.9.

The actual values for the IMFP of electrons in matter are a function of the energy of the electrons and nature of the sample through which they travel. Over the range of electron kinetic energies of most interest in ESCA, the IMFP increases with electron KE. The form of the dependence of IMFP on KE is described by KE^n where n has been estimated at 0.54–0.81 (0.7 is frequently used) [47, 48]. Equations that relate IMFP to electron energy and the type of material through which the electron is traversing have been developed by Seah and Dench [49]:

$$\text{IMFP} = \lambda = 538 KE^{-2} + 0.41(aKE)^{0.5} \quad \text{(for elements)} \tag{3.13}$$
$$\text{IMFP} = \lambda = 2170 KE^{-2} + 0.72(aKE)^{0.5} \quad \text{(for inorganic compounds)} \tag{3.14}$$
$$\text{IMFP} = \lambda_d = 49 KE^{-2} + 0.11 KE^{0.5} \quad \text{(for organic compounds)} \tag{3.15}$$

where:

λ is in units of monolayers
$a =$ monolayer thickness (nm)
λ_d (in mg m^{-2})
$KE =$ electron kinetic energy (in eV)

The equations were developed empirically based upon data from a large number of researchers. As a good fit to experimental data, they provide useful guidelines. However, precise values of IMFP in materials have been the subject of considerable controversy since the earliest days of the ESCA technique [47, 48, 50–55]. It is reasonable to say that the IMFP for photoelectrons of interest in ESCA probably falls in the range 1–4 nm. The actual IMFP values will depend on the density, composition and structure of the material being analyzed. Numbers calculated from the equations above can be used in most calculations if it is appreciated that, although precise values cannot be assigned, a reasonable estimate of sampling depths can be made. The sampling depth, as defined in Table 3.4, is three times the IMFP (i.e., the depth from which 95 % of the photoemission has taken place).

3.5 Quantification

As discussed in previous sections, the complete ESCA spectrum of a material contains peaks that can be associated with the various elements (except H and He) present in the outer 10 nm of that material. The area under these peaks is related to the amount of each element present. So, by measuring the peak areas and correcting them for the appropriate instrumental factors, the percentage of each element detected can be determined. The equation that is commonly used for these calculations is:

$$I_{ij} = KT(\text{KE})L_{ij}(\gamma)\sigma_{ij} \int n_i(z)e^{-z/\lambda(\text{KE})\cos\theta}\,dz \qquad (3.16)$$

where I_{ij} is the area of peak j from element i, K is an instrumental constant, T (KE) is the transmission function of the analyzer, $L_{ij}(\gamma)$ is the angular asymmetry factor for orbital j of element i, σ_{ij} is the photoionization cross-section of peak j from element i, $n_i(z)$ is the concentration of element i at a distance z below the surface, λ (KE) is the inelastic mean free path length, and θ is the take-off angle of the photoelectrons measured with respect to the surface normal. Equation (3.16) assumes that the sample is amorphous. If the sample is a single crystal, diffraction of the outgoing photoelectrons can cause peak intensities to deviate from values predicted by Equation (3.16) [56, 57]. By using large solid angle acceptance lenses (>20°) and either amorphous or polycrystalline samples, these diffraction effects can be neglected. Rarely are all of the quantities in Equation (3.16) evaluated. Typically, either elemental ratios (e.g. C/O atomic ratio) or percentages (e.g. atomic percentage carbon) are calculated. Thus, it is only necessary to determine the relative relationship, not the absolute values, of the quantities in Equation (3.16).

The instrumental constant, K, contains quantities such as the X-ray flux, area of the sample irradiated, and the solid angle of photoelectrons accepted by the analyzer. It is assumed not to vary over the time period and conditions used to acquire the ESCA spectra for quantification. Being a constant, it cancels when either elemental ratios or atomic percentages are calculated. The angular asymmetry factor $L_{ij}(\gamma)$ accounts for the type of orbital the photoelectron is emitted from and the angle γ between the incident X-rays and the emitted photoelectrons. The value of $L_{ij}(\gamma)$ for a particular peak can be calculated [58]. If only s orbitals are used for quantitation, $L_{ij}(\gamma)$ will be the same for all peaks and therefore will cancel. This situation is frequently encountered with polymeric samples since the 1s orbitals of many elements present in organic polymers (C, N, O, and F) are detectable by ESCA. Even for samples where different types of orbitals are used for quantification, the variation of $L_{ij}(\gamma)$ is typically small and is usually neglected for solids. However, it is always best to use orbitals of the same symmetry for calculating elemental ratios or atomic percentages.

The transmission function of the analyzer includes the efficiency of the collection lens, the energy analyzer, and detector. Most ESCA instruments are run in the constant-pass energy mode. This means that regardless of the initial KE of the emitted electrons, they will pass through the energy analyzer at a constant energy. This requires the collection lens to reduce the KE of the incoming electrons down to the pass energy. In this case, the only variation in the transmission function with KE of the photoelectrons is due to retardation in the lens system, which can be determined experimentally and usually has the form of KE^n. Most manufactures provide information about the transmission function of their instruments. Published data are also available for many instruments [59–61].

The photoionization cross-section σ_{ij} is the probability that the incident X-ray will create a photoelectron from the jth orbital of element i. Values for σ_{ij} are typically taken from the calculations of Scofield [62]. Selected values of the Scofield cross-sections are listed in Table 3.5. Empirically determined cross sections are also available [63, 64]. The IMFP, λ (KE), has been discussed in Section 3.4. For quantitative analysis, the values calculated from the equations published by Seah and Dench [49] (Equations 3.13–3.15) are commonly used. These equations show that λ depends both on the sample type (elemental, inorganic species, or organic species) and the KE of the photoelectron. Both quantities must be properly accounted for to obtain good quantitative results. The $\cos\theta$ term accounts for the decrease in sampling depth as the surface normal of the sample is rotated away from the axis of the acceptance lens. This is described in detail in Sections 3.4 and 3.9 and Figure 3.19.

3.5.1 QUANTIFICATION METHODS

The concentration of element i, n_i, is the unknown quantity in Equation (3.16). All other terms in this equation can either be measured (e.g. I_{ij}) or calculated (e.g. σ_{ij}). Therefore, Equation (3.16) can be solved for n_i. Once n_i is known for each element present in the ESCA spectrum, the atomic

Table 3.5 Select photoelectron binding energies (eV) and Scofield photoemission cross-sections from [62][a] with permission from John Wiley & Sons, Ltd.

Element	1s	2s	2p$_{1/2}$	2p$_{3/2}$	3s	3p$_{1/2}$	3p$_{3/2}$	3d$_{3/2}$	3d$_{5/2}$	4s	4p$_{1/2}$	4p$_{3/2}$	4d$_{3/2}$	4d$_{5/2}$	4f$_{5/2}$	4f$_{7/2}$
C	284 [1.00]															
N	399 [1.80]															
O	532 [2.93]	24 [0.141]														
F	686 [4.43]	31 [0.210]														
Al		118 [0.753]	73 [0.181]	73 [0.356]												
Si		149 [0.955]	100 [0.276]	99 [0.541]												
P		189 [1.18]	136 [0.430]	135 [0.789]	16 [0.112]											
S		229 [1.43]	165 [0.567]	164 [1.11]	16 [0.147]											
Ti		564 [3.24]	461 [2.69]	455 [5.22]	59 [0.473]	34 [0.276]	34 [0.537]									
Cu		1096 [5.46]	951 [8.66]	932 [16.73]	120 [0.957]	74 [0.848]	74 [1.63]									
Ag					717 [2.93]	602 [4.03]	571 [8.06]	373 [7.38]	367 [10.66]	95 [0.644]	62 [0.700]	56 [1.36]				
I					1072 [3.53]	931 [5.06]	875 [10.62]	631 [13.77]	620 [19.87]	186 [0.959]	123 [1.11]	123 [2.23]	50 [1.69]	50 [2.44]		
Au										759 [1.92]	644 [2.14]	546 [5.89]	352 [8.06]	334 [11.74]	87 [7.54]	84 [9.58]

[a]Cross-sections listed in brackets.

percentages can be calculated as:

$$\%n_i = 100\left(n_i \Big/ \sum n_i\right) \tag{3.17}$$

where $\%n_i$ is the atomic percent of element i. Atomic ratios (n_i/n_k) can also be calculated. To remove the integral from Equation (3.16) it is usually assumed that the elemental concentrations are homogeneous within the ESCA sampling depth. Equation (3.16) can then be integrated to obtain:

$$I_{ij} = KT(KE)L_{ij}(\gamma)\sigma_{ij}n_i\lambda(KE)\cos\ \theta \tag{3.18}$$

When n_i is not homogeneous with respect to z, the depth profiling experiments described in Section 3.9 are required to determine the form of $n_i(z)$.

3.5.2 QUANTIFICATION STANDARDS

Standard samples can be used to evaluate the validity of the quantification equations presented above. Four criteria for standard samples are they should have a known composition, be homogeneous with depth, be relatively stable, and be free of contaminants. Two polymer samples that meet these criteria are polytetrafluoroethylene (PTFE) and poly(ethylene glycol) (PEG).

PTFE is composed exclusively of chains of CF_2 units giving an F/C atomic ratio of 2.0. This polymer is known to be unreactive, and have low surface energy, so it is relatively easy to prepare a clean PTFE surface. The presence of oxidation or contamination can be readily determined by examination of the ESCA spectra. Since only F and C are present in PTFE, the detection of O by ESCA indicates the presence of surface oxidation or a surface contaminant. Likewise, any C_{1s} peaks other than the CF_2 peak at 292 eV indicate the presence of a surface contaminant. The most common contaminant is adsorbed hydrocarbon, which has a C_{1s} peak at 285 eV (Figure 3.9). Even small amounts of hydrocarbon contamination can be readily detected. The amount shown in Figure 3.9 represents $\sim$0.3 atomic percentage carbon. If the PTFE sample shows significant hydrocarbon contamination, sonicating the PTFE in acetone followed by methanol usually removes the contaminant. The results in Table 3.6 for cleaned PTFE samples were obtained using Equation (3.18). The excellent agreement between the experimental values and stoichiometric values support the use of the equations listed above for quantification. However, halogen-containing polymers such as PTFE are known to degrade with extended exposure to X-rays, especially non-monochromatic X-rays. So as a general practice it is recommended that monochromatized X-rays with minimal exposure times be used to analyze organic and biological samples.

Like PTFE, PEG is also a good candidate for a standard material. The formula for PEG is $HO-(CH_2-CH_2-O)_n-H$ so only C and O should be

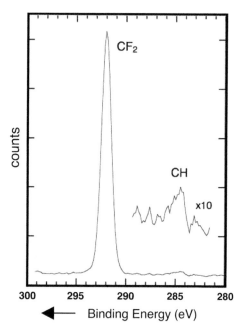

Figure 3.9 The ESCA C_{1s} spectrum of polytetrafluoroethylene. The peak at 292 eV corresponds to the CF_2 groups present in this sample. The weak peak at 285 eV corresponds to a small amount (~0.3 atomic percentage) of a hydrocarbon contaminant present on the surface of the sample

Table 3.6 Quantitation results for polytetrafluoroethylene[a]

Atomic percentage F $= 67.1 \pm 0.4$
Atomic percentage C $= 32.9 \pm 0.5$
F/C atomic ratio $= 2.04 \pm 0.04$

[a]Number of samples $= 22$.

present with a C/O atomic ratio of 2.0. The PEG C_{1s} spectrum should only have one peak at 286.5 eV (C—O species), so the presence of a hydrocarbon contaminant ($E_B = 285$ eV) can readily be detected. The results for PEG listed in Table 3.7, like the PTFE results, show excellent agreement between experiment and stoichiometry. This provides further support for the use of Equation (3.18) for quantification experiments.

3.5.3 QUANTIFICATION EXAMPLE

The results for a polyurethane sample presented in Tables 3.8 and 3.9 provide an additional example of the accuracy of quantitative analysis of polymeric

Table 3.7 Quantitation results for poly(ethylene glycol)[a]

Atomic percentage O $= 33.8 \pm 0.4$
Atomic percentage C $= 66.2 \pm 0.4$
C/O atomic ratio $= 1.96 \pm 0.03$

[a]Number of samples $= 12$.

Table 3.8 Quantitation results for a polyurethane sample

Element	Orbital	KE (eV)	σ	Peak area (counts $\times$ eV)	Atomic percentage
Carbon	1s	1200	1.00	26 557	76.9
Nitrogen	1s	1085	1.80	4478	7.7
Oxygen	1s	955	2.93	13 222	15.4

Table 3.9 Quantitation results for a polyurethane sample[a]

	Atomic percentage	
Element	ESCA	Stoichiometry
Carbon	76.6 ± 1.0	76.0
Nitrogen	7.9 ± 0.5	8.0
Oxygen	15.5 ± 0.8	16.0

[a]Number of samples $= 8$.

materials. Further information about the structure and identification of ESCA spectral features of this polyurethane are provided in the next section. In the quantification experiments only C, N, and O were detected, and the 1s peak areas were used for quantification. Therefore, $L_{ij}(\gamma)$ can be considered a constant. The spectra were acquired on a Surface Science Instruments X-probe spectrometer with the following characteristics: $T(KE)$ is constant over the range of detected photoelectron kinetic energies, $\lambda(KE)$ varies as $KE^{0.7}$, $h\nu = 1487$ eV, and $\theta = 55°$. Under these conditions Equations (3.17) and (3.18) can be combined to yield:

$$\%n_i = \left(I_{ij}/\sigma_{ij}KE^{0.7} \right) \Big/ \sum \left(I_{ij}/\sigma_{ij}KE^{0.7} \right) \tag{3.19}$$

Table 3.8 shows the values of I_{ij}, KE, σ_{ij}, and $\%n_i$ for analysis of one polyurethane sample. Table 3.9 summarizes the results from eight different analyses of this material with the calculated standard deviations. The composition expected from the polyurethane stoichiometry is also listed. The results show both good reproducibility and accuracy. As the atomic percentage of the element decreases towards the ESCA detection limits (~0.1 atomic percentage), the relative standard deviation will increase significantly. Near the detection limit the magnitude of the standard deviation is typically the same as the magnitude of the atomic percentage. Based upon the results in Table 3.9, the polyurethane sample has similar surface and bulk compositions. This is not always the case, and examples of variation in the surface composition with respect to the bulk composition are shown in Section 3.9. In addition to the examples in Section 3.9, the presence of contaminants is also often detected at the surface of samples. Oxidation of the sample and adsorption of hydrocarbons and silicones are common contamination processes. A more detailed discussion of quantification considering matrix effects and sensitivity factors has been published elsewhere [65].

3.6 Spectral Features

The understanding and analysis of ESCA spectra require an appreciation of the spectral features that are observed. ESCA analyses are typically performed by first taking a wide scan or survey scan spectrum, often covering a range of 1100 eV, and then looking in more detail over smaller ranges (perhaps 20 eV) at specific features found in the wide scan spectrum. A characteristic wide scan spectrum, energy referenced to compensate for sample charging as described in Section 3.3, is presented in Figure 3.10. High-resolution spectra of specific features observed in the wide scan spectrum are shown in Figure 3.11. First, let us consider the wide scan spectrum.

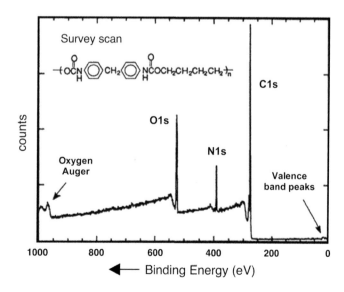

Figure 3.10 The ESCA survey scan of a hard-segment polyurethane

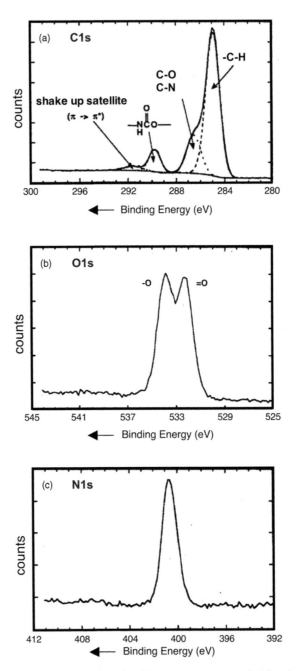

Figure 3.11   (a) The C_{1s} spectrum (resolved into component peaks) for the hard-segment polyurethane; (b) the O_{1s} spectrum for the hard-segment polyurethane; (c) the N_{1s} spectrum for the hard-segment polyurethane

The wide scan spectrum of a synthetic polymer, a polyurethane (Figure 3.10), has been annotated specifically for this example. The chemical structure of this polymer is also contained within this figure. First, note the x-axis. This axis is generally labeled 'binding energy.' However, from the Einstein equation, it is apparent that we could also plot it in terms of KE. As discussed in Section 3.3, the KE of the emitted photoelectron is a precisely measured value. The binding energy is a calculated value computed from the KE, the energy of the X-ray photon, the work function of the surface, and a correction term due to electrical charge accumulation on the surface. Still, when the ESCA instrument is in proper calibration, there is an inverse, linear relationship between KE and binding energy. Since binding energy has meaning for the chemistry and structure of the surface, it is most common to plot ESCA spectra in terms of binding energy. Typically the binding energies values on the x-axis decrease from left to right (i.e., KE increases from left to right). The y-axis is typically 'intensity' or 'number of counts'. For the presentation of ESCA data, it is usually linear rather than logarithmic.

Next, we can observe the background. The number of counts attributed to the background will typically first increase abruptly and then decrease slowly with increasing binding energy (decreasing kinetic energy) above the photoemission peak. This is the inelastic scattering, as suggested in Figure 3.7. After each photoemission event, there is a cumulative background signal associated with photoelectrons that have lost energy due to inelastic collisions in the solid, but that still have sufficient energy to escape the work function of the surface. The magnitude and dependence of the inelastically scattered background intensity with increasing E_B will depend on the composition and structure of the sample as well as the photoemission peak being analyzed [66]. There is a continuum of energies of the inelastic background electrons that range from the photoemission peak KE to zero KE, since the collision events reducing the KE of the photoelectron do not have discrete energies.

Rising prominently above the background signal we observe two types of peaks in Figure 3.10. There are photoemission peaks associated with core-level photoionization events and X-ray-induced Auger electron emission peaks. If binding energy referencing has been performed, peaks can be readily identified from their positions using tabulations of binding energy values [2, 67]. Where ambiguity exists as to a peak identity, it is useful to look for other photoemission lines from the same element. For example, iridium (irradiated by an aluminum anode X-ray source) should have reasonably strong emissions at 690 eV (4s), 577 eV ($4p_{1/2}$), 495 eV ($4p_{3/2}$), 312 eV ($4d_{3/2}$), 295 eV ($4d_{5/2}$), 63 eV ($4f_{5/2}$), and 60 eV ($4f_{7/2}$), with the latter five lines particularly intense. If all members of this series of lines (and especially the most intense of the set) are not observed in a spectrum, then iridium is probably not present. Table 3.5 contains the binding energies of a few select photoemission lines produced with Al Kα irradiation (1487 eV).

Auger lines are usually also listed in photoelectron peak tabulations. Examples of X-ray induced Auger lines for metallic Cu (Cu^0), Cu_2O (Cu^{+1}) and CuO (Cu^{+2}) are shown in Figure 3.4(b). Auger lines can be readily distinguished from photoemission lines by changing X-ray sources (e.g. using a Mg Kα source instead of an Al Kα source). The kinetic energies of all the Auger lines will remain the same, while the kinetic energies of the photoemission lines shift by the difference in energies of the two X-ray sources. Auger peaks can be used analytically, in conjunction with the photoemission peaks, to distinguish between different possible chemical species using the modified Auger parameter [68]:

$$\alpha' = h\nu_{\text{X-ray}} + KE_{\text{Auger}} - KE_{\text{photoelectron}} \qquad (3.20)$$

The final feature to be discussed in this wide scan spectrum is observed at low binding energies. The low-intensity features seen between 0 and 30 eV are due to photoemission of valence (outer shell) electrons. Interpretation of these spectral features is often more complex than the core-level lines, and has been presented elsewhere [69]. The valence band region will be discussed further in Section 3.12.

Much additional detail can be observed in the high-resolution ESCA spectra. Consider Figure 3.11(a), the high-resolution C_{1s} spectrum from the polyurethane sample used to generate Figure 3.10. From the peak shape, it is apparent that this spectrum is composed of a number of subpeaks. These subpeaks, attributed to chemical shifts from atoms and groups bound to the carbons (see Section 3.3), are identified in the figure. Methods and the rationale for resolving a peak envelope into subpeaks have been described [1, 28].

As well as peaks for each of the major carbon species, another feature is noted at 6.6 eV from the lowest binding energy (hydrocarbon) peak. This peak is referred to as a shake-up satellite. It represents photoelectrons that have lost energy through promotion of valence electrons from an occupied energy level (e.g. a π level) to an unoccupied higher level (e.g. a $π^*$ level). Shake-up peaks (also called 'loss peaks' because intensity is lost from the primary photoemission peak) are most apparent for systems with aromatic structures, unsaturated bonds or transition metal ions. Examples of the Cu_2O and CuO shake-up satellites are shown in Figure 3.4(a). In contrast to the continuum of reduced energies seen in the inelastic scattering tail, shake-up peaks have discrete energies (~6.6 eV higher binding energy than the primary peak in C_{1s} spectra of aromatic-containing molecules) because the energy loss is equivalent to a specific quantitized energy transition (i.e., the π → $π^*$ transition). If the departing photoelectron transfers sufficient energy into the valence electron to ionize it into the continuum, the photoemission loss peak is called a 'shake-off' peak. The shake-off satellite peaks of the photoemission peak can have a wide range of possible energies (of course, always with a lower KE than the photoemission peak). This energetically broad feature is typically hidden within the background signal and is usually not detected or used analytically.

Much additional information about the polymer is available by examining the other high-resolution spectra (Figures 3.11(b) and 3.11(c)). Specific interesting features are annotated on the spectra. Quantitative information comes from the ratio of the areas under the peaks in Figures 3.11(a−c).

As a general note, proper use of ESCA relies on taking advantage of all the information available in the spectra. Thus, the analysis does not end with the acquisition of the wide scan spectrum. High-resolution spectra of each of the features found in the wide scan spectrum are examined to extract maximum information. The information from a complete data set should be corroborative and not contradictory. For example, where significant levels of oxygen are seen in a wide scan spectrum of an organic polymer, subpeaks associated with carbons bound to oxygen should be found in the high-resolution C_{1s} spectrum. Furthermore, the subpeaks in the O_{1s} spectrum should also have binding energies appropriate for oxygen−carbon functionalities. When contradictions are found, further analysis and perhaps reacquisition of the data are in order. The theoretical and experimental understanding of ESCA spectra are very well developed, and contradictory evidence should have a sound explanation. For example, have artifacts been introduced into a spectrum due to surface charging? If so, exercising more care in charge compensation during reacquisition of the data can often resolve such problems.

In inorganic systems a number of other spectral features are observed that must be understood. These include spin−orbit doublets, multiplet splitting, and plasmon losses. Each will be described in the following.

In Figure 3.12, the initial state and final states (after photoemission) for a pair of electrons in a 3p orbital are schematically illustrated. Note that two energetically equivalent final states are possible, 'spin up' or 'spin down'. If there is an open shell (quantum number $1 > 0$, i.e. a p, d, or f orbital) with two states of the same energy (orbital degeneracy), a magnetic interaction between the spin of the electron (up or down) and its orbital angular momentum may lead to a splitting of the degenerate state into two components. This is called spin−orbit coupling or $j−j$ coupling (j quantum number $= 1 + s$). Figure 3.12 also shows common spin−orbit pairs. The ratio of their respective degeneracies, $2j + 1$, determines the intensities of the components. Figure 3.13 illustrates the $f_{5/2}$ and $f_{7/2}$ components of a gold 4f photoemission peak. The total 4f photoemission intensity for gold as used in quantitation is the sum of the two spin−orbit peaks. The trend for the doublet separation is p > d > f within a given atom.

A related phenomenon, referred to as multiplet or electrostatic splitting, is seen for the s orbital photoemission from some transition metal ions

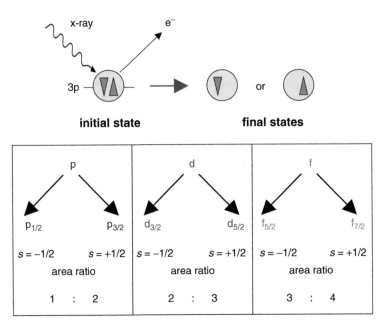

Figure 3.12   After electron emission from a 3p orbital subshell, the remaining electron can have a spin-up or spin-down state. Magnetic interaction between these electrons and the orbital angular momentum may lead to spin–orbit coupling

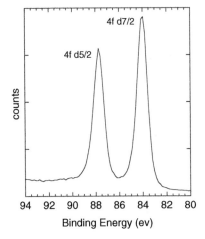

Figure 3.13   Spin–orbit coupling leads to a splitting of the gold 4f photoemission line into two subpeaks

(e.g. Mn^{+2}, Cr^{+2}, Cr^{+3}, Fe^{+3}). A requirement for this splitting of the s photoemission peak into a doublet is that there be unpaired orbitals in the valence shells. Complex peak splittings can be observed in transition metal ions and rare earth ions when multiplet splitting occurs in p and d levels. Additional detail on this splitting can be found elsewhere [26, 28].

The conduction electrons in metals, in contrast to being localized on each atom, have been likened to a 'sea' or continuum. Characteristic collective vibrations have been noted for this continuum of electrons and are referred to as plasmon vibrations. In some cases, the exiting photoelectron can couple with the plasmon vibrations leading to characteristic, periodic energy losses. An example of the plasmon loss series of peaks for the aluminum 2s photoemission is presented in Figure 3.14.

The features most often observed in ESCA spectra, independent of the type of instrument used, have been briefly reviewed. A feature that is associated with the type of instrument is the X-ray satellite. Non-monochromatized X-ray sources (see Section 3.7) may excite the sample with more than one X-ray line. From Equation (3.1), each X-ray line will lead to a distinct photoemission energy. Low-intensity X-ray lines, particularly the $K_{\alpha3,4}$, will produce low-intensity photoemission peaks with approximately 10 eV higher KE than the primary photoemission peak. These X-ray satellite peaks are not observed with monochromatic X-ray sources that are described in Section 3.7.

Table 3.10 summarizes all features that are important both to understand the spectra obtained, and for enhancing the information content of the ESCA experiment.

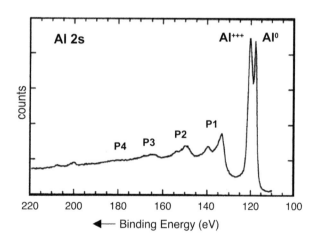

Figure 3.14 Plasmon loss peaks for the aluminum 2s photoemission peak

Table 3.10   Features observed in ESCA spectra

1. Photoemission peaks
   - Narrow
   - Nearly symmetric
   - Shifted by chemistry
   - Contain vibrational fine structure
2. X-ray satellite peaks
   - Not observed with a monochromatized source
   - Always the same energy shift from the photoemission peak
3. Shake-up satellites and shake-off satellites[a]
4. Photon-enduced Auger lines
5. Inelastic scattering background[a]
6. Valence band features
7. Spin–orbit coupling
8. Multiplet splitting
9. Plasmon loss peaks[a]

[a]Loss process.

3.7 Instrumentation

The ESCA experiment is necessarily tied to the complex instrumentation needed to stimulate photoemission and to measure low fluxes of electrons. A schematic drawing of a contemporary ESCA instrument is shown in Figure 3.15. The primary components that make up the ESCA instrument are the vacuum system, X-ray source, electron energy analyzer, and data system.

3.7.1 VACUUM SYSTEMS FOR ESCA EXPERIMENTS

Vacuum systems are described in detail in the appendix to this book, so only the aspects of the vacuum system that are pertinent to ESCA will be described here. The heart of the ESCA instrument is the main vacuum chamber where the sample is analyzed (analysis chamber). The ESCA experiment must be conducted under vacuum for three reasons. First, the emitted photoelectrons must be able to travel from the sample through the analyzer to the detector without colliding with gas phase molecules. Second, some components such as the X-ray source require vacuum conditions to remain operational. Third, the surface composition of the sample under investigation must not change during the ESCA experiment. Only a modest vacuum (10^{-6}–10^{-7} torr; 1 torr = 133 Pa) is necessary to meet the first two requirements. More stringent vacuum conditions are necessary to avoid contamination of the sample. The actual vacuum required will depend on

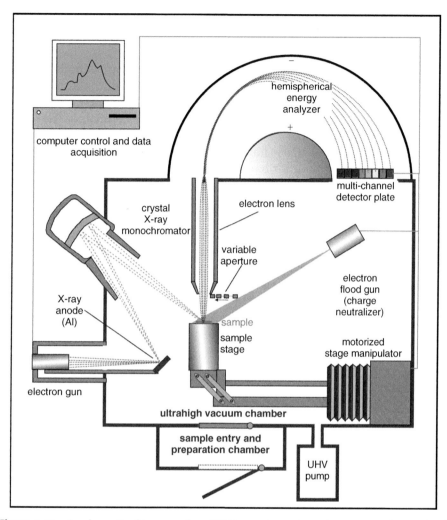

Figure 3.15 A schematic diagram of an ESCA spectrometer using a monochromatized X-ray source. The key components of a modern spectrometer are identified

the reactivity of the sample (e.g. metallic Na will require a better vacuum than PTFE). For most applications a vacuum of 10^{-10} torr is adequate. For studies on polymeric materials, good results can usually be obtained with a vacuum of 10^{-9} torr.

Samples are typically introduced into the analysis vacuum chamber via a load-lock or preparation chamber. In its simplest form, the load-lock is a small volume chamber that can be isolated from the analytical chamber and then backfilled to atmospheric pressure. One or more samples are placed in the load-lock chamber, which is then evacuated, typically with a turbomolecular pump. After the samples are pumped down, they are transferred into the analytical chamber. Depending on the vacuum require-ments and the type of samples, the pumpdown process can be as short as a few minutes or as long as several hours. In many cases it is desirable to do more elaborate sample processing before introducing the sample into the analytical chamber. For these cases, custom chambers providing an UHV environment with ion guns, deposition sources, sample heating and cooling, sample cleaving, gas dosers, etc., are available. The configuration of these sample preparation chambers depends on their intended use. Further details about sample handling and processing are described elsewhere [70].

After the samples have been placed in the analytical chamber, they must be properly positioned for analysis. This is accomplished with a sample holder/manipulator. Sample manipulators typically have the capability to translate a sample in three directions and to rotate it in one or two directions. Temperature control is also available on most manipulators. For spectrometers used for multisample analysis, the translation and rotation motions are computer controlled, so unattended operation of the instrument is possible. By coupling different sample mounting techniques with the manipulator capabilities and/or adding other components such as ion guns, a range of different ESCA experiments can be done (variable temperature, variable angle, multisample, destructive depth profiling, etc.).

3.7.2 X-RAY SOURCES

X-rays for an ESCA experiment are usually produced by impinging a high-energy ($\sim$10 keV) electron beam onto a target. Core holes are created in the atoms of a target material or anode, which in turn emits fluorescence X-rays and electrons (see Section 3.2 and Figure 3.2). It is the fluorescence X-rays that are used in the ESCA experiments. Common anodes along with the energies of their characteristic emission lines are listed in Table 3.11. A specific fluorescence line is used instead of the background emission (Bremsstrahlung) since its intensity is several orders of magnitude higher than the background emission. Thus, the X-ray emission energy is fixed for each anode. A multi-anode configuration is used to provide two or more X-ray energies. Most spectrometers use only one or two anodes, with Al and Mg the most common for non-monochromatic sources and Al the most common for monochromatic sources. Since most of the incident electron energy is converted into heat, anodes are usually water cooled. This allows operation at higher power loads without significant degradation (e.g. melting).

The emission from the anode can be allowed to strike the sample directly. Although this provides a high X-ray flux, it has several disadvantages. First, the energy resolution of the X-ray source is determined by the natural width of the fluorescence line (typically 1–2 eV). Second, the emission from weaker (satellite) X-ray fluorescence lines will also strike the sample, resulting in

Table 3.11 Characteristic energies and linewidths for common ESCA anode materials

Anode material	Emission line	Energy (eV)	Width (eV)
Mg	Kα	1253.6	0.7
Al	Kα	1486.6	0.85
Si	Kα	1739.5	1.0
Zr	Lα	2042.4	1.7
Ag	Lα	2984	2.6
Ti	Kα	4510	2.0
Cr	Kα	5415	2.1

the appearance of satellite peaks in the ESCA spectrum. Third, high-energy electrons, Bremsstrahlung, and heat will strike the sample, which can result in sample degradation. The flux of electrons and Bremsstrahlung can be minimized by placing a thin, relatively X-ray-transparent foil between the X-ray source and the sample. The presence of the foil will also minimize contamination of the sample by the X-ray source. For Al and Mg anodes, a ~2 μm thick Al foil is commonly used. The best way to optimize single energy production is to use an X-ray monochromator. The most popular monochromatized source combines an Al anode with one or more quartz crystals. The lattice spacing for the $10\bar{1}0$ planes in quartz is 0.425 nm, which is appropriate for the Al Kα wavelength (0.83 nm). For these wavelengths, the Bragg relationship ($n\lambda = 2d\sin\theta$) is satisfied at an angle of 78°. The geometry of a monochromatized X-ray source is illustrated in Figure 3.15. The quartz monochromator crystal and a thin Al foil to isolate the source from the sample will prevent electrons, Bremsstrahlung, satellite X-ray lines, and heat radiation from striking the sample. It will also narrow the energy spread of X-rays striking the sample. The disadvantages of a monochromator are the lower X-ray intensity that reaches the sample and the higher cost. The decrease in X-ray flux to the sample can be compensated for by using an efficient collection lens, energy analyzer, and multichannel detector system. Such a monochromatized instrument was successfully commercialized in the early 1970s [71]. In the mid 1980s, other manufacturers adopted this approach and now monochromatized ESCA instruments are widely used.

The area of the sample irradiated by the X-source depends on the geometry of the source and the type of electron gun used to stimulate X-ray emission. Most non-monochromatized sources illuminate a spot that is a few centimeters in diameter. In contrast, the monochromatized sources typically illuminate an area that is a few millimeters or smaller in diameter. With a focused electron gun and the quartz crystal used as both a monochromator and a focusing element, spot sizes <50 μm in diameter can be realized [72, 73].

The above discussion of X-ray sources deals with conventional instrumentation for individual laboratory experiments. The increased availability of synchrotron radiation in recent years has opened another avenue for ESCA experiments. The synchrotron provides a broad band of intense radiation (infrared to hard X-rays) that is highly collimated and polarized. When used with a suitable monochromator, synchrotron radiation can provide a tunable source of high intensity, focused X-rays for photoemission experiments. With the use of zone plates X-ray spot sizes <150 nm can be obtained [74]. The high X-ray flux typically focused into this small area can cause significant degradation of organic and biological samples. Complete degradation of these samples can occur with seconds if measures are not taken to reduce the X-ray brilliance (X-ray per unit area). However, the number of synchrotron facilities is far less than the number of stand-alone ESCA instruments. This often requires the investigator to travel extended distances to carry out experiments at synchrotron facilities. Further discussion of synchrotron facilities, instrumentation, and capabilities have been presented elsewhere [75–81].

3.7.3 ANALYZERS

The analyzer system consists of three components: the collection lens, the energy analyzer, and the detector. On most modern ESCA spectrometers, the lens system can collect photoelectrons from solid angles >20°. The higher the collection solid angle, the higher the number of photoelectrons collected per incident X-ray, which is generally advantageous. An efficient lens system can offset, in part, the decreased signal intensity encountered when using monochromatized and focused X-ray sources. The increased collection angle is particularly important for samples that degrade upon exposure to X-rays, since the more efficient the detection system is (e.g. the more photoelectrons collected per X-ray) the more data that can be collected before the sample is damaged. One type of experiment in which a large acceptance angle is a disadvantage, is non-destructive depth profiling. A large acceptance angle, by definition, contains a broad range of photoelectron take-off angles. This degrades the depth resolution obtainable in a variable take-off angle experiment. To improve the depth resolution, an aperture is placed over the entrance to the analyzer lens [82]. Recently an alternate approach has been developed where data from all photoelectron take-off angles are collected simultaneously using a two-dimensional detector [83].

In addition to collecting the photoelectrons, the lens system on most spectrometers also retards their KEs down to the pass energy of the energy analyzer. The entire ESCA spectrum is acquired by ramping appropriate voltages on the different lens elements. The range and retardation ratio used depends on the pass energy of the energy analyzer and the spectral range to be examined [28]. The lens system also projects the analyzed area a distance away from the entrance of the energy analyzer, which allows the sample to be positioned so that it is more readily accessible to the X-ray source and other components in the vacuum system.

The most common type of energy analyzer used for ESCA experiments is the electrostatic hemispherical analyzer. It consists of two concentric hemispheres of radius R_1 and R_2. A potential of ΔV is placed across the hemispheres such that the outer hemisphere is negative and the inner hemisphere is positive with respect to the potential at the center line, $R_0 = (R_1 + R_2)/2$. The center line potential is known as the pass energy. As noted previously, most ESCA experiments use a constant pass energy. This will maintain a constant absolute resolution, ΔE, for all photoelectron peaks, since the analyzer resolution is defined as $\Delta E/E$, where E is the energy of the electron as it passes through the analyzer. This ratio is a constant for a given analyzer, so if E is fixed (constant pass energy), ΔE will be fixed. This relationship shows that the lower the pass energy, the smaller ΔE will be. However, the signal intensity will also decrease at smaller pass energies. Typically 5–25 eV pass energies are used to acquire high-resolution ESCA spectra, while 100–200 eV pass energies are used to acquire survey scans.

Hemispherical analyzers are classified as dispersive analyzers, that is, the electrons are deflected by an electrostatic field. There is a range of electron energies that can successfully travel from the entrance to the exit of the analyzer without undergoing a collision with one of the hemispheres. The magnitude of this electron energy range depends on quantities such as the pass energy, the size of the entrance slits, and the angle with which the electrons enter the analyzer. In modern commercial analyzers, this range is ~10 % of the pass energy. Further information about hemispherical analyzers has been published elsewhere [28].

The electrons are counted once they have passed through the energy analyzer. Since the electrons arrive at the analyzer exit with a range of energies, the most efficient means of detection is to use a multichannel array to count the number of electrons leaving the analyzer at each energy. One method of accomplishing this is to use a channel plate to magnify the electron current and a resistive strip anode to monitor the position, and therefore energy, of the electrons. A less elegant method is to place a slit at the analyzer exit so that only electrons in a narrow energy range strike the detector. In this case, a device such as a channeltron is used to measure the number of electrons. Compared to a multichannel detection method using N channels, the single-channel detection method takes $N^{1/2}$ times longer to acquire the same spectrum.

Some of the analyzer systems used for ESCA experiments maintain the spatial relationship of the emitted photoelectrons throughout their transmission through the lens and energy analyzer. This means that the position where the photoelectrons strike the detector is related to their emission position from the sample. Thus, a position-sensitive detector can be used to image the sample. Depending on how the analyzer system is designed, spatial imaging can be done in one or two lateral directions. Spatial resolutions <10 μm have been achieved with imaging detectors [84–87].

Schematic diagrams illustrating how ESCA lens and analyzer systems work are shown in Figure 3.16. In both diagrams the analyzer pass energy is set to transmit electrons with a KE of 50 eV after retardation to the detector. The lens system in Figure 3.16(a) is set so only photoelectrons with KEs greater than 100 eV will be allowed into the analyzer and retarded. If a 150 eV photoelectron enters the lens it will be retarded to a KE of 50 eV, pass through the analyzer and strike the detector. Photoelectrons with KEs significantly great than 150 eV will pass through retardation lenses and strike the outer hemisphere of the analyzer since their KE after retardation in the lens is still greater than the 50 eV pass energy. Likewise photoelectrons with KEs significantly less than 150 eV, but still greater than 100 eV, will pass through the lenses and strike the inner hemisphere of the analyzer since their KE after retardation in the lenses is less than the 50 eV pass energy. In Figure 3.16(b) the lens setup has been changed so now only photoelectrons with KEs greater than 120 eV will be allowed into the analyzer and retarded. The 150 eV photoelectron will still pass through the lenses and be retarded. However, after retardation its KE will be significantly lower than 50 eV and it will strike the inner hemisphere. In Figure 3.16(b) photoelectrons with an initial KE of 170 eV will be passed through the lenses and be retarded to 50 eV so they can pass through the analyzer. The higher energy photoelectrons (e.g. 190 eV) will pass through the lenses, but after retardation will still have KEs greater than 50 eV and will strike the outer hemisphere.

3.7.4 DATA SYSTEMS

Modern computers provide a powerful means both for controlling instrument operation and performing data analysis. State-of-the-art ESCA spectrometers have virtually all aspects of their operation under computer control. Most accessories, components and status of the vacuum system (ion guns, electron guns, valves, pressures, etc.) can be controlled and monitored by the computer. The power supplies that control the analyzer functions (pass energy, scan rate, E_B range, etc.) are also under computer control. This, along with computer control of the sample positioning system, allows unattended, multi-sample runs to be executed. Since each sample may require several different types of scans, it is useful to be able to pre-select and store the desired scan parameters along with the sample position. Then, execution of these commands, which may take several hours, can be done automatically.

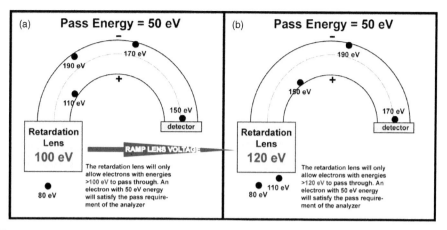

Figure 3.16 Schematic diagrams of the ESCA retardation lens and hemispherical analyzer. (a) The lens is set so photoelectrons with KEs greater than 100 eV will pass through the lens and be retarded by 100 eV, while the analyzer is set to allow electrons with 50 eV (after retardation in the lens) to pass through the analyzer and strike the detector. Photoelectrons that have initial KEs greater than or less than 150 eV (50 eV after retarding in the lens) will strike the outer or inner hemisphere, respectively. (b) The lens is now ramped to a higher voltage so photoelectrons with KEs greater than 120 eV will pass through the lens and be retarded, while the analyzer is still set to allow electrons with 50 eV after retardation in the lens to pass through the analyzer and strike the detector. Now photoelectrons that have initial KEs greater than or less than 170 eV (50 eV after retarding in the lens) will strike the outer or inner hemisphere, respectively. Note that with a multichannel detector, a spectrum of electron energies consistent with the dispersive nature of a hemispherical analyzer will impact the detector. Also note all electron are labeled according to their initial KE, not their KE after passing through the retardation lens

Since modern computers are multi-tasking, data acquisition and data analysis can be done simultaneously. Current software programs contain a wide range of data analysis capabilities. Complex peak shapes can be fit in seconds. Automatic peak finding, identification, and quantification for survey scans can also be accomplished in seconds. Numerous options for data scaling, smoothing, plotting, transferring, and transforming are readily available. Images, X–Y maps, and depth profiles can also be generated. Some software programs even include mathematical analysis packages (multivariant statistics, pattern recognition, etc.). Other software programs allow ESCA data to be directly transferred into word processing packages. In general, as the speed and power of computer systems increase, so do the capabilities for ESCA data acquisition and analysis. The recent improvements made in the ESCA software and hardware have dramatically increased the number of samples that can be run in a day.

3.7.5 ACCESSORIES

The types of accessories that can be added to an ESCA spectrometer are almost limitless. Common accessories include ion guns, electron guns, gas dosers, and quadrupole mass spectrometers. The accessories selected for a given system depend on the applications that are planned for the system. In many cases the ESCA instrument is part of a multicomponent surface analysis system that has one or more additional techniques (Auger, ion scattering, SIMS, LEED, EELS, etc.) mounted on the same vacuum chamber. In some cases, the ESCA analyzer is used in several different techniques (Auger, ion scattering, etc.). For monochromatized ESCA systems, the most important accessory is the low-energy electron flood gun, which is required to obtain high-quality spectra from insulating materials (described in Section 3.3.5).

3.8 Spectral Quality

Like all other spectroscopic techniques, the signal-to-noise (S/N) ratio and resolution are the most important properties to consider when evaluating spectral quality. Traditionally manufacturers have advertised the count rates for their ESCA spectrometers, which is not a particularly useful specification. More important is how noise-free the spectrum is, which determines the counting time needed to acquire a high-quality spectrum. Specifically, the length of time it takes to reach a given S/N ratio at a given energy resolution is the important criterion. As resolution is increased, the S/N will decrease. There are several ways to evaluate the S/N ratio. One convenient method is the peak-to-peak S/N ratio shown in Figure 3.17. A spectrum containing a photoemission peak is acquired for a known time period, then the channels with the highest and lowest number of counts are noted. The difference between these two channels (16738−48) is the peak-to-peak signal. The peak-to-peak noise is determined by using identical scan parameters (same acquisition time, number of data points, eV/point, pass energy, X-ray setting, etc.), except the scan window is shifted below the photoemission peak. The peak-to-peak noise, like the peak-to-peak signal, is the difference between the channels with the highest and lowest number of counts (67−38). The peak-to-peak S/N ratio is simply the ratio of the peak-to-peak signal to the peak-to-peak noise (16690/29 = 575). The energy resolution can be determined by measuring the FWHM of the peak. Thus, for the graphite sample shown in Figure 3.17, it was determined that three minutes of scanning produced a peak with an energy resolution of 0.65 eV and an S/N ratio of 575. The S/N of a spectrum can be increased by either increasing the scan time or decreasing the energy resolution. For a properly designed spectrometer, the S/N ratio will increase as $t^{1/2}$ where t is the scan time.

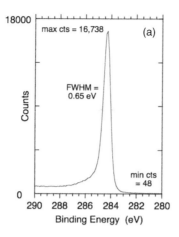
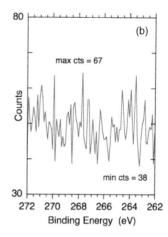

Figure 3.17 The (a) C_{1s} and (b) noise spectra from a graphite sample. The data in each region were acquired with the same scan parameters (time, eV/point, window, pass energy, etc.). These data, obtained after three minutes of scanning, have a peak-to-peak signal-to-noise ratio of 575 at an energy resolution of 0.65 eV

3.9 Depth Profiling

Although ESCA would seem to provide information from a highly surface-localized zone, in fact the surface zone has a finite thickness and often is composed of a vertical composition gradient. If we estimate that the sampling depth of ESCA is 10 nm and the atomic dimensions are 0.3 nm, then the surface region could be composed of ~30 atomic layers. Each of these layers may have a different composition. The ESCA spectrum we obtain will be a convolution of the information from all the layers. This problem is illustrated schematically in Figure 3.18. Depth profiling methods are used to deconvolute from the ESCA signal the composition as a function of depth. Three sampling depths are of interest here: 0–10 nm (the sampling depth of ESCA using conventional X-ray sources), 0–20 nm (the sampling depth of ESCA using special X-ray sources), and 0 to ~1000 nm using destructive depth profiling methods. Each of these zones will be separately described.

The information from the outermost ~10 nm of a surface is converted into a depth profile using data acquired in an angular dependent ESCA experiment. As the sample angle to the analyzer entrance is increased, with the X-ray source and the analyzer kept in a fixed position, the photoelectrons originate from an increasingly surface localized zone (Figure 3.19). If data are acquired at photoelectron take-off angles of, for example, 0°, 50°, and 80° from the surface normal, three sets of ESCA data can be obtained that contain information about the composition as a function of depth. Qualitatively, the shape of this composition versus take-off angle curve can

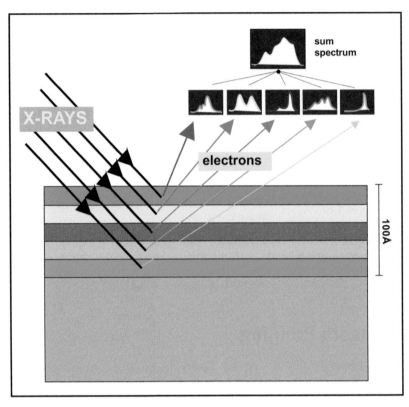

Figure 3.18 ESCA spectra are convolutions of the information from each depth within the sampling depth. In this model material, each colored layer represents material of a different composition. Overlayers attenuate the intensities of photoelectrons emitted from deeper layers and so contributions from the underlying layers to the final spectrum will be lower

reveal much about the compositional organization of a surface (Figure 3.20). To convert from a plot of 'angle versus composition' to a plot of 'depth into the surface versus composition' necessitates a deconvolution of the data set. The mathematical functions upon which the deconvolutions are based are forms of the equations given in Figure 3.8. A number of algorithms for performing such a deconvolution have been published [88–94]. Table 3.12 contains an ESCA data set taken at five electron take-off angles for a sample of a cast film of a fluorine-containing polyurethane [92]. Figure 3.21 shows the deconvolution of this data set using the algorithm developed by Tyler *et al.* [92]. Surface depletion of fluorine and nitrogen (hard segment components) is evident. For accurate and meaningful depth profiles from angular dependent ESCA data, we assume that the surfaces and interfaces studied are molecularly smooth and that overlayers are of uniform thickness. Effectively, an aspect ratio of 10:1 (length to height) for surface irregularities

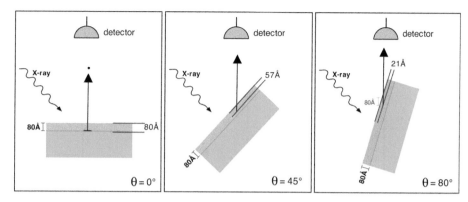

Figure 3.19   As the sample is rotated, maintaining the X-ray source and detector in fixed positions, the effective sampling depth decreases by a factor of cos θ. Note that the emitted photoelectron travels 80 Å through matter (the sampling depth) at all take-off angles. The sample angle, θ, is defined relative to the normal to the surface

Table 3.12   Angular dependent ESCA data from a fluorine-containing polyetherurethane (normalized signal intensities)[a]

Angle (degrees)	C	O	N	F
0	5456	1267	189	236
39	4341	979	118	157
55	3498	822	103	126
68	2736	642	68	70
80	1706	395	34	39

[a]Data taken from Tyler *et al.* [92].

(i.e., 'gently rolling hills') can be tolerated [95]. Other assumptions also apply for interpreting angular dependent ESCA data [92, 93, 95].

The ratio of the photoemission peak area to the inelastic background intensity at binding energies above the peak can also provide information about the depth profile of that element. For example, if a carbon substrate is covered with a few nm thick gold overlayer the background above the Au 4f peaks will initially rise slightly and then gradually decrease until the next photoemission peak is reached. In contrast, if a gold substrate is covered with a several nm thick carbon overlayer the Au 4f peak intensity will be decreased, while the background above the Au 4f will continue to increase with increasing E_B. The degree of reduction in the Au 4f peak intensity and the rate of increase in the background intensity will both depend on the thickness and structure of the carbon overlayer. The thicker the carbon overlayer, the smaller the elastic Au 4f peak intensity and the higher the

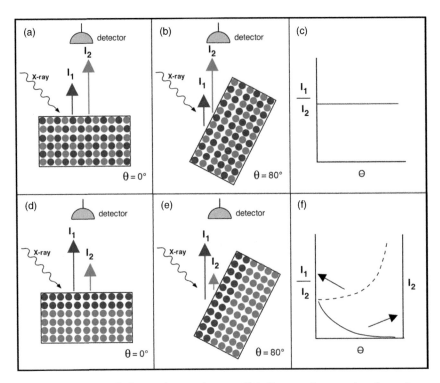

Figure 3.20 The morphology of a specimen will influence the angular dependence of the ESCA signal intensities. (a) For a specimen with homogeneously distributed atoms, note the ratio of the total intensity of the photoemission from the blue atoms and the red atoms; (b) the ratio as described in (a) will be constant at any sample angle; (c) because the ratio of intensities does not change with sample angle, for a sample homogeneous in depth, a plot of the ratio of photoemission intensities (or the ratio of atomic percents) with sample angle will show zero slope; (d) a sample is illustrated with an overlayer of red atoms on a substrate of blue atoms; (e) when this sample is rotated, the photoemission signal will localize closer to the outermost surface. Therefore, the intensity of the signal from the red atoms will increase relative to the intensity from the blue atoms with increasing angle; (f) a plot of the ratio of the red atom photoemission intensity to the blue atom photoemission intensity with sample angle will increase in an exponential fashion with sample angle. The photoemission from the blue atoms will decrease in intensity with increasing sample angle

intensity of the inelastic background intensity (i.e., the thicker the carbon overlayer, the higher the probability that the photoelectrons from the gold substrate will undergo an inelastic collision). Further details about how this information can be used to construct depth profiles has been described by Tougaard [96].

Non-destructive ESCA depth profiling can also be performed using X-ray sources of different energies. According to Equation (3.1) a higher energy X-ray source will liberate higher KE photoelectrons. These more energetic

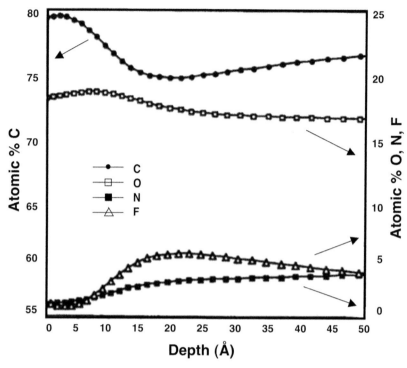

Figure 3.21 A depth profile diagram of a fluorine-containing polyurethane estimated using the regularization method [92] for deconvoluting the angular dependent data set in Table 3.12 (Reproduced from [92] with permission from John Wiley & Sons, Ltd.)

photoelectrons have a greater IMFP and, consequently, an increased sampling depth. If Al Kα (1487 eV), Ag Lα (2984 eV), and Cr Kα (5415 eV) X-ray sources are each used to generate ESCA spectra of the same sample, the C_{1s} electron sampling depths, using Equation (3.12) developed by Seah and Dench [49], can be estimated at 10.8, 16.2, and 22.4 nm, respectively. Thus, the information needed for a depth profile is acquired.

Depth profiles deep into the sample surface (to a micron or more) can be generated by an ion etching the surface, and then analyzing the bottom of the etching crater at regular time intervals using ESCA. Until recently monoatomic ions such as Ar^+ or Cs^+ were typically used for destructive depth profiling. These ion sources provide useful information from samples such as silicon wafers implanted with boron and other dopants [97]. However, for organic and biological materials, structural information will be lost due to the damaging effects of the monoatomic ion beams. Also, the ion beam can induce scrambling and knock-in of atoms at the bottom of the crater reducing the accuracy of the analysis – the longer the etching time (the deeper the crater), the more degraded will be the accuracy of the depth profile. Recently it has been shown that destructive ESCA depth profiles of some organic materials can be obtained by sputtering with a C_{60} cluster ion beam [98]. Further discussion of uses and benefits of C_{60} ion sources for etching organic and biological samples is provided in Chapter 4.

3.10 X–Y Mapping and Imaging

In the 1990s there were marked improvements made in the spatial resolution of commercial ESCA systems. Historically, spatial resolutions achieved with ESCA have been poorer than spatial resolutions achieved with other surface analysis techniques such as Auger electron spectroscopy and SIMS. This is because it is harder to focus an X-ray beam than an electron or ion beam. However, much can be gained from improving the spatial resolution of ESCA. One benefit is an improved ability to do spot analyses. As features on microelectronic chips and biological microarrays become smaller, improved spatial resolution is needed. Likewise, the analysis of surface defects requires good spatial resolution. The second benefit is the improved ability to construct images of a sample. By using the chemical specificity of ESCA, both elemental and functional group maps of a surface can be constructed.

As discussed in Section 3.7, there are two methods for obtaining spatial resolution in stand-alone ESCA spectrometers. One method is the microprobe mode where the X-rays are focused to a small spot on the sample [72, 73]. The best spatial resolution currently obtained with in the microprobe mode is <10 μm. The second method is the microscope mode where a position-sensitive detector is used to image the surface by mapping the position of the photoelectrons emitted from the sample [84–87]. The best spatial resolution currently obtained with the microscope mode is also <10 μm. The microscope mode has the advantage that it does not require a focused X-ray source. However, there are advantages to using focused X-rays in the microprobe mode. First, only the area being analyzed is irradiated. This can be important when working with samples that are prone to X-ray degradation. With a large-spot imaging detector system, parts of the sample can be degraded before they are even analyzed. However, the focused X-ray sources typically have higher X-ray fluxes per unit area, resulting in higher damage rates and more challenges for charge neutralization of electrically insulating samples. Second, positioning the X-ray beam for spot analyses is generally more straightforward with focused X-rays. After calibrating the X-ray position with a microscope, the sample is moved to the appropriate position and the analysis is started. However, it can be easier to acquire images at high spatial resolution over large areas using the microscope mode.

Spatial resolutions of 1 μm or better are desirable for many applications, requiring further improvements in currently obtainable spatial resolutions with laboratory-based ESCA systems. As mentioned previously synchrotron-based techniques can provide high spatial resolution ESCA data. One method uses a superconducting ESCA analyzer where the sample is placed in a superconducting magnet and then irradiated with X-rays [99]. The emitted photoelectrons follow the magnetic field lines as they leave the sample. A position-sensitive detector is then placed outside of the magnet to obtain a magnified image. With this system, a spatial resolution of a few microns has been obtained. Photoelectron microscopes using zone plates to provide X-ray spot sizes of <100 nm are also available at several synchrotron facilities [100]. Reviews of the progress made in this field have been published [74, 80, 81, 101].

An example of the spatial resolution of that can be obtained with commercial laboratory ESCA spectrometer in shown in Figure 3.22. The sample was a patterned polymer film on a silicon wafer substrate (10 μm photoresist lines separated by 15 μm silicon lines) prepared using standard photolithographic methods. Details of the sample preparation have been described elsewhere [102].

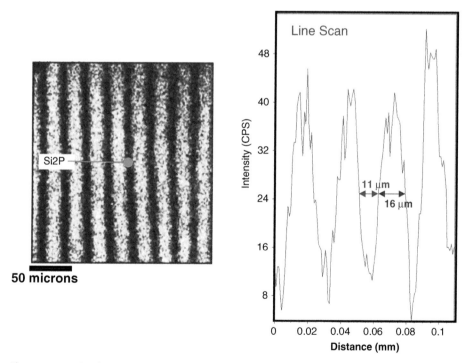

Figure 3.22　A silicon 2p image (peak – background) from a patterned sample with 10 μm photoresist lines separated by 15 μm lines of silicon (left). A line scan across the Si 2p image showing the FWHM widths of the photoresist and silicon lines are 11 and 16 μm, respectively. The image was acquired with a Kratos AxisUltra DLD ESCA system

3.11 Chemical Derivatization

The binding energy shifts in ESCA associated with specific functional groups often do not permit precise identification of the group. For example, ether carbons ($\underline{C}$—O—C) and hydroxyl carbons ($\underline{C}$—OH) both have observed E_{BS} of 286.5 eV. Also, carbons in carboxylic acid and ester environments are not readily distinguishable based upon binding energy shifts. There are many other examples of functional groups that are difficult to identify based solely upon their binding energies. To assist with the precise identification of chemical groups, a chemical reaction specific to only the functional group of interest can be performed. This reaction should uniquely identify the functional group in the ESCA spectrum by altering its binding energy position or by adding a tag atom not present in the specimen prior to reaction. This idea is illustrated schematically in Figure 3.23. There are many derivatization reactions that have been explored for this purpose [1,103–106]. A few common derivatization reactions are illustrated in Table 3.13. There are also many control studies that must be performed so derivatization reactions can be used with confidence. The concerns that must be addressed for derivatization studies are outlined in Table 3.14. After performing appropriate control studies, derivatization reactions can be used to enhance the understanding of the chemistry of complex surfaces [107–111].

3.12 Valence Band

Most ESCA experiments focus on the core level peaks. As discussed in previous sections of this chapter, the core level spectra are comprised of relatively sharp, intense peaks that can be used to identify oxidation states, molecular functional groups, concentrations, etc. of atoms in the surface region of a sample. For quantitfication each core level is assumed

Table 3.13 Four surface chemical derivatization reactions

Group	Reaction
Hydroxyl	—OH+ $\underset{\text{trifluoroacetic anhydride}}{(CF_3CO)_2O}$ $\longrightarrow$ —OCOCF$_3$
Carbonyl	$-\overset{\mid}{C}{=}O+$ $\underset{\text{hydrazine}}{NH_2NH_2}$ $\longrightarrow$ $-\overset{\mid}{C}{=}NNH_2$
Carboxylic acid	—COOH+ $\underset{\text{trifluoroethanol}}{CF_3CH_2OH}$ $\overset{\text{carbodiimide}}{\underset{}{\overset{C_6H_5N}{\longrightarrow}}}$ —CO$_2$CH$_2$CF$_3$
Unsaturated	—CH=CH— + Br$_2$ $\longrightarrow$ —CHBr—CHBr—

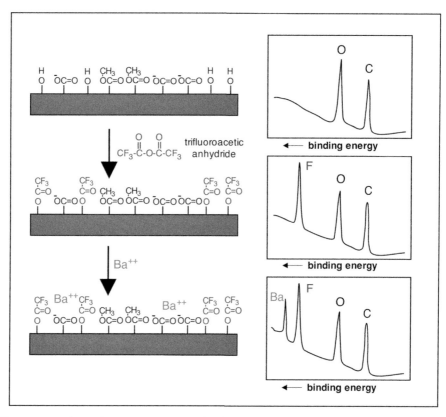

Figure 3.23 A surface many contain hydroxy, carboxylic acid and carboxylic ester groups. Unambigously identifying these different groups and quantifying their surface concentration can be challenging for ESCA. Derivatization reactions can be used to distinguish the different chemical species. If the surface is reacted with trifluoroacetic anhydride, only hydroxyl groups will pick up F. The intensity of the F peak in the ESCA spectrum will be proportional to the number of reacted hydroxyl groups. Similarly, carboxylic acid groups can bind a barium ion. Ester groups will not react with either of these chemistries

to have a specific photoionization cross-section that does not depend on the sample matrix or bonding. This is a good assumption for core electrons that are not involved directly in bond formation and molecular interactions, allowing ESCA to be used as a quantitative technique. This assumption is not valid for valance band analysis. The valence electrons are directly involved in bond formation and molecular interactions, so the intensity and energy of the valence band peaks depend on their bonding environment. Thus, quantitative interpretation of the valence band spectrum for most materials requires a full molecular orbital calculation. For multi-component materials or those containing a large number of atoms per structural unit (e.g. polymers) this requires computer calculations [112, 113].

Table 3.14   Concerns in derivatization studies

- Is the reaction stoichiometric?
- Has the reaction gone to completion (kinetics)?
- Does the derivatizing reagent cross-react with any of the other functional groups in the sample?
- The marker atom should be unique in the sample and have a high photoemission cross-section.
- The derivative formed must be stable under vacuum.
- The derivative formed must be stable over time, even under X-ray and electron flux.
- Surface rearrangement leading to migration of the marker atom from the interface is undesirable.
- The derivatizing reagent should not extract components out of the sample.

Valence band analysis is worth pursuing because it can provide electronic structure information that cannot be obtained from typical core level analysis. Also, it is sometimes possible to extract useful structural information from valence band spectra. In spite of these benefits, valence band spectra are not generally used in typical ESCA experiments. One reason is that the ESCA cross-sections for valence band peaks are significantly lower than core level cross-sections, which dictates that longer analysis times (up to several hours for older instruments) must be used to acquire valence band spectra with good signal-to-noise characteristics. This can be a problem for organic samples that degrade during analysis.

Although polymer valence band studies are not abundant, they do show the power of valence band analysis. Early studies showed that the combination of a monochromatized X-ray source, to minimize sample degradation, and theoretical calculations, to aid interpretation, resulted in an enhanced understanding of the polymer surface structure [69]. Small molecule models were also found to be valuable aids in interpreting valence band spectra, examples being linear alkanes for polyethylene and benzene for aromatic containing polymers (polyphenyl, polystyrene, etc.) [69]. The sensitivity of the valence band to polymer structure was highlighted in studies with polymers having the same elemental composition and core level spectra. These polymers were easily differentiated based on their valence band spectra. For example, poly(propylene oxide) and poly(vinyl methyl ether) both have C_3H_6O monomer units, but exhibit different valence band spectra [69]. Isomeric effects in pure hydrocarbon components can also be distinguished with valence band spectra. Examples include poly(3-methyl 1-butene) and poly(1-pentene), both having a C_5H_{10} monomer repeat unit [69], methyl substituted polystyrenes [114], and the normal, iso, and tertiary butyl side chains in methacrylates (C_4H_9 units) [115, 116]. Further details such as head-to-head versus head-to-tail linking of monomer units and the tacticity of the monomer units can also be differentiated with valence band spectra [69]. With the improvements in instrument performance, the acquisition time to obtain valence band spectra has been shortened considerably. A handbook that contains an extensive collection of polymer valence band spectra has been published [23].

In contrast to polymeric materials, more extensive use of valence band spectra has been made for metallic and semiconductor materials [117–119]. This is probably due to several reasons. First, experimental metallic and semiconducting samples are easier to study since they are less susceptible to degradation and charging problems than organic materials. Second, single crystalline samples are readily available for a wide range of metals and semiconductors. This allows one to perform a detailed investigation of the valence band electronic structure through angle-resolved photoemission experiments [119]. To overcome the low cross-section limitation in ESCA valence band experiments, many experimenters use synchrotron radiation for the incident photon source [120–122]. By using a synchrotron, the energy of the incident photons can be tuned to maximize a given valence band peak, since the photoemission cross-section depends on the energy of the excitation source. Thus, the combination of a tuneable X-ray source, a single crystal sample, and angle-resolved detection provides a powerful method for obtaining detailed information about the electronic structure of a material (work functions, band gap energies, band dispersions, band bending, etc). Valence band experiments can also be used to probe the development of the electronic structure of metal clusters as a function of cluster size [123, 124] and the electronic interactions in metallic alloys [125].

3.13 Perspectives

New instrumentation, technique development, and enhanced data analysis continue to expand the utility of ESCA for the analysis of surfaces. Two major instrumentation advances include (1) the development of efficient monochromatized x-ray sources and detectors (first introduced in the early 1970s, but not in wide spread use until the late 1980s) and (2) the introduction of imaging instrumentation in the 1990s. Now modern laboratory ESCA systems routinely provide high energy resolution spectra (e.g. C 1s peak widths <1 eV) and high spatial resolution (<10 microns) from a wide range of samples. Even higher performance in terms of both energy and spatial resolution can be obtained by using synchrotron radiation instrumentation. Although only small improvements in the spectral and imaging performance of ESCA instrumentation have been made in the past 10 years, there are other developments leading to further expansion of the ESCA technique. These advances include (1) the application of multivariate analysis methods to enhance the information obtained from ESCA imaging [126] and (2) the adaptation of C_{60} sputtering from the SIMS community for ESCA depth profiling experiments on polymers [98]. One future advance that could significantly impact the ESCA technique is the development of 'table-top' lasers. The lasers are actually quite large and now are approaching the ability to operate in the soft X-ray region. These sources would provide highly monochromatic, intense, polarized and focused X-rays. The advent of these sources would revolutionize ESCA.

The continuing expansion of ESCA usage can be documented by examining the number of publications on the Web of Science® that list ESCA/XPS in their abstract or keywords. The search was done using the topic words of ESCA, electron spectroscopy for chemical analysis, XPS and X-ray photoelectron spectroscopy. Similar topic searches were done for SIMS (topic words SIMS, ToF-SIMS and secondary ion mass spectrometry) and AES (topic words AES and Auger electron spectroscopy) for comparison. The number of publications using ESCA has nearly tripled from 1991 to 2006 (from ~1800 to ~4900), exhibiting a fairly steady increase of over 200 publications per year during this time period. In contrast, the increase in SIMS and AES publications were significantly lower over the same time period (increase of ~20 and ~10 publications per year for SIMS and AES, respectively). While the number of ESCA publications was higher than the SIMS and AES publications in 1991 (~1800 for ESCA compared to ~700 for SIMS and ~1100 for AES), the gap increased significantly by 2006 (~4900 for ESCA compared to ~1050 for SIMS and ~1200 for AES). These numbers provide strong support that ESCA is the major and most widely used surface analysis technique. We believe the reason for this is because ESCA can be used to determine quantitative elemental surface compositions from a wide range of samples. The extensive computer control of instrument operation, data acquisition and data analysis now make it straightforward to routinely and quickly obtain surface elemental compositions from multiple samples. The advantage of these advances is that ESCA instruments can be operated and used by a wide range of people, not just highly trained ESCA researchers. This has also lead to widespread use of ESCA instruments in corporate analytical laboratories. Development of expert systems for ESCA instruments should continue to expand the use and applications of ESCA [127]. The disadvantage of these advances is that many studies tend to just use ESCA for determining the surface elemental composition of a sample, neglecting the other detailed information that can be obtained with ESCA. Thus, while all indications point to the continued expansion in the use of ESCA, there are still significant opportunities to increase the impact of ESCA studies by using the full capabilities of the technique.

3.14 Conclusions

The ESCA technique has been a commercially available method since the late 1960s. In 40 years it has gone from a physicist's experiment to a practical and widely available surface analysis tool with thousands of published applications. The advantages of ESCA are its simplicity, flexibility in sample handling, and high information content. The heightened interest in materials science, biotechnology and surface phenomena in general, coupled with advances in ESCA technique and instrumentation, make it probable that ESCA will remain the predominant surface analysis technique in the foreseeable future. When used in conjunction with other surface analysis methods, ESCA will play a pivotal role in expanding our understanding of the chemistry, morphology, and reactivity of surfaces.

Acknowledgements

The authors acknowledge support from the National ESCA and Surface Analysis Center for Biomedical Problems (NIH grant EE-002027) and the University of Washington Engineered Biomaterials Program (NSF EEC-9529161) while revising and updating of this chapter. Support was received from NIH Grants RR01296 and HL25951 while writing the initial edition of this chapter, and for some of the experiments described in it. We thank Deborah Leach-Scampavia for acquiring some of the ESCA data used in this article. The ESCA data used in Figure 3.22 was acquired at Kratos Analytical in Manchester, UK.

📖 References

1. Andrade JD. X-ray Photoelectron Spectroscopy (XPS). In: Andrade JD (Editor). *Surface and Interfacial Aspects of Biomedical Polymers*. New York, NY: Plenum Press; 1985. pp. 105–195.

2. Carlson TA. *Photoelectron and Auger Spectroscopy*. New York: Plenum Press; 1975.

3. Clark DT. Some Experimental and Theoretical Aspects of Structure, Bonding and Reactivity of Organic and Polymeric Systems as Revealed by ESCA. *Physica Scripta* 1977; **16**: 307–328.

4. Dilks A. X-ray Photoelectron Spectroscopy for the Investigation of Polymeric Materials. In: Baker AD and Brundle CR (Editors). *Electron Spectroscopy: Theory, Techniques and Applications*. London: Academic Press; 1981. pp. 277–359.

5. Ghosh PK. *Introduction to Photoelectron Spectroscopy*. New York, NY: John Wiley & Sons, Inc.; 1983.

6. Ratner BD and McElroy BJ. Electron Spectroscopy for Chemical Analysis: Applications in the Biomedical Sciences. In: Gendreau RM (Editor). *Spectroscopy in the Biomedical Sciences*. Boca Raton, FL: CRC Press; 1986. pp. 107–140.

7. Siegbahn K. Electron-Spectroscopy for Atoms, Molecules, and Condensed Matter. *Science* 1982; **217**(4555): 111–121.

8. Siegbahn K, Nordling C, Fahlman A, Nordberg R, Hamrin K, Hedman J, Johansson G, Bergmark T, Karlsson SE, Lindgren I and Lindberg B. ESCA: atomic, molecular and solid state structure studied by means of electron spectroscopy. *Nova Acta Regiae Societatis Scientiarum Upsaliensis*, Series IV 1967; **20**: 5–282.

9. Swingle RS and Riggs WM. ESCA. *CRC Critical Reviews in Analytical Chemistry* 1975; **5**: 267–321.

10. Barr TL. Advances in the Application of X-Ray Photoelectron-Spectroscopy (ESCA).1. Foundation and Established Methods. *CRC Critical Reviews in Analytical Chemistry* 1991; **22**: 115–181.

11. Pijpers AP and Meier RJ. Core level photoelectron spectroscopy for polymer and catalyst characterisation. *Chemical Society Reviews* 1999; **28**: 233–238.

12. Castner DG and Ratner BD. Biomedical surface science: Foundations to frontiers. *Surface Science* 2002; **500**: 28–60.

13. McArthur SL. Applications of XPS in bioengineering. *Surface and Interface Analysis* 2006; **38**: 1380–1385.

14. Ratner BD and Castner DG. Advances in XPS Instrumentation and Methodology: Instrument Evaluation and New Techniques with Special Reference to Biomedical Studies. *Colloids and Surfaces B: Biointerfaces* 1994; **2**: 333–346.

15. Turner NH and Schreifels JA. Surface analysis: X ray photoelectron spectroscopy and Auger electron spectroscopy. *Analytical Chemistry* 1996; **68**: R309–R331.

16. Klein MJ. The Beginnings of the Quantum Theory. In: Weiner C (Editor). *History of Twentieth Century Physics*. New York, NY: Academic Press; 1977. pp. 1–39.

17. Pais A. *Inward Bound*. Oxford: Oxford Press; 1986.

18. Segre E. *From X-rays to Quarks*. San Francisco, CA: W.H. Freeman and Company; 1980.

19. Berkowitz J. *Photoabsorption, Photoionization, and Photoelectron Spectroscopy*. New York, NY: Academic Press; 1979.

20. Feldman LC and Mayer JW. *Fundamentals of Surface and Thin Film Analysis*. New York, NY: North Holland; 1986.

21. Hoffman R. *Solids and Surfaces. A Chemists View of Bonding in Extended Structures*, Volume 1. New York, NY: VCH Publishers; 1988.

22. Koopmans TS. Über die Zuordnung von Wellenfunktionen und Eigenwerten zu den Einzelnen Elektronen eines Atoms. *Physica* 1934; **1**: 104–113.

23. Beamson G and Briggs D. *High Resolution XPS of Organic Polymers*. Chichester, UK: John Wiley & Sons, Ltd; 1992.

24. Brundle CR, Chuang TJ and Rice DW. X-Ray Photoemission Study of Interaction of Oxygen and Air with Clean Cobalt Surfaces. *Surface Science* 1976; **60**: 286–300.

25. McIntyre NS and Cook MG. X-Ray Photoelectron Studies on Some Oxides and Hydroxides of Cobalt, Nickel, and Copper. *Analytical Chemistry* 1975; **47**: 2208–2213.

26. Shirley DA. Many-electron and Final-state Effects: Beyond the One-electron Picture. In: Cardona M and Ley L (Editors). *Photoemission in Solids*. Berlin: Springer-Verlag; 1978. pp. 165–195.

27. Anthony MT and Seah MP. XPS – Energy Calibration of Electron Spectrometers.1. An Absolute, Traceable Energy Calibration and the Provision of Atomic Reference Line Energies. *Surface and Interface Analysis* 1984; **6**: 95–106.

28. Briggs D and Seah MP. *Practical Surface Analysis*. Chichester, UK: John Wiley & Sons, Ltd; 1990.

29. Seah MP. Post-1989 Calibration Energies for X-Ray Photoelectron Spectrometers and the 1990 Josephson Constant. *Surface and Interface Analysis* 1989; **14**: 488.

30. Seah MP, Gilmore IS and Spencer SJ. Measurement of data for and the development of an ISO standard for the energy calibration of X-ray photoelectron spectrometers. *Applied Surface Science* 1999; **145**: 178–182.

31. Seah MP, Gilmore IS and Spencer SJ. XPS: Binding energy calibration of electron spectrometers 4 – Assessment of effects for different x-ray sources, analyser resolutions, angles of emission and overall uncertainties. *Surface and Interface Analysis* 1998; **26**: 617–641.

32. Seah MP. Instrument Calibration for AES and XPS. In: Briggs D and Grant JT (Editors). *Surface Analysis by Auger and X-Ray Photoelectron Spectroscopy*. Chichester, UK: IM Publications and SurfaceSpectra Ltd; 2003. pp. 167–189.

33. Lewis RT and Kelly MA. Binding-Energy Reference in X-Ray Photoelectron-Spectroscopy of Insulators. *Journal of Electron Spectroscopy and Related Phenomena* 1980; **20**: 105–115.

34. Dubey M, Gouzman I, Bernasek SL and Schwartz J. Characterization of self-assembled organic films using differential charging in X-ray photoelectron spectroscopy. *Langmuir* 2006; **22**: 4649–4653.

35. Havercroft NJ and Sherwood PMA. Use of differential surface charging to separate chemical differences in x-ray photoelectron spectroscopy. *Surface and Interface Analysis* 2000; **29**: 232–240.

36. Cohen H. Chemically resolved electrical measurements using X-ray photoelectron spectroscopy. *Applied Physics Letters* 2004; **85**: 1271–1273.

37. Suzer S. Differential charging in X-ray photoelectron spectroscopy: A nuisance or a useful tool? *Analytical Chemistry* 2003; **75**: 7026–7029.

38. Bryson CE. Surface-Potential Control in XPS. *Surface Science* 1987; **189**: 50–58.

39. Cardona M and Ley L. *Photoemission in Solids*. Berlin: Springer-Verlag; 1978.

40. Shirley DA. High-Resolution X-Ray Photoemission Spectrum of Valence Bands of Gold. *Physical Review B* 1972; **5**: 4709–4714.

41. Leclerc G and Pireaux JJ. The Use of Least-Squares for XPS Peak Parameters Estimation.1. Myths and Realities. *Journal of Electron Spectroscopy and Related Phenomena* 1995; **71**: 141–164.

42. Beamson G, Bunn A and Briggs D. High-Resolution Monochromated XPS of Poly(Methyl Methacrylate) Thin-Films on a Conducting Substrate. *Surface and Interface Analysis* 1991; **17**: 105–115.

43. Meier RJ and Pijpers AP. Oxygen-Induced Next-Nearest Neighbor Effects on the C-1s Levels in Polymer XPS Spectra. *Theoretica Chimica Acta* 1989; **75**: 261–270.

44. Pijpers AP and Donners WAB. Quantitative-Determination of the Surface-Composition of Acrylate Copolymer Latex Films by XPS (ESCA). *Journal of Polymer Science Part A – Polymer Chemistry* 1985; **23**: 453–462.

45. Ratner BD. The Surface Characterization of Biomedical Materials: How Finely Can We Resolve Surface Structure? In: Ratner BD (Editor). *Surface Characterization of Biomaterials*. Amsterdam: Elsevier; 1988. pp. 13–36.

46. Gelius U, Svensson S, Siegbahn H, Basilier E, Faxalv A and Siegbahn K. Vibrational and Lifetime Line Broadenings in ESCA. *Chemical Physics Letters* 1974; **28**: 1–7.

47. Powell CJ. The Quest for Universal Curves to Describe the Surface Sensitivity of Electron Spectroscopies. *Journal of Electron Spectroscopy and Related Phenomena* 1988; **47**: 197–214.

48. Jablonski A, Tanuma S and Powell CJ. New universal expression for the electron stopping power for energies between 200eV and 30keV. *Surface and Interface Analysis* 2006; **38**: 76–83.

49. Seah MP and Dench WA. Quantitative Electron Spectroscopy of Surfaces: A Standard Data Base for Electron Inelastic Mean Free Paths in Solids. *Surface and Interface Analysis* 1979; **1**: 2–11.

50. Brundle CR, Hopster H and Swalen JD. Electron Mean-Free Path Lengths through Monolayers of Cadmium Arachidate. *Journal of Chemical Physics* 1979; **70**: 5190–5196.

51. Cadman P, Evans S, Gossedge G and Thomas JM. Electron Inelastic Mean Free Paths in Polymers – Comments on Arguments of Clark and Thomas. *Journal of Polymer Science Part C – Polymer Letters* 1978; **16**: 461–464.

52. Clark DT, Thomas HR and Shuttleworth D. Electron Mean Free Paths in Polymers – Critique of Current State of Art. *Journal of Polymer Science Part C – Polymer Letters* 1978; **16**: 465–471.

53. Roberts RF, Allara DL, Pryde CA, Buchanan DNE and Hobbins ND. Mean Free Path for Inelastic Scattering of 1.2 keV Electrons in Thin Poly(methyl methacrylate) Films. *Surface and Interface Analysis* 1980; **2**: 5–10.

54. Wagner CD, Davis LE and Riggs WM. The Energy Dependence of the Electron Mean Free Path. *Surface and Interface Analysis* 1980; **2**: 53–55.

55. Jablonski A and Powell CJ. Electron effective attenuation lengths in electron spectroscopies. *Journal of Alloys and Compounds* 2004; **362**: 26–32.

56. Egelhoff WF. X-ray Photoelectron and Auger Electron Forward Scattering: A New Tool for Surface Crystallography. *CRC Critical Reviews in Solid State Materials Science* 1990; **16**: 213–235.

57. Fadley CS, VanHove MA, Hussain Z and Kaduwela AP. Photoelectron diffraction: New dimensions in space, time, and spin. *Journal of Electron Spectroscopy and Related Phenomena* 1995; **75**: 273–297.

58. Reilman RF, Msezane A and Manson ST. Relative Intensities in Photoelectron-Spectroscopy of Atoms and Molecules. *Journal of Electron Spectroscopy and Related Phenomena* 1976; **8**: 389–394.

59. Seah MP and Anthony MT. Quantitative XPS – the Calibration of Spectrometer Intensity Energy Response Functions.1. The Establishment of Reference Procedures and Instrument Behavior. *Surface and Interface Analysis* 1984; **6**: 230–241.

60. Seah MP, Jones ME and Anthony MT. Quantitative XPS – the Calibration of Spectrometer Intensity Energy Response Functions. 2. Results of Interlaboratory Measurements for Commerical Instruments. *Surface and Interface Analysis* 1984; **6**: 242–254.

61. Seah MP. A System for the Intensity Calibration of Electron Spectrometers. *Journal of Electron Spectroscopy and Related Phenomena* 1995; **71**: 191–204.

62. Scofield JH. Hartree–Slater Subshell Photoionization Cross-Sections at 1254 and 1487eV. *Journal of Electron Spectroscopy and Related Phenomena* 1976; **8**: 129–137.

63. Wagner CD, Davis LE, Zeller MV, Taylor JA, Raymond RH and Gale LH. Empirical Atomic Sensitivity Factors for Quantitative-Analysis by Electron Spectroscopy for Chemical Analysis. *Surface and Interface Analysis* 1981; **3**: 211–225.

64. Seah MP, Gilmore IS and Spencer SJ. Consistent, combined quantitative Auger electron spectroscopy and X-ray photoelectron spectroscopy digital databases: Convergence of theory and experiment. *Journal of Vacuum Science and Technology A – Vacuum Surfaces and Films* 2000; **18**: 1083–1088.

65. Seah MP. Quantification in AES and XPS. In: Briggs D and Grant JT (Editors). *Surface Analysis by Auger and X-Ray Photoelectron Spectroscopy*. Chichester, UK: IM Publications and SurfaceSpectra Ltd; 2003. pp. 345–375.

66. Tougaard S. Quantitative X-ray photoelectron spectroscopy: Simple algorithm to determine the amount of atoms in the outermost few nanometers. *Journal of Vacuum Science and Technology A – Vacuum Surfaces and Films* 2003; **21**: 1081–1086.

67. Wagner CD, Riggs WM, Davis LE, Moulder JF and Muilenberg GE. *Handbook of X-ray Photoelectron Spectroscopy*. Eden Prairie, MN: Perkin-Elmer Corporation; 1979.

68. Wagner CD and Joshi A. The Auger Parameter, Its Utility and Advantages – a Review. *Journal of Electron Spectroscopy and Related Phenomena* 1988; **47**: 283–313.

69. Pireaux JJ, Riga J, Caudano R and Verbist J. *Electronic Structure of Polymers*. ACS Symposium Series. Washington, D C: American Chemical Society; 1981. pp. 169–201.

70. Castner DG. Chemical Modification of Surfaces. In: Czanderna AW, Powell CJ and Madey TE, (Editors). *Specimen Handling, Beam Effects and Depth Profiling*. New York, NY: Plenum Press; 1998. pp. 209–238.

71. Kelly MA and Tyler CE. *Hewlett-Packard Journal* 1972; **24**.

72. Chaney RL. Recent Developments in Spatially Resolved ESCA. *Surface and Interface Analysis* 1987; **10**: 36–47.

73. Baer DR and Engelhard MH. Approach for determining area selectivity in small-area XPS analysis. *Surface and Interface Analysis* 2000; **29**: 766–772.

74. Escher M, Weber N, Merkel M, Kromker B, Funnemann D, Schmidt S, Reinert F, Forster F, Hufner S, Bernhard P, Ziethen Ch, Elmers HJ, Schönhense G. NanoESCA: Imaging UPS and XPS with high energy resolution. *Journal of Electron Spectroscopy and Related Phenomena* 2005; **144**: 1179–1182.

75. King DA. Looking at Solid-Surfaces with a Bright Light. *Chemistry in Britain* 1986; **22**: 819–822.

76. Margaritondo G. Synchrotron Radiation Photoemission Spectroscopy of Semiconductor Surfaces and Interfaces. *Annual Reviews of Material Science* 1984; **14**: 67–93.

77. Schuchman JC. Vacuum Systems for Synchrotron Light Sources. *MRS Bulletin* 1990; **15**: 35–41.

78. Winick H and Doniach S. *Synchrotron Radiation Research*. New York, NY: Plenum Press; 1980.

79. Kinoshita T. Application and Future of Photoelectron Spectromicroscopy. *Journal of Electron Spectroscopy and Related Phenomena* 2002; **124**: 175–194.

80. Ade H, Kilcoyne ALD, Tyliszczak T, Hitchcock P, Anderson E, Harteneck B, Rightor EG, Mitchell GE, Hitchcock AP and Warwick T. Scanning transmission X-ray microscopy at a bending magnet beamline at the Advanced Light Source. *Journal de Physique IV* 2003; **104**: 3–8.

81. Bluhm H, Andersson K, Araki T, Benzerara K, Brown GE, Dynes JJ, Ghosal S, Gilles MK, Hansen HC, Hemminger JC, Hitchcock AP, Ketteler G, Kilcoyne ALD, Kneedler E, Lawrence JR, Leppard GG, Majzlam J, Mun BS, Myneni SCB, Nilsson A, Ogasawara H, Ogletree DF, Pecher K, Salmeron M, Shuh DK, Tonner B, Tyliszczak T, Warwick T, Yoon TH. Soft X-ray microscopy and spectroscopy at the molecular environmental science beamline at the Advanced Light Source. *Journal of Electron Spectroscopy and Related Phenomena* 2006; **150**: 86–104.

82. Tyler BJ, Castner DG and Ratner BD. Determining Depth Profiles from Angle Dependent X-Ray Photoelectron-Spectroscopy – the Effects of Analyzer Lens Aperture Size and Geometry. *Journal of Vacuum Science and Technology A – Vacuum Surfaces and Films* 1989; **7**: 1646–1654.

83. Mack P, White RG, Wolstenholme J and Conard T. The use of angle resolved XPS to measure the fractional coverage of high-*k* dielectric materials on silicon and silicon dioxide surfaces. *Applied Surface Science* 2006; **252**: 8270–8276.

84. Coxon P, Krizek J, Humpherson M and Wardell IRM. ESCASCOPE – a New Imaging Photoelectron Spectrometer. *Journal of Electron Spectroscopy and Related Phenomena* 1990; **52**: 821–836.

85. Drummond IW, Street FJ, Ogden LP and Surman DJ. Axis – an Imaging X-Ray Photo-electron Spectrometer. *Scanning* 1991; **13**: 149–163.

86. Vohrer U, Blomfield C, Page S and Roberts A. Quantitative XPS imaging – new possibilities with the delay-line detector. *Applied Surface Science* 2005; **252**: 61–65.

87. Walton J and Fairley N. Characterization of the Kratos Axis Ultra with spherical mirror analyser for XPS imaging. *Surface and Interface Analysis* 2006; **38**: 1230–1235.

88. Baschenko OA and Nefedov VI. Depth Profiling of Elements in Surface-Layers of Solids Based on Angular Resolved X-Ray Photoelectron-Spectroscopy. *Journal of Electron Spectroscopy and Related Phenomena* 1990; **53**: 1–18.

89. Bussing TD and Holloway PH. Deconvolution of Concentration Depth Profiles from Angle Resolved X-Ray Photoelectron-Spectroscopy Data. *Journal of Vacuum Science and Technology A – Vacuum Surfaces and Films* 1985; **3**: 1973–1981.

90. Iwasaki H, Nishitani R and Nakamura S. Determination of Depth Profiles by Angular Dependent X-ray Photoelectron Spectra. *Japanese Journal of Applied Physics* 1978; **17**: 1519–1523.

91. Paynter RW. Modification of the Beer–Lambert Equation for Application to Concentration Gradients. *Surface and Interface Analysis* 1981; **3**: 186–187.

92. Tyler BJ, Castner DG and Ratner BD. Regularization – a Stable and Accurate Method for Generating Depth Profiles from Angle-Dependent XPS Data. *Surface and Interface Analysis* 1989; **14**: 443–450.

93. Yih RS and Ratner BD. A Comparison of 2 Angular Dependent ESCA Algorithms Useful for Constructing Depth Profiles of Surfaces. *Journal of Electron Spectroscopy and Related Phenomena* 1987; **43**: 61–82.

94. Cumpson PJ. Angle-Resolved XPS and AES – Depth-Resolution Limits and a General Comparison of Properties of Depth-Profile Reconstruction Methods. *Journal of Electron Spectroscopy and Related Phenomena* 1995; **73**: 25–52.

95. Fadley CS. Solid State and Surface Analysis by Means of Angular-dependent X-ray Photoelectron Spectroscopy. *Progress in Solid State Chemistry* 1976; **11**: 265–343.

96. Hansen HS and Tougaard S. Separation of Spectral Components and Depth Profiling through Inelastic Background Analysis of XPS Spectra with Overlapping Peaks. *Surface and Interface Analysis* 1991; **17**: 593–607.

97. Oswald S, Fahler S and Baunack S. XPS and AES investigations of hard magnetic Nd–Fe–B films. *Applied Surface Science* 2005; **252**: 218–222.

98. Sanada N, Yamamoto A, Oiwa R and Ohashi Y. Extremely low sputtering degradation of polytetrafluoroethylene by C-60 ion beam applied in XPS analysis. *Surface and Interface Analysis* 2004; **36**: 280–282.

99. King PL, Browning R, Pianetta P, Lindau I, Keenlyside M and Knapp G. Image-Processing of Multispectral X-Ray Photoelectron-Spectroscopy Images. *Journal of Vacuum Science and Technology A – Vacuum Surfaces and Films* 1989; **7**: 3301–3304.

100. Locatelli A, Aballe L, Mentes TO, Kiskinova M and Bauer E. Photoemission electron microscopy with chemical sensitivity: SPELEEM methods and applications. *Surface and Interface Analysis* 2006; **38**: 1554–1557.

101. Tonner BP. Photoemission Spectromicroscopy of Surfaces in Materials Science. *Synchrotron Radiation News* 1991; **4**: 27–32.

102. Wickes BT, Kim Y and Castner DG. Denoising and Multivariate Analysis of ToF-SIMS Images. *Surface and Interface Analysis* 2003; **35**: 640–648.

103. Batich CD. Chemical Derivatization and Surface Analysis. *Applied Surface Science* 1988; **32**: 57–73.

104. Povstugar VI, Mikhailova SS and Shakov AA. Chemical derivatization techniques in the determination of functional groups by X-ray photoelectron spectroscopy. *Journal of Analytical Chemistry* 2000; **55**: 405–416.

105. Kim J, Jung DG, Park Y, Kim Y, Moon DW and Lee TG. Quantitative analysis of surface amine groups on plasma-polymerized ethylenediamine films using UV-visible spectroscopy compared to chemical derivatization with FT-IR spectroscopy, XPS and TOF-SIMS. *Applied Surface Science* 2007; **253**: 4112–4118.

106. Pan S, Castner DG and Ratner BD. Multitechnique surface characterization of derivatization efficiencies for hydroxyl-terminated self-assembled monolayers. *Langmuir* 1998; **14**: 3545–3550.

107. Briggs D and Kendall CR. Derivatization of Discharge-treated LDPE: An Extension of XPS Analysis and a Probe of Specific Interactions in Adhesion. *International Journal of Adhesion and Adhesives* 1982; **2**: 13–17.

108. Chilkoti A, Ratner BD and Briggs D. Plasma-deposited Polymeric Films Prepared from Carbonyl-containing Volatile Precursors: XPS Chemical Derivatization and Static SIMS Surface Characterization. *Chemistry of Materials* 1991; **3**: 51–61.

109. Chilkoti A and Ratner BD. An X-Ray Photoelectron Spectroscopic Investigation of the Selectivity of Hydroxyl Derivatization Reactions. *Surface and Interface Analysis* 1991; **17**: 567–574.

110. Adden N, Gamble LJ, Castner DG, Hoffmann A, Gross G and Menzel H. Synthesis and characterization of biocompatible polymer interlayers on titanium implant materials. *Biomacromolecules* 2006; **7**: 2552–2559.

111. Hollander A. Labelling techniques for the chemical analysis of polymer surfaces. *Surface and Interface Analysis* 2004; **36**: 1023–1026.

112. Andre J-M, Dehalle J and Pireaux JJ. *Band Structure Calculations and their Relations to Photoelectron Spectroscopy.* ACS Symposium Series. Washington, DC: American Chemical Society; 1982. pp. 151–168.

113. Boulanger P, Pireaux JJ, Verbist JJ and Delhalle J. X-Ray Photoelectron-Spectroscopy Characterization of Amorphous and Crystalline Poly(Tetrahydrofuran) – Experimental and Theoretical-Study. *Polymer* 1994; **35**: 5185–5193.

114. Chilkoti A, Castner DG and Ratner BD. Static Secondary Ion Mass-Spectrometry and X-Ray Photoelectron-Spectroscopy of Deuterium-Substituted and Methyl-Substituted Polystyrene. *Applied Spectroscopy* 1991; **45**: 209–217.

115. Castner DG and Ratner BD. Surface Characterization of Butyl Methacrylate Polymers by XPS and Static SIMS. *Surface and Interface Analysis* 1990; **15**: 479–486.

116. Clark DT and Thomas HR. Applications of ESCA to Polymer Chemistry.11. Core and Valence Energy-Levels of a Series of Polymethacrylates. *Journal of Polymer Science Part A – Polymer Chemistry* 1976; **14**: 1701–1713.

117. Fadley CS. Elastic and Inelastic Scattering in Core and Valence Emission from Solids: Some New Directions. *AIP Conference Proceedings* 1990; **215**: 796–813.

118. Huefner S. *Photoelectron Spectroscopy: Principles and Applications*. Berlin: Springer-Verlag; 1995.

119. Smith KE and Kevan SD. The Electronic Structure of Solids Studied Using Angle Resolved Photoemission Spectroscopy. *Progress in Solid State Chemistry* 1991; **21**: 49–131.

120. Nilsson PO. Photoelectron-Spectroscopy by Synchrotron Radiation. *Acta Physica Polonica A* 1992; **82**: 201–219.

121. Olson CG and Lynch DW. An Optimized Undulator Beamline for High-Resolution Photoemission Valence-Band Spectroscopy. *Nuclear Instruments and Methods in Physics Research, Section A – Accelerators, Spectrometers, Detectors and Associated Equipment* 1994; **347**: 278–281.

122. Fleming I, Fulton CC, Lucovsky G, Rowe JE, Ulrich MD and Luning J. Local bonding analysis of the valence and conduction band features of TiO_2. *Journal of Applied Physics* 2007; **102**: 033707.

123. Eberhardt W, Fayet P, Cox DM, Fu Z, Kaldor A, Sherwood R and Sondericker D. Photoemission from Mass-Selected Monodispersed Pt Clusters. *Physical Review Letters* 1990; **64**: 780–784.

124. Bittencourt C, Felten A, Douhard B, Ghijsen J, Johnson RL, Drube W and Pireaux JJ. Photoemission studies of gold clusters thermally evaporated on multiwall carbon nanotubes. *Chemical Physics* 2006; **328**: 385–391.

125. Blyth RIR, Andrews AB, Arko AJ, Joyce JJ, Canfield PC, Bennett BI and Weinberger P. Valence-Band Photoemission and Auger-Line-Shape Study of AuxPd1-X. *Physical Review B* 1994; **49**: 16149–16155.

126. Walton J and Fairley N. Noise reduction in X-ray photoelectron spectromicroscopy by a singular value decomposition sorting procedure. *Journal of Electron Spectroscopy and Related Phenomena* 2005; **148**: 29–40.

127. Castle JE. Module to guide the expert use of x-ray photoelectron spectroscopy by corrosion scientists. *Journal of Vacuum Science and Technology A – Vacuum Surfaces and Films* 2007; **25**: 1–27.

Problems

1. The observed E_B values for the carbonyl C_{1s} and O_{1s} subpeaks in the high-resolution ESCA spectra of polyacrylamide, polyurea, and polyurethane samples are listed below. The atoms bound to the carbonyl group in each sample are also shown. All E_B values have been referenced by setting the hydrocarbon C_{1s} peak of each polymer to 285.0 eV. Based on the different structures of these functional groups, explain their different C_{1s} and O_{1s} E_B values. Make sure an explanation is given that is consistent for both the C_{1s} and O_{1s}.

Sample	Functional group	E_B values (eV)	
		C_{1s}	O_{1s}
Polyacrylamide	$-CH-C=O$ (with NH_2 on C)	288.2	531.4
Polyurea	$-NH-C-NH-$ (with O double bond on C)	289.2	531.7
Polyurethane	$-O-C-NH-$ (with O double bond on C)	289.6	532.2

2. An ESCA survey scan of a material detected the presence of carbon, nitrogen, and oxygen. High-resolution C_{1s}, N_{1s}, and O_{1s} scans of this material showed the presence of two, one, and one subpeaks, respectively. Assuming that the transmission function does not vary with KE, λ varies as $KE^{0.7}$, and that there is an Al $K\alpha$ X-ray source, use the data provided below to calculate the percentage of each component present in this sample. Also, propose a chemical structure for this sample and provide a consistent assignment of functional groups for the subpeaks. Remember, ESCA does not detect hydrogen. Correct the E_B values for sample charging by referencing C_{1s} hydrocarbon peak to 285.0 eV.

Peak	E_B (eV)	Area
C_{1s}	276.8	4000
C_{1s}	279.9	2000
N_{1s}	391.6	3355
O_{1s}	523.2	4995

3. An ESCA survey scan of a material detected the presence of only carbon and oxygen. High-resolution C_{1s} and O_{1s} scans of this material showed the presence of four and three subpeaks, respectively. Assuming the transmission function does not vary with KE, λ varies as $KE^{0.7}$, and an Al $K\alpha$ X-ray source, use the data provided below to calculate the C/O atomic ratio and the percentage of each component present in this sample. Also, propose a chemical structure for this sample and provide a consistent assignment of functional groups for the subpeaks. The E_B values have been corrected for sample charging.

Peak	E_B (eV)	Area
C_{1s}	285.0	1925
C_{1s}	286.6	675
C_{1s}	289.0	675
C_{1s}	291.6	100
O_{1s}	532.1	1600
O_{1s}	533.7	1685
O_{1s}	538.7	85

4. There is much interest in special properties of buckminsterfullerenes. The buckminsterfullerene consists of 60 carbon atoms arranged in an icosahedron (a soccer ball-like shape). The diameter of the icosahedral 'sphere' is 0.71 nm. Consider the cases where we have three close-packed layers, one close-packed layer, and a partial monolayer (covering 70 % of the surface area) of 'buckyballs' on molecularly smooth, contamination-free gold substrates. These specimens are examined by ESCA. Roughly sketch plots of the anticipated relative gold photoemission signal intensity (I) as a function of photoelectron (sample) take-off angle (θ) (note: in this example, θ is measured to the normal to the specimen) for the three specimens. Since the icosahedra pack closer than spheres, you can assume that the buckminsterfullerene molecules can be modeled as solid cubes, 0.71 nm on a side. The plots you sketch should approximately represent the anticipated characteristics of the signal intensity variation without the need for accurate x- or y-axis numbers – the functional relationship is being explored here. The y-axis should have two numbers on it: 100 % signal (relative to clean gold) and 0 % signal (no gold signal). The x-axis should have 0° and 80° on it.

5. Consider a molecularly smooth Teflon [—$(CF_2CF_2)_n$-] overlayer deposited on a molecularly smooth, contamination-free platinum surface. Using ESCA (the sample analysis angle relative to the surface normal is specified in parentheses) to study this type of specimen, arrange the following cases in order of decreasing relative platinum signal: [a] a 7.0 nm Teflon overlayer and an Al Kα X-ray source ($0°$), [b] a 0.5 nm Teflon overlayer and an Al Kα X-ray source ($0°$), [c] a 0.5 nm Teflon overlayer and a Ag Lα X-ray source ($0°$), [d] a 7.0 nm Teflon overlayer and an Al Kα X-ray source ($80°$), and [e] a 0.5 nm Teflon overlayer and a Ti Kα X-ray source ($0°$).

6. A polymeric surface rich in hydroxyl groups is derivatized with a vapour-phase reagent that converts the —OH groups to —OCF_3 groups. The specimen is studied in an ESCA instrument with a non-monochromatized Mg Kα X-ray source. The relative fluorine signal is observed to decrease with increasing time under analysis. Suggest three reasons why this might occur. Suggest two instrumentation strategies to make this derivatization analysis useful for analytically comparing specimens as to —OH content.

4 Molecular Surface Mass Spectrometry by SIMS

JOHN C. VICKERMAN

Manchester Interdisciplinary Biocentre, School of Chemical Engineering and Analytical Sciences The University of Manchester, Manchester, UK,

4.1 Introduction

Secondary ion mass spectrometry, SIMS, is the mass spectrometry of ionized particles that are emitted when a surface, usually a solid, is bombarded by energetic primary particles which may be electrons, ions, neutrals or photons. The emitted or 'secondary' particles will be electrons, neutral species, atoms or molecules or atomic and cluster ions. The vast majority of species emitted are neutral but it is the secondary ions that are detected and analysed by a mass spectrometer. It is this process which provides a mass spectrum of a surface and enables a detailed chemical analysis of a surface or solid to be performed.

At first sight the process is conceptually very simple. A pictorial representation of the process is shown in Figure 4.1. Basically when a high energy (normally between 10 and 40 keV) beam of ions or neutrals bombards a surface, the particle energy is transferred to the atoms of the solid by a billiard-ball-type collision process. A 'cascade' of collisions occurs between the atoms in the solid; some collisions return to the surface and result in the emission of atoms and atom clusters, some of which are ionized in the

Surface Analysis – The Principal Techniques 2nd Edition Edited by John Vickerman and Ian Gilmore
© 2009 John Wiley & Sons, Ltd

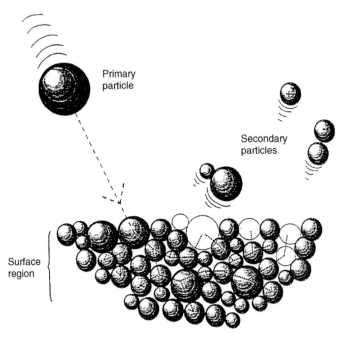

Figure 4.1 A schematic representation of the SIMS process. Reproduced with permission from N. Lockyer, *Ph.D. Thesis*, University of Manchester Institute of Science and Technology, 1996

course of leaving the surface. The point of emission of low energy secondary particles is remote (up to 10 nm) from the point of primary impact; the final collision resulting in secondary particle emission is of low energy (ca. 20 eV); over 95 % of the secondary particles originate from the top two layers of the solid. Thus the possibility of a 'softish' ionization mass spectrometry of the surface layer emerges.

Although the emission of secondary ions from surfaces was observed about 100 years ago [1], the *surface* mass spectrometry capability has developed since the late 1970s. By far the widest application of SIMS up to the early 1980s was to exploit its destructive capability to analyse the elemental composition of materials as a function of depth. The technique is known as *dynamic* SIMS (see Chapter 5). Dynamic SIMS has found extensive application throughout the semiconductor industry where the technique has had a unique capability to identify chemically the ultra-low levels of charge carriers in semiconductor materials and to characterize the layer structure of devices. Indeed, after instruments first appeared (Herzog [2]; Leibl [3]; Castang and Slodzian [4]) in the early 1950s and 1960s the technique developed rapidly under the impetus of this industry for the next two decades. Whilst of great importance, this variant of SIMS cannot be described as a surface *mass spectrometry*.

Static SIMS emerged as a technique of potential importance in surface science in the late 1960s and early 1970s as a consequence of the work of Benninghoven and his group in Münster [5]. Whilst the SIMS technique is basically destructive, the Münster group demonstrated that using a very low primary particle flux density ($<1\,nA\ cm^{-2}$) mass spectral data could be generated in a timescale that was very short compared to the lifetime of the surface layer. The information so derived would be characteristic of the chemistry of the surface layer because statistically no point on the surface would be impacted more than once by a primary particle during an analysis. The surface could be said to be essentially *static*. Obviously, the use of a very low primary flux density resulted in a very low yield of secondary particles and this imposed requirements of high sensitivity on the detection equipment. The fact that these experimental conditions could be used was due to advances in single particle detection equipment. Benninghoven and his team first demonstrated the surface analytical possibilities of static SIMS in a series of studies of the initial oxidation of metals [6]. Since those days a very large body of evidence from a wide range of chemistries, from model single crystal adsorbate systems [7] to complex polymer based materials, amply demonstrates that using *static* analysis conditions there is a clear relationship between static SIMS spectra and surface chemistry. Part of the positive ion spectrum of the surface of a film of the drug halperidol illustrates this well (Figure 4.2). It is distinctive and analytically definitive for this drug.

Developments at the turn of the 21st century have led to the use of cluster ions – Au_n^+, Bi_n^+, SF_5^+, C_{60}^+ primary ions instead of the atomic ions (Ar^+, Ga^+, Cs^+, etc.) used up to that time. As we shall see later these cluster ions generate higher secondary ion yields from molecular materials. Thus significantly increasing the sensitivity of the technique. However, the larger

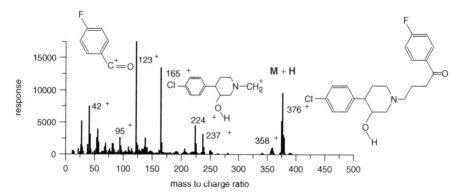

Figure 4.2 Positive ion spectrum of Halperidol – a neuroleptic drug (used to treat mental disorders, hyperactivity, agitation, etc.). Au^+ primary ion beam $\sim 10^{10}$ ions cm^{-2}

cluster ions – SF_5^+ and particularly C_{60}^+ seem to cause a great deal less bombardment induced chemical damage in many materials. The result is that the *static* requirement for analysis of these materials can be relaxed or even abandoned. Consequently in principle 100 % of the surface can be used for analysis, greatly increasing potential signal levels and detection limits. This variant of SIMS should perhaps now be termed *molecular* SIMS.

4.2 Basic Concepts

4.2.1 THE BASIC EQUATION

A more extensive introduction to the phenomenology of sputtering and secondary ion emission can be found elsewhere [8–11]. SIMS is concerned with the analysis of secondary *ions*. Ionization occurs at, or close to, emission of the particles from the surface with the consequence that the matrix participates in the electronic processes involved. This means that the yield of secondary ions is strongly influenced by the electronic state of the material being analysed (this phenomenon is termed the *matrix effect*) with consequent complications for quantitative analysis. The basic SIMS equation is as follows:

$$I_s^m = I_p y_m \alpha^\pm \theta_m \eta \qquad (4.1)$$

where I_s^m is the secondary ion current of species m, I_p is the primary particle flux, y_m is the sputter yield, $\alpha^\pm$ is the ionization probability to *positive or negative ions*, θ_m is the fractional concentration of m in the surface layer and η is the transmission of the analysis system.

4.2.2 SPUTTERING

The two fundamental parameters are y_m and $\alpha^\pm$ where y_m is the yield of sputtered particles of species m, neutral and ionic, per primary particle. It also increases with primary particle mass, charge and energy, although not linearly [4]. Figure 4.3 shows the variation of y for aluminium. The crystallinity and topography of the bombarded material will also affect the yield. The threshold for sputtering occurs at about 20 to 40 eV primary particle energy and y tends to maximize with energy at around between 5 and 50 keV. Beyond this energy yield drops away as the primary beam penetrates deeper into the solid and less energy returns to the surface region. For amorphous or polycrystalline materials sputter yield increases monotonically with increasing incident angle relative to the surface normal to a maximum at about 60 to 80°. The angular distribution of sputtered material tends to have a cosine distribution around the angle of reflection of the primary beam. Generally the larger the mass of the bombarding

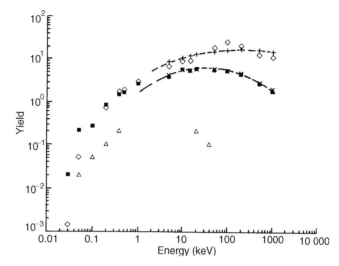

Figure 4.3 Experimental sputter yield data for aluminium as a function of primary ion energy for a number of different primary ions: △, He; ◇, Xe; ■, Ar; +, Xe (theoretical); ×, Ar (theoretical)

particle, the closer to the surface the energy will be deposited and hence the greater will be the yield. At a given bombardment energy the sputter yield for elements varies by a factor of 3 to 5 through the Periodic Table.

Sputtering is a damaging process, consequently it is more difficult to measure sputter rates for covalent organic materials. The yield of elemental carbon could be measured, but in static SIMS we are more interested in using the technique to detect and measure chemical structure. Sputtering of organic materials results in the removal of elements, structural fragments and molecular species. The loss of any of these entities from the surface will destroy the chemical structure within the area from which it is removed. If the material is molecular, every molecule impacted will be effectively destroyed, whether the whole molecule or only a small piece of the molecule is removed. If the material is a polymer, then that part of the monomer unit impacted will be destroyed. Thus instead of sputter rate the concept of *disappearance cross-section*, σ, has been found to be more useful. When a monolayer on a chemically different substrate is being studied some of the loss of signal will be due to the removal of intact molecules and some will be due to bombardment induced chemical damage. When the material being studied is in the form of a multilayer and there is a large supply of molecules the loss of structurally significant species from the SIMS spectrum as a function of bombardment time is taken to be a measure of the increasing *damage* and we speak of the *damage cross-section*. Clearly this measurement is in contrast to sputter yield measurements where the material removed from the surface is collected. The *disappearance cross-section, σ*, is related to the secondary ion intensity by the following equation:

$$I_m = I_{mo} \exp(-\sigma I_p) \tag{4.2}$$

Benninghoven *et al.* obtained *disappearance cross-sections* of around 5×10^{-14} cm^{-2} for amino acids and other small molecules on metal substrates [12]. In the case of polymers there is a decay in the intensity of characteristic fragment ions. It is usually observed that the larger the fragment, the greater the rate of damage. However, since removal of backbone parts of the polymer will require more than one scission point, there is frequently a rise in intensity of these fragments, before they start to decay [13]. The *disappearance cross-section* is an experimentally measured parameter which can be thought of as the average area per incident particle from which the emission of the particular species being analysed is excluded; 10^{-14} cm^2 is an area 10 Å square which is about the size of the fragments detected. The *disappearance cross-section* may be related to the *damage cross-section* depending on the precise form of material under study. Figure 4.4 shows the disappearance cross-section determined for a thick layer of cholesterol under 15 keV Au$^+$ bombardment.

As for elemental sputtering, secondary ion yields and damage cross-sections for organic materials increase with primary ion mass and energy and increasing angle of incidence away from normal [14]. It is also observed that there is an increased relative yield of high mass fragments and molecular species [15].

As mentioned in recent years there has been increasing interest in the use of polyatomic cluster primary ions, SF_5^+, Au_n^+, Bi_n^+, C_{60}^+. These ions have been shown to deliver significantly higher ion yields, particularly of high mass species [16] (see also Section 4.5.2). Much of the evidence suggests that the increased yields are because of very much increased sputter yields. In Table 4.1 the sputter yields of water molecules from ice under Au_n^+ and C_{60}^+ bombardment are compared [17]. There is a dramatic increase in yield

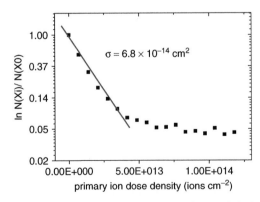

Figure 4.4 Disappearance cross-section plot for a thick layer of cholesterol under 15 keV Au$^+$ bombardment ($\sigma = 6.8 \times 10^{-14}$ cm^2)

Table 4.1 Sputter yields of water molecules from ice under bombardment by 20 keV Au_n^+ ions compared with C_{60}^+

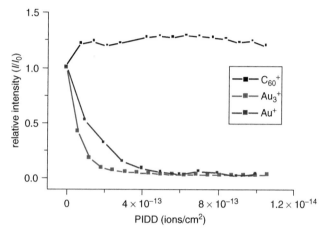

	Au^+	Au_2^+	Au_3^+	C_{60}^+
Removed # of H_2O Equivalents	100	575	1190	2510

Figure 4.5 Variation in relative intensity of the $(M-H)^+$ from a thick film of cholesterol as a function of ion fluence under 20 keV Au^+, Au_3^+ and C_{60}^+ bombardment ($m/z = 385$)

between the atomic projectile Au^+ and the cluster ions. The cluster breaks up as it hits the surface and the projectile energy is portioned between all the atoms (thus each atom from a 20 keV C_{60}^+ would have 666 eV energy). As a result they penetrate the material much less and generate much less chemical damage (see Section 4.4 for a pictorial representation). In addition the sputter rate is so high that most of any damage generated is removed by subsequent impacts so the apparent damage cross-section is greatly reduced, removing the need for the *static limit* for many materials. Figure 4.5 compares the loss of a molecular ion signal from a cholesterol film supported on silicon under Au^+ and C_{60}^+ bombardment. It can be seen there is rapid and almost complete loss under gold bombardment, but after an initial change there is a signal plateau until all the material is removed from the surface.

Atoms, molecules and molecular fragments sputtered from the surface of solids are emitted with a range of kinetic energies. The kinetic energy distribution is influenced indirectly by the primary ion energy, angle of incidence and atomicity which will determine the nature of the collision

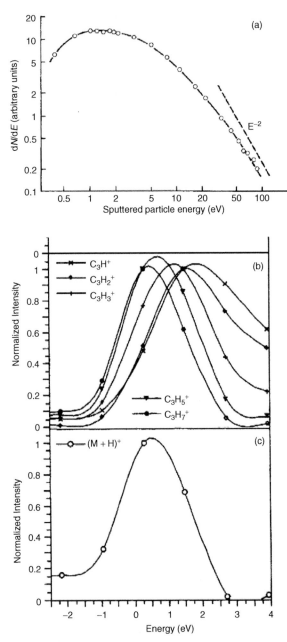

Figure 4.6 (a) The energy distribution of sputtered gold atoms due to the impact of $16\,keV\,Ar^+$ on polycrystalline gold (adapted from *J. Nucl. Mater*, **76/77**, 136 Copyright (1978), Elsevier). (b) and (c) Kinetic energy distributions of positive ion species sputtered from a layer of tricosenoic acid ($C_{22}H_{43}\backslash,COOH$) on gold (adapted from *Nucl. Instrum. Meth. Phys. Res. B* **115**, 246 Copyright (1996), Elsevier). Reproduced with permission from *ToF–SIMS: Surface Analysis by Mass Spectrometry*, John Vickerman and David Briggs (Eds), Chapter 1. Copyright 2003, SurfaceSpectra and IM Publications

cascades in the material. However, of more direct influence will be the binding of the species in the surface, the number of bonds to be broken, the degree to which internal energy can be stored by the emitted species. The kinetic energy distribution of an atomic secondary ion from a metal will generally be broad, typical of collisional sputtering (Figure 4.6(a)), whereas that of cluster ions will be very much narrower because they can lose energy by fragmentation or storing it in vibrations and rotations (Figure 4.6(b)). The kinetic energy distributions of large molecular species from organic materials are usually very narrow. If such a species fragments as it leaves the surface, the resulting fragment may show a negative tail in the distribution (Figure 4.6 (c)). Thus kinetic energy distributions can reveal a good deal about the mechanisms of the emission process.

4.2.3 IONIZATION

Secondary ion formation from inorganic materials is usually strongly influenced by electron exchange processes between the departing species and the surface. Thus the electronic state of the surface is critical. The yields of elemental secondary ions can vary by several orders of magnitude across the Periodic Table, Figure 4.7, and are very dependent on the chemical state of the surface. This phenomenon is known as the matrix effect. Thus the

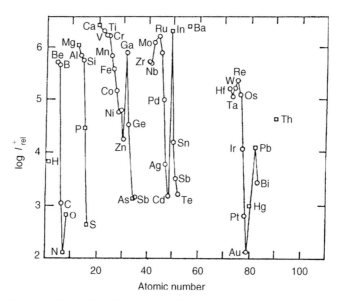

Figure 4.7   The variation of positive ion yield as a function of atomic number for 1 nA 13.5 keV O-bombardment: o, from elements; □, from compounds. Reproduced with permission from H.A. Storms, K.F. Brown, and J.D. Stein, *Anal. Chem.*, **49**, 2023 (1977). Copyright (1977) American Chemical Society

Table 4.2 Secondary ion yields from clean and oxidised metal surfaces

Metal	Clean metals M^+ yield	Oxide M^+ yield
Mg	0.01	0.9
Al	0.007	0.7
Si	0.0084	0.58
Ti	0.0013	0.4
V	0.001	0.3
Cr	0.0012	1.2
Mn	0.0006	0.3
Fe	0.0015	0.35
Ni	0.0006	0.045
Cu	0.0003	0.007
Ge	0.0044	0.02
Sr	0.0002	0.16
Nb	0.0006	0.05
Mo	0.00065	0.4
Ba	0.0002	0.03
Ta	0.00007	0.02
W	0.00009	0.035

ion yield for a particular element will vary dramatically, for example, for a metal as compared to its oxide, Table 4.2. It can be seen that oxidation changes the elemental ion yields to differing extents resulting in significant complications when absolute quantitative data is required.

Secondary ion formation from organic materials can occur by a number of mechanisms. Ejection of an electron to form an odd electron ion $M^{\cdot+}$; polar molecules may undergo acid base reactions to form $(M+H)^+$ or $(M-H)^{\pm}$ ions; cationization or anionization of neutral molecules may occur. These processes are mainly relevant to molecular type species while low mass fragments also provide important information for chemical structure determination. Ionization of these species probably occurs via a collision induced mechanism due to direct interaction with the primary ion or energetic recoil atoms within the material. The exact locus of these ionization processes is not known, but is likely in the emission region within or just above the surface. Matrix effects do influence secondary ion yields from organic materials and some cases can be quite severe (see Section 4.4.4). Ion yields from copolymers have been observed to be sensitive to identity of the components. Clearly cationization will be favoured when suitable cations (Ag, K, H) are available in the matrix.

4.2.4 THE STATIC LIMIT AND DEPTH PROFILING

Under atomic primary ion analysis we have seen that it is necessary to operate under so-called *static* conditions. This is to maintain the integrity of the surface layer within the timescale of the analytical experiment. This implies that a very low primary beam dose is used during analysis. It is estimated that each primary particle colliding with the surface disturbs an area of 10 nm^2, thus it would only require 10^{13} impacts cm^{-2} to influence all the atoms in the surface. We have seen that damage cross-sections for organic materials are about 5×10^{-14} cm^{-2}. If this value is entered into Equation (4.2), about 50 % of the signal intensity would be lost after a primary ion dose of 10^{13} ions cm^2. Traditionally a dose of 10^{13} ions cm^2 has been regarded as the static limit. However it is clear from the calculation that a value $\leqslant 10^{12}$cm^{-2} would be safer.

A distinction between dynamic and static conditions can be understood by computing the lifetime, t_m, of the topmost atomic layer as a function of the primary beam flux at the sample surface.

$$t_m = \frac{10^{15}}{I_p} \times \frac{Ae}{y} \tag{4.3}$$

where A cm^2 of the surface (surface layer atom density of 10^{15} atoms cm^{-2}) is bombarded by a primary beam of I_p in amps, e is the charge on an electron and the sputter yield is y (usually between 1 and 10 for atomic primary ions). The primary beam current is measured in amps (1 amp is equivalent to 6.2×10^{18} charged particles s^{-1}). Using this equation, assuming a sputter yield of 1 Table 4.3 has been assembled.

If an analysis requires say 20 min (1200 s) then static conditions can only be safely attained for atomic primary beam currents of about 1 nA cm^{-2} or less. For dynamic SIMS, high elemental sensitivity and rapid erosion rates are required, so high primary flux densities of 1 µA cm^{-2} or greater are

Table 4.3 The surface monolayer life-time as a function of primary beam flux density

I_p (A cm^{-2})	t_m (s)
10^{-5}	16
10^{-7}	1600
10^{-9}	1.6×10^5
10^{-11}	1.6×10^7

desirable (see Chapter 5). The time to complete a depth profile will be of principal interest.

We have seen that polyatomic cluster ion beams behave somewhat differently. The heavy metal cluster beams – Au_n^+ and Bi_n^+ – have higher sputter yields by factors 100 to 1000 compared to their atomic ions; however their damage cross-sections are comparable. Analysis using these ions is limited by the use of *static* conditions. The total ion dose used should not exceed 10^{13} ions cm^{-2}. However for ions such as SF_5^+ and C_{60}^+ the zone of material damaged is close to the surface and is approximately contained within the volume of material sputtered away. A consequence is the apparent damage cross-sections are very much less than $10^{-14} cm^2$ from multi-layers of many organic and bio-organic materials. For these materials the static limit can be ignored because chemically characteristic secondary ions continue to be emitted until all the material is removed. This phenomenon means that molecular analysis as a function of depth can be carried out (known as depth profiling). This is a very valuable capability when analysing heterogeneous synthetic or biological systems and was impossible with atomic primary ions because they destroyed the chemistry of the underlying layers before the top monolayer was removed. Examples of this type of analysis will be discussed later.

4.2.5 SURFACE CHARGING

Many of the important technological materials requiring surface analysis are insulators. When an insulating sample is bombarded by a positive ion beam the surface potential rises due to the input of positive charge and the emission of secondary electrons. The potential can rise very rapidly by several hundred volts in a few minutes, such that the kinetic energy of the emitted positive ions rises well beyond the acceptance window of the analyser [18]. The result is the loss of the SIMS spectrum. An early and very successful solution to this problem for positive ion quadrupole SIMS was to use a neutral atom beam (fast atom bombardment) [19]. The need for pulsed primary beams in ToFSIMS made this solution more difficult to apply and there are now two linked solutions to this problem. The first, widely used method, is to irradiate the sample surface with a beam of relatively low energy electrons. The theory is that the electrons will be attracted to the region of positive charge on the surface and hence the surface potential will return to neutral. This usually works quite well for positive ion SIMS; however for negative ion detection it is necessary to drive the surface potential negative in order for the negative ions to be released from the surface. This requires a higher flux of electrons, usually about 10 times that of the ion flux. The balance is sometimes difficult to attain, particularly if the material is rough or of small dimensions, e.g. fibres or granules. One disadvantage of this approach is that electron bombardment can also give rise to sample degradation and electron stimulated ion emission and it has been shown that the input of electrons needs to be kept below a total dose 6.3×10^{18} electrons m^{-2} if electron initiated damage is to be avoided [20]. The alternative or linked solution is to either to place a grid in close electrical contact with the sample, or deposit the material as a thin film on silver. The use of thin, incomplete polymer films supported on specially treated silver foils is particularly advocated by some workers. Because the film is very thin, little charging occurs and this allows the SIMS spectra to be acquired without any charge neutralization. The possibility of electron beam induced degradation effects is then obviated. However the spectra usually display considerable cationization by the silver support. This can be helpful or not depending on the analytical requirements.

4.3 Experimental Requirements

The basic arrangement for the SIMS experiment is shown in Figure 4.8. There are three main components: the primary particle source, the mass spectrometer and, since the secondary ions are emitted with a range of kinetic energies, an ion optical system which selects ions within a defined energy band compatible with the capability of the mass analyser.

4.3.1 PRIMARY BEAM

The range of primary beam source designs that have been used in SIMS can be classified into four basic types according to their mechanisms of primary beam production: (i) electron bombardment; (ii) plasma; (iii) surface ionization; (iv) field ionization. Each type of source offers different performance in terms of spatial resolution, ease/speed of use, sensitivity, coping with insulating materials, beam induced damage, etc. and all types have been pulsed successfully for time-of-light (ToF) mass spectrometer systems (see later).

The basic components of most types of ion beam source, Figure 4.9, are the source region/extraction zone, focusing and collimating regions, a

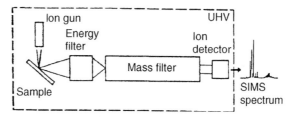

Figure 4.8 A schematic representation of a SIMS instrument. Reproduced with permission from J.C. Vickerman, *Chem. Brit.* 969 (1987)

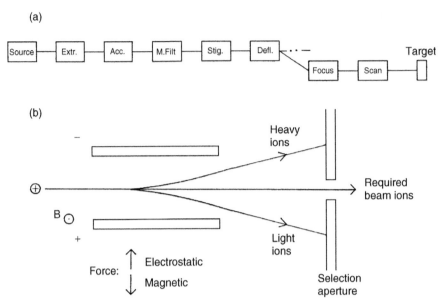

Figure 4.9 A schematic diagram of (a) the main components of a scanning, focused ion gun: Source, ion source; Extr, extractor; Acc, accelerator; M.Filt, mass/energy filter; Stig, stimator; Defl, neutral elimination bend; Focus, focusing lens; Scan, xy raster, reproduced from [7] with permission. (b) Operation of a Wien filter for mass selected ion beam: balance condition is $m/z = 2V(B/E)^2$. Reproduced with permission from *Secondary Ion Mass Spectrometry – Principles and Applications*, Oxford University Press (1989)

mass (Wien) filter for beam purification, a pulsing mechanism for systems with time-of-flight mass analysers, stigmation/focus lenses and finally scan rods. The mode of operation of the primary beam largely defines the type of SIMS information accessible. In static and dynamic SIMS it is common practice, to raster scan the primary ion beam across the surface region of interest (see Figure 4.9). For static (scanning) SIMS this enables sensitivity to be optimized by matching the analysed area to the field of view of the collection optics of the analyser. In dynamic SIMS the dimensions of the raster define the crater edges. Scanning SIMS with the mass spectrometer preset to detect certain masses also offers the possibility of mapping the distribution of secondary ions over the area of interest. An instrument with this capability is known as a scanning *SIMS microprobe* and using computer image storage and a colour coded graphics system it is possible to produce colour coded maps of elements and molecules in the surface.

A mapping of the distribution of surface elements can also be derived using an unscanned primary ion beam and a mass spectrometer with an ion optical arrangement such that the positional sense of the ions is retained throughout mass analysis process. This mode of mapping is known as *ion microscopy* and the ion image can be displayed directly onto a fluorescent screen or registered on a position sensitive detector for subsequent computer storage/manipulation [21].

The next section provides an overview of the basic aspects of some of the more popular primary beam sources. Critical parameters are the brightness and energy spread of the primary beam and the stability and reliability of the device.

Electron Bombardment. These ion sources are based upon the principle of using a high current density of electrons to ionise the primary beam gas, usually argon or xenon. Many types of source arrangement exist. For inert source gases generally a hot cathodic source of electrons (usually tungsten or iridium, frequently treated to increase its electron emission) is used. The electrons are accelerated towards the anode to give them the required energy to ionize the source gas (see Figure 4.10). Cross-sections for gas interaction (and therefore ionization yield or source efficiency) can be increased using electrostatic or magnetic fields to increase the path length of the travelling electrons. The beam is extracted from the ion source and accelerated and focused to produce a beam at the sample surface of between about 2–40 keV energy.

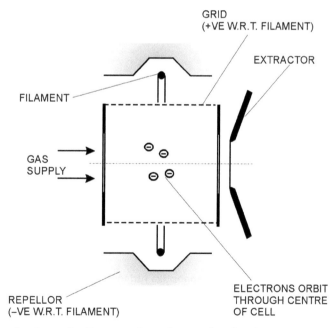

Figure 4.10 A schematic diagram of an electron bombardment source. Reproduced with permission from R. Hill in *ToF–SIMS: Surface Analysis by Mass Spectrometry*, John Vickerman and David Briggs (Eds), Chapter 4. Copyright 2003, SurfaceSpectra and IM Publications

A neutral beam can be generated by charge exchanging, say, an argon ion beam by passing it through a chamber (the Wien filter region is used in a number of cases) in the ion beam system, containing a pressure of gaseous argon (10^4 mbar) [22]. A proportion of the ions (10–30 %) lose their charge by capturing an electron from the atoms randomly moving in the chamber. Although these ions have lost their charge, they retain their velocity and direction and a fast atom beam is formed. The residual ion beam can then be deflected away.

Most electron bombardment sources are versatile, easy to use and comparatively reliable. They offer only moderate brightness (this is a measure of the current density available from the source) of about 10^5 A m^{-2} Sr^{-1}. In the past they were most commonly used de-focused over large areas (several mm^2) for static SIMS work. However with more efficient higher energy ion columns they are used now for cluster ion beam systems such as SF_5^+ or C_{60}^+ (see Figure 4.11); pA currents on target are accessible in micro-focused beams.

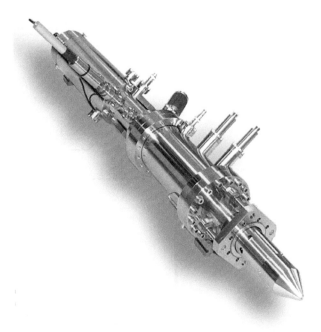

Figure 4.11   A 40 keV C_{60}^+ Ionoptika Ltd ion column. Reproduced with permission from Ionoptika Ltd

Plasma.   Duoplasmatron, RF and hollow cathode sources are all classified under this heading. The output of a simple electron bombardment source may be limited by the density of electrons which can be generated due to space charge effects. If the pressure of the source gas is raised, the ions and neutrals reduce the repulsions between the electrons and a much higher density of electrons can be sustained generating a higher ion beam current. Under such conditions a plasma is formed. The source will have an exit aperture through which the ions are extracted to form the ion beam. The source of electrons may be a hot filament; alternatively electrons may be generated by positive ion bombardment of a cathode to sustain the discharge. Reactive gases, such as oxygen, can be ionized using the cold cathode discharge method. Electric and magnetic fields are used to concentrate the discharge to increase output. The improved ionization efficiency of this method is reflected in the higher beam brightness ranging from 10^4 to 10^7 A m^{-2} Sr^{-1} attained with this type of source. However, this is achieved partly at the expense of reliability as the violence of the emission process tends to gradually destroy the source components by ion etching. Higher beam brightness renders this source type better suited for dynamic SIMS applications (μA into $\approx 50\,\mu$m) and for microfocused scanning analysis (nA into $\leqslant 5\,\mu$m).

As well as inert gases, this source type is used for oxygen ($O_2{}^+$) primary ion bombardment, the use of which provides improved sensitivities for the detection of electropositive species and thus is the primary source of choice for many semiconductor depth profiling applications (see Chapter 5).

Surface Ionization.   Sensitivity to electronegative species is enhanced when an electropositive primary beam is used. The availability of an alkali metal ion source is thus attractive to the depth profiling SIMS analyst and this is offered by the surface ionization source. In this case, ion emission is thermally stimulated by warming an adsorbed layer of, for example, cesium on the surface of a high work function metal (e.g. iridium) under vacuum conditions. The ionization potential of the Cs adlayer and the work function of the surface are such that electrons can move freely from adlayer to substrate and upon mild thermal excitation, ions of a very low and uniform energy spread are emitted.

Source brightness depends on the size of the emitting area and $>10^6$ A m^{-2} Sr^{-1} have been attained. There is the drawback of very careful handling and operational requirements of the source metal. However, the benefits they offer in electronegative ion yield for example, are such that this source is built in or retro-fitted to nearly all SIMS instruments used for depth profiling semiconductors. Cesium ion sources have been successfully adapted for ToFSIMS analysis where their use in the correspondingly lower ion dose regimes can lead to the generation of cationized secondary ions that are of diagnostic value.

Field Ionization Sources. These sources operate on the principle of stripping electrons off source atoms situated near to an extremely high local electronic field. A very fine tip with radius $< 1\,\mu m$ is used and source energies can be as high as $10-40\,kV$. Gas based field ionization sources have failed yet to produce sufficient beam currents for SIMS work; however, the liquid metal counterpart is widely used. In the latter case (the electrohydrodynamic ion source) a thin 'skin' of liquid metal (typically gallium, indium, gold or bismuth) is allowed to flow over a fine tungsten tip in a region of very high extraction field, Figure 4.12. This has the effect of distorting the skin towards the exit ring and setting up a (Taylor's) cone and (plasma) ball structure of liquid metal on the probe tip. Primary ions of the metal are stripped away from the plasma ball.

These are the highest brightness sources ($\approx 10^{10}\,A\,m^{-2}\,Sr^{-1}$) used for surface mass spectrometry and as such are the source of choice for work at the highest spatial resolution. The most commonly used liquid metal ion source has been based on liquid gallium which relies on field-ionizing Ga^+ from a tungsten tip to generate a very bright and highly focusable beam. Around the turn of the century liquid metal ion beams based on gold and

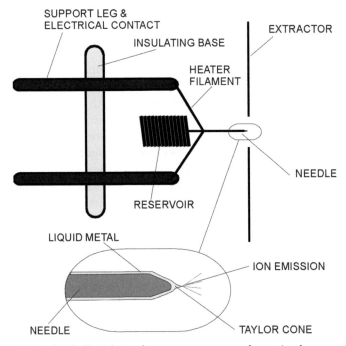

Figure 4.12 Principle of a liquid metal ion source: source schematic; close up of extraction region. Reproduced with permission from R. Hill in *ToF–SIMS: Surface Analysis by Mass Spectrometry*. John Vickerman and David Briggs (Eds), Chapter 4. Copyright 2003, SurfaceSpectra and IM Publications

bismuth were developed commercially. These metals generate quite high yields of cluster ions, e.g. Au_3^+, Bi_3^+, Bi_5^+ etc. (reviewed in Wucher [16]). As indicated earlier, such ions deliver significant increases in the yields of higher mass secondary ions and are attractive for the analysis of organic and bio-organic systems. Spatial resolutions down to the range 200–50 nm have been realized. It is, however, very difficult to maintain *static* conditions and obtain sufficient signal at these levels of spatial analysis (see Section 4.5.2).

Time-of-flight mass spectrometers require pulsed ion beams (see Section 4.3.2). Good mass resolution requires short (ns) pulses. The liquid metal ion beam systems have been adapted for pulsing by the introduction of deflection blanking plates that rapidly sweep the beam across an aperture. By appropriate design of deflection plates and lenses it is possible to motionlessly blank the beam. This can be very effective with liquid metal ion beams that have a very sharp source; however with gas beams it is more problematical and the minimum beam diameter attainable is degraded by the rapid beam movement. To obtain a very short pulse for very high mass resolution beam bunching is frequently used. The initial pulse width of the ions is approximately 20–50 ns; this pulse width is 'bunched' using an accelerating region producing a time-focused pulse at the surface, usually of less than 1 ns. Unfortunately, this process introduces an energy spread in the ions and chromatic aberrations in the lenses result in a degraded beam size at the sample. Consequently high spatial resolution is incompatible with high mass resolution.

4.3.2 MASS ANALYSERS

The different modes of SIMS have contrasting demands on the mass analyser. Where *static* conditions are required it is necessary to maximize the information level achieved per unit of surface damage. The analysis and detection system should be as efficient as possible for the total yield of secondary ions from the surface. In dynamic SIMS the requirement is usually to have the highest sensitivity possible for specific elemental ions. The preservation of surface structure is not important. In scanning or imaging SIMS where the spatial distribution of surface chemical information is being studied the requirements may be the same as for static SIMS. If the poly-atomic cluster beams are used, the static requirement may be lifted and more dynamic conditions can be tolerated.

The three most widely used mass analysers are the quadrupole RF mass filter, the magnetic sector and the time-of-flight instruments. The quadrupole analyser was widely used in the early work in static SIMS because it was easily incorporated in a UHV system due to its small size. Whilst a great deal of useful information has been obtained using this analyser in static SIMS, it is a low transmission device (less than 1 %). Furthermore, it is a scanning instrument so that it only allows the sequential transmission of ions, all other ions being discarded. The information loss is therefore very high. In recent years ToF analysers have been used for static SIMS because of their very high transmission and the fact that they are quasi-parallel detectors – they are not scanning instruments and collect all the ions generated. Consequently they are about 10^4 more sensitive than a quadrupole instrument. Historically, the magnetic sector analyser has been used for dynamic SIMS because of its high transmission, 10–50 %, high duty cycle and high mass resolution. Although it is usually a scanning device, since detection of only a few specific ions is required using a continuous high flux primary beam, it was traditionally preferred over the ToF instrument.

Magnetic Sector. This type of mass analyser was first used for conventional mass spectrometry and the principle of operation is well understood. Ions are extracted from the sample using a high extraction potential, circa 4 kV. Upon traversing a magnetic field, a charged particle experiences a field force in a direction orthogonal to the direction of magnetic flux lines and its original axis of travel; it thus adopts a circular path. The extent of force experienced by the particle and hence the radius of its path is directly related to its velocity and therefore since all ions are accelerated to a fixed potential before entering the magnetic field they can be readily separated according to their masses. The radius of curvature, R, for an ion of mass to charge ratio, m/z, traveling through a magnetic field, B, having been accelerated by potential V is given by:

$$R = \frac{1}{B}\left(\frac{2mV}{z}\right)^{\frac{1}{2}} \tag{4.4}$$

The dispersion of adjacent masses, i.e. mass resolution, is proportional to the radius of the magnets used and degrades with increasing mass. Secondary ions are emitted with a spread of kinetic energies. Elemental ions usually have a wider distribution (up to $\approx$100 eV) and peak around 10–20 eV whereas the multi-atomic or molecular ion distribution will peak between 1 and 5 eV and only have a width of a few tens of eV. A wide energy spread can degrade the mass resolution. Double sector instruments incorporating an electrostatic sector are often used in order to combat these resolution degradation effects. The electrostatic sector allows a small energy band of the ions to be selected and focused on the entrance slit of the magnet for analysis. The dispersed ions are commonly scanned across an exit slit of the magnet by scanning the field strength of the electromagnet. In a dynamic SIMS experiment where a few specific elemental ions are to be measured in a depth profile, rapid switching between masses is possible using the electromagnet (see Chapter 5).

An attractive feature of this form of mass spectrometry for surface analysis is that the positional sense of secondary ions can be retained throughout the analysis process such that secondary ion images can be projected in real time on to a fluorescent screen or directly into computer software via a position sensitive detector. This device can therefore be operated as an *ion microscope*. By irradiating the sample with a large diameter static beam the spatial distribution of the emitted secondary ions can be observed on the viewing screen with a spatial resolution in the 1–5 µm region [21].

More normally depth profiling is carried out in the *microprobe* mode. A highly focused ion beam 1–10 µm diameter is raster scanned across the sample surface to erode a crater with uniform edge and crater bottom. Emitted secondary ions are collected from the central area of the crater bottom.

Despite impressive performance characteristics, magnetic sector mass spectrometers are not the ideal. They can be large, cumbersome devices and pose considerable difficulties in the generation of true UHV. This is because these instruments cannot be easily baked to desorb chamber wall gases without seriously (irreversibly) modifying the magnetic properties of the magnet. Although transmission is affected, the most serious consequence of non-UHV conditions is in the sample region where interference effects from background gases raise the detection limits for residual gas elements in a matrix such as, for example, carbon in silicon. To minimize this problem instruments with a retractable magnet have been introduced which enables system baking. In another design the introduction of very high pumping rates in the sample region by the use of cryo-pumping reduces the local pressure.

The quadrupole Mass Analyser. A quadrupole mass spectrometer is so called since it makes use of a combination of a DC and a radio frequency (RF) electric field applied to four parallel rods, in order to separate ions according to their mass-to-charge ratio. A potential consisting of a constant DC (U) component plus an oscillating RF component ($V \cos \omega t$) is applied to one pair of rods whilst an equal but opposite voltage is applied to the other pair. The rapid periodic switching of the field sends most ions into unstable oscillations of increasing amplitude until they strike the rods and are hence not transmitted. However, ions with a certain mass to charge ratio, m/z, follow a stable periodic trajectory of limited amplitude and are transmitted to the detector (see Figure 4.13). By increasing the DC and AC fields whilst keeping a constant ratio between them, this resonant condition is satisfied for ions of each ascending m/z ratio in turn. The mass resolution and transmission of this device are interrelated by a complicated series of equations. The ion trajectories are a function of two dimensionless parameters:

$$a = \left(8U/r_0^2\omega^2\right)(z/m)$$
$$q = \left(4V/r_0^2\omega^2\right)(z/m) \tag{4.5}$$

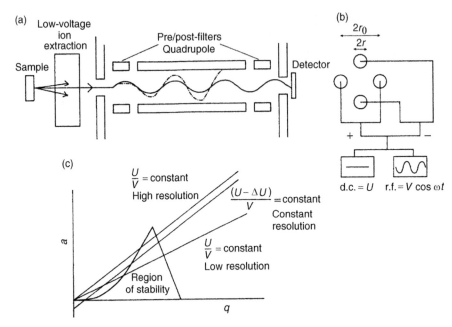

Figure 4.13 Operation of a quadrupole mass filter: (a) longtitudinal cross-section, showing stable and unstable trajectories; (b) radial cross-section, showing applied voltages; (c) ion trajectory stability diagram – ion trajectories are a function of two dimensionless parameters a and q (see text). Reproduced with permission from [7] and from *Secondary Ion Mass Spectrometry – Principles and Applications*, Oxford University Press (1989)

The interested reader is referred to other texts [23] for a more detailed description.

In practice the quadrupole mass analyser is tuned for transmission of secondary ions providing constant resolution $m/\Delta m$ throughout the mass range. Transmission usually falls with increasing mass $\approx m^{-1}$. Ion trajectory aberrations which can degrade performance may also be introduced by fringe field effects at the entrance and exit of the quadrupole rods. These problems can be compensated for by coaxially installing a miniature set of pre- and post-filter quadrupole rods to which is applied a proportion of the RF field.

This is a convenient device and was widely used for SIMS and other surface analysis applications since the electronics can be readily detached and replaced without degradation of performance to facilitate baking of an instrument. The same device can perform under both static and dynamic SIMS conditions. In static SIMS it is usually operated with low secondary ion extraction fields (10–100 eV). For dynamic SIMS, high fields (>1000 eV) can be used to improve transmission (see Chapter 5). Quadrupoles are often fitted with an ionizing filament attached to the 'front end' of a device which enables residual gas analysis. This is particularly convenient for monitoring the back-ground gasses in a vacuum system, but is also capitalized upon in surface science/adsorption studies for parallel thermal desorption studies (TPD) of the system e.g. Sakakini *et al.* [24].

Time of flight Mass Spectrometers.   Time-of-flight mass spectrometry is conceptually the simplest means of mass separation used in SIMS. In ToF analysis pulses of secondary ions are accelerated to a given potential (2 to 8 keV) such that all ions possess the approximately same kinetic energy; they are then allowed to drift through a field free space before striking the detector [25]. According to the equation of kinetic energy heavier masses travel more slowly through the 'flight tube' and so the measured flight time, t, of ions of mass-to-charge ratio, m/z, accelerated by a potential V down a flight path of length L provides a simple means of mass analysis.

$$t = L \left(\frac{m}{2zV} \right)^{\frac{1}{2}} \qquad (4.6)$$

The basic experimental requirement is for a precisely pulsed primary ion source, a highly accurate computer clock, a drift tube and considerable computing power for data acquisition. The flight times of all the ions to the detector are electronically measured and related to ion mass. Thus a mass spectrum of all the ions is generated from the flight time spectrum. Mass resolution is critically dependent upon the pulse length of the generated secondary ion pulse which should be precisely defined and very short. This in turn is dependent on the pulse length of primary beam which is typically of the order of nanoseconds (see Section 4.3.1).

The energy distribution of secondary ions (circa 20–100 eV) will also affect the mass resolution. This initial energy spread will cause ions of the same mass to enter the drift tube with slightly different velocities and thus degrade the resolution in the final spectrum. This is usually compensated for by an energy analyser in the flight tube. The most commonly used device is an *ion mirror* which consists of a series of precisely spaced rings to which is applied a gradually increasing retarding field. The more energetic ions will penetrate further into the mirror before they are reflected, whilst the less energetic ions will take a slightly shorter path. When tuned correctly all ions of the same mass will arrive at the detector at the same time, despite their small energy differences when leaving the sample surface.

Usually detection is by single particle counting using a microchannel plate detector. This is a flat plate device whose surface contains a multitude of miniature channel electron multipliers (10–100 μm diameter, length to diameter 40–100). The inner surface of the channels is a lead/glass matrix which when bombarded secondary ions generates an enormous cascade of electrons. In a common chevron arrangement two plates are placed with their channels at an angle to each other: $0°/15°$ or $8°/8°$ are common arrangements which result in high output gains and suppress ion feedback.

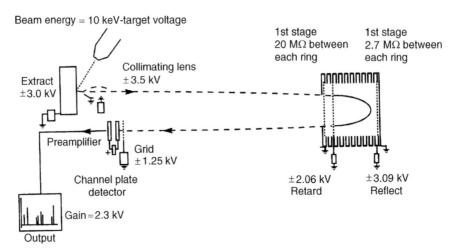

Figure 4.14 A schematic diagram of a ToF–SIMS instrument. Reproduced by permission of The Royal Society of Chemistry from J.C. Vickerman, *Analyst*, **119**, 513–523 (1994)

In some cases the heaviest ions travel so slowly that they do not register impact on the detector and this has been overcome by the introduction of an acceleration voltage immediately prior to detection.

The transmission of the ToFSIMS system is usually between 10 and 50 %, but the great advantage is that, because the analyser is a non-scanning device, none of the ions are discarded in the analysis method. Figure 4.14 shows a schematic outline of the main features of a ToFSIMS system.

Although the input of charged particles in ToFSIMS analysis is very much smaller than with the continuous beam analysers, sample charging is still a problem when analysing insulators. This has been successfully overcome by pulsing electrons onto the sample surface between each primary beam pulse.

The ToF analyser has further benefits for the analysis of organic materials. The more complex the organic materials being studied, the greater the mass range required and the possibility of mass spectral overlap can cause serious problems for interpretation. Whereas the quadrupole analyser is limited in its mass range to about 1000 amu and unit mass resolution, ToF instruments currently can provide mass resolution, $m/\Delta m$, in the region of 5000–20 000 with, in theory, a limitless mass range (usually in practice about 10 000 amu). Table 4.4 summarizes the performance of the mass analysers.

The sensitivity advantages of ToFSIMS suggested that *scanning* or *imaging* ToFSIMS would in principle enable sub $-\mu$m molecular ion imaging. However, the critical parameter now becomes the secondary ion yield per pixel and in practice spatial resolution is limited by the number of molecules in a pixel area and the yields of molecular secondary ions from the material being studied (see Section 4.5.2). Such yields tend to $\leqslant 10^{-4}$ via the SIMS

Table 4.4 Comparison of mass analysers for SIMS

Type	Resolution	Mass range	Transmission	Mass detection	Relative sensitivity
Quadrupole	$10^2 - 10^3$	$<10^3$	0.01–0.1	Sequential	1
Magnetic sector	10^4	$<10^4$	0.1–0.5	Sequential	10
Time-of-flight	$>10^3$	$10^3 - 10^4$	0.5–1.0	Parallel	10^4

process. The only way to increase the yield and hence the ultimate spatial resolution is to enhance the ionization of the vast number of neutral molecules in the sputtered plume (see Section 4.6).

Whilst ToF instruments have significant advantages for surface mass spectral analysis, until very recently they were not the chosen instruments for dynamic SIMS. Etching by a second ion source followed by static SIMS analysis is possible, but because the ToF instrument uses a pulsed analysis beam with a duty cycle of about 10^{-4} (i.e. the time the beam is on, divided by the time it is off) this is a time consuming process when analysing to a depth on the μm scale. Usually in a depth profile, perhaps six elements may be monitored, and there is no apparent advantage in collecting the whole spectrum. However, the developing requirement to analyse very shallow implants has made the ToFSIMS instrument attractive (see Chapter 5).

The advent of the wide use of cluster primary ion beams, particularly polyatomic ions such as C_{60}^+ has highlighted some of the drawbacks of the simple reflectron ToF–SIMS configuration. The most obvious is the requirement to use a very short pulsed primary beam to obtain good mass resolution. This greatly increases acquisition times because of the low duty cycle. Large chemical images can take many hours to acquire. The pulsed ion beam, if highly focused will also have a very low beam current. This will also extend acquisition times to generate sufficient ion signal. The action of pulsing the beam inevitably degrades the minimum spot size and the ultimate spatial resolution attainable. These issues can become more acute with a beam such as C_{60}^+. The ions are generated by electron bombardment so the 'brightness' of the ion source is limited when compared to a duoplasmatron ion source or a liquid metal field emission source. A fine probe can be produced by employing de-magnifying optics, but as the de-magnification increases the probe current falls. Often the short pulse is subsequently bunched to form a sub-nanosecond pulse but this further sacrifices spatial resolution in return for improved mass resolution in the spectrometer. The low velocity of the C_{60}^+ ions, even when accelerated to say 40 keV, and the presence of ^{13}C isotopes means that the shortest pulse formed will last tens of nanoseconds and the process of bunching destroys the spatial resolution which has been achieved by sacrificing ion current.

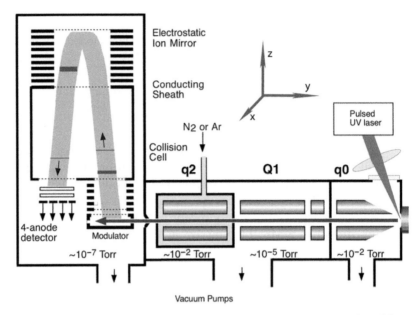

Figure 4.15 A schematic of a tandem QqToF mass spectrometer. Reproduced from A.V. Loboda A.N. Krutchinsky, M. Bromirski, W. Ens and K.G. Standing, *Rapid Commun. Mass Spectrom.*, **14**, 1047 (2000), with permission from John Wiley & Sons, Ltd.

The capabilities of ion beams such as SF_5^+ and C_{60}^+ require a different concept of mass spectrometer. We have seen that for many materials the static limit is not required and in principle molecular depth profiling should be possible. If a continuous beam could be used most of the drawbacks noted above could be eliminated. Mass spectrometers based on the principles of the ortho-ToF developed for MALDI are now being explored for cluster bombardment SIMS. These mass spectrometers have the added advantage that they enable tandem MS–MS experiments to be carried out that enable unknown compound identification to be carried out much more effectively. Figure 4.15 shows a schematic of the one form of the ortho-ToF configuration.

A continuous (or very long pulse) beam of ions is emitted from the sample by either ion bombardment (or photon bombardment in MALDI). Between the sample and the entrance optics there may be an extraction field, or in some cases a high gas pressure that serves the function of both collisional cooling and sweeping the ions into the mass spectrometer. The ions enter an RF only quadrupole which contains gas at $\sim 10^{-2}$ torr. This collisionally focuses the ion beam into a second quadrupole that may be operated in either RF only or RF and DC as a mass selector, the beam of ions then enters a quadrupole or cell which functions as a collisional activation chamber when MS–MS studies are to be carried out. Finally the ions enter a buncher or push-out region in which short packets of ions are injected orthogonally with several keV energy into a ToFMS for analysis. It will be clear that a 100 % duty cycle is not possible because as one packet of ions is being analysed some ions will be lost; however storing particular mass ranges of ions in the third quadrupole is possible using bunching techniques such that within certain mass ranges approaching 100 % duty cycles can be obtained. For MS–MS studies ions of interest are selected by quadrupole 2, activated with nitrogen or argon collision gas at a few 10s of eV in quadrupole 3 and the resultant fragments analysed by orthoganal injection into the ToFMS [26].

A variant on this configuration is to use a linear buncher, see Figure 4.16. The mass spectrometer collects secondary ions for a period of time, say 100 ms, and then mass analyses this collection of ions while the next group is being collected. This is achieved by collecting the secondary ions and introducing the ions after a collisional cooling quadrupole into a linear buncher cell at low velocity. It can be shown that ions of 500 amu will fill a linear buncher cell of length 300 mm with a high efficiency if they have a kinetic energy of a few tens of electron volts. Next the buncher fires by suddenly applying an accelerating field of kV cm^{-1}. The charged particles within the buncher cell are ejected rapidly downstream and the field is turned off so that the filling of the buncher with secondary ions can recommence. If the ejecting electric field is just the right shape the ejected bunch of charged particles will be brought into time focus down stream from the buncher and, at this intermediate time focus plane, ion bunches of increasing mass to charge ratio will be observed successively as time goes by. A bunch of ions of a given mass to charge ratio at the intermediate time focus plane will have a large range of kinetic energies. The high energy ions originate from the furthest point of the buncher and they are in the process of overtaking the lower energy ions which originate from nearby. These collections of highly chromatic ions are next admitted into a harmonic field reflectron. This reflecting device has the property that the time of flight in and out of the reflector depends on the mass to charge ratio only – not on the energy. So a group of ions with an energy spread of say 5 ns will be detected after passing through the reflector with the same temporal spread but a much longer time of flight. This enables high resolution mass spectrometry.

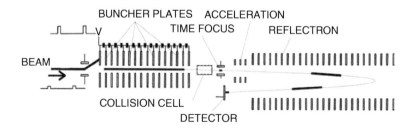

Figure 4.16 A schematic of the linear buncher with time of flight analysis

This intermediate time focus can also be used for MS–MS by positioning a timed ion gate in this plane. This gate can be made to admit only one precursor mass to charge ratio bunch into the next stage. Injecting a pulse of collisional gas at this stage will result in fragmentation and the resulting ions can be analysed with the ToF analyser.

In both these types of ToF mass spectrometer the mass spectral performance is not dependent on the performance of the primary ion beam and the full capability of the mass spectrometer can be exploited. Similarly since the ion beam can be operated in very long or continuous mode the full spatial resolution capability of the ion beam can be used. Of course the fact that the primary ion beam is operated in continuous or very long pulse mode means that sample may be consumed quickly. On the one hand this means that fast data acquisition is possible and this can bring challenges in coping with the large amounts of data being delivered especially in imaging mode; on the other hand where static conditions have to be observed, the static limit could be approached rapidly and care is required to optimize the operational parameters accordingly.

4.4 Secondary Ion Formation

4.4.1 INTRODUCTION

Experimental Evidence for the Mechanism of Secondary Ion Formation.
The mechanism of secondary ion formation from organics is far from fully understood. A range of experiments studied the process from different directions. Some of the earliest experiments directed at understanding the relationship between spectra and chemical structure investigated the adsorption of carbon monoxide and simple hydrocarbons on metal surfaces. The Vickerman group demonstrated that static SIMS could distinguish molecular from dissociative adsorption of CO on metal surfaces. Dissociative adsorption, which was observed on tungsten at 300 K, was characterized by M_xC^+ or M_xO^+ ions, whereas molecular adsorption, observed on Cu, Pd, Ni and Ru in the temperature range 100–300 K, was distinguished by M_xCO^+ ions [7]. Figure 4.17 shows a very early spectrum observed for CO adsorption on iron at 300 K. CO was known to adsorb in both modes at this temperature and the spectrum showed both types of ions. There was no evidence from the data, when the low primary flux conditions were used, that the static SIMS process modified or destroyed the surface state. The data were in complete agreement with what was known from other techniques. Subsequent studies demonstrated that the relative intensities of M_xCO^+ ($x = 1$–3) ions defined the adsorbate structure, whether linear, bridged or triply bridged to the metal surface atoms. It was further demonstrated that using ion ratios $\Sigma(M_xCO^+/M_x^+)$ the relative surface concentrations of the different adsorbate states could be quantitatively monitored and these

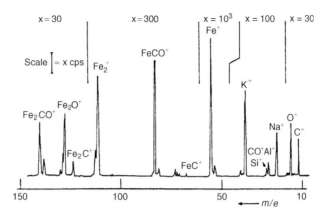

Figure 4.17 Static SIMS spectrum following the exposure of clean iron foil to 10^8 torr of carbon monoxide. Reproduced with permission from *J. Chem Soc., Faraday Trans. 1*, **71**, 40 (1976)

measurements used to determine the enthalpy of adsorption as a function of surface coverage [7]. This approach was ultimately extended to monitor surface reactions and more complex organic molecules [27].

Benninghoven *et al.* investigated the sputtering of organic molecules on metal substrates [28], Briggs studied the characteristics of polymer sputtering from damage studies [29], Leggett and Vickerman used MS–MS techniques to probe the mechanisms of fragment formation from polymers [30] and Delcorte and Bertrand studied the kinetic energy distributions of ions emitted from molecular and polymer materials [31]. From these various approaches a consensus reached that the SIMS spectrum does reflect the chemical structure of the surface. Furthermore a qualitative understanding of the overall process involved in the sputtering of organic surfaces has emerged. If the organic is a *thin* film supported on a metal substrate, close to the point of primary ion impact high energy events take place leading to the emission of atomic species and the fragmentation of the organic backbone. This will be followed by collision cascades in the metal substrate, the energy initially deposited by the primary particle falls off exponentially with successive collisions of recoiling atoms, transferring decreasing amounts of energy to adsorbed molecules such that some desorb with significant amounts of internal energy and fragment, while other molecules desorb without fragmentation [32]. The general concept of this model, formalized by Benninghoven *et al.*, is probably valid for almost any form of material. The energy may not be transferred by the type of collision cascade envisaged in homogeneous elemental substrates. In the case of covalent molecular solids with directed bonds energy will be transferred through vibrations. Thus in polymeric material the events occurring as the energy spreads out from the impact point can be envisaged as shown in Figure 4.18 [33]. The primary particle induces a physical scission in the

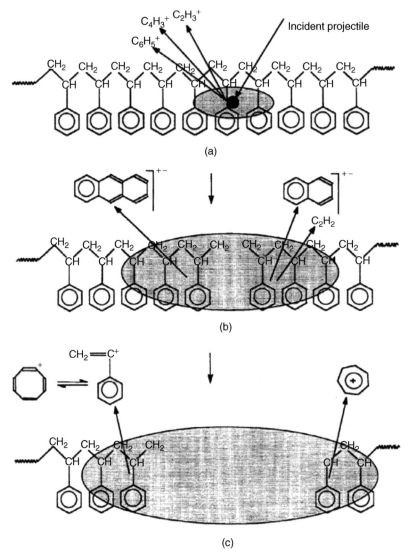

Figure 4.18 Model of sputtering of a polymer: (a) violent fragmentation in primary impact region; (b) unzipping to give large fragments in fingerprint region; (c) simple low-energy fragmentation in monomer region. Reproduced with permission from *The Static SIMS Library*, SurfaceSpectra Ltd, Manchester, 1998

polymer chain, which yields a macro-radical or ion. The primary ion energy is transformed into vibrational energy within the bonds of the molecule. As the energy is dissipated into the vibrational modes of the polymer, the polymer is unzipped from the point of fragmentation and successively larger lower energy fragments are emitted. Internal excitation leads to fragmentation via chemically determined pathways. Atomic species and small uncharacteristic organic fragments are thought to be emitted directly from the point of primary ion impact.

Ion formation in SIMS is a complex phenomenon. Simplistically the process can be divided into two components: the dynamical process by which atoms and multi-atomic clusters are desorbed and the ionization process in which a fraction of these sputtered particles become charged. Clearly, electronic factors are involved throughout the desorption event. Whilst a good deal of theoretical work has gone into understanding the process there is still some way to go before we have a comprehensive theory which fully explains the experimental observations. Here we briefly outline a few of the main approaches to date. Since we are mainly concerned with molecular analysis we will only consider the ideas which seek to explain the emission of molecular ions and fragments.

4.4.2 MODELS OF SPUTTERING

The simplest approach to sputtering of single component solids regards the atoms as hard spheres which obey Newtonian mechanics. Sigmund's linear cascade theory [34] has been the most successful model of the sputtering process so far. His model assumes that sputtering occurs by particle bombardment at small incident particle current and fluence. This excludes the situations where there is extensive heating and damage of the target and is close to the criteria for static SIMS. However, he also classifies sputtering events into knock-on sputtering and sputtering by electronic excitation. In his theory he disregards electronic excitation sputtering. This approximation may well be valid for high incident primary particle energies, but in the low energy (few keV) region typically used in molecular SIMS, electronic interactions between incident particles and target atoms may not be negligible and the hard-sphere model may not be appropriate. The theory is developed on the basis of elastic collisions between point particles. Particular predictive success has attended the linear cascade ideas. In this process the incident particle transfers its energy to the target atoms and thereby initiates a series of collision *cascades* between the atoms of the solid within about 30 Å of the surface. Some of these collisions return to the surface and cause the emission of sputtered particles (see Figure 4.1). When applied to medium-to-high energy particle bombardment of single component materials, the data match the experimental results in terms of the dependence of yields on primary particle mass and energy rather well (see Figure 4.3). However, at lower energies collision energy may be exchanged over greater distances than envisaged in the point mass collision cross-section and in complex multi-component materials (e.g. polymers) the transport of energy will not be isotropic but highly directional.

To understand the sputtering process from complex materials, whilst Sigmund's theory gives an important fundamental model of basic issues a different approach is required. In this regard various molecular dynamics (MD) simulations (particularly those due to Garrison and Winograd) have been very helpful in understanding the process occurring at the low primary flux densities encountered in static SIMS studies of inorganic materials [35]. An ensemble of a few hundred to a few thousand atoms is selected to model a crystal with a surface plane with specified initial conditions of atomic mass, position and velocity. An atomic interaction potential function is devised to account for the bonding of the crystal. The sample is then bombarded by a number of primary particles of specified mass, velocity and angle of incidence. The classical (Hamiltonian) equations of motion are solved in a sequence of iterative steps and the motions of the target atoms are determined as a function of time from the initial impact. The interaction potential is crucial in influencing the accuracy of the simulation. It not only determines the motions of the target atoms within the collision cascade volume, it also influences the nature of any bonding interactions between emitted atoms as polyatomic species are emitted. The embedded-atom potential method (EAM) has been very effective. It assumes that the total electron density in a metal can be approximated by a linear superposition of contributions from the individual atoms. The electron density in the vicinity of any atom is then the sum of the electron density contributed by the atom plus that from the surrounding atoms. The atom can be said to be *embedded* in this constant background electron density. For example in studies of Rh atom desorption from Rh (111) very good agreement has been obtained between the theoretical and experimental energy distributions [36]. Similarly the sputter yields from silver surfaces of Ag dimers relative to atoms together with their energy distributions also give good agreement between theory and experiment [37].

However in static or molecular SIMS analysis we are frequently more interested in the sputtering of organic films. These multi-element materials with highly directional properties are more challenging to model. In an early study Garrison and co-workers modeled the sputtering of an organic layer on a Pt (111) surface [38]. The first study was of a p(2×2) ethylidyne, C_2H_3 at coverages of 0.25 and 0.5 ML. Subsequent studies dealt with adsorbed C_5H_9 [39]. Many-body potential energy functions were required to allow for interactions among all of the three constituents of the target. As the collision cascade evolved it was necessary to be able to follow, via the integration of Hamilton's equations of motion, the reactions occurring among the substrate atoms, the substrate and adsorbate atoms and between the individual adsorbate atoms. The Pt {111} crystallite consisted of between 1500 and 2300 Pt atoms arranged in seven rows with C_2H_3 molecules placed in three-fold sites. The EAM approach was used to mimic the Pt crystallite. The C—C, C—H and H—H attractive interactions were well described by a reactive many-body potential energy function developed by Brenner [40]. The repulsive interactions were found to be improved by adding a Molière repulsive potential. The many-body interactions that could occur between Pt and C in the Pt—C complexes as sputtering proceeded were difficult to model with existing interaction potentials. A combination of the Brenner hydrocarbon potential and Pt—C and Pt—H Lennard–Jones pair potentials were used. Using this approach 'mass spectra' of the sputter yield following bombardment by 500 eV Ar were obtained, although this was a representation of the frequency of yields of particles as a function of mass with no account taken of ionization or stability. It was an encouraging start to the theory of surface mass spectra (Figure 4.19(a)). The most common mechanism of ejection observed was the fragmentation of the single C_2H_3 adsorbate. The main particles emitted were H and CH_3 with significant yields of Pt, and C_2H_3. Figure 4.19(b) shows one of the more common mechanisms for the formation of CH_3. About 50 fs into the collision event, a second layer Pt atom collides with a first layer Pt atom causing it to move outwards. As it tries to leave the surface in the 85 fs event it collides with the C_2H_3 adsorbate, which in the 125 fs event results in C—C bond rupture and the release of a CH_3 radical; by 200 fs the emission of a Pt atom and an intact C_2H_3 are also observed. This adsorbate was bound to the first layer Pt which was struck from below. The momentum of the first layer Pt is directed away from the adsorbate so the C_2H_3 *rolls off* the surface intact rather than fragmenting. Other fragments are formed by similar processes. If the emerging species has significant internal energy unimolecular fragmentations are observed. Other species such as H_2, CH_4 and HCCH were also observed although, with the exception of H_2 in relatively minor proportions. The simulations suggest that these arise from reactions between emerging particles with either of the adsorbates still bound to the surface or with other emerging fragments.

A more complex example is the sputtering of self-assembled monolayers of alkanthiolates on gold [41]. Using this approach 'mass spectra' of the sputter yield following bombardment by 500 eV Ar was obtained which although again no account was taken of ionization, was very similar to that obtained experimentally. By examining the MD emission sequences it is possible to obtain insights into the possible mechanisms of secondary particle formation. It is clear that intact molecules are emitted as a consequence of collision cascades emerging from below, sometimes attached to substrate atoms (Figure 4.20). Both experiment and MD models show that frequently such large species are emitted with significant amounts of internal energy that leads to unimolecular fragmentation above the surface [42]. However, smaller fragments can also be formed as a consequence of direct impact of the primary particle with the molecules at the surface. Thus the generation of an experimental spectrum can be rationalized.

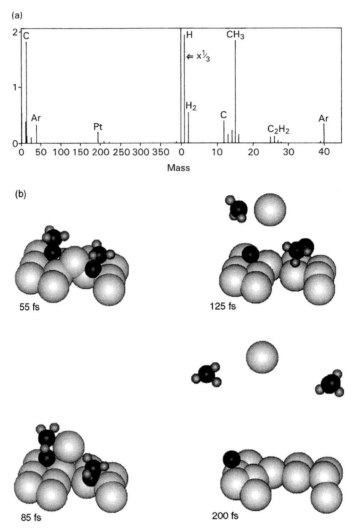

Figure 4.19 (a) Calculated mass distributions of sputtered particles from a C_2H_3 film adsorbed on Pr (111). (b) One of the more common mechanisms for the emission of CH_3 (see text for details). Reproduced with permission from *Langmuir*, **11**, 1220 (1995) [38]. Copyright 1995, American Chemical Society

These examples however did not tackle thick organic layers of relatively large organic molecules using primary ion beam energies close to those used in practice. Such simulations are extremely demanding both in terms of the multiple interaction potentials that need to be included, but also in terms of the computational time required. To approach these real systems Delcorte and Garrison started by modeling styrene tetramers on silver bombarded by 5 keV Ar [43]. These studies showed that the emission of the large tetramer

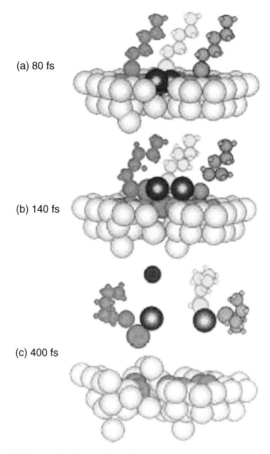

(a) 80 fs

(b) 140 fs

(c) 400 fs

Figure 4.20 Two MD mechanisms for the sputter ejection of intact alkanethiolate molecules from threefold sites on a gold surface. One results in the emission of AuM_2 species, the other in the emission of Au_2M species. Both are seen in significant yield in the experimental negative ion spectrum. Reproduced with permission from *J. Phys. Chem. B*, **103**, 3195 (1999) [41]. Copyright 1999, American Chemical Society

from the surface required co-operative action from a number of collision cascades impacting from below the molecule. The requirement for collective action was further underlined in a study of hexadecamers of polystyrene on silver (see Figure 4.21). Analysis of the trajectories involved showed that there were some high action/high yield trajectories that delivered most of the high mass yield.

This model has been extended to bulk polystyrene. Although the system is very different, the polystyrene molecules being part of an extensive 'soft' structure the modeling shows that the basic mechanistic features remain, namely the requirement for collective action to remove large molecular units and the emission of these large oligomers is due to high action/high yield trajectories [44].

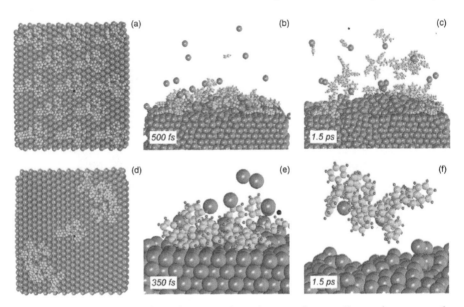

Figure 4.21   The emission of a polystyrene hexadecamer from a silver substratre under 5 keV argon impact. Reproduced with permission from reference [43], A Delcorte and BJ Garrison, *Nucl. Instrum. Meth. B* **180**, 37. Copyright 2001, Elsevier

The emergence of cluster primary ions as useful primary ions for molecular SIMS has stimulated some very informative MD studies on the mechanism by which they initiate sputtering. A study of the comparative reaction effects of the impact of 15 keV Au_3 and C_{60} on water ice (and organic substrates) shows that Au_3 fragments as it hits the surface and generates quite a significant crater, sputters many molecules, but because each of the gold atoms have 5 keV energy they deliver energy and molecular damage deep into the ice structure [45]. C_{60} also breaks up within fs of impact and generates a large shallow crater, causes a great deal of collective action and sputters a very large number of molecules, but almost all the energy is deposited close to the surface (each C only has ~ 250 eV) with none generated much below the crater depth (see Figure 4.22). This modeling provides a useful insight into the reason why C_{60} delivers high yields of larger molecules and for many materials shows much reduced bombardment induced chemical damage. The high degree of collective action will also help to remove large molecules from the surface. Because little damage is caused sub-surface, subsequent impacts uncover relatively undamaged material. In the case of the Au_3 ions, because sub-surface chemical damage has been caused by the high energy particles, subsequent ion impacts will uncover damaged material greatly reducing the yield of chemically significant secondary ions.

These calculations are beginning to approach the types of chemistry found in the analysis of real organic and bio-organic systems. They are yielding

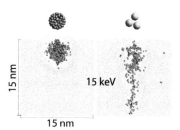

Figure 4.22 Time snapshots of 15 keV C_{60} and Au_3 bombardment of pure amorphous water ice. Grey and yellow spheres represent intact water molecules and projectile atoms, respectively, within a 2 nm slice through the center of the substrate at 0.5 ps. Orange, green, and blue spheres represent the water fragment species placed back in their initial positions and overlaid on the substrate at 0.5 ps. Reproduced with permission from K.E. Ryan, I.A. Wojciechowski, and B.J. Garrison, *J. Phys. Chem. C* **111**, 12822 (2007) [45b]. Copyright 2007, American Chemical Society

results that are providing very helpful insights into the mechanisms of molecule emission. However, the whole problem of ionization has yet to be addressed.

4.4.3 IONIZATION

The fraction of sputtered particles that are in the ionized state is in fact very small. In most cases over 99 % of the sputtered yield is neutral. Whether a sputtered particle escapes from the surface as an ion depends on the relative probabilities of ionization and de-excitation as it passes through the near surface region. Hence the high dependence of ion yield on the electronic/chemical properties of the matrix (the so-called *matrix effect*). For metals, the rapid electronic transitions (10^{14}–10^{16} s^{-1}) make de-excitation a high probability during the 10^{-13} s required for a sputtered particle to traverse the near surface region. The probability, P_a, of escape as an ion from a metal can be approximated by:

$$P_a \approx 2/\pi \, \exp[-\pi(\varepsilon_a - \varepsilon_F)/\hbar\gamma_N v_1]. \tag{4.7}$$

where ε_a and ε_F are the energies of the ionized state and the Fermi level, v_1 is the velocity of the emerging atom and γN^{-1} is the distance over which the level width decreases to $1/2.781$ of the bulk value.

However, the need to develop an understanding of secondary ion emission from adsorbate and organic materials demands that we take account of the 'molecular' covalent type of bonding and consider that ionization may take place at emission and also by subsequent fragmentation of vibrationally excited molecular units.

Although considerable progress has been made in using molecular dynamics to advance our understanding of the sputtering process, because of the complexity of the phenomenon very little real progress has been made in developing a theoretical understanding of the accompanying ionization processes. Two qualitative models developed quite some time ago are still helpful in describing the likely contributors to ionization of sputtered species from inorganic and organic molecular systems. They are briefly described here.

Nascent Ion Molecule Model.   Some early work by Benninghoven and Plog proposed the so-called Valence Model which predicted that the distribution of yields of these clusters (MO_x^+ and MO_y^-) from inorganic oxides, would be dependent on the cation valance if it was assumed the oxygen anion maintained its charge as -2 [46]. Empirically this approach has had some success; however it is highly doubtful that even before sputtering inorganic oxides can be regarded as purely ionic solids. Sanderson, many years ago demonstrated that inorganic oxides have a large degree of covalency and the actual partial charge on each ion is a small fraction of the nominal ionic charge [47]. Further it is highly unlikely that charge will be conserved when the bonds are broken during sputtering. A development of the Valence Model which makes no assumptions about the charge on oxygen takes this into account and demonstrates that indeed the static SIMS data suggested partial charges on the cations and anions which deviate from the pure ionic values [48].

Gerhard and Plog developed these ideas further into the Nascent Ion Molecule Model [49]. This suggested that the rapid electronic transition rates which occur in the surface region will neutralize any ions before they can escape. Secondary ions are thought to result as a consequence of dissociation of sputtered neutral molecular species some distance from the surface. In the terminology of the model, ions are formed by the non-adiabatic dissociation of *nascent ion molecules* (neutral molecules). For inorganic oxides most of the neutral molecules originate from direct emission of ion pairs such as MeO and keep their molecular character after leaving the surface. Only a few molecules have enough internal energy to dissociate into their constituents. The dissociation is considered to take place some distance from the surface where the electronic influence of the surface will be much smaller. The bond-breaking models that are used to explain emission from ionic materials consider the system solid–Me^+, whereas the nascent ion molecule considers the system $Me_xO_y^0$. It is clear that while the idea that the emission and fragmentation of nascent ion molecules is a major process is attractively straight-forward, and to first order may deliver helpful results, nevertheless the detail may be misleading if accepted at face-value.

The Desorption Ionization Model.   This is due to Cooks and Busch and introduces the concept that vibrational excitation may be important in understanding the emission of cluster or molecular ions from organic materials [50]. This model also emphasizes that the processes of desorption and ionization can be considered separately and however we understand the initial excitation process, the energy is transformed into thermal/vibrational motion as far as the molecules are concerned. A wide variety of ion emission processes are possible. Some pre-formed ions may be directly emitted. These are species that exist as ions within the material prior to bombardment and no ionization step occurs. It is suggested that neutral molecules are desorbed in high yield, but to be detected must undergo an ionization process such as cationization. To generate other ions the model suggests that desorption is followed by two types of chemical reaction: (i) in the selvedge or top surface layers fast ion/molecule reactions – or electron ionization can occur; (ii) in free vacuum, unimolecular dissociations may occur, governed by the internal energy of the parent ion giving rise to fragment ions.

According to these ideas the desorption event is of relatively low energy. The linear cascade ideas are not wholly appropriate when considering molecular solids. It is more helpful to think of energy being transferred to the vibrational modes of the molecule thus leading to fragmentation and ionization. This is consistent with the observation that molecular SIMS is a relatively soft-ionization phenomenon: for many materials, there is relatively little low mass fragmentation and large yields of molecular ions are observed.

4.4.4 INFLUENCE OF THE MATRIX EFFECT IN ORGANIC MATERIALS ANALYSIS

A complication which affects all desorption mass spectrometries is the suppression or enhancement of ion formation due to the matrix effect. As we have already seen ionization depends on either electron or proton transfer during sputtering. The relative capabilities of the desorbing species to capture or lose electrons or protons can inhibit or enhance ionization such that in extreme cases even though a particular chemistry is present the ions that characterize it may be totally suppressed by the presence of other chemistry in the system.

As indicated earlier within SIMS the matrix effect has been known for many years, the ionization probability of a given element varying greatly depending upon the composition of its immediate environment (see Section 4.2.3). This concept of ionization probability being dependant upon the chemical environment has in fact been exploited by the desorption MS techniques, fast atom bombardment (FAB) and MALDI, where the analyte is incorporated into an excess of a suitable matrix. In the case of FAB this is typically a liquid such as glycerol, in MALDI the analyte is typically co-crystallized onto the target plate with an excess of an organic acid molecule. Although mainly used for the analysis of isolated species, the use of MALDI for MS and MS imaging has become a rapidly expanding application; however, great care must be taken with the application of matrix to the sample surface in order to obtain accurate results. A major benefit of cluster ion sources for SIMS analysis is that they offer secondary ion yield enhancements without the need for chemical modification of the surface, unlike the techniques mentioned here. However, the matrix effect can be just as important when analysing samples in an unaltered state, as one compound may strongly influence the detection of another that it is co-localized with. An understanding of the matrix effect with respect to organic molecules is therefore essential to allow the results obtained to be interpreted correctly. The example shown below highlights the importance of having some understanding of the effect and of the situations where it may become critical in the analysis. In the course of a study of the incorporation of drugs into the brain of rats a control study involved spraying the drug haloperidol over the surface of a slice of the brain to check the signal response from different areas of the brain [51]. The slice exposed different domains – white matter and grey matter. The drug was deposited evenly over the two domains. Figure 4.23 shows three ToF–SIMS images obtained using C_{60} primary ions.

The distribution of the drug [M+H]+ signal ($m/z\,376$) across the two different domains of the tissue is shown. The signal from the phosphatidyl-choline lipids which predominate in the grey matter are indicated by the PC headgroup ($m/z\,184$), whilst the localization of the white matter is characterized by the peak from the cholesterol ($m/z\,369$). Although the drug species covered the whole area visible in the image, the molecular signal is only detected from the cholesterol rich areas. This model system demonstrates the severity of the suppression/enhancement effects that can

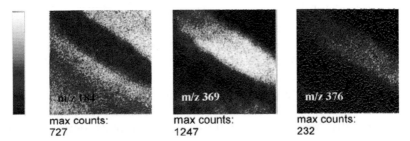

max counts: max counts: max counts:
727   1247   232

Figure 4.23   Distribution of the molecular signal of the drug haloperidol ([M+H]$^+$ signal at $m/z\,376$) spun cast onto a section of brain with respect to the chemical domains of the tissue. The signal from cholesterol ($m/z\,369$) and phosphatidylcholine ($m/z\,184$) are shown to indicate the different chemical domains within the tissue surface; the analysed area is 800 µm × 800 µm with a dose of 8×10^{10} ion/cm^2. Reproduced with permission from E.A. Jones, N.P. Lockyer and J.C. Vickerman, *Int. J. Mass Spectrom.*, **260**, 146 (2007) [51]. Copyright 2007, Elsevier

be encountered across a two-domain system such as brain tissue sections. Without prior knowledge of the system in question it would be easy to assume that the peak at m/z 376 was linked to the constituents of the white matter along with the cholesterol.

The precise mechanisms by which this signal enhancement and suppression occurs are not fully understood, but for organic ions that rely on the formation of M+H or M−H ions for detection it is clear that proton transfer processes are involved either in the sample before emission or in the course of the sputtering process. Zenobi *et al.* have shown quite convincingly that in the MALDI process the competitive formation of M+H ions occurs in the desorbing plume of analyte and matrix and that a quasi-equilibrium is set up in which the relative gas phase basicities of the various molecules and ions strongly influences which ions are detected and which are suppressed [52]. However the density of the sputtered plume in the SIMS process even under bombardment by cluster beams such as C_{60}^{+} is orders of magnitude less dense than under laser irradiation. A quasi-equilibrium plasma does not seem likely. However, a study of a series of mixtures of model compounds whose gas phase basicities are known has shown that matrix suppression and enhancement can be understood in terms of the drivers of proton transfer and can be correlated with the gas phase basicity of the molecules involved [53]. The systems were formed by mixing 2,4,6-trihydroxyacetophenone (THAP) with one of the DNA bases cytosine and thymine, or the structurally similar barbituric acid (BA) (see Figure 4.24).

They were studied using Au^{+} primary ions. This figure shows the chemical structure of the molecules studied and their gas phase basicities. Inspection of the spectra shown in Figure 4.25 shows that in the case where cytosine has a significantly higher gas phase bascity (gpb) than THAP, the M+H ion is seen for cytosine, but not for THAP, whereas the negative M−H ion is not seen for cytosine, whereas it is for THAP. On the other hand where the mixture consists of THAP and barbituric acid where THAP

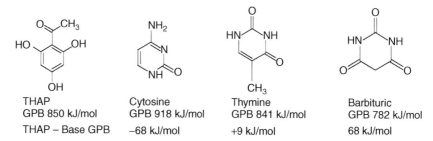

Figure 4.24   The structures and gas phase basicities of 2,4,6-trihydroxyacetophenone (THAP), the DNA bases cytosine and thymine and the structurally similar barbituric acid (BA)

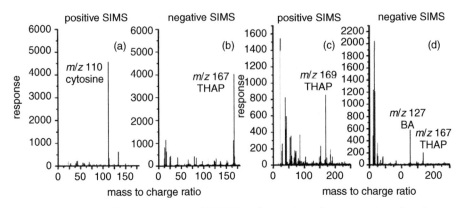

Figure 4.25 (a) and (b) 1:1 mixture of THAP and cytosine, demonstrating that the gas phase basicities of the molecules within the mixture dictate the ion polarity in which the quasi-molecular ion will be detected. (c) and (d) A 1:1 mixture of THAP with barbituric acid, demonstrating that by mixing the THAP molecule with a compound of lower gas phase basicity the $[M+H]^+$ ion which is suppressed in the previous example can become the favoured ion. Reproduced with permission from E.A. Jones, N.P. Lockyer, J. Kordys and J.C. Vickerman, *J. Am. Soc. Mass Spectrom.* **18**, 1559 (2007) [53]. Copyright 2007, Elsevier

has a greater gpb than BA, the positive M+H ion is seen for THAP, but not for BA, whereas the negative ion mode M−H is seen for BA but only as a very small peak for THAP. In the mixture of THAP and thyamine in which the gpbs are very close, both M+H and M−H ions are seen for both components. The clear conclusion is that there is proton mobility as the sputtering event occurs and the protons tend to be mopped up by the species with the highest relative basicity.

This observation provides an explanation for the suppression of the haloperidol drug M+H ion in the brain tissue. The phospholipids head group ion requires protons to be formed and must have a high basicity, thus the formation of the haloperidol M+H ion must be suppressed. On the other hand cholesterol is a proton donor and the formation of the haloperidol M+H ion is enhanced in the white matter domain.

The matrix effect could have a serious effect on the validity of analyses and analytical images. Although it can be understood and factored into analysis when there are only a few components in a system and perhaps even the gpbs of some or all the components are known, the analysis of real systems is a different matter and the absence of a signal may not indicate the absence of the corresponding chemical. It has been suggested that the matrix effect may be ameliorated by providing a matrix of very low gpb, such that most other molecules in the system would be able to draw protons from it to form M+H ions. Indeed in the above experiments if BA, that has the lowest gpb, is added to the THAP-cytosine mixture, the M+H ion for THAP is restored. Water is a molecule with an even lower gpb of

$650\,kj\,mol^{-1}$. It has the advantage of being the majority component in biological systems. There is some good evidence that the yield of M+H ions from water matrices may indeed be significantly higher [54], but the possibilities of exploiting it to reduce the matrix effect in biological systems has yet to be demonstrated.

4.5 Modes of Analysis

4.5.1 SPECTRAL ANALYSIS

The most widely applied mode of SIMS analysis as a surface mass spectrometry has been to use the spectra obtained to characterize the chemistry of the surface in terms of determining what is there. Questions as to how the surface changes as a consequence of some process or treatment are clearly issues that the analyst would hope and expect to answer using this technique. Similarly the ability to determine the surface chemistry differences that give rise to the different behaviour of products or biological systems would also be a hoped for capability. The ability to deal with such analytical needs by studying the SIMS spectra from the samples of interest has been demonstrated in many important analytical applications.

Usually in the spectral mode analysis is carried out on the sample of interest with minimal sample pretreatment. Analysis may be carried out in spot mode where the ion beam is defocused to cover most of the area from which the analyser will extract ions. Charge compensation with a defocused low energy electron beam may be required if the sample is an electrical insulator. If a monatomic primary beam is used the spectra are acquired using a primary ion dose below the static limit, which for organic samples the dose should be less than 10^{12} primary ions cm^{-2}. An alternative method to spot mode, is to raster scan the focused ion beam over a precise area of the sample, say $300\,\mu m \times 300\,\mu m$. The fluence requirements will be the same. The spectra obtained are then interpreted and the analysis follows. Where the sample only has a few components it should be relatively easy to interpret the spectra using the general approach to mass spectral analysis outlined in references such as Briggs [55]. However for many biological samples the spectra will be very complex and spectral interpretation of all the components may be very difficult. In such cases tandem mass spectrometry, MS–MS, and computer assisted multivariate analysis may be required (see Chapter 10).

The ability of static and molecular SIMS to characterize complex chemical systems has been widely applied in many materials areas – from synthetic polymers, catalysts and optoelectronics to bio-materials, biological tissue and cell studies. There are a number of reviews [56]. In the following sections three very different examples illustrate the range of application.

Characterization of Technical Catalysts.   Matrix effects and inhomogeneous sample charging seriously hinder quantitative analysis of SIMS on technical catalysts. Although full quantitation is almost impossible in this area, the interpretation of SIMS data on a more qualitative is basis nevertheless offers unique possibilities. Molecular cluster ions may be particularly informative about compounds present in a catalyst. A ToF–SIMS study on the automotive exhaust or three-way catalyst by Oakes and Vickerman [57] illustrates the type of information that SIMS provides.

The three-way catalyst consists of a ceramic monolith made of cordierite, which serves as the structural support and is covered by a porous washcoat with alumina and ceria as major constituents, onto which small amounts of platinum and rhodium are deposited. Platinum efficiently oxidizes CO and unburnt hydrocarbons to CO_2 and H_2O, while rhodium catalyzes the reduction of NO to N_2 by CO, H_2 and hydrocarbons. During its active life the catalyst may easily travel up to 2×10^5 km, during which poisons such as lead, sulfur, and phosphorous accumulate on the surface. In addition, the catalyst may deteriorate due to thermal and mechanical damage.

Figure 4.26 shows positive and negative ToF–SIMS spectra from a used three-way catalyst. One readily recognizes contributions from the alumina–ceria washcoat and from contaminants (originating from lubricants and fuel additives), which are obviously present in the form of lead sulfate and calcium phosphate. In addition, elements such as Fe, Cu and Zn, are present, due to wear of engine parts. Note that secondary ions from the

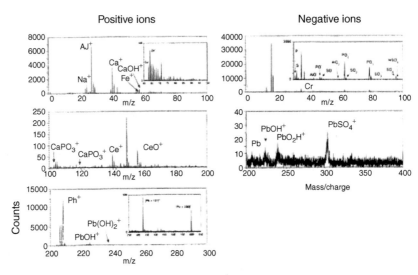

Figure 4.26   Positive and negative ToF–SIMS of an automotive exhaust catalyst after 100 000 km (adapted from Oakes and Vickerman [57])

noble metals are not prominently visible, because they are to a large extent covered by the contamination layer. Similar spectra from a fresh catalyst straightforwardly show aluminium, cerium and combinations with oxygen, some chlorine, and rhodium and rhodium–oxygen clusters. Interestingly, cerium is only observed in clusters containing one cerium, whereas a CeO_2 reference compound emits clusters up to $Ce_2O_3^+$ and $Ce_2O_5^-$, indicating that the cerium ions in the washcoat are highly dispersed through the alumina and certainly not present in the form of CeO_2. Rhodium, at mass 103, overlaps with several secondary cluster ions in a SIMS spectrum. The high resolution of a ToF–SIMS instrument is useful here, as Figure 4.27 illustrates.

Platinum is not easily detected in SIMS when an argon primary beam is used. Cesium guns offer an alternative for the detection of noble metals. A cesium layer deposited through the experiment alters the electrostatic state of the surface and favours the emission of negative ions, enabling the detection of a series of platinum ions, including Pt^-, PtO^-, $PtCl^-$ and $Pt(OH)_3^-$, and combinations with hydrocarbon fragments, all showing the isotopic distribution pattern characteristic of platinum.

In addition to static SIMS spectra, as shown in Figures 4.26 and 4.27, depth profiles of catalysts were measured that had travelled different distances in cars. Basically, these depth profiles show that the poisons deposit on the surface, and that poisons such as lead, sulfur and zinc penetrate further into the material than phosphorous does. Imaging SIMS has been used to visualize the distribution of the poisons species over the washcoat, making

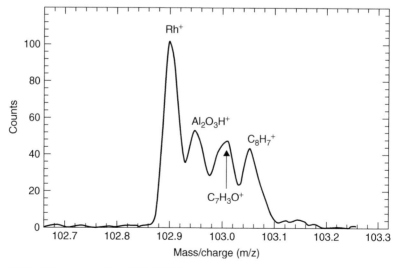

Figure 4.27   High-resolution ToF–SIMS spectrum in the mass region of rhodium (103 amu) of a fresh automotive exhaust catalyst (adapted from Oakes and Vickerman [57])

the study a complete illustration of the capabilities of ToF–SIMS in catalyst characterization.

Surface Analysis of an Adhesive System.   A good illustration of the application of static SIMS to the surface characterization of a complex material is a study from some years ago on an adhesive system. The surface and interfacial chemistry of adhesives and organic coatings can have a major effect on their properties and performance. Whilst XPS and AES have been used widely for adhesion studies, these techniques inherently lack the specificity to provide the level of molecular information that is essential for a full understanding of their interfacial chemistry. In this study, four epoxides were investigated using ToF–SIMS [58]. They are Epikotes 828, 1001, 1007 and 1009. Epikote 828 is commonly used in adhesive formulations, whilst Epikotes 1001, 1007 and 1009 are used in lacquers, coatings and thermosetting agents. The Epikote resins are based on the diglycidyl ether of bisphenol-A epoxide structure (Figure 4.28).

Two sets of studies were carried out. First, silver cationization was used to obtain detailed information about the oligomer distributions and composition of the polymers. In these experiments, thin layers of the Epikote were deposited on silver from butan-2-one solutions. In another set of experiments, to obtain data relevant to the study of the real adhesive and coating systems, thick films were laid down on aluminium foil.

Part of the silver cationized spectrum for the 1007 sample is shown in Figure 4.29. At intervals corresponding to the mass of the single monomer unit of the di-epoxide-terminated diglycidyl polyether, 284, ions are seen due to the oligomer plus the two silver isotopes 107 and 109 for $n = 2$ to

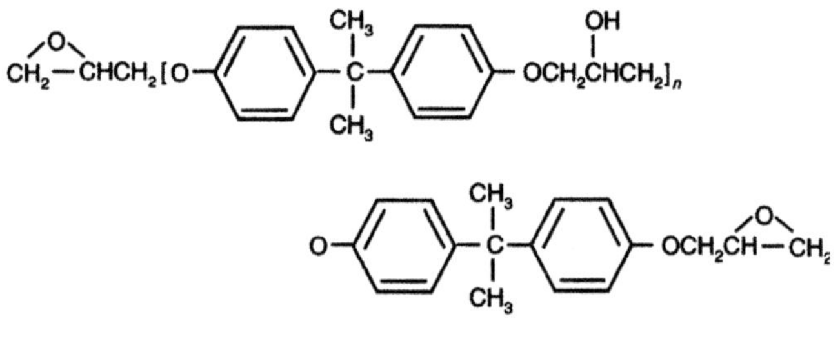

Resin	Mean molecular mass	n distribution
Epikote 828	450	mainly $n = 0$
Epikote 1001	880	centred $n = 3$
Epikote 1007	2870	centred $n = 10$
Epikote 1009	4000	centred $n = 14$

Figure 4.28   General structure of Epikote samples, listing their mean molecular masses and the central value of n. Reproduced by permission of the Royal Society of Chemistry from J.C. Vickerman, *Analyst*, **119**, 513–523 (1994)

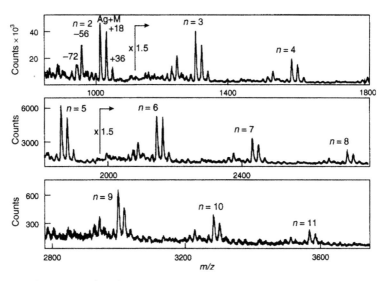

Figure 4.29 Spectrum of Epikote 1007 deposited on a silver substrate. Reproduced with permission from Treverton *et al.* [58]. Copyright (1993) John Wiley & Sons, Ltd

$n = 11$. Due to the wide mass range of the spectrum, the peak splitting due to the two isotopes cannot be seen. Each of the Ag cation peaks is accompanied by a signal at $m/z = 18$ higher. This corresponds to the presence of epoxide-*glycol* oligomers formed by hydrolysis of one of the two terminal epoxy groups. Signals at m/z 56 higher may correspond to the presence of *propanol* derivatives. In contrast to the 828 and 1001 Epikotes, in the 1007 and 1009 samples the epoxide-*glycol* terminated oligomer signals were higher relative to the normal di-epoxide terminated species. In addition, the 1007 and 1009 resins show other prominent Ag cationized species at m/z values 36 higher, 56 lower and 72 lower than the normal di-epoxide peaks. These additional signals correspond respectively to the presence of *diglycol-*, *phenol-* and *phenyl*-terminated oligomers. Thus it can be seen that there is a significant amount of information about the detailed chemical state of polymers. The preparation of the silver supported samples means that the analysis may not reflect the actual *surface* state of the resins; nevertheless the composition of the resin mixture is qualitatively indicated.

The positive ion spectra from the thick films generally did not give rise to signals above $n = 2$ oligomers. Fragments from the $n = 1$ and $n = 2$ oligomers were observed for Epikote 828. In the higher mass region of the negative ion spectra, fragments from $n = 1$ and $n = 2$ oligomers could be identified. However, the most useful signals were fragments of the monomer that could be attributed to terminal epoxide or bisphenol-A components. Figure 4.30 shows part of the positive ion spectra for the samples 1001, 1007 and 1009. The ions at m/z 191, 252 and 269 can be assigned to fragments containing terminal epoxide groups (Scheme 4.1).

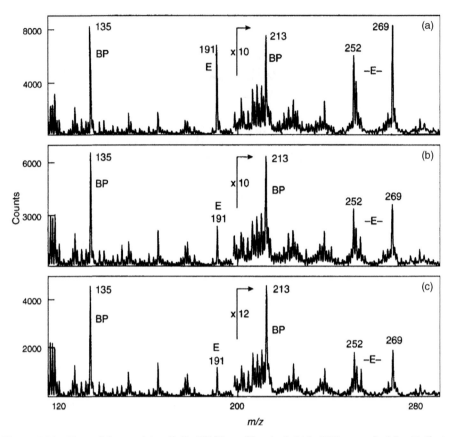

Figure 4.30 Part of the positive ToF–SIMS profiles (m/z 115–290) recorded for Epikote 1001: (a) 1007; (b) 1009: (c) BP, bisphenol-A component; E, epoxide end group. Reproduced with permission from Treverton *et al.* [58]. Copyright (1993) John Wiley & Sons, Ltd

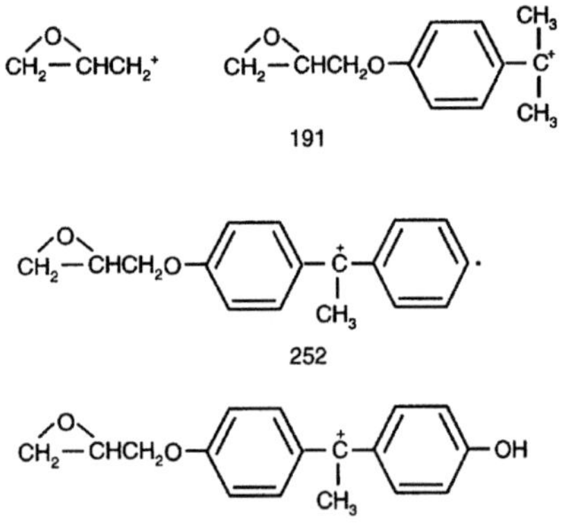

Scheme 4.1

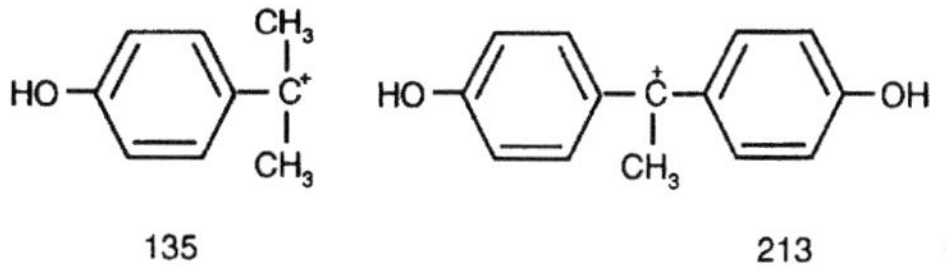

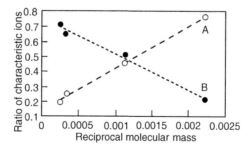

Scheme 4.2

In addition, positive ion signals specifically diagnostic of the bisphenol-A component were observed at m/z 135 and 213 corresponding to the structures shown in Scheme 4.2. In the negative ion spectra, similar assignments can be made.

Examination of these spectra makes it clear that changes in the molecular weight have a marked effect on the relative intensities of the signals specific to the terminal epoxide groups on the one hand, compared to those characteristic of the bisphenol-A groups on the other. As the molecular weight of the polymers increases, the ratio of epoxide end group to the bisphenol-A component would be expected to fall. A plot of the component peak ratios as a function of the reciprocal of the mean molecular weight shows a straight-line decrease for the epoxide ratio and an increase for the bisphenol-A (Figure 4.31). There is a clear quantitative relationship between the ToF–SIMS spectral data and the composition of the Epikote resins.

This study demonstrates some of the information accessible from a static SIMS study of the surface of a complex organic system. The cationized spectra showed that a range of glycol and phenol terminated oligomers were present for the higher molecular weight resins. The array of secondary ions specifically diagnostic of terminal epoxide and bisphenol-A obtained from the spectra of the thicker films have been shown to be quantitatively related to the concentration of terminal epoxide and bisphenol-A groups. This latter observation should be useful in probing the effect of cross-linking in cured

Figure 4.31 Plot of peak area ratios 191(epoxide)/[191(epoxide) + 135(bisphenol-A)] and 213(bisphenol-A)/[269(epoxide) + 213(bisphenol-A)], the Epikote resins. A, epoxide ratio; B, bisphenol-A ratio. Reproduced by permission of the Royal Society of Chemistry from J.C. Vickerman, *Analyst*, **119**, 513–523 (1994)

adhesives. The bisphenol-A part of the molecule is relatively unaffected by cross-linking. Thus most of the spectral features associated with this group should continue to be observed, whereas new features would be expected due to changes in the rest of the molecule. Preliminary studies of fully cross-linked adhesives indicated that the bisphenol-A fragments are observed in the spectra. Bisphenol-A fragments have also been observed in ToF–SIMS spectra of cured epoxy paint systems [59].

Analysis of Complex Biological Systems. The analysis of complex biological systems such as bacteria using SIMS would seem an attractive application. The ability to discern the differences in the chemical state of cell surfaces and perhaps depth profile into the cell interior is an exciting possibility. It might be thought that it would be possible to distinguish between different bacteria on the basis of chemical differences detected by mass spectrometry. However, these systems are multicomponent and the spectra turn out to be very complex. A good example is the various bacteria that are the source of urinary tract infection (UTI) in adult women. This is a considerable problem in general practice and on average leads to a consultation rate of approximately 63.5 consultations in every 1000 women each year. With this high incident rate of bacteria (counts of above 105 organisms per ml urine) there is a growing need to identify the causal agent prior to treatment. The bacteria typically associated with UTI include *Escherichia coli* (in over 50 % of cases) and *Klebsiella* species that can be resistant to antibiotics. In addition, other *Enterobacteriacea* are implicated, including *Proteus mirabilis* and *Citrobacter freundii*, whilst the Gram-positive *Enterococcus spp.* also often causes infection (10 % of cases). In this study nineteen strains of UTI bacteria, previously identified by conventional biochemical tests, were investigated using ToF–SIMS with a C_{60} primary ion beam [60]. The isolates consisted of *E. coli* (five isolates coded 'Eco'), *Klebsiella oxytoca* (one isolate coded 'Kox'), *Klebsiella pneumoniae* (three isolates coded 'Kp'), *C. freundii* (two isolates coded 'cf'), *Enterococcus spp.* (four isolates coded 'Ent') and *P. mirabilis* (four isolates coded 'Pm'). ToFSIMS spectra of four of the isolates are shown in Figure 4.32.

The individual strains give rise to spectra containing many common ions and identification of an individual specimen by visual inspection of the spectrum would be extremely difficult since separation would be based on the changes in the relative intensities of a number of these common peaks. Thus it would be impossible to distinguish unknown samples of these bacteria on the basis of a 'stare and compare' approach. The multivariate methods of analysis described in Chapter 10 are required if we are to have any success in identifying the differences between these bacteria on the basis of their ToF–SIMS spectra.

The analysis of these data was performed by the cluster analysis method of principal components-discriminant function analysis (PC-DFA). With

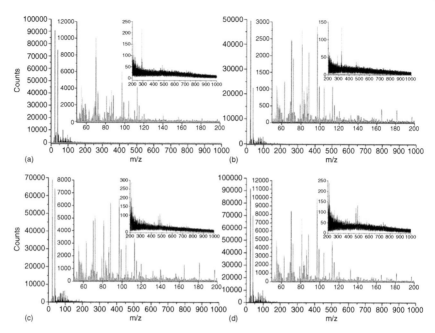

Figure 4.32 ToF–SIMS spectra of (a) *Citrobacter freundii* [cf102], (b) *Escherichia coli* [Eco13], (c) *Enterococcus spp.* [Ent93] and (d) *Klebsiella oxytoca* [Kox105]. Reproduced with permission from J.S. Fletcher, A. Henderson, R.M. Jarvis, N.P. Lockyer, J.C. Vickerman and R. Goodacre, *Appl. Surf. Science* **252**, 6869 (2006) [60]. Copyright 2006, Elsevier

this method the PC-DFA multivariate algorithm seeks 'clusters' in the data, in order to group objects together on the basis of their perceived closeness in two or three-dimensional PC-DFA ordination space. The initial step in cluster analysis involved the reduction of the dimensionality of the data by principal components analysis (PCA). PCA is a well-known technique for reducing the dimensionality of multivariate data whilst preserving most of the variance. Plots of the first two principal components scores represent the best 2D representation of natural variance in the data. Discriminant function analysis (DFA) then discriminated between groups on the basis of the retained principal components (PCs) and the *a priori* knowledge of which spectra were acquired from which biological replicates. Following the PCA step, no significant clustering of the data was observed and inspection of the loadings plots indicates that most of the variance between spectra was related to the changes in the intensities of the Na^+ and K^+ signals that dominated the spectra. However, the subsequent application of DFA, produced loadings that show the inclusion of a large number of higher mass organic fragments, with a relatively small influence from Na^+ and K^+. PC-DFA of the entire data set, where the classes used in the DFA represented individual strains, was performed and the ordination plots are presented in Figure 4.33.

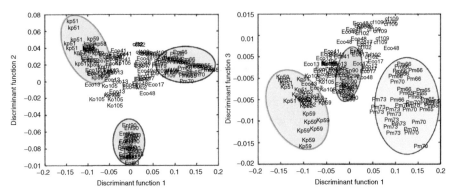

Figure 4.33   PC-DFA ordination plots of for the entire data set. For clarity three species that are separated from the main cluster have been highlighted: *Enteroccocus spp.* (red); *Proteus mirabilis* (blue); *Klebsiella pneumoniae* (green). Reproduced with permission from J.S. Fletcher, A. Henderson, R.M. Jarvis, N.P. Lockyer, J.C. Vickerman and R. Goodacre, *Appl. Surf. Science* **252**, 6869 (2006) [60]. Copyright 2006, Elsevier

The PC-DFA utilized three principal components. Although clear separation of the individual strains is not evident, the isolates have clustered at the species level with the enterococci isolates (highlighted in red) forming a cluster that is well separated from the remainder of the group. The recovery of the *Enterococcus spp.* from the other bacteria is expected, since this is the only species of bacteria in this study to be Gram-positive. These bacteria would therefore have a distinctly different cell membrane in comparison with the Gram-negative *Enterobacteriacea*, incorporating a thick outer layer of peptidoglycan. Isolates of the *P. mirabilis* also cluster into a well-defined group and this result can also be explained by the phenotype of these organisms. The *P. mirabilis* cell walls contain a polysaccharide-rich capsule layer which includes N-acetyl-D-glucosamine that is also present in peptidoglycan. As the SIMS analysis is expected to probe only the outer surface of the bacterial sample, it is understandable that such biological differences would have a strong influence on the spectra and hence the clustering observed in this multivariate analysis. For further analysis all the Gram-positive enterococci isolates were removed from the data set, the PC-DFA was repeated and again, three principal components were used to generate the PC-DFA model. The resulting ordination plots (Figure 4.34) show extensive clustering of the remaining isolates. The clustering observed is no longer at the species level but now shows good strain level separation.

The *P. mirabilis* isolates are separated from the main group, as seen in the analysis of the entire data set; however distinct strain level separation is observed. The Pm70 and Pm73 strains are clearly isolated, although there is a small overlap of the Pm65 and Pm66 strains. The above results clearly show that ToF–SIMS spectra are information rich and that the chemical

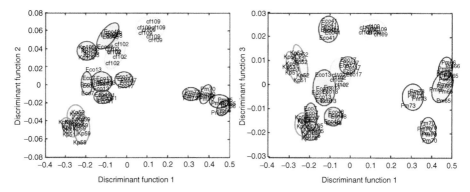

Figure 4.34 PC-DFA ordination plots of the data set following the removal of the *Enterococci*. Rings have been placed around individual strains as a visual aid only. Rings of the same colour indicate strains belonging to the same species: *E. coli* (red); *C. freundii* (yellow); *P. mirabilis* (blue); *K. pneumoniae* (blue); *K. oxytoca* (pink). Reproduced with permission from J.S. Fletcher, A. Henderson, R.M. Jarvis, N.P. Lockyer, J.C. Vickerman and R. Goodacre, *Appl. Surf. Science* **252**, 6869 (2006) [60]. Copyright 2006, Elsevier

data therein allow for the classification of these UTI isolates to the species level using this multivariate analysis approach. The next stage was to assess whether the application of PC-DFA to ToF–SIMS spectra could provide strain level discrimination of the bacteria. A single species subset of the data was chosen for this exercise. The *E. coli* spectra were selected since this group contained the highest number of individual isolates ($n = 5$). Three spectra of each strain were chosen randomly and removed from the group. PC-DFA was then performed on the remainder 'training set' to generate a model as before. The excluded 'test' data were then projected into the model (first into the PCA-space and then the resultant PCs projected into DFA-space) and the resulting ordination plot is shown in Figure 4.35 where the training set is labeled in red and the projected test data appear in blue. As with the previous analyses, and for consistency, three principal components were used for the DFA model. Figure 4.35 clearly shows that for each of the five isolates, the test data were recovered very close to their corresponding training data clearly showing that ToF–SIMS does provide the mass spectral sensitivity and reproducibility for sub-species discrimination. In principle the loadings plots provide an indication of the mass spectrum peaks that give rise to the discrimination. From these it should be possible to unravel the chemistry that distinguishes one bacterial strain from another. However these loading plots are still quite complex and high level mass spectrometry will be required to go the next step. The new generation of the mass spectrometers with high mass accuracy combined with MS–MS capabilities will be needed to realize this potential.

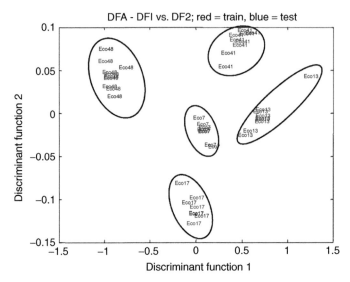

Figure 4.35   PC-DFA ordination plot of five strains of *E. coli*. The training data is labelled in red and the projected, test data in blue. Three principal components were used to generate the DFA model. Reproduced with permission from J.S. Fletcher, A. Henderson, R.M. Jarvis, N.P. Lockyer, J.C. Vickerman and R. Goodacre, *Appl. Surf. Science* **252**, 6869 (2006) [60]. Copyright 2006, Elsevier

4.5.2 SIMS IMAGING OR SCANNING SIMS

In principle, combining one of the mass analyser systems with a liquid metal ion source enables surface analysis with high spatial resolution to be carried out. Indeed there is the potential for scanning electron microscopy-type images to be generated with the very considerable added facility of full chemical sensitivity. The highest spatial resolution is in principle accessible when operating in microprobe mode.

As indicated earlier, liquid metal ion beam systems can be operated with beam diameters at the sample down to 50 nm, although the more usual range is 200 nm to 1 μm. The beam is digitally rastered across the surface of interest such that there are, say, 256 × 256 pixels in an image. At each pixel point it is possible either to collect ions of a single m/z, of a few specified m/z or a whole mass spectrum, dependent on the detail required and the sophistication of the analyser and data system. Elemental or chemical state images can then be generated of the areas of interest.

The images in Figure 4.36 illustrate the basic capability. They are images of a single frozen hydrated yeast cell [61]. This has been freeze-fractured and can be seen to be surrounded by water ice. The lipid content of the cell is detected by the head group ion m/z 184 and the molecular ion m/z 734 from dipalmatoylphosphatidylcholine (DPPC). The high potassium

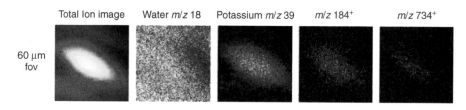

Figure 4.36 Freeze-fractured single cell of *Candida-glabrata*

concentration within the cell is characteristic of a viable cell. Although quite a simple analysis this example shows the potential of ToF–SIMS to provide chemical maps of biological and indeed many other complex materials. Analysis of complex chemistry with good spatial resolution is possible.

To date there have been two approaches to biological imaging. One utilizes an approach described in Figure 4.36 using a ToF–SIMS instrument below the static limit. The aim is to acquire molecular spectra of the biochemistry of the biological system in the so-called discovery mode, i.e. without any previous detailed knowledge of the sample. The other approach is to use a dynamic SIMS approach. In this approach some specific analytical aim is in view such that some specific chemistry in the sample is labeled with an elemental isotope. The analysis then probes elements or small cluster ions to identify the location and quantification of the specific labeled chemistry. In this mode of operation the analysis requires the destruction of the molecular chemistry to detect labeled fragments that can be related to the chemistry being studied. High sensitivity combined with high spatial resolution are possible.

A Dynamic SIMS Study of Cochlear Hair Cells of a Mouse. Figure 4.37 shows an impressive application of this approach in the imaging of protein renewal in the cochlear hair structures of a mouse. The mice had been fed with a diet of ^{15}N-L-leucine. The presence of protein generated as a consequence of this diet is detected via the $^{12}C^{15}N^-$ fragment ion. This is imaged relative to $^{12}C^{14}N^-$ to identify the locus of the renewed protein. The spatial resolution is on the order of 30 nm using a cesium primary ion beam. A magnetic sector mass spectrometer has been used. The technique has been given the name multi-isotope imaging mass spectrometry (MIMS). A full description of the technique is provided in an excellent review [62].

It should be emphasized that this approach requires that the chemistry to be studied is known so that appropriate labeling can be carried out.

A study of the mating of tetrahymena protozoa. This chapter's main focus is the analysis of unknown chemistry by the mass spectrometric detection of molecular and related fragment ions. An example that nicely illustrates the

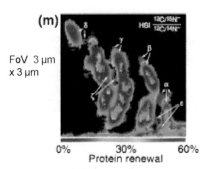

Figure 4.37 A study of protein renewal in the cochlear structures of the mouse. Quantitative MIMS images of mice cochlear hair cells after 9 days on a diet of ^{15}N-L-leucine. Reproduced with permission of BioMed Central from C. Lechene, F. Hillion, G. McMahon, D. Benson, A.M. Kleinfeld, J.P. Kampf, D. Distel, Y. Luyten, J. Bonventre, D. Hentschel, K.M. Park, S. Ito, M. Schwartz, G. Benichou and G. Slodzian, *J. Biol.*, 5, 20 (2006) [62]

power of imaging SIMS to solve a biological problem was an investigation of the lipid content in the conjunction region during the mating of tetrahymena protozoa [63]. Mating involves formation of many fusion pores in a ~8 mm membrane junction region. It was thought that the entire junction region may have a different lipid composition from the cell body. Fusing cells were captured onto a silicon wafer covered with a shard of silicon and rapidly frozen in liquid nitrogen cooled propane. This fast freezing ensures that the water forms amorphous ice so that the cell structure is not ruptured by ice crystals. The frozen cells are rapidly transferred to the cold stage (<150 K) of the spectrometer and the top shard removed to fracture frozen cell assembly and expose the cells. ToF–SIMS analysis takes place while the sample is still frozen. Figure 4.38 shows the process involved.

Analysis of the fusing cells showed that m/z 69 was uniform throughout the two cells and the fusion region. This ion is common to all the lipids in the cells and can be representative of the total lipid content. However, m/z 184 from the phosphocholine head group is depleted in the fusion region (see Figure 4.39).

A PCA analysis of the cell bodies compared to the fusion region suggested that m/z 126 head group ion characteristic of the 2-aminophosphonolipid distinguished the cell bodies from the fusion region. It was present in the cell bodies *and* the fusion region. The conjugation junction contained elevated amounts of 2-AEP. This lipid is conical and forms the curved structures required for the fusion region., whereas PC is cylindrical and forms only planar surfaces and is only found in the cell bodies.

SIMS Imaging Beyond the Static Limit. While surface analysis at high spatial resolution is an attractive proposition, as the magnification increases

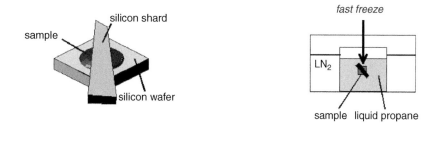

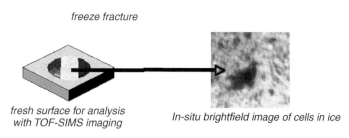

Figure 4.38 A schematic of the handling of frozen-hydrated samples. Reproduced by permission of N. Winograd

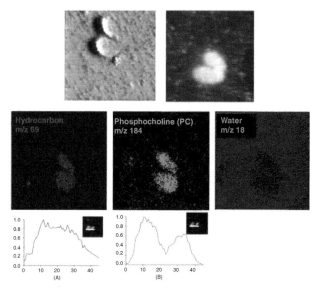

Figure 4.39 SEM and SIMS images of the fusing cells. Row 1: figures of SEM and brightfield images. Row 2: m/z 69 SIMS image; m/z 184 SIMS image; m/z 18 water image. Row 3: (A) line scan of m/z 69 intensity across fusion region; (B) line scan of m/z 184 intensity across fusion region. Reproduced with permission from S.G. Ostrowski, C.T. Van Bell, N. Winograd and A.G. Ewing, *Science*, **305**, 71 (2004) [63]. Copyright 2004, American Association for the Advancement of Science

Table 4.5   Estimation of the number of molecules and atoms per pixel area

Imaged area (µm)	Pixel size	Pixel area	Molecules per pixel	Atoms per pixel
100	10µm × 10µm	10^{-6} cm^2	4×10^8	2.5×10^9
10	1µm × 1µm	10^{-8} cm^2	4×10^6	2.5×10^7
5	500 nm × 500 nm	2.5×10^{-9} cm^2	1×10^6	6.25×10^6
1	100 nm × 100 nm	1×10^{-10} cm^2	40 000	2.5×10^5
0.2	200 Å × 200 Å	4×10^{-12} cm^2	1600	10 000

it becomes more and more difficult to obtain images with adequate dynamic range in the molecular ion or fragment signal and still maintain static conditions. SIMS is a destructive technique; as magnification increases, the number of atoms or molecules in a pixel area decreases (see Table 4.5).

This raises a severe problem. In a 500×500 nm pixel area there are about 1×10^6 molecules. The static limit imposes the restriction that less than 1 % of all surface chemistry, θ, should be removed. Therefore the maximum number of molecules for analysis is 10^4 molecules. The ionization probability, $a^{+/-}$, is usually much less than 10^{-3} and the yield of high mass ions is even lower, so there will be less than 10 ions per pixel for analysis. Since at least 100 ions per pixel are required to deliver a useful spectrum per pixel, there is little prospect of useful imaging at this spatial resolution. It is clear that to have any real hope of chemical state analysis at high spatial resolution (<1 µm pixel size), secondary ion yields have to be improved dramatically. This is only possible if either the static limit can be removed and/or the ionization probability can be raised. The removal of the static limit could potentially increase yields by 2 orders of magnitude and since about 99 % of the particles sputtered from a surface are neutral, if some effective method can be devised to ionize most of these neutrals there is the possibility of a startling improvement in sensitivity.

As indicated in Section 4.3 cluster ions generated by liquid metal ion sources such as gold or bismuth, Au_3^+, Bi_3^+ or Bi_5^+, deliver increased secondary yields mainly as a consequence of increasing the sputter yield. Figure 4.40 compares the yield of molecular ions and fragment ions from PET from Au^+, Au_3^+ and C_{60}^+ [64]. These increases frequently favour the higher mass ions. Liquid metal ion beams are relatively easy to focus.

A 100 nm Au_3^+ or Bi_3^+ beam can be obtained routinely. Thus some improvement in the minimum useful pixel size might be hoped for. However, these metal clusters contain relatively few rather heavy atoms. As discussed in Section 4.4.2, upon collision with the surface the 20 or 25 keV energy is partitioned amongst only 3 to 5 atoms. Each atom will have 5 to 8 keV of energy that will penetrate deep into the material generating chemical damage as it goes. Measurements show that the damage cross-sections

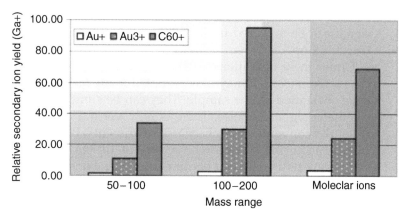

Figure 4.40 Comparison of the positive ion yield from bulk poly(ethylene terephthalate) (PET) under 10 keV Au^+, Au_3^+ and C_{60}^+ bombardment. Reproduced with permission from D. Weibel, S. Wong, N. Lockyer, P. Blenkinsopp, R. Hill, and J C. Vickerman, *Anal. Chem.* **75**, 1754 (2003) [64]. Copyright 2003, American Chemical Society

are very similar to atomic primary ions so the static limit has to apply to the use of these ions. Secondary ion yields are higher, so efficiencies are a little higher, but the number of useful ions to be accessed from a 500 nm pixel is only increased by less than a factor of 10. This is demonstrated in Table 4.6. Using the damage cross-sections for Au^+, Au_3^+ and C_{60}^+ determined for a cholesterol film the efficiencies and numbers of useful ions in a 500 nm pixel area have been computed. It can be seen that the metal cluster ion does improve things, but not enough. Nevertheless some very impressive imaging investigations have been reported using metal cluster ions. However, for many organic systems the polyatomic cluster ions C_{60}^+ and SF_5^+ can operate beyond the static limit and significant yield increases result. As Table 4.6 shows in principle C_{60}^+ should be able to provide useful images down to less than 100 nm pixel area [51]. This is because the damage cross-section is so low the static limit does not apply and the beam is able to sample well beyond the first monolayer and essentially use up a volume

Table 4.6 Comparison of ion yields, damage cross sections and resulting ion formation efficiencies from a thick cholesterol film bombarded by 20 keV Au^+, Au_3^+ and C_{60}^+ primary ions

	Secondary ion per primary ion, $Y(m/z385)$	Damage cross section, $\sigma\,(cm^2)$	Efficiency $E(cm^{-2})$	Number of m/z 385 ions in $(500\,nm)^2$ pixel	Imaging possible
Au^+	5.5×10^{-6}	4.5×10^{-14}	1.2×10^8	0.3	×
Au_3^+	6.5×10^{-5}	7.3×10^{-14}	9.0×10^8	2.3	??
C_{60}^+	4.8×10^{-4}	3.9×10^{-16}	1.2×10^{12}	3×10^3	✓

cube (sometimes known as a *voxel*) as its sample. If there is plenty of sample, signal can be acquired until adequate signal to noise for the analysis has been built up.

Under these conditions a 2D image may have sampled more than just the surface layer – it may have sampled many layers into the sample, even have consumed all the sample. They can be summed to give a total image of greater signal to noise. There is however a drawback to this approach using the conventional ToF–SIMS instruments. Because the ion beams are pulsed, acquiring spectra with a dose of 10^{13} to 10^{14} ions cm^{-2} will take very many hours and can become impractical and inefficient. Thus to take advantage of the possibility of acquiring spectra at high ion dose levels it is necessary to move to the use of a DC primary beam and an ortho-ToF or buncher type instrument (see Section 4.3.2). This is a rapidly moving field. Early examples of the possibilities presented by this approach are shown in Figure 4.41. Using a buncher-ToF type instrument images of cheek cells have been obtained using a focused C_{60} DC beam. The images have been acquired after a total primary ion dose of 5×10^{13} ions. Whereas previously the signal in each pixel might lie between 1 and at best 100, in this case thousands of secondary ions are obtained per pixel. Because a DC beam is used the spatial and mass resolutions are not compromised. These are early

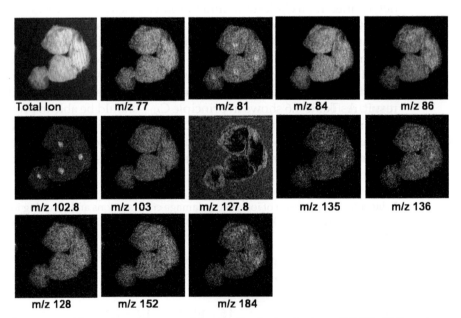

Figure 4.41   Mass selected images of cheek cells obtained using a DC 40 keV C_{60} primary beam on an Ionoptika J105 imaging mass spectrometer. Total C_{60} ion fluence was 5×10^{13} ions/cm^2 following a pre-etch of 1×10^{13} ions/cm^2 : $160 \times 160 \,\mu m^2$ field of view

results but they show that nucleus of the cells are clearly distinguished at m/z 81 and 102.8. The cell membranes are evident from m/z 127.8 and 184.

4.5.3 DEPTH PROFILING AND 3D IMAGING

The ability to analyse and image beyond the static limit opens up the possibility of molecular depth profiling and imaging in three dimensions. Both of these capabilities have been available to dynamic SIMS for a long time (see Chapter 5). In the latter the instrumental and materials issues that influence the quality of depth profiles are outlined. These issues are pertinent to molecular depth profiling. The experimental procedure is much the same as that used in dynamic SIMS. A focused DC ion beam is rastered over a defined rectangular area on the surface of the sample to be analysed. The aim is to sputter etch a precise crater down several nm to µm below the surface. The x/y position of the beam on the surface will be controlled by digital electronics. As the beam steps across the surface, secondary ions are generated and are collected by the mass analyser. Usually the beam energy will be in the range 1–20 keV while the beam flux will usually be in the nA cm^{-2} region; the beam will be focused dependent on beam flux in the sub µm to tens of µm diameter range (the smaller the ion beam flux, the smaller the minimum beam diameter possible); the beam will be rastered over an area ranging from 10 µm × 10 µm–100 s µm ×100 s µm. In carrying out a depth profile analysis there are two main parameters for which the instrumentation and experimental procedure have to be optimized. First, the *dynamic range* of concentration sensitivity for the chemistry (elemental or molecular) to be analysed and second, the *depth resolution*. In dynamic SIMS the main application is the analysis of electronic devices on silicon wafers or similar. Compared to the samples that are likely to be encountered in other areas of chemistry and biology they are ideal because are usually geometrically very flat. This is fortunate because accurate elemental compositions as a function of depth are required. The *dynamic range* accessible will be very dependent on the experimental procedure. It is usual to collect the ions for analysis from a smaller area than the etch area. This is because even in the ideal case of a flat sample as the crater is formed ions will be emitted from the crater edges as well as the crater bottom. To eliminate the collection of ions other than from the crater bottom, two instrumental procedures are used. First, a lens is incorporated in the collection optics that cuts down the field of view such that only the crater bottom is 'seen'. This is the so-called *optical gate*. This helps to reduce the collection of ions from the crater edges. To improve matters further an *electronic gate* may also be added. This permits the detection system to be switched off other than when the primary beam is located within a small defined area in the centre of the crater. Thus ion detection only occurs in the *gated area*. For further details see Chapter 5.

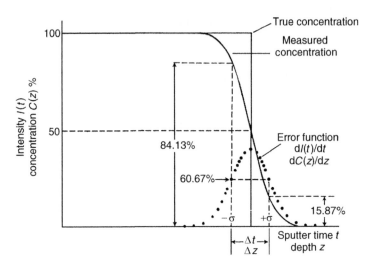

Figure 4.42   The definition of depth resolution

Depth resolution defines how accurately we can measure the depth relative to measured concentration. If, as shown in Figure 4.42, the true concentration in a material changed abruptly, the actual measured change would deviate from it to a greater or lesser extent. The depth resolution is defined from the measured intensity/sputter time curve or the concentration/depth curve. It is either Δt or Δz which is the depth or time difference corresponding to the 84.13 % and 15.47 % levels on the concentration or intensity scale. It measures the degree to which the experiment is able to measure an abrupt interface. The parameters that affect the depth resolution are (a) instrumental effects, particularly the quality of the ion beam, (b) the sputter rates of the components of the sample, (c) the surface topography of the material to be analysed and (d) radiation induced effects as a consequence of the ion bombardment process.

The most important *instrumental effect* on depth resolution is the uniformity of the ion beam over the analysed area. The beam intensity profile does not have a flat top to it, it is usually approximately Gaussian, thus as it scans across a surface it will sputter more in the centre section of the beam compared to the edges. The beam edges will overlap at each pixel point. Dependent on the ratio of the width of the beam profile, d, to the adjacent pixel distance, D, microroughness can be developed in the crater bottom. It is clear that this will significantly degrade the depth resolution of the profile. A squarer beam profile can be obtained by passing it through an aperture which cuts off the edges. Then by optimizing d/Δ this effect can be minimized. The influence of *sputter rate* on depth resolution arises from the possibility of different components in a sample sputtering at significantly different rates. Clearly the crater depth attained by the ion beam will be dependent on the rate at which the sample is sputtered by the ion beam used. If the sample is multi-component as is frequently the case in materials chemistry and biology, it is possible that the different components may sputter at widely different rates. Clearly depth resolution would be compromised. This would be a difficult situation to allow for. Fortunately organic molecules of about the same mass are likely to sputter at similar rates so although the precision of profiling found in lightly doped single component electronic materials will not be attained, useful qualitative profiles should be expected.

The starting *surface topography* will have a very important influence on depth resolution. If the surface of the material to be analysed is not flat to start with, it will not be possible to etch a well defined crater and good depth resolution will be impossible to attain. The rougher the sample surface at the start, the more complicated things will be. Some materials develop rough structures as etching proceeds. Spectacular cones and columns can be formed in some inorganic materials as a consequence of various types of preferential etching. This may be chemically determined or a consequence of the angle of the ion beam to the material surface. Obviously the development of such structures has a terminal effect on depth resolution! In dynamic SIMS one effective way around this in many cases is to rotate the sample while it is being depth profiled [65].

The ion bombardment process itself can induce changes due to what are termed *radiation effects*. When a high energy ion collides with the atoms of the material in the sputtering process, some of the atoms at the bottom of the crater can be moved deeper into the material, while some recoil upwards. Thus bombardment-induced mixing occurs. This effect occurs over a depth similar to the range of the primary ions. For a 2 keV beam this would be around 7 nm.

All of these issues will be important in determining how useful in a quantitative or qualitative sense the depth profiles obtained from chemical and biological systems are. Clearly biological cells and tissue are quite rough to start with; they are multicomponent so high depth resolution cannot be expected. Nevertheless by cross-referencing the profiling process with other depth sensitive techniques such as AFM or ellipsometry useful information about the variation of chemistry with depth should be accessible. Examples of depth profiling and 3D imaging applications are described below.

Depth Profiling a Multilayer Structure.   Model multi-layer structures were grown using Langmuir–Blodgett methods. The layers consisted of multilayers of dimyristoyl phosphatidic acid (DMPA) and arachidic acid (AA). A number of systems were constructed having different thicknesses of films built up of blocks of multilayers of each component [66]. Figure 4.43 shows one example consisting of alternating films of barium salts of AA and

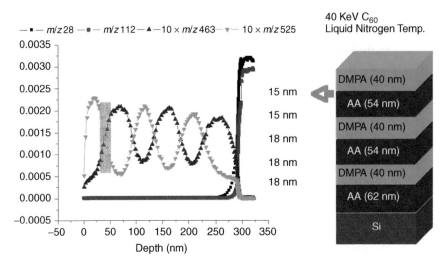

Figure 4.43 C_{60} depth profile of Langmuir–Blodgett formed multilayer films of barium salts of dimyristoyl phosphatidic acid (DMPA) and arachidic acid (AA). Reproduced with permission from L. Zheng, A. Wucher, and N. Winograd, *J. Am. Soc. Mass Spec.*, **19**, 96–102 (2008) [66]. Copyright 2008, Elsevier

DMPA. The thickness of each film equals the thickness of one monolayer multiplied by the number of layers applied to the substrate. The monolayer thicknesses for AA and DMPA are 2.7 nm and 2.2 nm, respectively, as measured by ellipsometry. The films were depth profiled using a 40 keV C_{60} ion beam and the crater depths at the end of a profile were determined by AFM. This figure shows the obtained profile when the sample was held at liquid nitrogen temperature; $m/z\,28$ (Si^+) and 112 (Si_4^+) were used to follow the substrate silicon signal, $m/z\,463$ for AA and $m/z\,525$ for DMPA. The depth resolution was calculated using the same approach as that above. It can be seen that it lies between 15 and 18 nm. It is significant to note that when the experiment was carried out at room temperature the depth resolution ranged from 18 nm at the first interface to 30 nm at the fourth. The surface topography had a roughness of about 20 nm and this was not degraded significantly after profiling. These films may be susceptible to some radiation effects in terms of molecular mixing. The fact that temperature seems to have an effect suggests that the films may be subject to thermally induced mixing. Computer simulations of 40 keV C_{60} sputtering suggest that the best depth resolution that could be expected is about 4.2 nm. As well as topography the angle of incidence of the beam may have an effect. These are early results and there is clearly much to be learnt about molecular depth profiling, but at least it is now a viable study for some molecular systems.

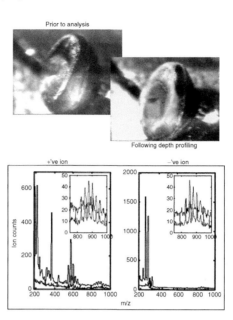

Figure 4.44 *Xenopus laevis* Oocyte optical images before and after 40 keV $C_{60}{}^+$ depth profile analysis together with the positive and negative ion spectra used to generate the data. Reproduced with permission from J.S. Fletcher, N.P. Lockyer, S. Vaidyanathan and J.C. Vickerman, *Anal. Chem.*, **79**, 2199 (2007) [67]. Copyright 2007, American Chemical Society

Depth Profiling and 3D Imaging of Biological Cells. The ability to depth profile molecular chemistry automatically opens up the possibility of 3 dimensional imaging. The possibility of carrying this out on biological samples is really exciting because although there are a number of optical techniques that allow 3D imaging of the inside of bio-systems, in most cases one has to know what one is looking for in order to tag the molecules with active markers to make them detectable. In principle mass spectrometry does not require this and spatially resolved chemistry may be accessible without changing the chemistry of the system. The first cell system using C_{60} depth profiling to be published was a study of a large frogs' egg, *Xenopus laevis* Oocyte [67]. This was followed by a similar study of normal rat kidney cells [68]. The frogs' egg is about 1 mm diameter and although this makes for easy manipulation on the sample holder it does mean that the topography effects mentioned above intrude in assessing the data generated. However, as a proof of concept it was successful. It demonstrated that after the removal of more than 175 μm of cell it was still possible to detect molecular information and that the spatial distribution of chemistry of a cell could be described. Figure 4.44 shows the cell before and after depth profile analysis together with the positive and negative ion spectra used to generate the data.

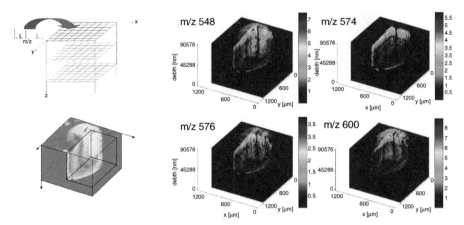

Figure 4.45 Three-dimensional imaging of part of the *Xenopus laevis* Oocyte. The left-hand figures show how assembling a stack of 2D images enables a 3D image to be produced. The figures on the right-hand side show the distribution of four different lipids in the Oocyte section. Reproduced with permission from J.S. Fletcher, N.P. Lockyer, S. Vaidyanathan and J.C. Vickerman, *Anal. Chem.*, **79**, 2199 (2007) [67]. Copyright 2007, American Chemical Society

This study used a conventional pulsed ion beam ToF–SIMS system and as a consequence took many days of sputter etching and analysis. A total ion dose of 10^{16} ions cm^{-2} was used. The spectral features used in the analysis were the groups of peaks in the range m/z 540–700. These are mainly assigned as phosphatidylcholine lipids following loss of the PC head group, e.g. m/z 548 = PC 16:0–16:1 [(M-Head Group) + H]$^+$; m/z 574 = PC 16:0–18:2 [(M-Head Group) + H]$^+$; m/z 576 = PC 16:0–18:1 [(M-Head Group)+H]$^+$. These correspond to the three most abundant lipids found in analysis of lipid extractions. There is a similar set of peaks between m/z 800 and 1000. The intense peak at m/z 369 is cholesterol [M+H−H$_2$O]$^+$. Then there are negative ion fatty acid peaks, i.e. peaks at m/z 255, 279 and 281. (Composition C16:0, C18:2 and C18:1, respectively.) By generating a stack of 2D images for a particular m/z or range of m/z values it is possible to put together a 3D image as shown on the left-hand side of Figure 4.45. The right-hand side of Figure 4.45 shows the distribution of four lipid molecules with the volume of oocyte analysed.

Three dimensional images while useful in providing pictures of the distribution of chemistry are not always very useful in giving an idea of relative signal intensities and hence hopefully concentrations. One way to help assess relative intensities is to use the *isosurface* technique. The isosurface is used to highlight regions of pixels with intensity values above a specified threshold. By selecting a high threshold, regions of high intensity can be clearly visualized. Low thresholds, just excluding noise, can be used to visualize the sample as a whole. Thus for this particular example

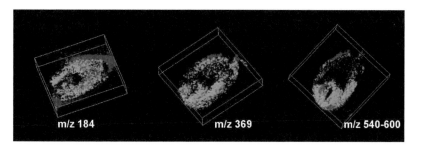

Figure 4.46   Isosurface representation of C_{60} depth profile of Oocyte (see text for details)

Figure 4.46 shows isosurface plots for the lipid headgroup m/z 184 with a low threshold so the whole cell is seen, the cholesterol peak at m/z 369 with a median threshold and the lipids, m/z 540 to 600 less the headgroup with a high threshold showing they are at high concentration in the outer membrane region.

Correcting for Topography in 3D Imaging.   It is clear that there is a difficulty when trying to construct real 3D images of biological cells using only the data obtained from 2D images of topographically rich systems such as biological cells. One way around the problem is to combine SIMS analysis with a technique that measures topography such as AFM. Figure 4.47

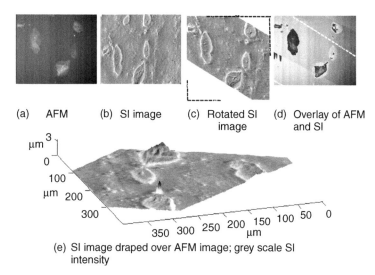

Figure 4.47 C_{60} 3D secondary ion (SI) imaging of cheek cells, combined with AFM measurements to correct for topography and provide a z-scale measurement (see text for details)

illustrates the approach on a cheek cell. Before analysis in a SIMS system the topography of the cell is measured with an AFM (a). Then the cell is analysed using SIMS and a secondary ion image is generated (b). This image is rotated to correspond with the AFM image (c) and then overlaid on that image to give images (d) and (e). Image (e) shows the topography along the x-, y- and z-axes while the secondary ion intensity is provided as a grey scale. In this case the z-direction data is provided by the AFM. If the cell was then chemically profiled using SIMS, the z stack of images could be off-set using the initial AFM image. Ideally there should be an AFM depth measurement during the profile but that would entail taking the sample out during the profile which would not be practical. However it is possible to envisage some sort of optical measurement of depth as the analysis proceeds.

Tissue Depth Profiling.   The analysis of tissue provides the same challenges as those encountered with the cell analysis. In biology tissue is frequently stored for some time at low temperature and then freeze dried prior to analysis. The application of SIMS in this area is developing. However, it is clear that the matrix effect alluded to in Section 4.4.4 can be a problem. There is evidence that unless samples are kept cold, components of the tissues, e.g. cholesterol, can be lost. Tissue contains salts and these can interfere with SIMS depth profile analyses. In the following example a rat brain section was depth profiled at the interface between the white and grey matter to see if the differing composition of the two regions could be monitored [69]. The first attempts were not successful because as the profile proceeded the lipid, cholesterol and amino acid signals disappeared very rapidly to be replaced by Na and K related peaks that dominated the spectra. Studies of biological systems with mass spectrometry are frequently distorted by the presence of salt. The tissue samples were washed in ammonium formate, a treatment known to remove salt from biological materials and the analysis repeated. The data obtained are shown in Figure 4.48. The plots at the top of the figure report data from the whole sample and show in plot (a) that the signal from m/z 44 (alanine immonium ion) and m/z 70 (proline) are stable through the sample, unlike the rapid fall of cholesterol ion signals at m/z 369 and 385; m/z 56 from the steel substrate rises in both plots at the end of the profile when the brain tissue has been etched right through. There appears to be two different regimes in the behaviour of the m/z 184 ion, the phosphocholine head group ion. This is supported by 3D imaging experiments that show the first 70 layers of a depth profile through a white/grey matter boundary region (Figure 4.48). In Figure 4.48(c) an AFM analysis of the sputtered tissue region allows the crater depth to be measured, and this was shown to be 4 μm. In Figure 4.48(d) the cholesterol signal at m/z 369 is shown to be confined to the upper layers of the white matter only, and in Figure 4.48(e) the m/z 184 ion is initially intense in the grey matter but then decreases

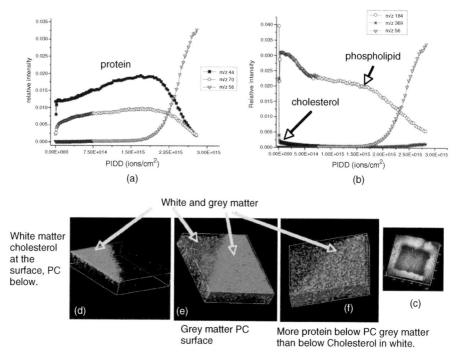

Figure 4.48 40 keV $C_{60}{}^{+}$ depth profile of rat brain tissue at the interface between white and grey matter (see text for details). Reproduced with permission from E.A. Jones, N.P. Lockyer and J. C. Vickerman, *Anal. Chem.* **80**, 2125 (2008) [69]. Copyright 2008, American Chemical Society

to nothing, whilst its intensity increases in the white matter region. The signal from the immonium ions (Figure 4.48(f)) is only detected from the grey matter region and appears to increase following the first few layers removed.

It is curious that the cholesterol concentration decreased so rapidly from the surface. There was concern that under vacuum conditions at room temperature cholesterol would diffuse and could even be removed from the tissue altogether. Such effects could affect any relatively low molecular weight chemical in biological samples. A depth profile study of a similar sample of tissue held at − 120 C showed that the cholesterol concentration was constant with depth providing confirmation of the need to analyse biological samples at low temperatures. However a depth profile imaging study of the type shown in the above figure took over two days with a conventional ToF–SIMS instrument, precluding a low temperature study.

Molecular Biological SIMS Analysis. It is clear that using polyatomic primary ions such as C_{60} and perhaps even larger ions, the potential for

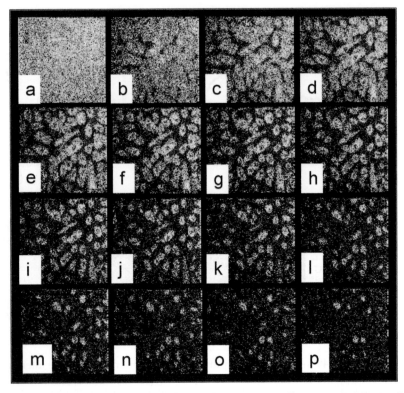

Figure 4.49   A 3D analysis of benign prostatic hyperplasia cells using a DC C_{60} ion beam. This figure shows the m/z 184 signal variation through the cell sample using a logarithmic intensity scale. Images (a–p) were acquired using $C_{60}{}^{+}$ (40 kV) with an accumulated ion dose of 2×10^{13} ions/cm^2 for each image: field of view is $\sim$200 $\times$ 200 μm^2. The upper membrane is removed (a–e), low signal areas where the nucleus is located become visible (f–j) and then the lower membrane is removed (k–p)

biological analysis using SIMS is bright. Exploiting the capabilities of such ion beams requires mass spectrometers able to cope with DC ion beams and providing all the facilities of modern mass spectrometry, particularly tandem mass analysis. Figure 4.49 shows some early results of a 3D analysis of benign prostatic hyperplasia cells using a DC C_{60} ion beam and a buncher-ToF–SIMS instrument. This figure shows a series of images of the m/z 184 ion, where each image represents the signal accumulated after a dose of 2×10^{13} ions cm^{-2}. This contrasts with the 'etch, then analyse' approach previously that wasted all the signal potentially available during the etch period. In the present analysis all the signal generated during sputter etching is acquired.

The figure shows the m/z 184 signal variation through the cell sample using a logarithmic intensity scale. Images (a)–(p) were acquired using $C_{60}{}^{+}$

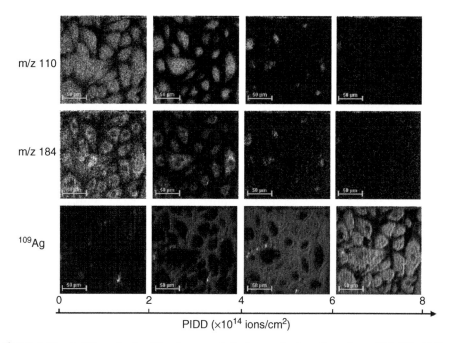

Figure 4.50 A 3D analysis of benign prostatic hyperplasia cells using a DC 40 keV C_{60} ion beam. Each image is the accumulated signal in the etch period. The m/z 110 signal is from histidine, an amino acid signifying the presence of protein and m/z 184 from the phospholipids of cell membrane. The cells were supported on silver foil

(40 kV) with an accumulated ion dose of 2×10^{13} ions cm^{-2} for each image. The field of view is $\sim 200 \times 200 \, \mu m^2$. The upper membrane is removed (a–e), low signal areas where the nucleus is located become visible (f–j) and then the lower membrane is removed (k–p). In a second set of images, Figure 4.50, the m/z 110 signal from histidine, an amino acid signifying the presence of protein and m/z 184 from the phospholipids that predominate in the cell membrane are followed. The cells were supported on silver foil and a ghost signal showing where silver was protected from sputtering by the cells is evident at the end. Again the nucleus area is evident by the absence of m/z 184 at the centre of the cell.

These results show that the prospects for mass spectrometric imaging of biological and other chemically complex systems using SIMS are excitingly favourable. Rapid advances are to be expected in coming years.

4.6 Ionization of the Sputtered Neutrals

Over 95 % of the particles emitted during sputtering are neutral. It is clear therefore that the SIMS experiment ignores the vast majority of the information generated. If this could be accessed there is the possibility of increased sensitivity and further types of information regarding the surface of materials. There is, however, a further potential attraction from accessing the neutral component. Quantification by SIMS is bedevilled by the matrix effect. This arises because the particle emission and ionization processes occur 'simultaneously'. If we can decouple ionization from emission, ionization occurring after the neutrals have moved away from the surface, the ion yield would be independent of the matrix and quantification would be easier.

Thus the principle of the so-called sputtered neutral mass spectrometry (SNMS) is to suppress the sputtered secondary ion component and post-ionize the sputtered neutrals some distance above the solid surface. A number of different approaches were tried from the late 1980s. Commercial instruments were produced but the technique has not been very widely applied. The aim was easier quantification, higher sensitivity and augmented information.

Experimentally post-ionization can be accomplished by either electron or photon bombardment. The possibility of a type of 'chemical ionization' is also being explored in which cesium bombardment leads to the formation of MCs^+ ions which also appear to be less influenced by the matrix effect [70]. Electron post-ionization was developed primarily for elemental analysis under dynamic SIMS conditions [71, 72]. This is not the focus of this chapter, so we will not consider it further. Photon ionization has been explored with ToF–SIMS instruments to overcome the matrix effect and also to try to increase ion yield. Although a commercial instrument was launched in the 1990s, the complexities and limitations of the technique prevented its development. However, the approach is still used in some research laboratories to explore the fundamentals of ion formation in the sputtering experiment. We will therefore consider photon ionization briefly.

4.6.1 PHOTON INDUCED POST-IONIZATION

The most elegant and most efficient method of post-ionization is to use pulsed high-energy laser photons to ionize the neutrals as they leave the surface. In this mode of operation a ToF–MS must be used for mass analysis. Figure 4.51 shows the basic arrangement. There are two basic mechanisms by which the photons can ionize the neutrals. First, *resonant* multi-photon ionization (REMPI) in which one or more photon wavelengths are chosen to match the energy differences between electronic levels lying between the ground state and the vacuum level of the element or species which is to be detected. An electron is thus stimulated to climb the energy level ladder to ionization. Second, *non-resonant* MPI (NRMPI) uses a very high power laser, usually in the UV, to stimulate electrons from the ground state to ionization via *virtual* energy levels.

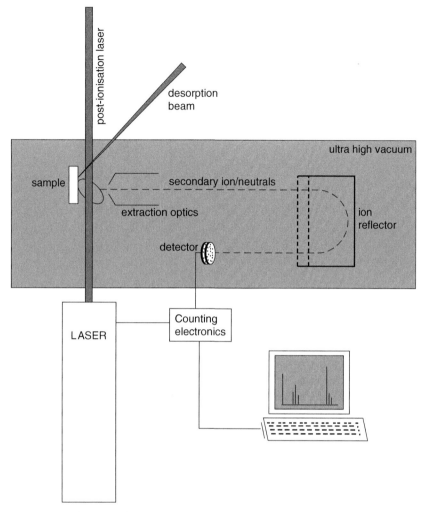

Figure 4.51 A schematic diagram of a ToF–SIMS system with laser postionization. The potentials on the various electrodes are set so that only post-ionized ions are transmitted to the detector. Reproduced with permission from N. Lockyer, *Ph.D. Thesis*, University of Manchester Institute of Science and Technology, 1996

Sputtered Neutral Ionization by Resonant Multi-Photon Ionization (REMPI). This approach tunes the laser photons to the energy levels of the atom or species of interest. Thus if we want to analyse for indium, we consult the energy level diagram, e.g. Figure 4.52, and we find that we could ionize indium using two photons of 410.2 nm or one of 303.9 nm followed by one of 607.9 nm. Clearly the former arrangement would be more convenient since only one laser wavelength would be required. This example highlights the fact that REMPI is element or species specific. To ensure that the correct

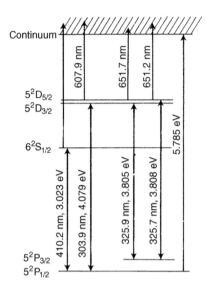

Figure 4.52 Partial electron structure of indium. Reproduced with permission from F.M. Kimock, J.P. Baxter, D.L. Pappas, P.H. Kobrin and N. Winograd, *Anal Chem.*, **56**, 2782 (1984) [72]. Copyright 1984, American Chemical Society

photons are available one has to know beforehand which elements are to be analysed. Some elements require three, four or even five photons. REMPI–SNMS has the great advantage that it is extremely efficient. The ionization probability for neutrals in the ionization volume can be close to 100 %. To attain these sensitivity levels, the overlap between the laser photon field and the sputtered plume of neutrals is crucial. The laser is pulsed about 1 µs after the ion beam pulse to give the neutrals time to fly from the surface into the region of the laser beam. The factors which will affect the overlap between the laser field and the sputtered plume will be the primary pulse width, the energy of the sputtered neutrals, the neutral spatial distribution, the laser photon timing and the laser beam diameter. If all these factors are tuned in well, very high sensitivities are attainable down to sub-ppb levels [73]. Figure 4.53 shows an analysis of ^{56}Fe in silicon. There is a difficult mass interference between ^{56}Fe and ^{56}Si$_2$ making the use of SIMS difficult. Using REMPI, selective ionization of ^{56}Fe is possible enabling a depth profile down to 2 ppb to be carried out [74].

The high sensitivities attainable mean that static conditions of ion bombardment can be used. The analysis of complex organic and inorganic materials is also possible. However, whilst the resonant requirements are not so stringent for many organic species because of the many vibrational levels associated with each electronic state, nevertheless some knowledge of the electronic absorption characteristics of the species to be ionized are required. Unknown analysis requires a less specific ionization process.

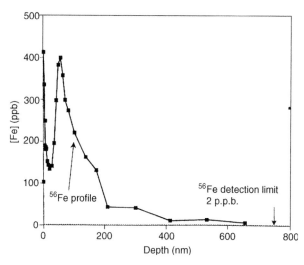

Figure 4.53 A depth profile of ^{56}Fe-implanted silicon performed by REMPI-SNMS of sputtered Fe atoms. Reproduced with permission from C.E. Young, M.J. Pellin, W.F. Calaway, B. Jorgenson, E.L. Schweitzer, and D.M. Gruen, *Nucl. Instrum. Meth. Phys. Res.*, **B27**, 119 (1987) [73]. Copyright 1987, Elsevier

Sputtered Neutral Ionization using Non-Resonant Multi-Photon Ionization. Non-resonant multi-photon ionization is not species specific. Ionization occurs via virtual energy levels. Electron lifetimes in virtual levels are very short so it is essential that the rate of the photon arrival is high enough to elevate the electron to the vacuum level before it returns to the ground state. Thus, the process is not so efficient and high power densities are required: 10^9–10^{10} W cm^{-2} compared to $\approx 10^7$ W cm^{-2} for REMPI. Usually UV photons are used, e.g. 193 nm or 248 nm from ArF or KrF eximers or 266 nm from a frequency quadrupled Nd–YAG. The technique is very effective for elemental analysis of unknowns using a ToF–MS analyser. Sensitivities are not quite as high as with REMPI but very close. Figure 4.54 shows the spectrum of a standard steel using NRMPI [75]. From such a material, tables of relative sensitivity factors (RSFs) can be derived to use the instrument for unknown analysis. The RSFs vary by less than an order of magnitude across the Periodic Table. Sensitivities in the low ppm to ppb region can be attained. The technique offers easier quantification, rather uniform sensitivity for all elements and hence higher sensitivities than SIMS for many elements.

Static analysis is clearly obtainable with NRMPI–SNMS. This offers the possibility of higher sensitivity for surface chemical structure analysis of organic and inorganic materials. Although a number of studies of organic compounds, desorbed from glass substrates using a CO_2 laser followed by ionization by UV–MPI, have been successful in generating molecular ions,

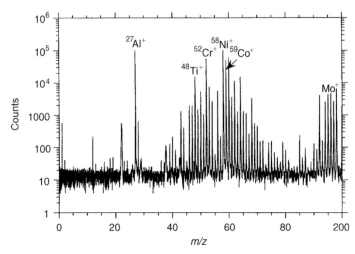

Figure 4.54   NRMPI-SNMS spectrum of NIST standard SRM 1243 steel. Reproduced with permission from E. Scrivenor, R. Wilson and J.C. Vickerman, *Surf. Interface Anal.*, **23**, 623 (1995) [75]. Copyright 1995, John Wiley & Sons, Ltd

early studies into the post-ionization of sputtered PMMA were not encouraging [76]. The spectrum showed only carbon fragments (see Figure 4.55(a)). If the power level is reduced there is no evidence of ion production. The observation was attributed to excitation of emitted clusters by a number of photons whose energy was not quite enough to ionize the cluster, but which increase the vibrational energy so that the molecule falls apart before it can absorb another photon to be ionized. There are two solutions to this problem. The first is to generate VUV photons which have sufficient energy to ionize most molecules and clusters with one photon. The other is to use fs photons so that the photon energy can be input more rapidly than the molecular vibration. Ionization should then occur before the molecule has a chance to fragment. Becker and co-workers [76] demonstrated that the first approach is viable. They generated 118 nm radiation by frequency tripling the 455 nm Nd–YAG radiation in a xenon–argon gas mixture (Figure 4.55(b)). The tripling process is of low efficiency ($\approx 10^{-4}$), but with a 20 mJ input pulse, 10 ns pulses of 118 nm radiation containing 1.3×10^{12} photons were obtained. This was sufficient to produce a good spectrum of PMMA (see Figure 4.55(c)). The spectrum was somewhat different from the SIMS spectrum. Whilst m/z 59 and 69 are strong SIMS peaks the m/z 100 peak due to the monomer ion does not appear in SIMS. The sensitivity is at least on a par with SIMS and there is augmentation of the data. The technique was commercialized under the acronym SALI – surface analysis by laser ionization. As with REMPI, sensitivity is very much dependent on the efficiency of the interaction of the laser beam with the sputtered plume.

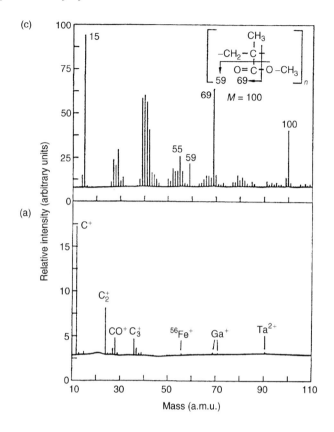

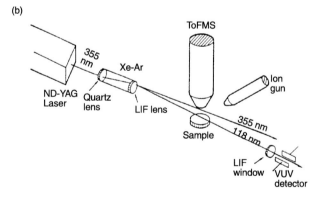

Figure 4.55 (a) Laser post-ionization spectrum of PMMA obtained with pulsed Ar^+ sputtering with multi-photon ionization (258 nm at 1×10^7 W cm^{-2}). Spectra were obtained with 1000 pulses. (b) Schematic diagram of the photon induced post-ionization arrangement using VUV (118 nm) radiation for ionization. (c) Spectrum of PMMA obtained with pulsed Ar^+ sputtering with single photon ionization (118 nm, 3×10^3 W cm^{-2}). Spectra were obtained with 1000 pulses. Reproduced with permission from U. Schühle, J.B. Pallix and C.H. Becker, *J. Am. Chem. Soc.*, **110**, 2323 (1988) [76]. Copyright 1988, American Chemical Society

Furthermore, the lasers used had very low repetition rates, so data acquisition was slow. The technique had great potential but was not a commercial success. Further research in the area has not developed.

One of the attractions of photon post-ionization is to greatly increase sensitivity in molecular species analysis so that high resolution analysis in the imaging mode becomes possible under a close to static regime. Because the ion yield in SIMS is so low, in theory there is the possibility of an increase by at least 100, perhaps 1000. Since in imaging, the experiment has to interrogate say $(256)^2$ points on the surface there is also the requirement of high laser repetition rate if the acquisition of an image is to occur in a reasonable time. The fs lasers based on Ti-sapphire technology have the advantage of kHz repetition rates, so taken with their capability to rapidly inject energy into molecular species, they would seem to be the laser of choice for photon induced post-ionization of molecular species. Early data were very promising. Compared to ns photons, fs photons do result in much higher ionization efficiency and less fragmentation [77, 78]. In the example shown in Figure 4.56 benzo[a]pyrene has been sputtered and ionized in SIMS and by photon post-ionization. Only using photon post-ionization is the molecular ion at m/z 252 clearly evident. However, there is significantly more fragmentation when the ns radiation is used as compared to the fs. The yield is also significantly higher for the fs case.

4.6.2 PHOTON POST-IONIZATION AND SIMS

All the photon ionization methods when applied to organic molecular systems are at best only able to generate radical cations. These are different from the ions normally generated by sputtering. These are usually M±H ions. The hope that somehow photon post-ionization would generate higher yields of molecular ions or that the yields would be independent of the chemistry was probably misplaced. Although organic molecules have closely spaced vibrational levels associated with the electronic levels that mean that photon wavelength is not so critical, nevertheless other than in the VUV the chemistry of the molecules to be detected is important in determining the exact experimental conditions used. As a consequence photon post-ionization is really successful for a quite limited set of molecules. It is not easily used as part of a 'discovery' mass spectrometry where one does not know what one will find in any presented analyte. There are however specific problems where it has been useful and can be seen to have real potential in the future. An example is shown in Figure 4.57. As outlined in Section 4.4.4 the matrix effect can suppress the formation of secondary ions in tissue samples. DPPC in tissue has been found to be a very strong proton acceptor such that it inhibits the formation of M+H ions from drug molecules in the tissue. Cholesterol also found in tissue is more of a proton donor and enhances the formation of M+H ions from many drug

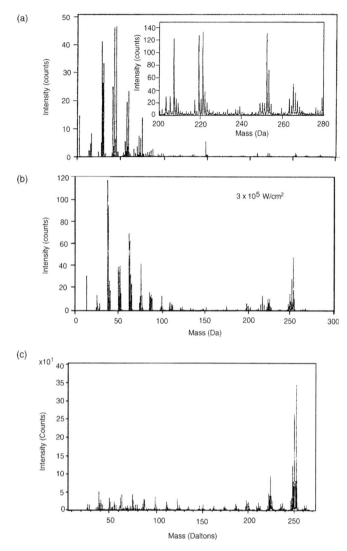

Figure 4.56   A comparison of spectra of benzo[a]pyrene (a toxic polyaromatic hydrocarbon) sputtered by a 25 keV gallium beam from a silicon surface: (a) a SIMS spectrum; (b) SNMS by 280 nm, ns photons; (c) SNMS by 266 nm, 250 fs photons (1.5×10^{12} W cm^{-2}). Reproduced with permission from C.L. Brummel, K.F. Willey, J.C. Vickerman and N. Winograd, *Int. J. Mass Spectrom. Ion Phys.*, **143**, 257 (1995) [77]. Copyright 1995, Elsevier

molecules. Thus it can appear as if the drug is co-located in those parts of the tissue rich in cholesterol (e.g. white matter of brain tissue), but absent from those rich in DPPC (e.g. grey matter of brain tissue), even though it may be evenly distributed. To see if using post-ionization could overcome this effect a model sample was prepared consisting of the drug atropine in

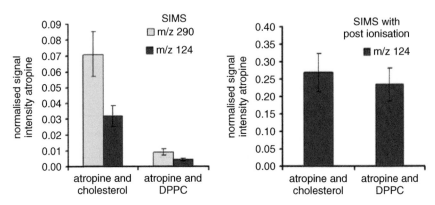

Figure 4.57 The analysis of a drug molecule, atropine, mixed with two abundant biological lipids, cholesterol and DPPC using SIMS analysis and SIMS with photon post-ionization (see text for details). Reproduced with permission from E.A. Jones, N.P. Lockyer and J.C. Vickerman, *Int. J. Mass Spectrom.*, **260**, 146 (2007) [51]. Copyright 2007, Elsevier

1:1 mixtures with both cholesterol and DPPC. The samples were analysed using C_{60}^+ SIMS in the conventional manner and laser post-ionization of the sputtered neutrals was performed using 5 mJ, 266 nm laser pulses from a Nd–YAG laser. The results presented in Figure 4.57 clearly show a great difference in the intensity of the peaks associated with the atropine within the SIMS experiment, with the intensity of the $[M+H]^+$ at m/z 290 and a major fragment at m/z 124 differing by an order of magnitude between the two lipid matrices. When the same samples are analysed using laser post-ionization the most abundant representative ion seen from the drug is the fragment peak at m/z 124, the $[M+H]^+$ ion of course is not generated by laser PI. When the intensity of this peak is compared across the two samples the difference is negligible within experimental error. This suggests that the same amount of the drug molecule is present at the surface of both samples, and the same number are being sputtered into the vacuum; however the nature of the sample has a great effect upon the percentage of these molecules that enter the vacuum in a charged state.

A further example using SPI–VUV post-ionization is one that has been successfully developed for use with laser desorption, but would seem to be applicable to ion desorption too. It is the use of aromatic tags to lower the ionization potentials of bio-molecules to bring them within the photon energy range of the only commercially available high intensity VUV source – the F_2 laser that delivers photons at 157 nm, or 7.87 eV. Most bio-molecules have IPs above this energy, but tagging them with aromatic molecules frequently brings the IP down below 7.87 eV enabling post-ionization with this laser. Hanley and co-workers have combined a nitrogen laser for desorption with aromatic tagging and F_2 – laser post-ionization to detect peptides and other bio-molecules [79]. An advantage of what they term '7.87-eV SPI' of tagged molecular analytes is that the selective ionization of only those species whose IPs are below the photon energy also reduces background and interference peaks in the mass spectrum. This reduction in mass spectral chemical noise combined with tagging of the target analyte improves the ability to identify the target. Again it is not a 'discovery' technique, but in cases where it is possible to specify the molecules of interest and they are chemically susceptible to tagging *in situ* this technique has much to recommend it. A good illustrative example is the detection and imaging of signaling peptides in the bacteria in biofilms [80]. Individual bacteria communicate within biofilms by using signaling molecules in a process referred to as quorum sensing. Both Gram-positive and Gram-negative bacteria use quorum sensing to regulate sporulation, biofilm formation and other developmental processes. Small peptides that behave as quorum-sensing species are found on the external cell surface within Gram-positive bacterial biofilms. In order to contribute to understanding the identity, distribution and activity of quorum sensing species that will help develop strategies for controlling health problems that arise from biofilms, the authors showed that one such signaling pentapeptide, ERGMT in *B. subtilis*, could be tagged with anthracene or quinolene and detected and imaged with 7.87 eV LDPI–MS. This peptide could not be readily detected with straight MALDI–MS, so the combination of aromatic tagging with either laser or ion desorption can be envisaged to offer benefits where ion formation is difficult.

To date, the main interest in SNMS has been in elemental quantification and it does seem that matrix effects are greatly reduced. They are not wholly eliminated, however, because sputter yield is sensitive to surface bond strengths and to the angular distribution of emission. There is therefore some variation from matrix to matrix, but probably only by a factor of two or three. Accurate quantification can be disturbed by two other parameters. First, if secondary ion emission is high, as it can be for alkali metals, the neutral yield will be significantly reduced and will affect the post-ionized yield. Second, a significant yield of atom clusters (as can happen with metals such as silver) may also distort the elemental yield. Despite these provisos electron beam SNMS is being exploited by a number of industrial concerns world-wide for elemental analysis. Elemental analysis by laser-SNMS is much less widely used as yet. The cost and complexity of the equipment is inhibiting its application other than in one or two specialized contract laboratories. The analysis of complex chemistry by laser-post-ionization is still very much in the research and development laboratory. It is unlikely to form part of the arsenal of 'discovery' surface analysis techniques, but it is clear that where the target analyte can be defined there are great benefits in its exploitation. The potential is enormous, but much has to be done to realize it.

4.7 Ambient Methods of Desorption Mass Spectrometry

Classically most methods of surface analysis have been carried out in a vacuum system. Where the detection of ions or electrons is involved this has been essential to ensure the particles were not lost or modified on the way to the detector. It was also thought that a vacuum would keep the surfaces to be studied free from contamination. Of course a vacuum can also interfere with analysis by removing important weakly held components before they can be detected. As has been seen cryo-methods have been developed for biological systems to try to preserve the samples in their native state. During the early years of the 21st Century a new set of techniques analogous to SIMS have emerged that enable desorption mass spectra to be obtained from surfaces while under ambient conditions. The mass spectrometer and its detector still require to be in a vacuum system, but the desorption and ion formation process can take place at atmospheric pressure. It has been shown that ions can be desorbed from surfaces by electrospray jets, by various types of plasma discharges and by laser photons followed by entrainment in electrospray jets [81] (see Figure 4.58). The ions so formed are collected in an ion transfer tube, which may be heated to desolvate the ions as is traditionally done in electrospray mass spectrometry. The ions pass through a small aperture that provides an interface between atmospheric pressure and the vacuum of the mass spectrometer. As the ions travel further into the spectrometer the pressure is reduced until the vacuum is sufficient for effective mass analysis. A variety of skimmer plates, focusing optics and quadrupoles are used to condition the ion beam. The ions may be analysed with ToF, quadrupole or ion trap mass spectrometers. The various modes of ion formation all appear to result from a relatively soft ionization process. The mechanisms of desorption and ion formation are still the subject of investigation and discussion, but their apparent ability to deliver analytically useful data directly from everyday surfaces has generated considerable excitement. Demonstrator studies showing the ability to detect drugs, explosives and suggested biomarkers for diseases have ensured rapidly rising interest. We have seen that the ion formation probability in SIMS in common with MALDI is of less than 10^{-4}. These techniques have very efficient electrostatic ion collection systems to optimize the transmission of ions from the sample surface to the mass spectrometer. For a ToF–SIMS instrument transmissions as high as 50 % have been measured. The ambient pressure instruments collect ions from a desorption plume by placing the entrance to the capillary in the plume and collecting the ions by gas flow into the capillary. It is difficult to envisage that the transmission can be very high and yet detection capabilities in the femtomole are claimed.

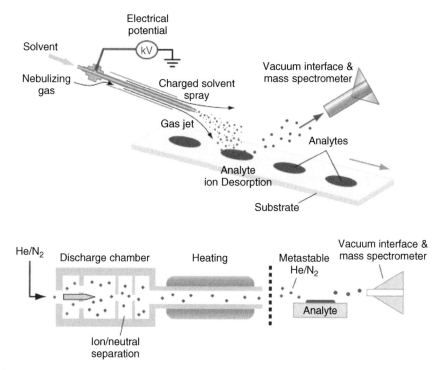

Figure 4.58 Schematics of the principles of operation of the DESI (upper) and DART (lower) methods for ambient mass spectrometric analysis of samples without any pre-treatment. Reproduced with permission from R.G. Cooks, Z. Ouyang, Z. Takats and J.M. Wiseman, *Science* **311**, 1566 (2006) [81]. Copyright 2006, American Association for the Advancement of Science

The most widely applied technique is desorption electrospray ionization mass spectrometry (DESI) developed in Cooks' laboratory [81]. In this arrangement charged droplets and ions produced in an electrospray jet usually composed of water and methanol are directed at the surface of the analyte. Surface compounds are removed as either ions entrained in liquid droplets or as bare ions. The vast majority of material removed will probably be uncharged. The pressure differential around the interface capillary draws ions and gas into the mass spectrometer where the ions are analysed. The angle of impact of the spray and the position of the collection capillary are also important in determining the overall detection capability. The efficiency of the desorption process is dependent on how strongly molecules are bound to the surface. The composition of the spray can be varied to aid the desorption and ionization process. It can be seen that the whole process is very soft and this is reflected in the mass spectra that show very little fragmentation. Mainly molecular ions are detected. Analysis requires accurate mass measurement and tandem MS–MS methods to be applied. Multiple mechanisms of ionization are likely to be involved, including chemical sputtering involving gas-phase ions generated by electrospray ionization (ESI) or corona discharge with subsequent charge transfer between these primary ions and sample molecules on the surface. The occurrence of gas phase ion–molecule reactions has also been suggested together with a droplet splashing or pick-up mechanism, which is at present the generally favoured model. This involves the impacting of multiply charged solvent droplets dissolving sample molecules from the surface leading to the formation of secondary charged droplets carrying sample molecules and resulting in ion formation mechanisms similar to that of electrospray ionization.

Some very impressive results have been obtained from DESI and it is a very actively developing technique. Because it can apparently be used without any sample pre-preparation its application to forensic investigations, to security situations, e.g. airport baggage screening, to the detection of drugs and the diagnosis of diseases are some of the main areas where it is hoped it will make a real contribution. Figure 4.59 shows the spectrum obtained from a dry urine spot on paper (2 ml of urine), showing the complex nature of this mixture. Minor components can be identified by exploiting the use of MS–MS spectra; for example, the isolation of the ion with $m/z\,214$ and the measurement of its product spectrum allow its identification as aspartyl-4-phosphate. It has been shown that experiments of the type illustrated can be performed at a rate of one per second. There is no preparation of the biological fluid other than its deposition on the surface. While an electrospray jet doesn't have the spatial capability of ion beams used in SIMS, nevertheless spatially resolved analysis can be usefully performed. The simplest approach to DESI imaging is to use a microprobe beam of solvent microdroplets and to raster the surface.

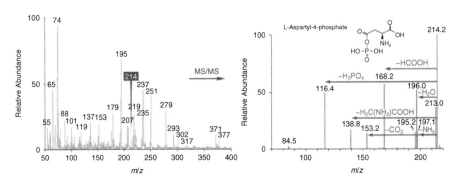

Figure 4.59 A DESI mass spectrum for dried 2 µl raw urine spots on paper, sprayed with 1:1 methanol containing 1 % acetic acid. The product ion MS/MS spectrum identifies one of the minor components, $m/z\,214$ as aspartyl-4-phosphate. Reproduced with permission from R.G. Cooks, Z. Ouyang, Z. Takats and J.M. Wiseman, *Science* **311**, 1566 (2006) [81]. Copyright 2006, American Association for the Advancement of Science

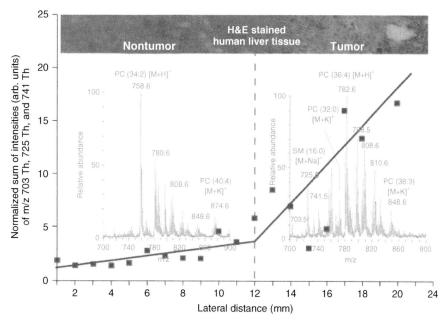

Figure 4.60 Direct tissue profiling of human liver adenocarcinoma using DESI in the positive ion mode. The tissue was sectioned and untreated and was subject to a spray of 1:1 methanol:water containing 0.1 % ammonium hydroxide. Reproduced with permission from R.G. Cooks, Z. Ouyang, Z. Takats and J.M. Wiseman, *Science* **311**, 1566 (2006) [81]. Copyright 2006, American Association for the Advancement of Science

Tissue imaging by DESI shows only modest spatial resolution (spot sizes 0.5 to 1.0 mm), but it removes the constraints of the high vacuum met in SIMS imaging and that of sample preparation, which is a requirement for MALDI imaging. This is illustrated in Figure 4.60 where an analysis has been carried out by scanning the DESI beam across the interface between the non-tumour and tumour regions of human liver tissue, revealing differences in the distributions of compounds, including elevated levels of certain phospholipids in the tumor region as compared with the non-tumor region.

DESI can be thought of as an atmospheric version of SIMS. However the mechanism of molecule removal from the surface is rather different, less energetic and perhaps more dependent on chemical factors, for example the solubility or basicity of the analyte in the spray liquid [82]. Unlike SIMS, DESI can be used to detect proteins and even it is claimed to sequence them. Matrix effect type issues will affect ion formation and hence need to be factored into data interpretation. For example in the tissue analysis shown above it is unlikely that all the compounds in the tissue are showing up in the spectrum and although elevated phospholipid content in the tumour region is likely to be significant it probably does not reflect all the chemical changes due to the cancer. As we have seen this is an issue common to all desorption mass spectrometries. In the case of DESI to try to ensure that the detection of more compounds, the ionization can be influenced and enhanced by changing the composition of the spray liquid to provide a type of chemical ionization.

As indicated above there are a number of methods that employ plasma discharges. One quite widely applied is DART (direct analysis in real time). As shown in Figure 4.58 an electrical potential is applied to a gas with a high ionization potential (typically nitrogen or helium) to form a plasma of excited-state atoms and ions, and these desorb low-molecular weight molecules from the surface of a sample. As may be perceived DART is very good for the analysis of gas phase samples. The mechanism of ion formation would be thought to be rather different from DESI. Penning ionization has been suggested, in which ionization of the sample occurs by energy transfer from an excited atom or molecule of energy greater than the ionization energy of the sample. It has been observed that when helium is used as the gas, the mechanism involves the formation of ionized water clusters followed by proton transfer reactions. However similar spectra are obtained although the degree of fragmentation is sometimes higher than with DESI [83]. In practice because it is difficult to actually see the plasma discharge it is more challenging to obtain the optimum configuration for sampling from solid samples.

Another variant is electrospray enhanced laser desorption mass spectrometry (ELDI) [84]. Material is desorbed from the analyte surface using a pulsed nitrogen laser, an electrospray jet is directed through the desorbing plume towards the mass spectrometer atmospheric capillary inlet. This approach has been shown to enable small proteins to be analysed without the need to apply a matrix. The laser alone does not produce any ions, but the ESI source and the laser together are shown to generate good spectra.

To be able to apply all the chemical characterization capability of mass spectrometry to materials in the ambient atmosphere is a great advance and it is clear that these techniques potentially have many applications in the modern world. A great advantage is that these techniques are directly compatible with a wide variety of mass spectrometers traditionally used in analytical chemistry. Consequently this may make surface chemical analysis very much more accessible. However, analysis in the ambient brings its own special challenges. It is frequently difficult enough to ensure that spurious contaminants do not interfere with the quality of analysis when carried out in vacuum; this will be even more of a challenge when the atmosphere and environment of the analysis cannot be controlled.

All the methods of surface mass spectral analysis introduced in this chapter have their own special capabilities and challenges. It can also be seen from our survey that their capabilities are all still developing. No one variant of desorption mass spectrometry will be able to solve every problem. The researcher or analyst needs to consider the data to be acquired in order to specify the experimental approach likely to provide the most useful answers to the questions posed.

📖 References

1. J.J. Thomson, *Phil. Mag.*, **20**, 252 (1910).

2. R.F.K. Herzog and F.P. Viebock, *Phys. Rev.*, **76**, 855L (1949).

3. H.J. Liebl and R.F.K. Herzog, *J. Appl. Phys.*, **34**, 2893 (1963).

4. R. Castaing and G. Slodzian, *J. Microscopy*, **1**, 395 (1962).

5. A. Benninghoven, *Z. Physik*, **230**, 403 (1970).

6. (a) A. Müller and A. Benninghoven, *Surf. Sci.*, **39**, 427 (1973); (b) *Surf. Sci.*, **41**, 493 (1973); (c) A. Benninghoven, in *ToF–SIMS: Surface Analysis by Mass Spectrometry*, J.C. Vickerman and D. Briggs (Eds), Chapter 2, SurfaceSpectra and IM Publications, Chichester, UK (2003).

7. M. Barber, J.C. Vickerman and J. Wolstenholme, *J. Chem. Soc., Faraday Trans. I*, **72**, 40 (1976) and subsequent papers reviewed in J.C. Vickerman, *Surf. Sci.,* **189/190**, 7 (1987).

8. J.C. Vickerman, A. Brown and N.M. Reed (Eds), *Secondary Ion Mass Spectrometry, Principles and Applications*, Oxford University Press, Oxford (1989).

9. R. Behrisch (Ed.), *Sputtering by Particle Bombardment, I*, Springer Series Topics in Applied Physics **47**, Springer-Verlag, Berlin (1981); (b) R. Behrisch (Ed.), *Sputtering by Particle Bombardment, II*, Springer Series Topics in Applied Physics **52**, Springer-Verlag, Berlin, Heidelberg (1983); (c) R. Behrisch and K. Wittmaack (Eds), *Sputtering by Particle Bombardment, III*, Springer Series Topics in Applied Physics, **64**, Springer-Verlag, Berlin (1991).

10. A. Benninghoven, F. Rüdenauer and H.W. Werner, *Secondary Ion Mass Spectrometry*, John Wiley & Sons, Ltd, Chichester, UK (1987).

11. J.C. Vickerman and D. Briggs (Eds), *ToF–SIMS: Surface Analysis by Mass Spectrometry*, SurfaceSpectra and IM Publications, Chichester, UK (2003).

12. W. Sicthermann and A. Benninghoven, *Int. J. Mass Spectrom. Ion Phys.*, **40**, 177 (1981).

13. I.S. Gilmore and M.P. Seah, *Surf. Interface Anal.*, **24**, 746 (1996).

14. R. Galera, J. Blais and G. Bolbach, *Int. J. Mass Spectrom. Ion Proc.*, **107**, 531 (1991).

15. D. Briggs and M.J. Hearn, *Int. J. Mass Spectrom. Ion Proc.*, **67**, 47 (1985).

16. A. Wucher, *Appl. Surf. Sci.,* **252**, 6482 (2006).

17. C. Szakal, J. Kozole, B.J. Garrison and N. Winograd, *Phys. Rev. Lett.,* **96**, 216104, 1–4 (2006).

18. (a) N.M. Reed, P. Humphrey and J.C. Vickerman, in *Proceedings of the 7th International Congress on SIMS*, A. Benninghoven, C.A. Evans, K.D. McKeegan, H.A. Storms and H.W. Werner (Eds), p. 809, John Wiley & Sons Ltd, Chichester, UK (1990); (b) A. Brown and J.C. Vickerman, *Surf. Interface Anal.*, **8.**, 75 (1986).

19. D. Surman, J.A. van den Berg and J.C Vickerman, *Surf. Interface Anal.*, **4**, 160 (1982).

20. I. Gilmore, in *ToF–SIMS: Surface Analysis by Mass Spectrometry*, J.C. Vickerman and D. Briggs (Eds), Chapter 10, SurfaceSpectra and IM Publications, Chichester, UK (2003).

21. S.L. Luxembourg, T.H. Mize, L.A. McDonnell and R.M.A. Heeren, *Anal. Chem.*, **76**, 5339 (2004).

22. A. Brown, J.A. van den Berg and J.C. Vickerman, *Spectrochim. Acta B*, **40**, 871 (1985).

23. K. Wittmaack, *Rev. Sci. Instrum.*, **47**, 157 (1976).

24. B Sakakini, A.J Swift, J.C Vickerman, C Harendt and K Christmann, *J. Chem. Soc., Faraday Trans. I*, **83**, 1975 (1987).

25. (a) K. Tang, R. Bevis, W. Ens, F. Lafortune, B. Schueler and K.G. Standing, *Int. J. Mass Spectrom. Ion Proc.*, **85**, 43 (1988); (b) E. Niehuis, T. Heller, F. Feld and A. Benninghoven, *J. Vac. Sci. Technol.*, **A5**, 1243 (1987); (c) A.J. Eccles and J.C. Vickerman, *J. Vac. Sci. Technol.*, **A7**, 234 (1989).

26. I.V. Chernushevich, A.V. Loboda and B.A. Thomson, *J. Mass Spectrom.* **36**, 849 (2001).

27. (a) B.H. Sakakini, C. Harendt and J.C. Vickerman, *Spectrochim. Acta*, **43A**, 1613 (1987); (b) A.J. Paul and J.C. Vickerman, *Proc. R. Soc. London A, Math. Phys. Sci.*, **330**, 147 (1990).

28. D. Greifendorf, P. Beckmann, M. Schemmer and A. Benninghoven, in *Ion Formation from Organic Solids (IFOSIII)*, Springer Series in Chemical Physics, **25**, A. Benninghoven (Ed.), p. 118, Springer-Verlag, Berlin (1983); (b) D. Rading, R. Kersting and A. Benninghoven, in *Proceedings of the 11th International Conference on SIMS (SIMS XI)*, G. Gillen, R. Lareau, J. Bennett and F. Stevie, p. 455, John Wiley & Sons, Ltd, Chichester, UK (1998).

29. M.J. Hearn and D. Briggs, *Surf. Interface Anal.*, **11**, 198 (1988).

30. (a) G.J. Leggett and J.C. Vickerman, *Int. J. Mass Spectrom. Ion Phys.*, **122**, 281 (1992); (b) G.J. Leggett and J.C. Vickerman, *Annu. Rep. R. Soc. Chem.*, *C*, 77 (1991).

31. (a) A. Delcorte and P. Bertrand, *Nucl. Instrum. Meth. Phys. Res. B*, **117**, 235 (1996); (b) A. Delcorte and P. Bertrand, *Nucl. Instr. Meth. Phys. Res. B*, **135**, 430 (1998; (c) A. Delcorte and P. Bertrand, *Surf. Sci.*, **412/413**, 97 (1998).

32. D. Rading, R. Kersting and A. Benninghoven, *J. Vac. Sci. Technol. A*, **18**, 312 (2000).

33. G. Leggett, in *The Static SIMS Library*, J.C. Vickerman, D. Briggs and A. Henderson (Eds), SurfaceSpectra Ltd, Manchester, UK (1999).

34. P. Sigmund, in *Sputtering by Particle Bombardment*, Springer Series Topics in Applied Physics, **47**, R. Behrisch (Ed.), p. 9, Springer-Verlag, Berlin (1981).

35. N. Winograd, *Prog. Solid State Chem.*, **13**, 285 (1981).

36. N. Winograd, in *Fundamental Processes in Sputtering of Atoms and Molecules*, P. Sigmund (Ed.), Mat. Fys. Medd. Dan. Vid. Selsk., **43**, 223 (1993).

37. (a) A. Wucher and B.J. Garrison, *Surf. Sci.,* **260**, 257 (1992); (b) A. Wucher and B.J. Garrison, *Phys. Rev., B,* **46**, 4855 (1992).

38. R.S. Taylor and B.J. Garrison, *Langmuir,* **11**, 1220 (1995).

39. R.S. Taylor, C.L. Brummel, N. Winograd, B.J. Garrison and J.C. Vickerman, *Chem. Phys. Lett.*, **233**, 575 (1995).

40. D.W. Brenner, *Phys Rev., B,* **42**, 9458 (1990).

41. K.S.S. Liu, C.W. Yong, B.J. Garrison and J.C. Vickerman, *J. Phys. Chem., B,* **103**, 3195 (1999).

42. A. Delcorte, X. Vanden Eynde, P. Bertrand, J.C. Vickerman and B.J. Garrison, *J. Phys. Chem., B,* **104**, 2673 (2000).

43. A. Delcorte and B.J. Garrison, *Nucl. Instr. Meth. Phys. Res. B,* **180**, 37 (2001).

44. A. Delcorte and B.J. Garrison, *J. Phys. Chem., B.,* **108**, 15652 (2004).

45. (a) M.F. Russo, Jr, I.A. Wojciechowski and B.J. Garrison, *Appl. Surf. Sci.,* **252**, 6423 (2006); (b) K.E. Ryan, I.A. Wojciechowski, and B.J. Garrison, *J. Phys. Chem., C,* **111**, 12822 (2007).

46. C. Plog, L. Wiedermann and A. Benninghoven, *Surf. Sci.,* **67**, 565 (1977).

47. R.T. Sanderson, in *Solid State Chemistry and its Applications*, A.R. West (Ed.), p 291. John Wiley & Sons, Ltd, Chichester, UK (1987).

48. N.M. Reed and J.C. Vickerman, in *Practical Surface Analysis*, Volume **2**, *Ion and Neutral Spectroscopy*, D. Briggs and M.P. Seah (Eds), p 332, (John Wiley & Sons, Ltd, Chichester, UK (1992).

49. W. Gerhard and C. Plog, *Z. Phys. B, Condensed Matter,* **54**, 59, 71 (1983).

50. R.G. Cooks and K.L. Busch, *Int. J. Mass Spectrom. Ion Phys.,* **53**, 111 (1983).

51. E.A. Jones, N.P. Lockyer and J.C. Vickerman, *Int. J. Mass Spectrom.* **260**, 146 (2007).

52. (a) R. Zenobi and R. Knochenmuss, *Mass Spectrom. Rev.,* **17**, 337 (1998); (b) K. Breuker, R. Knochenmuss, J. Zhang, A. Stortelder and R. Zenobi, *Int. J. Mass Spectrom.,* **226**, 211 (2003).

53. E.A. Jones, N.P. Lockyer, J. Kordys and J.C. Vickerman, *J. Am. Soc. Mass Spectrom.,* **18**, 1559 (2007).

54. X.A. Conlan, N.P. Lockyer and J.C. Vickerman, *Rapid Commun. Mass Spectrom.,* **20**, 1327 (2006).

55. D. Briggs, in *ToF–SIMS: Surface Analysis by Mass Spectrometry*, J.C. Vickerman and D. Briggs (Eds), Chapter 16, SurfaceSpectra and IM Publications, Chichester, UK (2003).

56. See, for example, various chapters in *ToF–SIMS: Surface Analysis by Mass Spectrometry*, J.C. Vickerman and D. Briggs (Eds), SurfaceSpectra and IM Publications, Chichester, UK (2003).

57. A.J. Oakes and J.C. Vickerman, *Surf. Interface Anal.*, **24**, 695 (1996).

58. J.A. Treverton, A.J. Paul and J.C. Vickerman, *Surf. Interface Anal.*, **20**, 449 (1993).

59. W.J. van Ooij, A. Sabata and A.D. Appelhans, *Surf. Interface Anal.*, **17**, 403 (1991).

60. J.S. Fletcher, A. Henderson, R.M. Jarvis, N.P. Lockyer, J.C. Vickerman and R. Goodacre, *Appl. Surf. Sci.*, **252**, 6869 (2006).

61. B. Cliff, N. Lockyer, H. Jungnickel, G. Stephens and J.C. Vickerman, *Rapid Commun. Mass Spectrom.*, **17**, 2163 (2003).

62. C. Lechene, F. Hillion, G. McMahon, D. Benson, A. M. Kleinfeld, J. P. Kampf, D. Distel, Y. Luyten, J. Bonventre, D. Hentschel, K. M. Park, S. Ito, M. Schwartz, G. Benichou and G. Slodzian, *J. Biol.*, **5**, 20 (2006).

63. S.G. Ostrowski, C.T. Van Bell, N. Winograd and A.G. Ewing, *Science*, **305**, 71 (2004).

64. D. Weibel, S. Wong, N. Lockyer, P. Blenkinsopp, R. Hill and J.C. Vickerman, *Anal. Chem.*, **75**, 1754 (2003).

65. M.R. Houlton, O.D. Dosser, M.T. Emeny, A. Chew and D.E. Sykes, in *Proceedings of the 8th International Conference on SIMS (Amsterdam, 1991)*, A. Benninghoven, K.T.F. Janssen, J. Tümpner and H.W. Werner (Eds), p. 343, John Wiley & Sons, Ltd, Chichester, UK (1992).

66. L. Zheng, A. Wucher, and N. Winograd, *J. Am. Soc. Mass Spectrom.*, **19**, 96 (2008).

67. J.S. Fletcher, N.P. Lockyer, S. Vaidyanathan and J.C. Vickerman, *Anal. Chem.*, **79**, 2199 (2007).

68. D. Breitenstein, C.E. Rommel, R. Möllers, J. Wegener and B. Hagenhoff, *Angew. Chem. Int. Ed. Engl.*, **46**, 5332 (2007).

69. E.A. Jones, N.P. Lockyer and J.C. Vickerman, *Anal. Chem.*, **80**, 2125 (2008).

70. H. Gnaser, *Surf. Interface Anal.*, **24**, 483 (1996).

71. H. Oechsner, W. Ruhe and E. Stumpe, *Surf. Sci.*, **85**, 289 (1979).

72. R. Wilson, J. van den Berg and J.C. Vickerman, *Surf. Interface Sci.*, **14**, 393 (1989).

73. F.M. Kimock, J.P. Baxter, D.L. Pappas, P.H. Kobrin and N. Winograd, *Anal Chem.*, **56**, 2782 (1984).

74. C.E. Young, M.J. Pellin, W.F. Calaway, B. Jorgenson, E.L. Schweitzer and D.M. Gruen, *Nucl. Instrum. Methods Phys. Res.*, *B*, **27**, 119 (1987).

75. E. Scrivenor, R. Wilson and J.C. Vickerman, *Surf. Interface Anal.*, **23**, 623 (1995).

76. U. Schühle, J.B. Pallix and C.H. Becker, *J. Am. Chem. Soc.*, **110**, 2323 (1988).

77. C.L. Brummel, K.F. Willey, J.C. Vickerman and N. Winograd, *Int. J. Mass Spectrom. Ion Phys.*, **143**, 257 (1995).

78. N. Lockyer and J.C. Vickerman, in *Proceedings of the 10th International Conference on SSIMS (Münster, 1995)*, A. Benninghoven, B. Hagenhoff and H.W. Werner (Eds), p. 783, John Wiley & Sons, Ltd, Chichester, UK (1997).

79. P.D. Edirisinghe, J.F. Moore, W.F. Calaway, I.V. Veryovkin, M.J. Pellin and L. Hanley, *Anal. Chem.*, **78**, 5876 (2006).

80. P.D. Edirisinghe, J.F. Moore, K.A. Skinner-Nemec, C. Lindberg, C.S. Giometti, I.V. Veryovkin, J.E. Hunt, M.J. Pellin and L. Hanley, *Anal. Chem.*, **79**, 508 (2007).

81. R.G. Cooks, Z. Ouyang, Z. Takats and J.M. Wiseman, *Science*, **311**, 1566 (2006).

82. (a) Z. Takats, J.M. Wiseman, B. Gologan and R.G. Cooks, *Science* **306**, 471 (2004); (b) Z. Takats, J.M. Wiseman and R.G. Cooks, *J. Mass Spectrom.*, **40**, 1261 (2005).

83. J.P. Williams, V.J. Patel, R. Holland and J.H. Scrivens, *Rapid Commun. Mass Spectrom.*, **20**, 1447 (2006).

84. J. Shiea, M.-Z. Huang, H.-J. HSu, C.-Y. Lee, C.-H. Yuan, I. Beech and J. Sunner, *Rapid Commun. Mass Spectrom.* **19**, 3701 (2005).

Problems

1. Define the terms 'sputtering' and 'surface sensitivity' and 'surface damage' as they apply to SIMS.

2. The secondary ion signal intensity from element m in a SIMS experiment is given by:

$$I_{ms} = I_p \theta_m y_m \alpha^+ \eta \qquad (4.8)$$

   Identify and explain the importance of each parameter. Describe and explain the variation of the Al^+ signal as a function of depth, as a depth profile analysis is performed through an aluminium oxide film on aluminium. What are the implications of this observation for *quantitative* elemental analysis using SIMS?

3. Explain why in secondary ion mass spectrometry (SIMS) the yield of secondary ions arising from different chemical species in a surface is not directly proportional to the chemical composition of the surface.

4. Explain why the *static* SIMS primary particle bombardment dose limit has been set at 10^{13} ions cm^{-2} for atomic primary ions. Outline why this may be too high for the analysis of organic surfaces.

5. An X-ray photoelectron spectrum of poly(ethylene terephthalate) shows three peaks in the C 1s region with binding energies of about 284, 287 and 290 eV. In the static SIMS positive ion spectrum there are three significant peaks at m/z 193, 149 and 104. Provide an interpretation for these spectral data and discuss the relative advantages and disadvantages of XPS and SSIMS for the surface analysis of polymers.

6. Outline the analytical benefits for molecular surface analysis from the use of polyatomic cluster primary ion beams as compared to atomic primary ions. Distinguish the advantages of small cluster metal ion beams compared to the benefits of large polyatomics such as C_{60}.

7. There is a requirement to analyse a 1 µm × 1 µm area of a sample within the static limit using a liquid metal primary ion beam. Carry out a calculation to determine whether this is possible, assuming the sputter yield of all the components is 1, the ionization probability is 10^{-3}, the transmission of the mass analyser is 0.1 and a yield of at least 100 secondary ions are needed across the spectrum for analysis. How could the calculated yield be increased whilst keeping within the static limit?

8. A bio-organic system requires to be analysed in 2 and 3 dimensions to sub-micron resolution. Outline the issues to be considered in choosing the ion beam to be used.

9. What difference would the use of a polyatomic beam make in question 7, if the sputter yield is raised to 100, the ionization probability remains the same and the requirement to stay within the static limit is removed allowing the whole surface layer to be consumed in the analysis. What would the minimum area for analysis be under these circumstances if 100 ions are required in the spectrum?

10. There is a requirement to depth profile completely through a 50 µm diameter biological cell using a ToF–SIMS system requiring a pulsed polyatomic ion beam. Assume a sputter yield of 100, a primary beam in DC mode of 1 pA with a beam diameter of 500 nm raster scanned over a 70 µm×70 µm area and a pulse length of the ion beam of 20 ns with a repetition rate of 10kHz. If an ortho-ToF instrument was used that allowed a DC beam to be used calculate the profiling time in this case.

11. If the analysis of Fe in silicon shown in Figure 4.53 were to be carried out by SIMS rather than REMPI–SNMS what mass resolution would be required? Consider this if the transmission of the mass analyser was 0.05, the sputter yield of both Fe and Si was 3, the ionization probability of Fe was 10^{-3}, the minimum acceptable count level was 5 counts, the analysis area was 500 µm × 500 µm and the primary beam current was 1 µA cm^{-2}. What

would the minimum detectable Fe concentration be? How quickly would the depth profile take to be completed to the depth of the minimum detectable concentration of Fe; to 600 nm?

12. In Section 4.6.1, in producing VUV radiation, Becker *et al.* generated 10 ns pulses containing 1.3×10^{12} photons of 118 nm radiation from 20 mJ input pulses of 355 nm radiation. What was the efficiency of the conversion?

5 Dynamic SIMS

DAVID MCPHAIL

Department of Materials, Imperial College, London, UK

MARK DOWSETT

Department of Physics, University of Warwick, Coventry, UK

5.1 Fundamentals and Attributes

5.1.1 INTRODUCTION

Dynamic secondary ion mass spectrometry (SIMS) is a powerful mass spectrometric technique for the chemical microanalysis of (usually) solid materials in which the source consists of ions removed from the analyte by sputtering. Since the mid-1960s the principal applications of the technique have been in microelectronics and geological sciences but the range of applications has steadily grown to encompass glasses, ceramics, metals, plastics, pharmaceuticals, bio-materials, materials from museums and even materials from space. The technique now enjoys sub-nanometre depth resolution (with ultra low energy primary ions), parts per billion sensitivity (for high yield elements) and lateral resolution in the tens of nm range. It should be noted though that the ultimate performance in any one parameter usually excludes such in another – e.g. high sensitivity and the highest lateral resolution are mutually exclusive. Nevertheless, these attributes are useful in the study of a very wide range of processes in materials including diffusion, corrosion, oxidation segregation, dating and origin studies (e.g. in the case of pre-solar material).

Surface Analysis – The Principal Techniques 2nd Edition Edited by John Vickerman and Ian Gilmore
© 2009 John Wiley & Sons, Ltd

Over the last forty years dynamic SIMS has contributed heavily to basic semiconductor material science, new device and process development, and failure analysis. Conversely, advances in SIMS have been stimulated by the demand for analytical methods for the measurement of part or fully processed semiconductor materials and structures. Since electronic materials can be produced to extremely high standards of purity and reproducibility they form ideal samples for SIMS technique characterization. As active device volumes continue to shrink, the demands of microelectronics continue to stimulate the development of instrumentation and methods. In parallel, demands for high precision measurement from microvolumes in geological and astrophysical contexts have lead to the development of high mass resolution spectrometers capable of exceptionally accurate isotope ratio measurements. The advent of ultra-low energy depth profiling and the associated sub-nanometre depth resolution has facilitated the analysis of thin layers such as oxides and nitrides. Slow corrosion processes can now be studied in real time without the need for accelerated ageing. For example the decay of museum glass can be followed by studying the rate of ingress of moisture and egress of mobile sodium cations. SIMS depth profiling can also be combined with stable isotope labelling to facilitate the measurement of diffusion processes. For example, the diffusion of oxygen through an oxide ceramic can be measured by heating the ceramic in an environment containing an enhanced level of the gas $^{18}O_2$. The diffusion of the ^{18}O atoms through the ceramic are then easily measured. This approach is being used in research for Solid Oxide Fuel Cell Technology.

In this chapter we describe the application of dynamic SIMS to a variety of materials problems, with an emphasis on depth profiling. Section 5.1 describes the overall capabilities and limitations of dynamic SIMS. Section 5.2 outlines some of the areas in which SIMS analysis is used. Section 5.3 deals with the quantification of 1- and 2-D data, the use and fabrication of standards, and sources of error. Some novel ways of applying the technique are described in Section 5.4 and aspects of the instrumentation are discussed in Section 5.5. Sections 4.1 to 4.3 of Chapter 4 will also provide useful background to the basic concepts and instrumentation of SIMS. An exhaustive list of the parameters important in dynamic SIMS can be found in Benninghoven et al. [1] (pp. xxvii–xxxv).

In dynamic SIMS the sample surface is eroded by sputtering with a mono-energetic beam of primary ions in the energy range 0.25 keV to 50 keV. It is likely that the routinely achievable lower end of the range will be extended to 0.1 keV shortly as the column designs to do this already exist. In general below 0.1 keV one enters the regime of ion beam deposition, rather than sputtering (indeed, heavy ions such as Au^+ deposit below ~0.3 keV). The top end of the range is used by liquid metal ion sources (LMIS) and other heavy ion guns to overcome chromatic aberration and space charge effects to achieve sub-micron spot sizes.

Some of the sputtered atoms or molecular clusters are themselves ionized directly in the sputtering process and a significant fraction of these is collected in an electrostatic field and transported into a double focusing magnetic sector (DFMS) [2], quadrupole (QMS) [3] mass spectrometer or time-of-flight (TOF) mass spectrometer [4, 5]. The type of analysis obtained (micro-volume analysis, depth profile, image or image depth profile) depends on the shape and location of the sample volume contributing to each data point.

Dynamic SIMS and static SIMS are distinguished by the primary ion dose acceptable during analysis. For static SIMS, it is essential that the probability of secondary ions being detected from a region of the sample surface already modified by previous ion impact is $\ll 1$. In practice, this limits the acceptable primary ion dose density to $<10^{13}$ ions cm^{-2} per experiment [1]. In contrast, in dynamic SIMS, the objective is to establish steady state conditions of erosion rate and surface chemistry, and the primary ion dose retained in the (receding) near surface region is intended to attain a stationary value. The minimum dose in a dynamic SIMS measurement [6] is $\sim 10^{17}$ ions cm^{-2}. Where experimental conditions exist which lead to such a steady state for a particular primary beam/matrix/impurity combination, SIMS can be used as a high precision depth profiling and three dimensional characterization technique. For ion doses between these two limits, the sample's surface chemistry, the erosion rate and the ion yields may change drastically as irradiation proceeds. This is termed the pre-equilibrium region and persists for the erosion of a minimum of several nanometres. Here, accurate quantification is impossible, although qualitative information may be derived under some circumstances. The analysis of very shallow buried features therefore requires special treatment (see Section 5.2.3).

SIMS has four fundamental attributes which give it its high sensitivity and dynamic range and establish its limitations:

(i) Mass spectrometry is inherently background free (unlike electron spectroscopies and ion scattering spectroscopies) because the mass spectrum is discrete and not superimposed on a continuum. Even at a modest mass resolution (e.g. $M/\Delta M = 1000$) a well set up magnetic or quadrupole SIMS spectrometer should be capable of a rejection ratio $(I_M/I_{(M\pm1)})$ of $>10^8$, where $I(M)$ is the intensity recorded at mass M, etc. The optics of the ToF generally preclude abundance sensitivities better than $\sim 10^5$. Table 5.1 shows some typical mass interferences (after Balake *et al.* [6]).

(ii) Secondary ion yields are quite high (often in the range 10^{-1} to 10^{-4}), giving useful quantitative precision for a small probe dose. They are also, however, very matrix and species dependent and vary by a factor of 10–10^7 across the matrix/species combinations of interest. Measured intensities can therefore change because of a change in

Table 5.1 Common mass interferences (after Balake *et al.* [6])

	Interfering ion	Analytical ion	Required resolution	ΔM
Error	$^{28}\text{Si}^+$	$^{14}\text{N}_2{}^+$	960	0.0146
Matrix				
Ions	$^{16}\text{O}_2{}^+$	$^{32}\text{S}^+$	1800	0.0178
	$^{28}\text{Si}_2{}^+$	$^{56}\text{Fe}^+$	2960	0.0189
	$^{47}\text{Ti}^{28}\text{Si}^+$	$^{75}\text{As}^+$	10 940	0.0069
	$^{46}\text{Ti}^{29}\text{Si}^+$	$^{75}\text{As}^+$	10 500	0.0091
Matrix + Primary	$^{29}\text{Si}^{30}\text{Si}^{16}\text{O}^+$	$^{75}\text{As}^+$	3190	0.0235
Hydrides	$^{30}\text{Si}^1\text{H}^+$	$^{31}\text{P}^+$	3950	0.0078
	$^{27}\text{Al}^1\text{H}^-$	$^{28}\text{Si}^-$	2300	0.0120
	$^{54}\text{Fe}^1\text{H}^+$	$^{55}\text{Mn}^+$	6290	0.0087
	$^{120}\text{Sn}^1\text{H}^+$	$^{121}\text{Sb}^+$	19 250	0.0062
Hydrocarbons	$^{12}\text{C}_2\text{H}_3{}^+$	$^{27}\text{Al}^+$	640	0.0420
	$^{12}\text{C}_5\text{H}_3{}^+$	$^{63}\text{Cu}^+$	670	0.0939

chemical environment, rather than a change in concentration of the analyte species.

(iii) The technique consumes sample material and the analytical precision that can be achieved in a measurement is determined by the volume of material consumed to make the measurement. In the absence of a background signal, the volume consumed will determine the detection limit. The fractional atomic concentration C_X of an atomic species X in a measurement which consumes N sample atoms is determined from a SIMS or similar analysis from:

$$C_X = \frac{n_X}{NT_X\alpha_X} \tag{5.1}$$

where n_X is the number of secondary particles (positive or negative ions, or neutrals) of X detected; T_X is the product of the collection, transmission and detection efficiencies of the spectrometer for species X; α_X is the emission or production probability for the charge state detected, and is sometimes also referred to as the ionization probability.

$$\alpha_X^\phi = \frac{\text{number of } X \text{ produced in charge state } \phi}{\text{total number of } X \text{ sputtered}} \tag{5.2}$$

It is important to note that α implicitly contains factors due to impurity migration occurring during analysis (e.g. segregation effects) and similar phenomena; T_X and α_X are difficult to measure independently. Their product, $\tau_X = T_X\alpha_X$, known as the useful yield of X, is easily obtained, and is used as a measure of instrumental quality.

(iv) Finally, the interaction between the primary ion beam and the sample leads to complex mass transport effects (Kirkendall effects [7]), atomic mixing [8], radiation enhanced diffusion and segregation etc. [9–15] in the near surface region, because of energy deposition by the beam and the accretion of probe atoms. These processes distort the sample's internal three dimensional chemical distribution prior to measurement and can lead to the development of surface topography. The effects can be minimized, but not eliminated, by the correct choice of experimental conditions.

The combination of the attributes summarized above leads to a technique with a working dynamic range (D_w) of around 10 orders of magnitude in concentration (i.e. from 5×10^{22} to 5×10^{12} atoms cm^{-3}). However, at the lower end of this range, for species with average useful yields, it is necessary to sputter about 100 µm^3 of material rapidly to get acceptable measurement precision, so high spatial resolution is out of the question. Dynamic SIMS has the following mutually exclusive ultimate specifications: detection limits between 10^{13} and 10^{16} atoms cm^{-3} for most impurities in semiconductors, a lateral resolution [16] of 20 nm and sub-nanometre depth resolution [17–19]. The combination of lateral and depth resolution achievable for a given detection sensitivity depends, of course, on the statistical precision required from the analysis and the shape of the material volume selected to satisfy this.

5.1.2 VARIATIONS ON A THEME

Dynamic SIMS is capable of five basic modes of operation, namely: micro-volume analysis, depth profiling, imaging, image depth profiling and surface spectroscopy (at sub-keV energies). Additionally, the combination of these with samples specially prepared *in-* or *ex-situ* gives rise to powerful extensions of the technique such as two-dimensional profiling and *in-situ* sectioning and imaging.

Micro-volume analysis is common in geological applications for isotope ratio measurements, for example, Betti [20], Becker and Dietze [21] and Long and Gravestock [22]. The primary ion beam is used to erode a limited sample volume, typically ~1 µm^3, and secondary ions are collected whilst ignoring their position of origin within the volume. The spectrometer may scan over a mass range to perform an impurity survey or concentrate on a few masses to compare impurity ratios from different parts of the sample. Owing to the significant species to species variation in ion yields, survey data are hard to quantify unless a reference material containing the same impurities is available. In microelectronics, because of the growing importance of polycrystalline materials in device fabrication, and the shrinkage in device size to the point where one device contains insufficient material for an internal analysis with high spatial resolution, micro-volume analysis is likely to increase in importance.

Historically, depth profiling of semiconductor samples was by far the most common and best developed mode of analysis, because most materials of interest sputter extremely well, and samples with high lateral homogeneity over large ($>1\,\text{mm}^2$) areas such as implanted wafers are available. The primary ion beam is rastered over an area typically 0.1–0.5 mm square, eroding a certain depth increment per scan, and producing a flat bottomed crater. The edges of the eroded region are carefully excluded from the analysis by ion optical means [2] or electronic gating [3] to ensure that the secondary ions collected originate only from a crater bottom. The variation in intensity of a restricted number of species as a function of ion dose is recorded, and may be converted to a profile of concentration vs. depth using suitable quantification procedures (however, in ToF–SIMS analysis the whole spectrum is collected). Depth profiling has now reached the stage where in favourable cases data can be quantified with precisions of better than 1 % in concentration and 4 % in depth. However, this requires both that accurate standards are available and that the sample contains no sharp spikes or abrupt interfaces. The development of ultra-low energy ion beam columns has meant that depth resolutions better that one nanometre can now be achieved [17–19]. Rutherford backscattering, Auger profiling and related techniques compete favourably with SIMS above 10^{20} atoms cm^{-3}, whilst secondary neutral mass spectrometry (SNMS) [21, 23, 24] and resonant ionization mass spectrometry (RIMS) [25, 26] are useful from 10^{18} atoms cm^{-3} upwards. Below this level SIMS has no competition from other chemical analysis techniques in terms of sensitivity per unit sampling volume, but may be usefully compared with electrical characterization methods such as electrochemical capacitance/voltage profiling (eCV) [27], spreading resistance profiling (SRP) [28] and deep level transient spectroscopy (DLTS) [29].

The acquisition of images in SIMS may be carried out by two different means: stigmatic imaging in an ion microscope, where the spatial information on the region of origin of the secondary ions is preserved by the mass spectrometer and mass filtered images may be projected directly onto a screen or channel plate [1] and scanned imaging in an ion microprobe, where a micro-focused primary beam is scanned over the sample surface and the secondary ion signal intensity is recorded and displayed as a function of beam position [3, 30]. With the ion microscope mode, the whole sample area may be illuminated at once and an image may be displayed rapidly. The lateral resolution achievable is limited by chromatic and spherical aberrations in the secondary ion optical system to around 0.5 µm. The dynamic range in an image element (pixel) is limited by the detector (usually a micro-channel plate combined with a resistive anode encoder), which also influences the time taken to record (rather than merely display) an image as the total incident flux for the whole image cannot exceed about $10^6 \, s^{-1}$. In the microprobe mode, the ultimate lateral resolution so far achieved is 20 nm [26, 30], with 100–200 nm as a more routine performance. Since each pixel is illuminated individually, the time taken to acquire an image of acceptable statistical precision is determined by the probe current. The image acquisition is not limited by the specification of the detector which (for a channeltron) will count reasonably linearly up to an incident flux of $10^6 \, s^{-1}$ per pixel. Most SIMS instruments can operate in the microprobe mode, but the Cameca range of magnetic sectors and the TRIFT ToF instrument from Ulvac Phi will also run in the microscope mode. Image quantification is complex and in its infancy because many contrast mechanisms exist. For example: ion yield variations, topographic contrast and chemical concentration contrast. In addition, instrumental factors such as sensitivity variations across channel plates must be accounted for [31]. Because of the fundamental limitations imposed by the sputtered volume and the number of atoms contained within it, it is only possible to combine high spatial resolution with dopant sensitivities under special circumstances (see Section 5.5.3). Competing or complimentary techniques are Auger [32], XPS imaging [33] and SNMS imaging [34].

The availability of fast microcomputers with large amounts of mass storage is progressively releasing SIMS from the need to constrain the type of analysis performed at the time the data are collected. The extension of the technique to three dimensions is a logical step that combines lateral resolution with the successive removal of layers by sputtering. Data are stored as successive image planes, preserving both the lateral and depth information for emitted ions [26, 35]. Rudenauer [35] has reviewed technical developments in imaging secondary ion mass spectrometry (SIMS) analysis.

3D SIMS analysis is now being applied to a wide range of materials problems, such as patterned polymer films [36], solder joints [37] and biological materials. Chandra [38], for example, has analysed human glioblastoma T98G cells with 3D SIMS imaging used in order to study the chemical composition of specialized sub-cellular regions, like the mitotic spindle, hidden beneath the cell surface. Pharmaceutical samples have also been studied [39].

Ultra-low energy dynamic SIMS spectroscopy [40] exploits a low energy beam to get the most stable conditions possible close to the surface, and also with the objective of sputtering out clusters which are chemically characteristic of the surface. The removal of the constraint imposed by the limiting dose for static SIMS means that adsorbed species on a surface can be cleaned away by the primary beam to reveal, for example, inorganic surface composition. This technique is being pioneered for corrosion and surface modification studies on metals and other materials.

5.1.3 THE INTERACTION OF THE PRIMARY BEAM WITH THE SAMPLE

Primary beams in common use are oxygen, caesium, gallium, and to a lesser extent, argon and these beams were the 'work-horses' in the early days of the technique [41–43], reflecting the high secondary ion yields offered with oxygen and caesium beams for positive and negative ions respectively, and the high lateral resolution offered by gallium beams. Many other beam types have been reported [44–46] with a C-60 beam developed by Hill and Blenkinsopp at Ionoptika a very recent development offering the potential for the in-depth analysis of polymeric films and biomaterials [47]. This C-60 beam was designed and tested at the same time as a gold cluster beam and was originally intended primarily for imaging. The C-60 was found to sputter very efficiently and possibly lead to higher ion yields and Hill and colleagues [48] have now developed a range of C-60 ion beam systems for use in static SIMS, dynamic SIMS and SIMS imaging, including intracellular imaging of biological compounds, and surface cleaning and depth profiling in electron spectroscopy. Recent examples of applications have included C-60(+) depth profiling for high depth resolution SIMS analysis of silicon semiconductor materials [49] and depth profiling analysis of organic and inorganic multilayer films [50]. Other beams include gold clusters [51], SF_5 and indium and bismuth clusters. Wagner [52], for example, has studied molecular depth profiling of some spin-cast polymer films by secondary ion mass spectrometry (SIMS) using an SF_5 beam. There seems to be a consensus emerging that cluster ions such as C-60 and SF_5 offer the possibility of more benign analysis so that damage sensitive materials such as polymers and biomaterials may be studied more effectively in the SIMS depth profiling mode [53, 54]. However, it is also the case that alternatives such as ultra low energy oxygen and inert gas analyses are relatively untested in this area, and a major source of damage can be electron beams used for charge compensation.

When a solid is bombarded by a beam of ions (or neutrals), the accompanying sputtering, beam induced mixing and probe atom incorporation give rise to an altered layer in the surface of the material [8, 55, 56]. The altered layer will, in general, be a disordered mixture of the original matrix elements and probe atoms. Except for favourable combinations of primary beam (species, energy, angle) and target material, the surface of the sample will become textured on the nano- to micro-scale, and this will impose a significant, and usually depth-dependent, limitation on the achievable depth resolution and quantitative accuracy. All the processes which give rise to distortions in SIMS data take place in the altered layer or at its boundaries, and the ion yields observed will be determined by the chemistry of its top 2–3 monolayers. Many studies of the chemistry and morphology of altered layers have been carried out [42,57–67], including the occurrence of profile distorting beam-induced diffusion [3, 6, 69]. The oxygen bombardment of silicon at near normal incidence has been studied in some detail because of its applications importance. Figure 5.1 shows under focus transmission electron microscopy (TEM) cross-sections of altered layers formed on silicon and silicon germanium alloy by $8\,keV\ O_2^+$ bombardment at normal incidence [67]. In the silicon case, the back interface and the surface are flat to about 1 nm which has important consequences for the ultimate depth resolution of the technique. A Fresnel fringe is consistently observed towards the back of the layer where the transition from SiO_2 stoichiometry through sub-oxides to silicon begins. Over a wide primary ion projected range (Rp) the steady state altered layer consists of a uniform surface layer with the stoichiometry of SiO_2 extending to about $2.5Rp$, followed by a layer with the stoichiometry graded smoothly from SiO_2 to Si about $0.5Rp$ thick and a relatively sharp back interface with the silicon. The TEM data show the boundaries between these regions clearly. The transition zone is not laterally homogeneous [68]. For the silicon germanium, it can be seen that the Ge exhibits a complicated segregation/precipitation behaviour in the altered layer, because the silicon oxidises preferentially. The thickness of altered

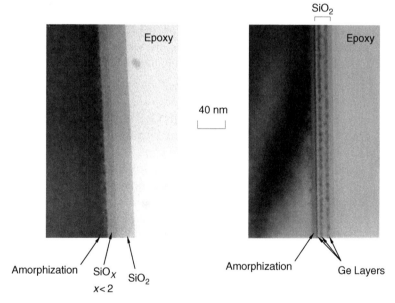

Figure 5.1 Off-focus TEM cross-section of the altered layer formed in the surface of a (100) silicon wafer and a thick SiGe alloy layer by bombardment at normal incidence with $8\,keV\ ^{16}O_2^+$ ions. Layers of similar appearance are formed over the full energy range investigated (2 keV to 12 keV). The surface and the back interface with the silicon or SiGe are flat to better than 1 nm, but in the case of the SiGe a complex internal structure has developed (TEM provided by courtesy of P.D. Augustus at GEC–Marconi Research Centre, Caswell, UK)

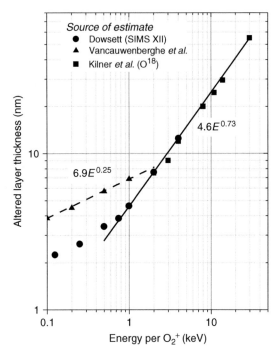

Figure 5.2 Altered layer thickness plotted as a function of incident O_2^+ energy for normal incidence on silicon (100) from the work of Kilner *et al.* [70], Vancauwenberghe *et al.* [71] and Dowsett *et al.* [72]. Note that there is considerable disagreement below 2 keV

layers in silicon bombarded with O_2^+ at normal incidence as a function of primary ion energy, is shown in Figure 5.2 which summarizes data from Kilner *et al.* [70], Vancauenberge *et al.* [71] and Dowsett *et al.* [72]. There is considerable disagreement as to what happens below 2 keV.

At high beam energies (14 keV), Beyer *et al.* [73] have shown that full oxidation is not achieved for angles greater than 20° to normal. However, as the primary beam energy is decreased below 1 keV, the angular range over which full oxidation occurs (in silicon) is extended, probably reaching beyond 45° at 250 eV (per O_2). This brings the benefit that higher erosion rates can be achieved compared to normal incidence, whilst maintaining maximum positive ion yields and the planarizing effect of full oxidation.

With the advent of devices such as the floating low energy ion gun (FLIG) [74] invented by Dowsett, it has become possible to erode the sample surface at a reasonable rate with beam energies down to below 250 eV. Near normal incidence sub-keV O_2^+ bombardment minimizes the widths of the transient regions at the surface of a shallow profile in addition to providing a very high depth resolution. At 300 eV, for example, the transient width in silicon is ~1 nm, and the ion yields are relatively insensitive to the presence of native oxide. This suggests that ideal conditions for the profiling of very shallow implants are available at 300 eV and below. Nevertheless, the transient signals reflect differences in the thickness of native oxide, as well as differences in primary beam energy and could, in principle, be used to measure process and wafer age related differences in the top few nanometres.

It should be noted that practical experience shows that the depth resolution attained using Cs^+ ions at low beam energy (impact angles between 45 and 60° to normal) is typically worse that that with O_2^+, and that at energies below 400 eV Cs tends to accumulate on the surface, depressing negative ion yields and further degrading the depth resolution. A potential solution to this problem is co-bombardment with Xe^+ to control the surface concentration of Cs.

5.1.4 DEPTH PROFILING

Most depth profiling in microelectronic materials takes place in part processed wafers where the objective is to assess the results of process steps such as ion implantation, thermal annealing, co-diffusion and epitaxial growth. There is also a growing requirement for highly localized device profiling. In these applications, precise quantification in depth and concentration is required, with a depth resolution that, as a rule of thumb, needs to be about 3 % of the active device thickness for implanted material but ideally better than 3 nm for epitaxially grown layers. Generally, a dynamic range of 10^4 is sufficient. Higher dynamic ranges can be achieved (see Section 5.1.4 below) but extreme care is required or the 5th and successive decades will be instrumental/technique responses rather than representative of concentration. Some of the parameters are now discussed in more detail. Because semiconducting materials can be grown very precisely they are the best available test structures for calibration of SIMS instruments and most of the work to date on depth resolution has used multilayer and delta-doped semiconductor structures although some work has been conducted on metallic multi-layers. Although oxygen bombardment in particular can give very high depth resolution in silicon, silicon germanium and gallium arsenide, extreme roughening limits its use in metallic structures. Winograd and co-workers [75] have pioneered the use of C_{60}^+ in this application.

Depth Resolution. Depth resolution is a measure of the ability to localize a concentration measurement at a depth and distinguish between features at different depths. It is usually expressed as a parameter extracted from the measured profile of a thin plane of impurity (a response function $R(z)$) grown by molecular beam epitaxy (MBE) or chemical vapour deposition

(CVD). (The logical definition of depth resolution as the ability to resolve two adjacent features according to, say Rayleigh's criterion, is not used partly because the response function parameters make the technique look better than it really is.) Figure 5.3 shows such a layer labeled with parameters in common use. They are decay length (λ_d) usually quoted in nm, assuming the slope to be of the form $K\exp(-z/\lambda_d)$ where z is depth and K is a constant, the decay slope (Λ_d) which is the depth over which the signal changes by a factor of 10 (nm per decade) and the standard deviation σ (nm) correctly defined as the square root of the second moment of the response function:

$$\sigma = \sqrt{\int_{-\infty}^{+\infty} z^2 R(z)\, dz} \tag{5.3}$$

but sometimes referred to as the depth over which the apparent concentration varies from 16–84 % on an interface profile (a definition valid only for a Gaussian response, and of little use in SIMS). The standard deviation of the response as measured and other parameters such as the exponential parameter from the *shallow* edge of the profile are misleading if used as resolution parameters because they contain a significant (indeed, dominant in the case of the shallow edge parameter) contribution from the true distribution. Dowsett *et al.* [76, 77] have explained this point in more detail. As a result, the decay parameters are the only ones which can be directly measured and assumed to be representative of the experimental conditions, rather than the sample.

It is important to note that in SIMS the ion and sputter yield variations across interfaces between different materials (e.g. silicon and silicon dioxide)

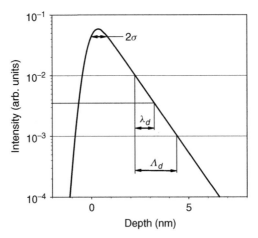

Figure 5.3　SIMS depth profile of an MBE grown narrow boron dopant distribution in silicon (nominally a single doped atomic plane or 'delta layer'). The depth resolution parameters in common use are displayed on the profile (see text for definitions)

will dominate the apparent concentration change, and resolution parameters measured on such interfaces are meaningless because the measurement violates the assumption of linearity implicit in the definition of resolution. Figure 5.4 shows the behaviour of some resolution parameters as a function of beam energy for boron in silicon and normally incident O_2^+ ions. The standard deviation shown has been corrected for its intrinsic sample-related width [76].

For a single matrix, the depth resolution is impurity species dependent since the mass transport (mixing) effects are species dependent. Depth resolution is also matrix dependent for a given set of experimental conditions. Figure 5.4 represents one of the best impurity matrix combinations. Similar performance can be achieved for silicon in gallium arsenide, but other cases are typically worse.

Early SIMS instrumentation often failed to produce a uniform primary beam dose across the analysed area, resulting in uneven erosion and consequent crater macrotopography [78, 79]. In turn, this resulted in a linear loss of depth resolution with depth. When attempting to achieve high depth resolution one still has to bear this problem in mind and beware of effects due to the trapezoidal projection of the primary beam scan when working away from normal incidence [80], high extraction fields and sparse pixelation where the overlap of the beam between dwell points in a digital raster is less than 25 %.

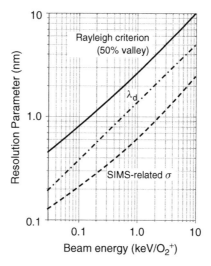

Figure 5.4 Primary ion energy dependence of resolution parameters for normal incidence O_2^+ analysis of boron in silicon. The 'SIMS-related σ' is that of the Gaussian component of the resolution function extracted according to Dowsett *et al.* [76]. A depth resolution based on a Rayleigh-like criterion (separation at 50 % valley) for two adjacent delta layers of equal areal concentration is also plotted (solid line)

More subtle problems occur because many combinations of analytical conditions (probe, angle, energy) and sample type result in the development of microtopography, often in the form of ripples. This causes sudden or progressive loss of depth resolution with depth. With high beam energies (>1 keV) catastrophic increases in surface roughness were observed for profiles deeper than 1–2 μm [81], which caused the depth resolution to deteriorate abruptly to ~1 μm (see Section 5.3.4). These effects can be alleviated by using near-normal incidence for Si and GaAs, rotating the sample in-situ, and smoothing the starting surface [82]. Cirlin [83, 84] has investigated the effects of sample rotation and sputtering conditions on the depth resolution and ion yield during (SIMS) sputter depth profiles on bulk GaAs and a GaAs (5 nm) per $Al_{0.3}Ga_{0.7}As$ (5 nm) super-lattice. Profiles without sample rotation with 1.0–7.0 keV O_2^+ show a rapid degradation of the depth resolution with increasing sputter depth. Profiles with Ar^+ show only slight degradation. Scanning electron microscope (SEM) studies indicate that degradation is associated with development of periodic surface ripples. The wavelength of the ripples is energy dependent and increases with increasing ion impact energy. With sample rotation, no degradation of the depth resolution is observed and SEM micrographs indicate that surfaces sputtered with rotation are smooth.

At low beam energies, nano-scale roughening can start right from the surface. Figure 5.5(a) shows a comparison between 500 eV O_2^+ profiles at 0° and 60° to normal for a 10 layer boron doping layer structure in silicon. The slight broadening of the deeper layers in the former is due to diffusion of boron at the elevated growth temperature. The loss of depth resolution at 60° is obvious, and is due to severe ripple formation. The surface topography of the 60° bombardment is shown after erosion of '50 nm' in Figure 5.5(b). Had the beam energy been 1 keV, the depth resolution would have remained constant with depth. Jiang [85] has discussed how, through the use of a deceleration electrode in the primary beam line of a magnetic sector SIMS instrument, an oxygen primary beam of variable energy and angle can be produced – a strategy now adopted in commercial instrumentation. The SIMS measurements of ultra-thin Ge and B layers in silicon were performed with low-energy (0.7–2 keV) and grazing incidence (50–75 degrees) ions. The depth resolution was measured on a Ge delta layer and very good decay lengths of 0.25 nm and 0.9 nm, respectively, together with a 1.6 nm full width at half maximum (FWHM) were obtained. A depth profile analysis without any appreciable loss in depth resolution was achieved down to a depth of 1 μm with a 1 keV oxygen beam at 60 degrees. Furthermore, a good dynamic range, acceptable detection limit and moderate sputtering rates were achieved in this ultra-high depth resolution mode. Juhel [86] has recently analysed an epitaxial Si multi-layer stack consisting of boron delta-doped layers separated by 6.4 nm thick un-doped films using a Cameca IMS 6f magnetic sector SIMS instrument. Using a

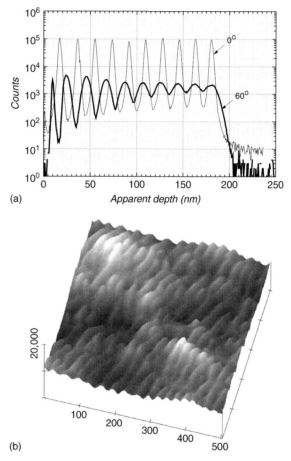

(a)

(b)

Figure 5.5 (a) A set of 10 boron delta layers in silicon with nominal 17 nm spacing profiled at normal incidence and 60° using 500 eV O_2^+. The slight decrease in modulation amplitude at normal incidence is due to real boron diffusion during growth. The distortion of the profile is due to a combination of roughening and consequent ion yield and erosion rate variations. (b) An AFM image of a crater base made at 60° at an intermediate depth of 50 nm (all dimensions in nm). The development of ripples (aligned in a direction normal to the direction of incidence of the beam) is clear. (AFM data provided by courtesy of Alan Pidduck, DERA, Malvern, UK)

low energy oblique O_2^+ beam, the boron depth resolution was improved from 1.66 nm per decade at 500 eV down to 0.83 nm per decade at 150 eV. Very low impact energy O_2^+ bombardment induced almost full oxidation of silicon and oxygen flooding was not needed in the analytical chamber to get smooth sputtering of silicon at 45° incidence. Chanbasha and Wee [87] have used an Atomika 4500 SIMS tool with O_2^+ primary ions at an ultra-low energy (E_p) of 250 eV and incidence angles between 0 and 70 degrees without oxygen flooding. A sample with 10 delta layers of

$Si_{0.7}Ge_{0.3}$ nominally grown 11 nm apart was used. For characterization of ultra-shallow junctions a beam incident normally provided the narrowest surface transient of 0.7 nm. The depth resolution denoted by the full width at half maximum (FWHM) of the $^{70}Ge^+$ peaks was comparable for both incident angles close to 0° and 40° at 1.6 nm and 1.4 nm, respectively. However, in the case of MQW profiling, where the quantum wells are normally located deeper, an incident angle of 40° was preferable. At this angle, the average sputter rate of 47 nm min^{-1} nA^{-1} cm^{-2} was significantly higher, more than double that at close to normal incidence, and a better depth resolution was achieved with a decay length λ_d of 0.64 nm compared to 0.92 nm at normal incidence. Moreover, the dynamic range possible was also better close to 40°. Incident angles close to 60° were not ideal, even though there was no sign of the onset of roughening. Although the higher sputter rate was an advantage, the depth resolution deteriorated. Further useful studies include work on semiconductors [88], polymer multilayers [89] and development on nanostructured layers for SIMS tests [90]. General discussions of sub-keV depth profiling are to be found in Hofmann [91] and Dowsett [92].

It is obvious that the depth resolution cannot be better than the sampling depth (the depth sputtered to accumulate a single data point) so erosion and sampling rates need to be matched to the sample. To achieve the highest depth resolution, it is necessary to start with an atomically flat surface and work with a primary beam/matrix combination which avoids the development of beam induced surface topography and the occurrence of segregation. For rough materials, or those where the surface roughens unacceptably during sputtering, it is worth considering the use of chemical and mechanical sectioning techniques to approach or expose the layer of interest. Some alternative techniques for dealing with this situation are described in Section 5.4.

Dynamic Range and Memory Effects.   Two criteria give an indication of an instrument's ability to profile through high to low concentrations for a particular species. They are the surface to background dynamic range (D_s) and the peak to background dynamic range (D_p), and are defined as the difference in concentration (or measured signal intensity) between the surface or peak levels respectively, and the steady state background level. Either the latter is simply caused by the bulk impurity concentration in the material, or it is due to intrinsic performance limitations in the instrument. Trivial, but frequently encountered limitations are shot noise in the detection system (which should not exceed 0.1 ions s^{-1} equivalent signal), and mass interferences. Another important factor, especially when gaseous and associated impurities (e.g. H, C, N, O) are examined, is adsorption onto the instantaneous crater base from the residual gas phase [93, 94] which may be alleviated by improving the vacuum and increasing the erosion rate. Ultimately, D_s and D_p are a measure of two fundamental factors: the quality of the gating (crater wall and surface concentration rejection) and re-deposition on the instantaneous crater base by species sputtered from the crater walls and surrounding electrodes [95–97]. For the latter effect, Wittmaack [96] has derived a model based on approximating the electrodes by a hemisphere radius R centred on the analysed area. Re-deposition effects may be alleviated by replacing the local electrode surfaces between analyses, or even during an analysis [98].

Amongst the first authors to study the effects which limit the dynamic range were Huber *et al.* [93], Wittmaack and Clegg [99] and McHugh [100]. Wittmaack and Clegg observed that energetic neutral atoms striking the target surface were largely responsible for limiting the dynamic range to 3–4 orders of magnitude. They achieved the six decade profile shown in Figure 5.6 by bending the primary beam by 4° just prior to the final lens. The dotted line shows the recovery of an extra 0.5 decades by subtraction of

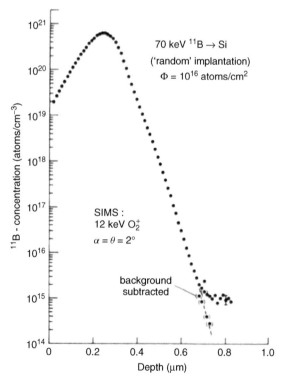

Figure 5.6   Six order of magnitude depth profile obtained by Wittmaack and Clegg in an electronically gated Quadrupole SIMS instrument (Atomika DIDA) using an ion column with a small bend just before the final lens to suppress neutrals and scattered ions in the beam (reproduced from McHugh [100] by kind permission of the authors and the American Institute of Physics)

the residual background. Nowadays routine results are more typically 5–6 decades. Von Criegern reported a value of seven decades using specially prepared samples [101]. The dynamic range can sometimes be improved by reducing the crater area by collapsing the scan at an appropriate point during the profile. This increases the erosion rate, diluting the effects of residual gas adsorption [94] and re-deposition, and reduces the current density falling on the original surface and crater walls. Wangemann and Langegieseler [102] in their studies of high dose arsenic and antimony implants in silicon, which have high surface concentrations of dopants, have shown that special sample preparation techniques can improve the dynamic range by up to three orders of magnitude. McKinley [103] has described optimization of the low energy analysis of boron and arsenic implants with either oxygen or caesium low energy beams on a Cameca IMS 6F and Napolitani [104] has studied how best to analyse ultra-shallow boron implants.

5.1.5 COMPLIMENTARY TECHNIQUES AND DATA COMPARISON

Analytical science has always required a multi-technique approach and there are many potential companion techniques for SIMS. There are also numerous reasons for using them: (a) extra information is required to inter-pret the SIMS data (obvious examples of this are the various means for calibrating depth and quantifying reference materials and the parallel use of ion scattering techniques to assess the surface and internal sample com-position during analysis); (b) structural information is required in addition to chemical information (to measure, for example, crystallography, strain, etc.); (c) SIMS is inappropriate in some part of the concentration range (for example, a non-linear or even multi-valued relationship between sig-nal and concentration can make SIMS data impossible to interpret at high concentrations – signals can go up when concentrations are going down, etc.); (d) in-situ chemical information specific to time dependent changes in a controlled environment is required (SIMS needs the sample to be placed in a vacuum, so at best it can only measure snapshots of environmen-tal chemistry, and these may be distorted or destroyed by the vacuum requirement) – ultimately, non-destructive analysis is required so that SIMS must help to establish the parameters of a non-destructive method or sensor using simulated samples (although a typical dynamic SIMS analysis only requires pico-litres of material, just cutting up samples to put them into the instrument can be too destructive to contemplate); (e) SIMS measures atomic or molecular concentrations but the level of electrical activity of an impurity is required (it is the carrier concentration in a semiconductor which matters – donor or acceptor atoms are a necessary inconvenience, but the spatial distributions of carriers and their associated impurity atoms can dif-fer significantly). In general different analytical techniques are combined so that their mutual foibles and imperfections are accounted for. Typical combinations with SIMS include: (a) Rutherford backscattering spectroscopy (RBS) [106] and medium energy ion scattering (MEIS) [107, 108] (structure, high concentration profiles, internal profiles during SIMS, possible air-side operation for RBS) – see Chapter 6; (b) low energy ion scattering [109] (surface structure and composition, e.g. during SIMS analysis by scattering the primary ions or an auxilliary probe); (c) TEM [80] (layer thickness, separation and crystallography, precipitate structure, delta-layer studies); (d) energy dispersive X-ray analysis [110] (EDX) (micro-scale bulk composition); (e) Auger and photoelectron spectroscopies [111] (matrix level depth profiling with chemical shifts supplying some binding information – but beware of calibration problems due to electron backscattering from interfaces) – see Chapters 2 and 3; (f) X-ray diffraction [112] (XRD) (layer separation and thickness calibration, bulk composition and crystallography, strain, all available without sectioning (unlike TEM) and in air, or even in a controlled environment); (g) finally in this non-exhaustive list – carrier profiling and imaging techniques such as scanning spreading resistance microscopy [113] (SSRM) and electrochemical capacitance voltage profiling [114] (eCV).

Figure 5.7 shows a high resolution TEM of a boron delta layer structure in silicon and a superimposed (inverted) density scan. Because the magnification of the microscope can be absolutely calibrated with reference to the silicon lattice, the density scan provides a highly accurate method of calibrating the SIMS depth scale for the same sample, and qualifying the sample as a depth reference [76, 80].

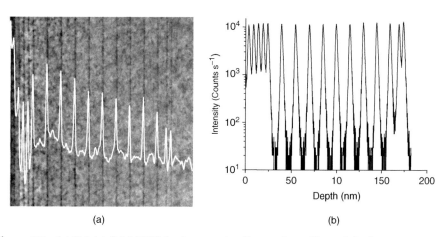

(a)   (b)

Figure 5.7   (a) Bright-field XTEM micrograph of boron in a silicon delta layer structure 611/08 (nano-Silicon group, University of Warwick) with inverted intensity scan. (b) SIMS depth profile taken using 500 eV normal incidence O_2^+ and depth calibrated from the XTEM data using the protocol in Dowsett *et al.* [76] and Kelly *et al.* [80] (XTEM data provided by courtesy of P.D. Augustus and R. Beanland, at GEC–Marconi Research Centre, Caswell, UK)

5.2 Areas and Methods of Application

5.2.1 DOPANT AND IMPURITY PROFILING

Dopant and impurity profiling of semiconductors reveal the strengths of the technique. For a single matrix sample, a depth resolution of 1 nm, an analytical precision of 10–50 % and a detection limit of 10^{17} atoms cm^{-2} over a depth of about 1 µm, the technique generally provides an extremely rapid analysis for almost the entire Periodic Table (<10 min per sample). However, many different strategies are required to obtain the full performance of the technique and other materials can be far more challenging.

Mass Interference.   Mass interferences, i.e. limitations in dynamic range due to the presence in the sample of more than one species with the same nominal mass, are common in SIMS analysis. The classic examples [115] are arsenic and phosphorous in silicon and silicon dioxide where ^{75}As$^+$ interferes with ^{29}Si^{30}Si^{16}O$^+$ and ^{31}P$^+$ interferes with ^{30}Si^{1}H$^+$, and arsenic in silicon germanium alloy profiled using oxygen where ^{75}As$^+$ interferes with ^{74}Ge^{1}H$^+$. A variety of different countermeasures are available; use of high mass resolution in the DFMS (and now the TOF) is one, and using energy discrimination to distinguish between atomic and molecular ions [115] is another. Alas, this latter method will not be effective with low primary beam energies, as the secondary ion energy spectra become narrower and indistinguishable from one another.

Use of Molecular Ion Detection.   Detection limits can in some cases be improved by monitoring a molecular ion which is a combination of either the matrix or the primary beam ion and the appropriate impurity ion. This strategy may both solve mass interference problems and increase sensitivity as in the analysis of ^{74}Ge implanted into InP, where the Ge profile is limited by a background at ^{74}(P$_2$C)$^-$. Choosing the molecular ion ^{105}(PGe)$^-$ improves the detection limit by a factor of 100, made up of a ten-fold increase in the ion yield and a ten-fold decrease in the background level, this time of (P$_3$C)$^-$.

The existence of microstructure (e.g. precipitates) in an otherwise discrete random impurity distribution can sometimes be inferred from the relative behaviour of single and cluster ion profiles [116].

The method also facilitates the detection of both electropositive and electronegative dopants when using Cs bombardment and an important development has been the use of molecular secondary MCs$^+$ ions during SIMS depth profiling with a caesium beam. It has been demonstrated that secondary ions such as MCs$^+$ and MCs$_2$$^+$ can exhibit a reduced (and occasionally an insignificant) dependence on matrix composition. Gnaser

[117], one of the earliest to investigate this phenomenon, originally stated that *the yields of these rather ubiquitous species exhibit little or no dependence on sample composition (matrix effect) even in the presence of electronegative elements and are thus well suited for quantitative SIMS evaluation. Specifically, for series of binary and ternary systems (a-Si:H, a-SiGe:H, a-SiC:H, and HCN) composition independent relative sensitivity factors could be established; thus, quantification by means of a single standard is feasible* which was perhaps a little optimistic. Nevertheless, much subsequent work has demonstrated the usefulness of the method and established that the mechanism is effectively postionization by means of Cs^+ attachment.

Multilayer Samples. All of the problems that are present in the analysis of single matrix samples occur and indeed are often exacerbated in multi-matrix samples. Non-equilibrium regions similar to that at the surface will occur at interfaces. Spurious concentration spikes may also occur; for example, as segregating impurities from one layer meet a barrier formed by another, or because an ion yield enhancing impurity is trapped at the interface. Background levels may vary from one matrix to another, as different mass interferences come into play in each matrix. If one is pursuing a particular analyte species across matrices, then its ion yield will differ in each. The best experimental conditions and quantification method will be strongly problem dependent. A typical example of a two matrix quantification scheme for dopants in the SiO_2/Si system is given by Spiller and Davis [118] and for boron in $SiO_2/Si/SOS$ and aluminium in Si/SOS by Dowsett *et al.* [119, 120]. Development of the $Si_{(1-x)}Ge_x$ system for device applications over the last 10 years has lead to several protocols for measuring the Ge content [121, 122], just as earlier work focused on protocols for determining x in Al_xGa_{1-x} As layers [123, 124]. Companion techniques, notably double-crystal X-ray diffraction, and RBS are used to establish reference levels in a range of samples.

Techniques such as RIMS and SNMS are ideal companions to SIMS where quantitative data from interface regions are to be obtained. Figure 5.8 shows an early example: SIMS and RIMS profiles, obtained by Downey [125] from a GaAs/AlGaAs heterojunction base transistor structure (HBT). The beryllium SIMS profile (bold line) shows a sharp spike at the emitter/base interface but the RIMS profile (thin line) shows no such feature and demonstrates that SIMS is misleading in this case.

5.2.2 PROFILING HIGH CONCENTRATION SPECIES

In the dilute concentration regime (<1 %) the linear relationship between the amount of dopant and its measured intensity makes quantification straightforward. However, when levels are in excess of 10^{20} atoms cm^{-3} this

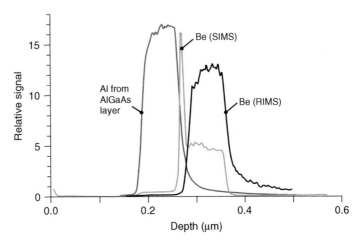

Figure 5.8 Aluminium and beryllium profiles from a GaAs/$Al_{0.3}Ga_{0.7}$As HBT structure. The bold line shows a Be^+ SIMS profile obtained using O_2^+ ions, which exhibits a sharp concentration spike at the emitter/base interface, but the RIMS profile (thin line) has no such spike. The spike is therefore a SIMS transient and not a measure of the true chemical profile for Be (after Downey *et al.* [122])

linear relationship breaks down, leading to the so-called 'matrix effect' [126, 127]. Secondary ionization probabilities are determined by the perturbation of the electronic configuration of the sputtered particle during sputtering, and the starting point for this process is the local electronic configuration just prior to sputtering. For an isolated impurity, this, in turn, is determined by the matrix equilibrium with the probe, and stays constant if the matrix stays constant. However, at high concentrations, impurity atoms will both influence each other, and modify the matrix. Therefore, ion yields are determined by the instantaneous chemistry of the top 2–3 monolayers of the sample, and may, for example, show a dependence of the form [57]:

$$Y_i^+ = K_i C_j^n(\phi) \tag{5.4}$$

where Y_i is the ion yield for species i, K_i is a constant, $C_j(\phi)$ is the primary ion dose dependent surface concentration of species j (where j may equal i) and n may be as high as 4 (but is also dependent on C). It is not surprising, therefore, that a change of matrix composition, intrinsically altering the surface chemistry and the degree of retention of the probe atoms [128] and surface bond energies [129] have a strong influence on ion yields. The variability of ion yields represents a major stumbling block in quantifying data from multi-layer samples and from samples in which dopant concentrations are sufficiently high that the assumption of a dilute system no longer holds true.

High impurity levels ($>10^{20}$ atoms cm^{-3}) frequently occur in devices containing ultra-shallow implants. The dopant itself may be at sufficiently high

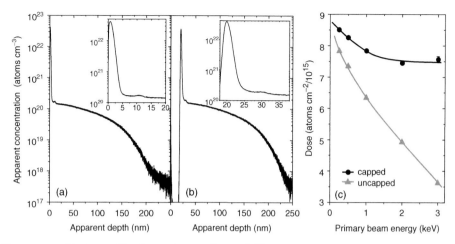

Figure 5.9 (a) A profile through a 10^{16} cm^{-2} ^{11}B implant into (100) Si after rapid transient annealing at 1100°C. Neither the depth nor concentration scales have been corrected for transient effects. The extremely sharp surface spike is boron silicide which has precipitated during the thermal process. (b) The same sample profiled through an amorphous silicon cap. Note that the spike (inset) is closer to its true width, as the erosion rate had stabilized in the cap. (c) Apparent doses measured as a function of beam energy for the capped and uncapped samples. The capped sample tends to the correct dose at higher beam energies as the silicide matrix is diluted by atomic mixing before it is sputtered. (Sample courtesy of Eric Collart, Applied Materials Ltd.)

concentration to cause matrix effects, especially after thermal processing and at the surface. Figure 5.9 shows 250 eV O_2^+ depth profiles of a high dose boron implant into silicon [130], after rapid thermal processing at 1100°C. Both the concentration and depth scales are labelled 'apparent' because no corrections have been made for erosion rate changes in the transient region, or ion yield changes. Figure 5.9(a) shows a direct profile, whilst Figure 5.9(b) shows the same sample profiled through a ~20 nm cap of amorphous silicon. By comparing the inset figures one can see the effect of the distortion due to the high transient erosion rate. Figure 5.9(c) is the apparent dose measured as a function of beam energy. XTEM shows that the boron has precipitated to form silicon boride at the surface. For the capped sample, low beam energies overestimate the dose because the silicide has a different ionization probability for sputtered boron. Higher beam energies get the dose correct, because the silicide is remixed and diluted before it is sputtered (cascade dilution, first described by Williams *et al.* [131]). The uncapped dose is always wrong, because of both matrix and transient effects.

Materials science presents the SIMS analyst with many other examples of multi-layer structures. For example, Montgomery [132] has shown how a multilayer superconducting structure can be accurately depth calibrated by correcting for the sputter rates of the individual components. The film was a multilayer composed of $YBa_2Cu_3O_7$ – delta/10% cobalt-doped $YBa_2Cu_3O_7$ – delta/$YBa_2Cu_3O_7$ – delta (YBCO/Co–YBCO/YBCO) which had been laser-ablated on $LaAlO_3$. The thicknesses of the layers were also validated by using a focused ion beam (FIB) instrument to mill a cross-section, which in turn was then imaged.

SNMS and RIMS techniques, employing post-ionization of the neutral component of the sputtered flux, have potential advantages over conventional SIMS for the analysis of matrix or high concentration species. Because the processes of sputtering and ionization are decoupled in SNMS the sputtered neutral signal of an element X shows much less dependence on matrix composition than the corresponding sputtered ion signal [125, 133].

Post-ionization techniques include lasers [133], electron impact ionization [134, 135] and the use of a hot electron gas [136, 137], multi-photon resonance [138, 139], non-resonant multi photon ionization [140] and thermal post-ionization [141].

5.2.3 USE OF SIMS IN NEAR SURFACE REGIONS

The near surface is a very important functional part of many materials systems and it will generally be different from the bulk in terms of chemical composition, defect chemistry and electronic structure due to processes such as oxidation, diffusion and segregation, combined with any contaminating layers due to sample handling, preparation for analysis and adsorption from the vacuum. For example, many of the organic solvents used for cleaning surfaces prior to analysis contain sodium (and oils, water, etc.) at low levels. When the solvent film evaporates it leaves behind at least a monolayer or so of concentrated contamination. The extent of the near surface will vary from system to system but will generally span the length metric from nanometres to microns. At the nanometre end, adsorbates and handling contamination become increasingly significant. Examples of the near surface include the thin oxide film that develops on most metals (native oxide), the alkali-depleted layer on most glasses and the doped region in modern semiconductor devices.

Primary ion energies used in routine depth profiling lie in the range 0.2–14 keV at angles from 60° to normal incidence. Under these conditions, a few nanometres are usually sputtered before equilibrium is achieved and the variations in secondary ion signal, known as surface transients, are at best due to changes in sputter rate and the degree of ionization, but can also occur because of monolayer levels of contamination. Special care must be taken if accurate quantifiable data are to be extracted from the top 50 nm, but there is no known method for quantification in the transient region.

The most elegant method for recovering a profile from the near surface region is to coat the sample with a few tens of nanometres of similar material [130, 142, 144, 145] as demonstrated in Figure 5.9. This side-steps the problem of quantifying transient signals by physically separating the surface transient from the distribution of interest and allows the erosion rate to equilibrate before the instantaneous surface reaches the now buried impurity distribution. Although Clegg [145] has suggested that contamination spikes introduced by the evaporation of the over-layer might provide a convenient method of identifying the interface, in practice one finds that interfacial spikes in different species do not necessarily line up, and it is not obvious which part of the spike profile coincides with the original surface. Valizadeh [146] has used an 10–15 nm isotopically pure Si-28 layer to conduct some fundamental SIMS studies on the Cs beam induced altered layer in silicon. Ng [147] has studied the effectiveness of the silicon capping technique in ultra-shallow SIMS depth profiling of silicon using a 1 keV O_2^+ beam at 56 degrees incidence angle. It was shown that a capping layer of at least 20 nm was needed to ensure a steady-state erosion rate when profiling the doped region. Miwa [148] has capped silicon with amorphous silicon and has successfully measured shallow depth profiles for ultra-shallow B and P.

Analysis of shallow samples at different primary energies can reveal the magnitude of profile broadening effects. If the profile remains unchanged, it is probable that beam-induced broadening is small. The profile shape which would be obtained by using primary energies below the practical minimum ($\sim$200 eV) can be estimated by extrapolating some parameters of the distribution to estimate the 'true' or undistorted profile shape. Figure 5.10(a) shows profiles of a 5 keV As implant analysed using 1 and

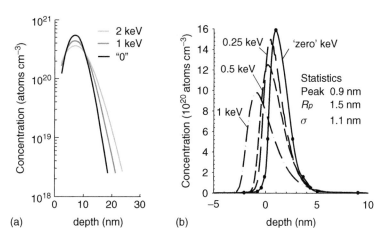

Figure 5.10   (a) A 'zero beam energy SIMS' shallow arsenic profile in silicon (solid line) synthesized as a Pearson IV distribution by extrapolation of the measured moments from distributions fitted to profiles taken with 1 keV and 2 keV O_2^+ ions at 45° incidence (after Clegg [145]). (b) Similar treatment [17] of a 200 eV 5×10^{14} cm^{-2} ^{11}B implant, with a 20 nm cap

2 keV O_2^+ ions at 45° obtained by Clegg [145]. The solid line represents a Pearson IV distribution constructed on the basis of the extrapolated moments; the true As profile probably lies between this curve and the curve for 1 keV O_2^+. Figure 5.10(b) shows this 'energy sequencing' idea applied to the peak region of a capped 250 eV boron implant profiled with 0.25, 0.5 and 1 keV O_2^+ at normal incidence [17]. Note how the apparent depth of the profile increases as the beam energy decreases.

SIMS measurements of thin surface layers have been extended into many different areas of materials science. Rees [149] has shown how it is even possible, using ultra-low energy SIMS depth profiling, to measure the distribution of elements within an oxide film on stainless steel. She showed that the chromium and iron were partitioned in such an oxide. Fearn [150] has used ultra-low energy SIMS depth profiling to determine sodium diffusion profiles in museum glass – of critical importance in conservation studies to save irreplaceable glass collections from 'glass disease' (see Figure 5.11). When glass is stored, it is exposed to repeated cycles of humidity and temperature which lead to the leaching out of alkali metals from the glass network. The resulting surface depletion, especially of sodium, produces a sol–gel layer which cracks when it dries out allowing humid air to reach deeper into the glass causing further depletion. The cracks multiply and propogate ('crizzling') and the artefact ultimately disintegrates. Venetian and similar glasses are especially prone to this

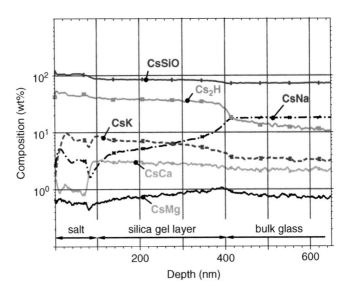

Figure 5.11 Positive ion TOF–SIMS depth profile of a 'crizzled' Venetian glass surface showing sodium depletion and the formation of a salt crust. The sample was sputtered with 1 keV Cs^+ and analysed with 25 keV Bi_3^+ (depth profile provided by courtesy of Ion-TOF GmbH)

problem. Work like that shown in Figure 5.11 is a first step to correctly diagnosing and controlling the problem. The very high depth resolution in ultra low energy SIMS allows ageing of the glass to be conducted under normal storage conditions, rather than through accelerated ageing [151].

5.2.4 APPLICATIONS OF SIMS DEPTH PROFILING IN MATERIALS SCIENCE

In the 1970s and 1980s SIMS depth profiling was mainly used to solve semiconductor problems, but in the last two and a half decades the uses have become far more diverse. The high sensitivity and excellent depth resolution of the technique have been used to investigate processes such as oxidation, diffusion and segregation in a very broad range of materials including functional ceramics, biomaterials, geological materials, metals (including aerospace alloys), plastics and polymers and glasses. SIMS depth profiling has also been used to look at surface processes in museum objects (cultural heritage). The broad range of application is best appreciated by looking at the proceedings of the more recent SIMS conferences [152–157]. The range of applications of SIMS in materials science has been reviewed recently by McPhail [158].

5.3 Quantification of Data

5.3.1 QUANTIFICATION OF DEPTH PROFILES

A depth profile is obtained as a set of discrete ordinates (usually a count) almost always obtained at uniform increments of primary ion dose $\Delta\varphi$. Each ordinate is obtained by accumulating counts across an interval $\Delta\varphi'$ where $0 < \Delta\varphi' \leq \Delta\varphi$. For example, in a depth profile with two mass channels one might have $\Delta\varphi' = \Delta\varphi/2$ in a QMS or DFMS. In a TOF using the dual beam technique [5] $\Delta\varphi' << \Delta\varphi$ but data for different channels is obtained simultaneously. For a stable primary ion source, time intervals Δt etc. may be substituted for dose intervals. Then, the raw data in a SIMS depth profile consist of set of n ordinates $Y_X(t_i)$ where Y is the detected count for some species X at time t_i. For many purposes, for example qualitative comparisons between process steps, this format is adequate. For more rigorous applications it is necessary to convert t to z where z is depth and $Y_X(t)$ to $C_X(z)$ where C is concentration. This is almost invariably done by means of two constants: the erosion rate $\Delta z/\Delta t$, which is used to convert time to depth, and a calibration constant derived from the useful ion yield τ_X which is used to convert Y_X to C_X. This technique results in a small error in the depth (the differential shift) if the erosion rate in the pre-equilibrium region differs from the bulk. It must be adapted to multi-matrix samples by using values for τ_X and $\Delta z/\Delta t$ appropriate to each layer and there will be a cumulative error if the erosion rate varies at each interface. These depth errors can usually be minimized by using low primary ion energies and/or non-normal incidence. The technique does not quantify the concentration in the pre-equilibrium region or at matrix interfaces, but careful choice of experimental conditions can minimize problems here. Most of the variation in quantification procedures is in the method used for the determination of the τ_X.

Use of Reference Materials. A reference sample for accurate SIMS quantification should be as close to the analyte in matrix stoichiometry and morphology as possible and the total concentration of impurities to be quantified should not exceed 1 % (unless it is matrix species which are to be quantified). The accuracy of the technique depends critically on the close matching of the analysis conditions for the standard and the unknown. Apart from obvious factors such as the use of identical probe conditions (energy, current, crater size, bombardment angle), it is vital that constant spectrometer transmission is achieved. The mounting and orientation of the sample and reference with respect to the extraction system should be identical. For example, in instruments which analyse through windows in a thin foil mask on the sample surface, it is important that the foil is flat and that the same part of the window (preferably the centre) is used for each analysis [159]. This ensures that the effect of the distortion of the extraction field by the edges of the mask is reproduced.

Ion implants of known dose are amongst the most commonly used reference materials. They permit the measurement of τ_X, background levels and dynamic range; they can act as a useful guide to data quality and instrumental performance. An implanted sample will contain damage and amorphized regions which may not be present in the analyte matrix and which can, therefore, potentially limit the quantitative accuracy [160]. The parameter τ_X is implicitly determined from the Y_X profile of the implant after depth calibration:

$$\tau_X = \frac{\Delta z}{\Delta z'} \frac{1}{AD} \sum_{i_{min}}^{i_{max}} Y_X(z_i) \tag{5.5}$$

where A is the area of the crater from which ions are detected (the gated area), D is the implanted dose, i_{min} is the index of the shallowest ordinate, chosen to avoid problems with transient surface spikes, i_{max} is the index of the deepest ordinate and $\Delta z/\Delta z' = \Delta\phi/\Delta\phi'$ accounts for the fraction of the sputtering spent in the X mass channel. This expression is commonly written as an integral, but that is not correct, except in the case $\Delta\phi' << \Delta\phi$, as the accumulation of counts is already an integration process. The useful yield can be explicitly used in Equation (5.1) as an absolute sensitivity factor to obtain the fractional concentration C_X^u of X in the unknown sample at a specific depth:

$$C_X^u(z_i) = \frac{Y_X^u(z_i)}{A \Delta z_u' \rho \tau_X} \tag{5.6}$$

where ρ is the atomic density (atoms per unit volume) for the sample and $\Delta z_u'$ is the depth eroded recording $Y_X^u(z_i)$. However, a more usual approach is to take the ratio of $Y_X(z)$ to a matrix element signal, to give a relative sensitivity factor S_X ostensibly free from the effects of changes in the primary current and other instrumental drifts:

$$S_X = \frac{\Delta z''}{\Delta z'} \frac{1}{AD} \sum_{i_{min}}^{i_{max}} \frac{Y_X(z_i)}{Y_M(z_i)} \tag{5.7}$$

Here, Y_M is the matrix signal recorded as close as possible to depth z_i (bearing in mind that in most spectrometers the signals are measured sequentially) and $\Delta z''$ is the depth increment over which Y_M is recorded. Then, for the unknown sample:

$$C_X^u(z_i) = \frac{1}{A \Delta z_u' \rho S_X} \frac{Y_X^u(z_i)}{Y_M^u(z_i)} \tag{5.8}$$

Samples which have been ion implanted as part of their processing are 'self-standardizing', provided that none of the implant is lost during annealing, and precipitation and segregation effects do not result in regions with different degrees of ionisation. It is occasionally convenient to implant the standard directly into the analyte [161]. This method is useful for samples doped with a nearly uniform concentration of a particular element; the implant distribution then stands out clearly. The modified sample is profiled to a depth enabling the uniform signal to be accurately measured. A variation of this 'standard addition' method involves the use of minor isotopes [162]. This can be particularly useful in avoiding mass interferences, high instrumental backgrounds or to quantify high concentration levels. It is possible to use the primary column of the SIMS instrument as the implanter itself [163–166]. A particular advantage for microelectronic samples is the ability to implant small areas at low energies (0.2–30 keV) with large areal doses.

Uniformly Doped Materials. Uniform samples such as substrates and epitaxial layers make useful and inexpensive reference materials. They are easy to characterize using electrical measurements, XRD, RBS, photoluminescence and bulk chemical analysis etc., as appropriate. For example, sheet resistance can be used to establish the carrier concentration n_c in a uniformly doped semiconductor, and if the level of activation σ is known, the dopant concentration (atoms/unit volume) is:

$$C_X = n_c/\sigma \tag{5.9}$$

From the above, τ_X is then obtained from:

$$\tau_X = \frac{\displaystyle\sum_i Y_{X_i}}{C_X A \Delta z} \tag{5.10}$$

where the sum is taken over all the isotopes of X present in the sample and Y_{X_i} is the count for the ith isotope of X accumulated for the sputtering of a depth Δz. Note that if the isotopic abundance is known, useful yields for particular isotopes can be determined. However, it is not safe to assume that tabulated isotopic abundances hold good for a particular sample and detection efficiencies can also vary for different isotopes, especially at low mass. As before, τ_X may be used directly as an absolute sensitivity factor. Otherwise, by ratioing the measured signal for the impurity to that for a matrix channel Y_M with fractional concentration C_M (taking care to avoid detector saturation with the latter), a relative sensitivity factor (RSF) S_X may be determined:

$$S_X = \frac{C_M}{C_X} \frac{\displaystyle\sum_i Y_{X_i}}{Y_M} \tag{5.11}$$

so that the concentration C_X^u of the species in the unknown sample is

$$C_X^u = \frac{C_M}{S_X} \frac{\displaystyle\sum_i Y_{X_i}^u}{Y_M^u} \tag{5.12}$$

Use of an RSF allows the standard and unknown to be analysed using different primary ion currents and analysed areas, provided that there are no current density or charging effects, and the ratio of the volumes sputtered in the impurity and matrix channels is the same for both. Simons [167] has shown that RSFs can be transferred to some extent between instruments, with errors in quantification of up to a factor of two.

Delta Layers, Response Functions and Deconvolution.   Techniques such as MBE and CVD can produce impurity layers which approach a single atomic plane in thickness. Where the layers are dilute, their SIMS depth profile can be shown to be the convolute of sample-dependent and SIMS-dependent parts [76, 77]. The SIMS dependent part is known as the response function. Where the layers are not dilute, the SIMS profile will still contain sample and SIMS dependent elements, but the inherently non-linear nature of the sputtering and ionization processes means that separating these is non-trivial or even impossible. Delta layers can be used as references in three different ways. (i) Where the sheet concentration of the delta is known, it can be used as a concentration reference in exactly the same way as an ion implanted sample. (ii) If the depth of the delta (or of each delta in a set) is known then it can be used to establish a close to absolute depth scale [76, 80, 168] – as shown in Figure 5.7. However, it is essential to have some way of associating some feature of the sample-related part of the profile with its true position – a feature depth indicator (FDI) [76]. (iii) If the response function can be extracted, then it can be used to establish the depth resolution parameters and for deconvolution in linear systems (see below).

The original concept of the response function in SIMS depth profiling, and its potential for deconvolution was largely due to Clegg *et al.* [169, 170]. A useful functional form, R_X, which appears to fit many measured deltas in different materials is found by convolving two truncated exponentials and a Gaussian [171]:

$$R_X(\varphi) = e^{\varphi/\lambda_g} \Big|_{\varphi=-\infty}^{0} * e^{-\varphi/\lambda_d} \Big|_{\varphi=0}^{\infty} * e^{-\varphi^2/2\sigma^2} \tag{5.13a}$$

$$= k\big[(1 - \operatorname{erf} \xi_g)e^{\zeta_g} + (1 + \operatorname{erf} \xi_d)e^{\zeta_d}\big] \tag{5.13b}$$

where

$$\xi_g = \frac{1}{\sqrt{2}}\left(\frac{\varphi}{\sigma} + \frac{\sigma}{\lambda_g}\right) \quad \xi_d = \frac{1}{\sqrt{2}}\left(\frac{\varphi}{\sigma} - \frac{\sigma}{\lambda_d}\right) \tag{5.13c}$$

and

$$\zeta_g = \left(\frac{\varphi}{\lambda_g} + \frac{\sigma^2}{2\lambda_g^2}\right) \quad \zeta_d = \left(-\frac{\varphi}{\lambda_d} + \frac{\sigma^2}{2\lambda_d^2}\right) \tag{5.13d}$$

where φ is dose, time or depth, λ_g and λ_d are the exponential parameters and σ is the standard deviation of the Gaussian (see Figure 5.12(a)). The cusp of the double exponential forms a convenient FDI, φ_δ. The position of the centroid is easily found by integrating Equations (5.13a,b) to find the first moment:

$$\langle\varphi\rangle = \varphi_\delta + \lambda_d - \lambda_g \tag{5.14}$$

Figure 5.12(b) shows the use of Equations (5.13a,b) in the construction of an energy sequence for the measured response. The extrapolation to zero beam energy is then as close as one can come to the true impurity distribution in the delta layer. For boron in silicon (at least) it has been shown that R_X can be analytically deconvolved into sample and SIMS related contributions by studying the dependence of the parameters on the SIMS and growth conditions [76, 77, 172]. The growth exponential is found to be sample related, whilst the decay exponential is entirely SIMS related (λ_g is due to segregation during growth and λ_d is the SIMS decay length). Then the chemical distribution R_C and the SIMS response R_S are:

$$R_C(\varphi) = k_C\big[(1 - erf\xi_C)e^{\zeta_C}\big] \tag{5.15a}$$
$$R_S(\varphi) = k_S\big[(1 + erf\xi_S)e^{-\zeta_S}\big] \tag{5.15b}$$

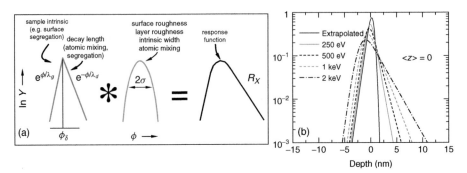

Figure 5.12 (a) The measured SIMS response function for a delta layer. Note that this form contains both sample and SIMS-related contributions, and is not suitable as it stands for e.g. deconvolution. (b) Measured and fitted responses for boron delta layers grown by MBE in silicon with beam energy for O_2^+ at normal incidence as a parameter. The '0' energy profile is extrapolated from the energy sequence and is the pure sample related profile

where the parameters are found from equations like Equations (5.13c) and (5.13d) and

$$\sigma^2 = \sigma_C^2 + \sigma_S^2.$$

It follows that the best estimate of the true delta position (the centroid of R_C) is:

$$\langle \varphi_C \rangle = \varphi_\delta - \lambda_g \tag{5.16}$$

The availability of a purely SIMS-related response function R_S gives rise to the possibility of 'improving' the depth resolution in a SIMS measurement by deconvolution as originally proposed by Clegg. Historically, pressure to do this increased at the beginning of the 1990s when the then attainable beam energies were insufficiently low to satisfy the depth resolution requirements of the semiconductor industry, and again from around 2002 onwards as history repeats itself. Firstly, one must comment that it is *always* better to improve the estimate of the truth by improving the experiment wherever possible, and that deconvolution, which is inherently unverifiable, must be a last resort. There is still scope to reduce the routine energies in use towards the sputtering threshold [86] and the beam transport of ion guns such as the FLIG can be further improved to make this attractive in practice. Another technique which should be mentioned, both in a context of achieving high depth resolution, and highly localized analysis is the 'shave-off' method of Nojima *et al.* [173] which uses a FIB beam parallel to the surface for sectioning. Secondly, if deconvolution is to be attempted, a forward method such as maximum entropy (pioneered in SIMS depth profiling by Collins, Dowsett and co-workers [77, 172, 174]) is much to be preferred to an inverse method. Unfortunately, a survey of the deconvolution literature shows that there is a lack of appreciation of the fact that SIMS data are undersampled and subject to Poisson noise, and that deconvolution is only a valid process if convolution is a valid model – i.e. if the analytical process is linear [175] and the data can be synthesized by superposition [176]. Deconvolution cannot be used close to the surface of a sample, because the surface transition could not have come through the Fourier window of the response function, even in the absence of transient behaviour. Data treatments which purport to control this behaviour amount to producing an artist's impression of the true distribution. Finally, energy sequencing can be used to check the results of deconvolution, and may be a superior method in many cases.

5.3.2 FABRICATION OF STANDARDS

Ion Implantation.   Ion implantation is well suited to the fabrication of standards for semiconductor analysis as well as many other systems. Any solid matrix can be used and a wide range of isotopes can be implanted. The total implanted fluence can be accurately monitored and beam/sample rastering ensures lateral homogeneity. Concentrations from a few atoms in 10^9 up to $>10\%$ can be obtained. Ideally the implanted standard should reflect as closely as possible the unknown sample, both in terms of matrix composition and dopant concentration. The dose and energy should be chosen so that at the peak of the implant there are no ion yield variations due to the formation of a non-dilute system or due to precipitation. More than 95 % of the material implanted should be beyond the transient behaviour at the surface for the SIMS conditions envisaged. A rough estimate of the peak concentration can be obtained by assuming a Gaussian profile and using the expression:

$$C_p = \frac{0.4\Phi}{\sigma} \tag{5.17}$$

where C_p is the peak concentration, σ is the standard deviation and Φ is the dose (atoms cm^{-2}). In general the peak concentration should not exceed 1 %, and it should be at least two orders of magnitude above the instrumental background. It is also desirable to avoid mass interferences which may arise from the matrix, an impurity or a combination of the matrix, primary beam and impurity ions. This can be sometimes be accomplished by a judicious choice of isotope. For example, implanting ^{70}Ge rather than ^{72}Ge into silicon avoids mass interference with the molecular ion Si_2O^+ at 72–76 dalton. Instrumental backgrounds can be similarly avoided.

The major source of inaccuracy in the fabrication of ion implanted standards arises from the uncertainty in the implanted flux. In general the implanted dose can only be measured with an accuracy of 3 %. Implanter dosimetry errors are directly transferred from the implant standard to all subsequent analyses. RBS, NRA, and PIXE are useful techniques for checking the dosimetry of implanted standards. Studies [177, 178] have also shown that contamination of the implant, by species adjacent in mass, or recoil implanted from the surface is common. Recently Singh [179] has discussed changes in implant process control and the effect of metallic and cross-contamination. For standards of gaseous species such as H, C, N and O it is important to keep the projected range $Rp > 0.1\,\mu m$ to avoid the near surface concentration from the latter effect.

Channelling can be reduced by pre-amorphization of the sample using a high energy self or heavy implant such as ^{28}Si or ^{70}Ge for silicon. The implant energy is then chosen so that it is captured within the amorphized layer. This technique has been shown to reduce the depth inhomogeneity of B implants to 2% over a 5 cm wafer [180] and is a frequently used VLSI process step (e.g. for the fabrication of shallow emitters). Capello [181] has described how a pre-amorphization implant (PAI) is commonly used in industrial processing in order to avoid unfavourable profile broadening and channelling tails during dopant atom implants in the ultra-low energy ion implantation regime (<5 keV).

Multiple implants can be used if several species in a matrix are to be sought. Careful examination of such samples is necessary to ensure that the initial implant profiles are not modified by the later implants. Clegg [182] observed that radiation enhanced out-diffusion of ^{52}Cr occurred during ^{56}Fe implantation of a Cr + Fe + Zn standard for GaAs. At low doses $(1 \times 10^{13}$ atoms cm^{-2}) this created a sizeable surface peak of Cr (50% of total Cr) making the implant unsuitable for SIMS calibration. Mitra [183] has described formation of deep levels during double implantation procedures used to form pn junctions in SiC and the SIMS analyst needs to be cognisant of the potential consequences of such issues, for example, the possible change in ion yield that may arise as the local electronic environment is changed.

At the time of writing, there are five certified reference materials (CRMs) for SIMS available from NIST. These are 103.2 (synthetic glasses), 2133 (phosphorous implanted into silicon), 2134 (arsenic implanted into silicon), 2137 (boron implanted into silicon) and 2135c (Ni–Cr thin film multi-layer structure). Use of this material is almost essential in qualifying local references. For further information see www.nist.gov/.

Extreme care should be taken when qualifying references for ultra low energy implantation. Many of the problems described in Section 5.2.2 can occur, and if significant material is within the transient, qualification is inherently impossible.

Thin Film References. Techniques such as MBE and chemical vapour deposition (CVD) are capable of producing very thin films with carefully controlled interfacial properties. Generally, low growth temperatures should be used to minimize impurity migration during growth. An independent check (e.g. XTEM) on the morphology of the layer is useful if unexpected effects are to be avoided. Undesirable impurities may also accumulate at interfaces when growth is temporarily halted. The incorporation of oxygen in this way is particularly important since it may enhance the ion yield from the impurity during analysis.

A particularly useful thin film material is a surface layer of some SiGe alloy some 20 nm thick, on silicon. This can be used to tune the analytical beam parameters and others in the SIMS instrument to achieve high dynamic range depth profiles with good rejection of surface species.

5.3.3 DEPTH MEASUREMENT AND CALIBRATION OF THE DEPTH SCALE

SIMS depth profile craters are typically between 10 nm and a few μm deep. The ideal depth calibration method would inform the user of the mass of matrix material removed as a function of time. This could be converted to a depth scale from a knowledge of the original density. No such method exists to the knowledge of the authors. The depth scale is usually quantified by measuring the SIMS crater with a surface profilometer after analysis. It must be remembered that the profilometer measures to the surface of the altered layer, and will introduce a systematic error in the 1–20 nm range according to the degree of density renormalization of the matrix due to incorporated probe atoms, and the primary beam species, energy and angle of impact. An alternative is to use an internal reference (e.g. impurities at a layer-substrate interface). Zalm was an early proposer of this strategy [184]. This too will introduce a small error because the FDI associated with the feature is a defined entity. Recently, following an original proposal by Kempf [185], Cameca have introduced a built-in interferometer for real time depth measurement in the IMS WF instrument [186]. This too will have associated systematic errors, for example, due to the transparency of some layers (and the altered layer, perhaps).

Surface profilometry is the most commonly used method for establishing the depth scale in SIMS. The accuracy of the measurement is limited by surface roughness and particulate contamination, sample curvature and other factors such as instrumental levelling corrections.

Depth calibration is usually based on the assumptions that the primary current remains stable during the course of the profile and that the erosion rate is constant. The apparent depth of an ordinate in a profile D_{app} is then given by:

$$D_{app} = z_{meas} \frac{\varphi}{\varphi_{tot}} \tag{5.18}$$

where z_{mes} is the measured depth of the crater (i.e. to the top of the altered layer), φ is the dose or time corresponding to the ordinate and φ_{tot} is the total dose or time. If the transient erosion rate differs from that in the bulk, then Equation (5.18) will require correction on the scale of nanometres. A similar correction will be required if retained probe atoms in the crater cause z_{meas} to differ from the depth of matrix sputtered.

In principle, samples like the one shown in Figure 5.7, could be used to establish absolute erosion rates for particular combinations of matrix and primary ion beam. Unfortunately, most SIMS instruments lack the means to measure or reproduce parameters such as primary ion current with sufficient accuracy. Even more difficult to reproduce is the current density distribution in a focused beam. Quite small changes in beam shape can change the erosion rate for a given current and scan size by several percent; hence the continued use of surface profilometers.

5.3.4 SOURCES OF ERROR IN DEPTH PROFILES

Transient Behaviour.   There are many combinations of probe and sample where the sputtering system never achieves steady state. SIMS may be a useful comparative spectroscopy tool in these circumstances, but it will be no use for depth profiling. Even where steady state is achieved, highly non-linear effects occur during its establishment and a number of phenomena occur at the commencement of sputtering (and also at buried interfaces between different matrices) amongst which are:

(i) The gradual dilution of the sample with incorporated probe atoms which are implanted from the primary beam, and then continuously remixed in the top nanometres of the surface. For steady state to be achieved, the flux of probe atoms leaving the sample through sputtering and effusion must equal the arriving flux in the beam. Initially though, the surface and sub-surface concentration of probe atoms increases with two linked effects: The erosion rate of the sample changes from one characteristic of the pure sample to one characteristic of the chemical combination of sample and probe atoms. An average erosion rate reduction of more than a factor of 2 can be expected, for example, for near-normal oxygen bombardment of silicon, increasing as the beam energy decreases, but persisting over a smaller depth. At the same time, the ionization probability for emitted species can change by many orders of magnitude (some increasing, some decreasing).

(ii) Preferential sputtering and segregation will change the relative surface concentrations of impurity and matrix atoms and of the components of an alloy matrix. Relative yields will change as the surface concentrations change, and there will also be an effect on ionization probabilities.

Therefore ion yields can vary by several orders of magnitude and sputter yields by up to an order (typically) and making quantitative use of the data from this region a distant goal. One of the major applications of ultra-low energy beams and other strategies such as oxygen flooding is to reduce the depth range of these effects. However, if quantitative information is required right from the surface, capping is the only reliable option, unless the concentrations are sufficiently high to use some form of ion scattering.

The change of erosion rate across the transient region, first described by Wittmaack and Wach [61], contributes to an error in the depth calibration known as the profile, or transient, shift. This is dependent on the primary beam species, energy and angle of incidence, as well as the matrix and impurity species. The fact that it is impurity species dependent shows that it is not entirely due to the transient behaviour, and it is probably compounded from the transient erosion rate and anisotropic bombardment induced relocation of impurity atoms, the latter leading to a shift in the centroid of the measured feature [187]. For example, for oxygen at near normal incidence on silicon, it amounts to [188] $- 1\,nm\,(keV/O_2^+)^{-1}$ (where the $-$ sign denotes a shift towards the surface), whereas for similar bombardment conditions on diamond [189], it is between 0 and $- 0.5\,nm\,(keV/O_2^+)^{-1}$. Note that these figures are derived from measurements of the change in apparent depth (centroid) of a delta layer (from Equation (5.18) as a function of beam energy, and different magnitudes are obtained depending on the FDI chosen (e.g. peak, centroid, etc.), because the bombardment conditions also change the profile shape. The way in which the apparent depth of a feature increases as beam energy decreases under oxygen bombardment of silicon can clearly be seen in Figure 5.12(b). The profile shift has not been investigated in detail for anything except boron in silicon under oxygen bombardment, but the factors which should be investigated include the following: (i) The effect due to the sputter rate change at the surface, (ii) a shift due to the dominant mass transport processes occurring as the altered layer engulfs the buried feature and (iii) a contribution arising from the swelling of the matrix arising from the incorporation of probe atoms which causes the final crater depth to be slightly underestimated. Effect (iii) is cancelled out if the measured crater depth remains proportional to the primary ion dose throughout the profile [61]. The contribution from (ii) may be negative (towards the surface), or positive, so that (i) may be reinforced or compensated. In other matrices the incorporation of chemically active probe atoms may cause densification (e.g. for oxygen in an alkali metal), so the sense of (iii) may be reversed.

It has also been shown that the centroid positions measured from deltas which are too closely spaced are incorrect, and can lead to the illusion of a depth and structure dependent profile shift [76].

In general, near surface behaviour has been investigated by authors such as Wittmaack, Vandervoorst, Van de Heide and the present authors, but much remains to be done.

Specimen Selection, Preparation and Surface Condition. Because SIMS is capable of high sensitivity, dynamic range and depth resolution, the data are affected adversely by poor sample handling and preparation. Unnecessary surface contamination (especially in the form of dust) and surface scratches will restrict the dynamic range and depth resolution whilst causing serious profile distortion. As a general rule, clean room and wafer preparation standards of cleanliness should be observed during sample selection, preparation and mounting for analysis. If the surface contains sputter-resistant or masked regions (e.g. by dust), or if the sputtering conditions lead to the nucleation of defects in the material, or if the material itself contains voids, the profiled surface may develop mesas or pinholes. The effects are shown schematically in Figure 5.13. Raised features such as mesas lead to features (e.g. shoulders) on the deep edge of a profiled layer because some of the surface still lies within the layer after the majority has passed through, whilst pinholes give rise to features on the shallow side of the profile (as buried material is reached 'too soon'). Complex effects can occur in multi-layer structures where part of the surface is in one layer and part in another. These include gradients on regions that should be at constant concentration (Figure 5.13(d)), and the observation of 'beats'. The effects of sample quality on depth resolution in particular have been discussed by Wittmaack [105].

Figure 5.14 also demonstrates how particulate surface contamination will give rise to poor depth resolution, profiles distorted in a manner

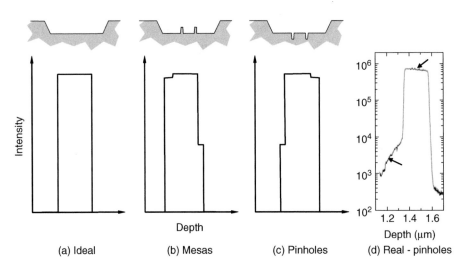

Figure 5.13 Effects due to crater topography forming due to masking of the surface by dust, or pinholes in the sample: (a) ideal crater bottom and profile; (b) mesas; (c) pinholes; (d) pinholes forming in an SiGe layer due to nucleation of defects during analysis, or voids in the material. Note shoulder on shallow side of profile, and slope on top

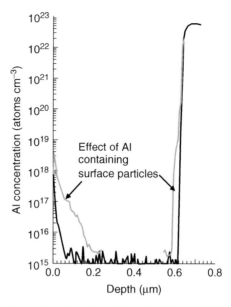

Figure 5.14 SIMS depth profiles for aluminium in 0.6 μm SOS layers (after Dowsett *et al.* [119]). Al_2O_3 particles embedded in the soft surface of the silicon epilayer during wafer dicing are impossible to remove by subsequent pre-treatment and lead to profile distortions shown by the grey line. The effect at the interface is due to surface topography caused by the low sputter rate of the particles (see also Figure 5.13). The black line shows the profile obtained from a sample properly protected during preparation

characteristic of the surface topography and high background levels (especially if surface debris contains the species of interest). SIMS depth profiles for aluminium in silicon on sapphire (SOS) with and without alumina surface debris due to the sawing of the wafer prior to analysis are shown. In the presence of the alumina a high surface Al^+ level is recorded, there is apparently a high level of Al^+ in the silicon epilayer and the interface response is seriously distorted in a manner which is suggestive of outward migration of Al from the substrate. The problem was avoided by breaking wafers for analysis, rather than sawing them, and by protecting the relatively soft silicon surface with a layer of spin on oxide until just prior to analysis [185].

The following factors should be considered in advance of analysis. What depth resolution is required, and is it compatible with the surface condition (roughness, cleanliness) of the material? Is the sample representative of the process under examination (for example, did it come from the centre or extreme edge of a wafer)? What impurities are to be sought, at about what levels? Have mass interferences been avoided in any implantation? Are suitable standards available? Is special preparation of the sample or instrument required to meet the demands of the analysis?

Erosion Dependent Surface Topography.   The effect of surface topography on depth resolution has already been discussed in Section 5.1.4. Three types of surface topography develop during a profile. Firstly, there is the basic roughening and pitting of the surface by individual ion impacts. This has been directly observed by scanning tunnelling microscopy (STM) [190, 191], and is believed to place a fundamental limit on leading edge resolution for sharp features when combined with the statistical nature of sputtering [7, 169]. Secondly, surface topography may evolve due to fundamental mass transport processes occurring as sputtering proceeds [192, 193], and preferential sputtering [194]. The degree to which this occurs can be linked to the initial state of the surface as originally described by Wehner and Hajicek [194] and recently by Fares [195]. Catastrophic increases in surface roughness will influence both the erosion rate and ion yield. Thirdly, macroscopic topography can develop due to poor beam scanning electronics [78] and the influence on the beam impact angle and focus of a high extraction field [79, 80].

In general, delta layer or other multi-layer samples give a good guide to the prevalence of ion beam related topography generation for a particular ion beam/material combination, and loss of depth resolution with depth combined with relatively sudden changes in matrix intensities (early observations by Stevie *et al.* [81]) can be taken as first evidence of roughening. Apparent loss of depth resolution with depth can also be caused by diffusion of deeper layers during growth, however [196]. In this case, in the absence of topography, the SIMS decay length should stay constant, even if the standard deviation increases with depth.

5.4 Novel Approaches

The way in which sample consumption specifically links spatial resolution and sensitivity, the need to analyse rough or non planar samples and the loss of depth resolution due to ion beam induced mass transport effects have stimulated the development of SIMS techniques to circumvent these limitations. A selection of these is described in this section.

5.4.1 BEVELLING AND IMAGING OR LINE SCANNING

In direct depth profiling, the depth resolution will be limited by numerous effects, the worst of which are usually surface roughness and segregation, if they are present. If it is possible to produce a bevel on the sample surface, then buried layers which are thinner than the achieved depth resolution may be converted into surface stripes wider than the lateral resolution (see Figure 5.15(a)). SIMS imaging, or line-scanning of these stripes using a static SIMS dose yields a depth profile. A variety of methods

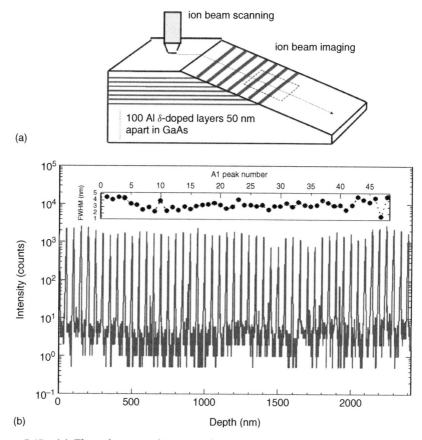

(a)

(b)

Figure 5.15 (a) The schematic diagram of ion beam line-scanning and imaging of a bevelled sample containing one hundred Al ∂-layers nominally 50 nm apart. (b) The depth calibrated linescan profile of the corresponding Al image; the inset is the FWHM of Al peak vs. Al peak number. It can be seen that there is no indication of a loss of depth resolution with depth with an average FWHM of 3.3 nm

are available for bevelling, from mechanical lapping with chemical polishing [197], *in-situ* ion milling [198–200] and chemical etching [201]. The bevel may be produced with the SIMS probing beam, in which case the interaction between the altered layer and buried chemical distributions, including segregation effects themselves, may be investigated. In order to produce sufficient magnification, bevel angles of the order of 10^{-3}–10^{-4} radians are required. This gives a magnification factor sufficient to transform a 1 nm thin layer into a 1–10 μm wide stripe at the surface. It also has the consequence that a complete structure contained within a depth of $\sim$1 μm requires an instrument with a field of view >1 mm for a complete analysis.

Ion beam milling naturally leaves a damaged and chemically altered layer on the sample surface. It is easy to show that the redistribution of material normal to the surface due to atomic mixing is magnified by the bevel and results in a broadening of the surface stripe [198]. Two methods for alleviating this problem have been suggested. For silicon milled with oxygen at near normal incidence, McPhail [198] has shown that the SiO_2 altered layer can be removed along with the damage it contains by etching in 5–10 % hydrofluoric acid. This method has the advantage that the sub-nm native oxide formed on the bevel surface after etching stimulates secondary ion emission, and the disadvantage that it is an *ex-situ* process. An alternative, suggested by Skinner [200] for AES applications, is to make the bevel using a relatively high energy beam (with a high current for speed), and then sputter away the damage *in-situ* with a very low energy beam.

Hsu [201–203] has developed an alternative method of bevelling, in which the sample is simply lowered into (or raised out of) an appropriate etchant. The resulting bevel surface is then imaged or line scanned in the SIMS instrument. In this method the bevel magnification is determined by the ratio of the etch rate to the dipping rate. Very high depth resolutions may be achieved, that are independent of depth. For example [201], a depth independent depth resolution from an $Al_{0.3}Ga_{0.7}As/GaAs$ multi-quantum well of 3.1 nm (FWHM) down to a depth of 1.6 μm was achieved. Sub-nanometre depth resolution can be achieved with sufficiently high bevel magnifications (Figure 5.15(b)) and it may be shown that the depth resolution depends only upon the magnification, the beam width and the information depth of the SIMS technique [202]. The approach developed by Hsu is also useful when conventional SIMS depth profiling introduces very large profile distortions. For example, the analysis of copper in silicon with an oxygen beam can lead to gross distortions due to beam induced segregation. This can be eliminated by the bevelling approach in which the beam can be line-scanned both up and down the bevel to check for beam induced problems. Fearn [204] has recently shown that high magnification linear bevels may be introduced into silicon and accurate profile shapes recovered using a bevel and image approach. She analysed a shallow boron implant in silicon and the profile shape produced by the bevel and image approach was very similar to that from a 1 keV depth profile (Figure 5.16).

The sample volume considerations are not as favourable for bevelling as for direct depth profiling as only the top nanometres of the bevel may be consumed before beam induced effects destroy the 'depth' resolution. Detection limits will be typically 10–1000 times worse than in depth profiling but the situation can be optimized by co-adding lines in an image [204].

Bevelled materials may be analysed by many other microscopic techniques. Srnanek has analysed chemically etched semiconductor samples by SEM, Raman and AES [205–207], and more recently micro-Raman [208].

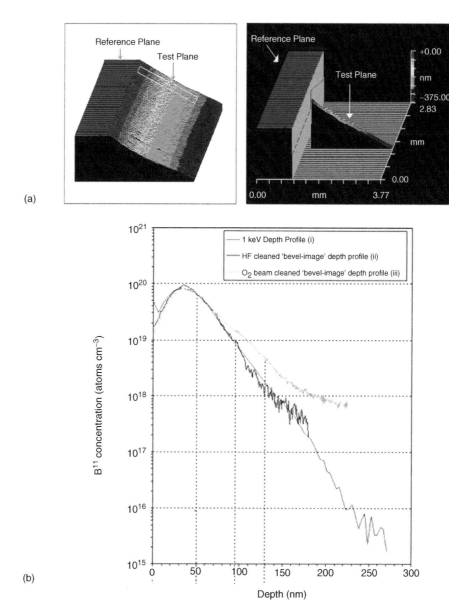

Figure 5.16 (a) A ZYGO map of a bevel produced in silicon by Fearn [204]. The bevel is smooth and linear making depth-calibration of the subsequent line-scan data very straightforward. (b) Comparison of the boron concentration depth profiles obtained by (i) 1 keV conventional SIMS depth profile of a 10 keV boron implant in silicon with (ii) data produced by imaging an HF cleaned bevel; (iii) SIMS ion imaging an O_2^+ ion beam cleaned bevel

5.4.2 REVERSE-SIDE DEPTH PROFILING

Long tails observed in SIMS depth profiles are often caused by mass transport processes such as segregation occurring during analysis, and significant backgrounds can be simply due to high near-surface concentrations. If the sense of the profile can be reversed, i.e. if profiling can proceed up the concentration gradient, or without the surface concentration, more accurate information can be recovered. This can be achieved using reverse-side depth profiling [209–211], in which a conventional profile is compared with one acquired by profiling from the back of the wafer after it has been thinned. Tails due to segregation and similar effects observed when profiling from a high concentration to a low one are immediately apparent. Figure 5.17, after the work of Lareau [210], shows forward and reverse profiles through a complex alloy ohmic contact on GaAs. In the reverse side profile, the interface responses for the alloy constituents all coincide, demonstrating that the separation observed in the forward profile is due to differential segregation during analysis, or a similar effect. The technique is also appropriate for samples whose front surface is not flat or which roughen severely during

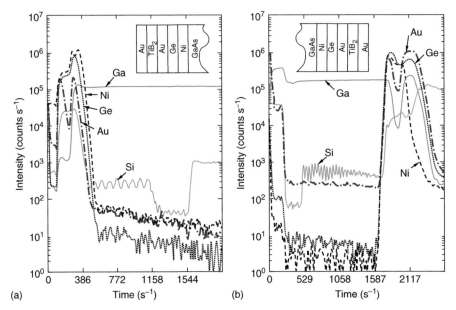

Figure 5.17 Forward and reverse side profiles of an ohmic metal/marker layer sample after alloying of the Ni/Ge/Au/TiB2/Au contact (after Laureau [210]). (a) Forward profile. (b) Reverse profile. Note that the Au, Ge and Ni signals coincide at the GaAs interface, showing that their separation in the forward profile was due to segregation or a similar effect occurring during analysis. The reverse profile shows the effect (in the Si channel) of some topography on the polished back starting surface

sputtering. However the difficulties associated with sample preparation should not be minimized. The wafer must be thinned over a limited area ($1\,mm^2$) through about 0.5 mm, maintaining strict parallelicity between the back surface and the buried feature of interest.

5.4.3 TWO-DIMENSIONAL ANALYSIS

The separation and size of modern semiconductor devices is now less than the possible extent of lateral impurity migration during fabrication. This situation is exacerbated by the frequent use of a polycrystalline matrix, especially in multi-layer devices, where grain boundary diffusion may be especially rapid. Coincident measurement of lateral and vertical impurity distributions has become an important objective for SIMS. The limitations imposed on SIMS by the necessarily restricted volumetric consumption imposed by the need for high spatial resolution are discussed in Section 5.1 and elsewhere [212, 213]. In particular, if high lateral resolution direct imaging or image depth profiling of a sample must be used, then AES may be more sensitive than SIMS [212]. SIMS lateral resolutions of the order of a few tens of nanometres can be achieved; for example, Nojima [214] has reported 22 nm with good sensitivity and in the FIB SIMS 5 nm is possible [158]. However under these conditions the analytical volume removed is so small that detection limits will generally be worse than 1 % atomic, even for a high yield element such as boron.

Analytical sensitivity may be recovered if one dimension of lateral resolution is sacrificed and the analyte volume element becomes a long cuboid. However, it is still necessary to use a yield enhancing probe such as Cs^+ or O_2^+ for the ultimate useful yield and the lateral resolution is then limited to 0.1−1 µm at best. Hill and co-workers [215, 216] developed a sample and data processing techniques based on multiple stripes and using a rotated mask to produce vertical bevels. This sample combined the attributes of high lateral resolution at modest probe diameters and high spatial resolution with a large consumed volume. Ukraintsev [217] has improved on this general idea by using a moire pattern so that conventional processing steps can be used. He showed that the approach significantly simplified 2D SIMS test chip manufacturing, data acquisition and analysis. As a result, 2D dopant profiling with a lateral resolution limited by the photomask pixel size (10 nm) and sensitivity of $3 \times 10^{17}\,cm^{-3}$ were realized on commercial equipment and the 2D dopant profiles were reproducible to a precision of 10 nm. The measured profiles were compared with 2D Monte-Carlo and calibrated TSUPREM4 simulations and showed good agreement.

More recently 2D SIMS has been used by Rosner [218] to look at hot-pressed steel particles.

5.5 Instrumentation[1]

5.5.1 OVERVIEW

Three types of mass spectrometry system, based on the double focussing magnetic sector mass spectrometer (DFMS) [2, 22, 219–222], the quadrupole mass spectrometer (QMS) [223–225] and the time of flight mass spectrometer (ToF) [5, 226] are widely used for dynamic SIMS. In some ways, these instruments are complimentary. Some comments on their different advantages and disadvantages are given below. It is also worth noting that, whilst one spectrometer type may have a clear advantage in a particular application *in principle*, the manufacturer may not have emphasized that side of the design, or the specification of the actual hardware used may fall short of the ultimate.

Both the DFMS (with multiple collectors or a strip detector [222]) and the ToF are capable of parallel detection – i.e. detecting different masses from sputtering events falling within the same time period. This increases the efficiency of the instrument with respect to the amount of material consumed if (as is usually the case) more than one mass is of interest. The ToF has parallel collection across 100s or 1000s of Da, but its injection must be pulsed. The DFMS, due to practical limitations in the size of discrete detectors, or the width of its image plane usually collects up to 10 parallel channels or a continuous mass range of a few 10s of da. Most commercial DFMS designs are not equipped for parallel detection, however.

The DFMS has the highest mass resolution, and spectrometers for SIMS have been built with achieved $M/\Delta M$ of 20 000 and a target performance of least 100 000 [22, 221]. The most common DFMS designs will achieve $M/\Delta M$ of 5000 fairly easily and this is sufficient to resolve many important mass interferences in SIMS. A ToF with a long optical path and aberration correcting optics can also achieve $M/\Delta M$ in the 3000–5000 range but the abundance sensitivity (see next paragraph) is nowhere near as good as for the DFMS.

Abundance sensitivity is the ability to detect $M \pm 1$ in the presence of a large flux of mass M (e.g. to be able to detect Al as an impurity in bulk Si). High abundance sensitivity is important in the measurement of minor isotopes and in impurity profiling, for example. Abundance sensitivities in excess of 10^8 can be achieved by a large and well set-up QMS and the DFMS [120, 220]. The ToF is limited, so far, to around 10^5 because scattering inside the open geometry, and high order aberrations in devices like the reflectron lead to a broadening of the peaks about 5 orders down on the maximum.

---

[1]This section can be usefully read in conjunction with Chapter 4, Section 4.3, which provides more background.

When used in spectroscopic as opposed to depth profiling mode (see below) the ToF has easily the highest useful yield of the set when this is averaged across the mass range. This is because the DFMS (used in peak switching or scanning modes) and the QMS (which has no parallel capability) simply throw away sputtered ions other than the ones they are tuned to. However, if the spectrometry system must be operated with an effectively continuous sputtering beam, as in almost all dynamic SIMS, the advantages of the ToF disappear: The analytical primary beam must be pulsed, typically with a width of a few nanoseconds and repetition rate of microseconds. To achieve quasi-continuous sputtering, dual beam methods can be employed (see Section 5.5.4), but then very little of the sputtered material is sampled. In continuous sputtering mode, the DFMS can achieve useful yields which can exceed 0.1 in some cases, and a modern quadrupole can approach this. Practical detection tests show that the ToF is comparable.

5.5.2 SECONDARY ION OPTICS

The three types of spectrometer differ greatly in the optical systems used to extract the secondary ions and this impacts on their possible range of applications. The DFMS has, ultimately, a high field extraction system (keV mm^{-1}) and electrodes within a few mm of the sample [2]. The sample must be part of a low aberration optical system and modern DFMS designs place it parallel to the entrance optics. Although strategies such as split extraction fields allow some flexibility between energy and angle of impact of the primary beam, these two parameters remain strongly coupled. A significant achievement in recent years, however, has been the development of optical systems which retain the beneficial properties of the DFMS (high useful yields, mass resolution and abundance sensitivity) whilst allowing sub-keV beams access to the sample [226, 227]. The confined space around the sample makes it difficult to add extra techniques and vacuum can be a problem requiring extra measures in some cases.

The ToF has a more open optical system, but still requires pass energies in the keV range, and so tends towards a high extraction field. However, this may be pulsed and may not be present when the sputtering beam is on, removing potential problems with low energy bombardment. However, the ToF may still require a strict relationship between the sample position and the entrance optics, so that some inflexibility remains.

The QMS combines (or can combine) an open apertured optical system with a low extraction field of $\sim$10 V mm^{-1}. Sample position is usually very flexible and the instrument can bombard with any charge state at any angle whilst collecting any charge state, for a primary ion energy from $\sim$100 eV upwards. There can be an open volume of several cc over the sample making it simple to introduce additional techniques. The QMS has a large depth of field, especially compared with the DFMS and copes easily with rough samples (also a potential problem in the ToF).

5.5.3 DUAL BEAM METHODS AND TOF

With a pulsed primary ion beam, and parallel detection across a wide mass range, the ToF is the archetypal spectrometer for static SIMS. However, it is superficially not well suited to dynamic SIMS applications which require a high and continuous flux of primary ions. A major development on ToF–SIMS was the advent of dual beam instruments, originally developed in Benninghoven's group at Münster [5]. Here, a sputtering beam with a high flux density and a long pulse length is interleaved with a sampling beam with a dose per cycle in the static SIMS range and a short pulse length. The ToF is gated so that it only collects secondary ions from the sampling pulse, but sputtering of the sample with the sputtering beam continues whilst the collected ion plume is in transit around the ToF. Figure 5.18 shows the timing scheme. The sputtering beam is generally oxygen or caesium, and loads the sample with emission promoting probe species. The sampling beam may come from an LMIG and be tightly focused, improving the instrumental gating (see next section). Recently, the power of dual beam methods has been extended to the use of two sputtering beams, one which controls the surface concentration of the other (caesium and xenon). This may pave the way to the use of Cs beams in the sub-500 eV range, and be appropriate outside the ToF application.

5.5.4 GATING

The most important single parameter of a high performance depth profiling instrument is its ability to reject sputtered material from the walls of the crater and the sample surface. This determines the dynamic range and

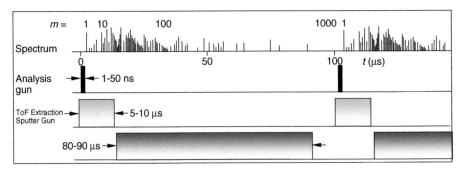

Figure 5.18   The timing scheme for the interleaved operation of a sputtering and an analysis beam in dual beam ToF–SIMS (after K. Iltgen et al. [5])

profile shape in the low orders, and is the key to high performance depth profiling. The fundamental problem arises because, however well designed the primary ion column, the final spot will contain aberrations and scattered ions which lead to a low but finite current density several FWHM multiples from the centre of the beam. One might describe three methods of gating: optical gating – which is only really effective in a high field extraction system [2], electronic or digital gating [228, 229] – which requires a very high quality scanned primary ion beam and probe gating where, as in the dual beam ToF, the sampling beam is microfocused, and can be confined entirely within the crater bottom. Optical and digital gating are illustrated in Figure 5.19.

Optical gating requires that the sample be a well controlled part of the optical system and because of the relatively large energy and angular spread of the secondary ion emission, it also requires a high extraction field. The objective lens of the secondary ion optical system is used to form a low aberration magnified image of the sample in secondary ions on an aperture in the field image plane. The aperture then defines that area of the sample which is 'seen' by the spectrometer. If this is confined to the bottom of the crater, then one has a high dynamic range system – almost independent of the quality of the primary ion beam. It is not necessary to scan the beam, although this can improve the dose uniformity across the crater, to produce

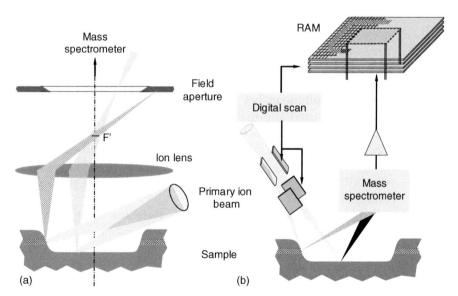

Figure 5.19 (a) Optical gating – a feature of the ion microscope type of instrument which can form magnified images of the sample on an aperture which then limits the field of view. (b) Digital gating as used in a scanning ion microprobe. Any part of the image stack can be used (in principle) to form the recovered information, and transverse sections as well as depth profiles may be extracted

a flat bottom and high depth resolution. (A high quality focus and beam scanning may also be required for other reasons). An extension of optical gating is known as dynamic emittance matching [2, 230]. This allows the stationary optical gate size to be small, promoting high transmission and resolution in the DFMS, but the image of the gate on the sample plane is scanned so that a fine scanned probe lies within it for all areas where ion collection is desired.

Electronic or digital gating uses a scanned probe which must be sharp and relatively free from scattered halo (this last requirement means that the vacuum in the instrument must be in the 10^{-10} mbar range for the highest dynamic range). Either the ion counting system is inhibited when the beam is scanning the crater walls or the whole scan is collected as an image and a depth profile reconstructed from those parts of the image corresponding to the bottom of the crater. Clearly, this technique allows other anomalous regions, such as regions of high intensity caused by dust or other inhomogeneities, to be rejected [231].

Probe gating as used (implicitly) in the dual beam ToF is similar to digital gating, but the beam need not actually reach the crater walls.

5.6 Conclusions

Dynamic SIMS has some very important attributes including high depth resolution, high lateral resolution, excellent sensitivity and high dynamic range. Sub-nanometre depth resolution may be achieved with sub-keV primary ion beams and it is now possible to obtain quantitative data within a few nanometres of the surface. Lateral resolutions of a few tens of nanometres are possible and sensitivities down to parts per billion can be achieved (but not simultaneously). All elements and isotopes in the Periodic Table may be measured and mass interferences may be resolved with high mass resolution instruments. Dynamic SIMS is complementary to techniques giving structural information (RBS, XTEM, X-ray diffraction), electrical profiles (SSRM, e-CV) and quantitative analysis at high concentrations (RBS, AES, XPS). SIMS, XTEM, and e-CV or SSRM form a particularly powerful combination.

Dynamic SIMS was developed mainly for applications in electronic materials and in geology but it is now enjoying a far wider range of applications covering many different types of materials used in many different technological areas. Overall, it is proving an increasingly versatile and sophisticated technique, but one might also comment that it lags behind in some application requirements. The profit margins on SIMS instrumentation are rather small and there has always been a lack of willingness on behalf of major users of the technique to invest in its future development. At the same time, one feature of modern SIMS research is reinvention of the wheel – often with some spokes missing – inexcusable with modern literature search methods. In this chapter we have retained many older references to show that there is a significant history to check before claiming novelty.

📖 References

1. A. Benninghoven, F.G. Rudenauer and H.W. Werner, *Secondary Ion Mass Spectrometry*, John Wiley & Sons, Inc., New York, NY, p. 671 (1987).

2. G. Slodzian, in *Applied Charged Particle Optics*, Advances in Electronics and Electron Physics, Supplement 13B, Ed. A. Septier Academic Press, New York, NY, p. 1 (1980).

3. K. Wittmaack, *Vacuum*, **32**, 65 (1982).

4. J. Bennett and J.A. Dagata, *Journal of Vacuum Science and Technology*, **B12**, 214 (1994).

5. K. Iltgen, C. Bendel, A. Benninghoven and E. Niehuis, *Journal of Vacuum Science and Technology*, **A15**, 460 (1997).

6. (a) D.K. Balake, B.N. Colby and C.A. Evans, *Journal of Analytical Chemistry*, **47**, 1532 (1975); (b) R. Hernandez, P. Ianusse, G. Slodzian and G. Vidal, *Recherche Aerospace*, **6**, 313 (1972).

7. (a) D.G. Armour, M. Wadsworth, R. Badheka, J.A. Van den Berg, G. Blackmore, S. Courtney, C.R. Whitehouse, E. A. Clark, D.E. Sykes and R. Collins, in *Proceedings of SIMS VI*, John Wiley & Sons, Ltd, Chichester, UK, p. 399 (1988); (b) R. Badheka, M. Wadsworth, D.G. Armour, J.A. Van den Berg and J.B. Clegg, *Surface and Interface Analysis*, **15**, 550 (1990).

8. U. Littmark and W.O. Hofer, *Nuclear Instruments and Methods*, **168**, 329 (1980).

9. P.R. Boudewijn, H.W.P. Ackerboom and M.N.C. Kemperers, *Spectrochimica Acta*, **39B**, 1567 (1984).

10. M.G. Dowsett, R.D. Barlow and P.N. Allen, *Journal of Vacuum Science and Technology*, **B12**, 186 (1994).

11. S.A. Schwarz, B.J. Wilkens, M.A.A. Pudensi, M.H. Rafailovich, J. Sokolov, X. Zhao, W. Zhao, X. Zheng, T.P. Russell and R.A.L. Jones, *Molecular Physics*, **76**, 937 (1992).

12. E.H. Cirlin, *Thin Solid Films*, **220**, 197 (1992).

13. B.A. Pint, J.R. Martin and L.W. Hobbs, *Oxidation of Metals*, **39**, 167 (1993).

14. K. Wittmaack, *Philosophical Transactions of the Royal Society A – Mathematical, Physical and Engineering Sciences*, **354**, 2731 (1996).

15. W. Vandervorst, T. Janssens, B. Brijs, T. Conard, C. Huyghebaert, J. Fruhauf, A. Bergmaier, G. Dollinger, T. Buyuklimanli, J.A. Vandenberg and K. Kimura, *Applied Surface Science*, **231**, 618 (2004).

16. E. Napolitani, A. Carnera, V. Privitera and F. Priolo, *Materials Science in Semiconductor Processing*, **4**, 55 (2001).

17. M.G. Dowsett, *Applied Surface Science*, **203**, 5 (2003).

18. R. Liu and A.T.S. Wee, *Applied Surface Science*, **231**, 653 (2004).

19. P.C. Zalm, *Mikrochimica Acta*, **132**, 243 (2000).

20. M. Betti, *International Journal of Mass Spectrometry*, **242**, 169 (2005).

21. J.S. Becker and H.J. Dietze, *Spectrochimica Acta – Atomic Spectroscopy*, **B53**, 1475 (1998).

22. J.V.P. Long and D.C. Gravestock, *Proceedings of SIMS VI, Vacuum*, **34**, 903 (1984).

23. A. Wucher, *Fresenius Journal of Analytical Chemistry*, **346**, 3 (1993).

24. H. Oechsner, in *Thin Film and Depth Profile Analysis*, Topics in Current Physics, H. Oechsner (Ed.), Springer-Verlag, Berlin, p. 63 (1983).

25. H.J. Mathieu and D. Leonard, *High Temperature Materials and Processes*, **17**, 29 (1998).

26. S.W. Downey, A.B Emerson and R.F. Kopf, in *Proceedings of the 1st International Conference on Measurement and Characterization of Ultra Shallow Doping Profiles in Semiconductors*, eds. C. Osburn and G. McGuire, Microelectronics Centre of North Carolina, Research Triangle Park, NC, p. 172 (1991).

27. J.J. Kopanski, J.F. Marchiando and J.R. Lowney, *Materials Science and Engineering B – Solid State Materials for Advanced Technology*, **44**, 46 (1997).

28. P. De Wolf, R. Stephenson, T. Trenkler, Y. Clarysse, T. Hantschel and W. Vandervorst, *Journal of Vacuum Science and Technology*, **B18**, 361 (2000).

29. L. Dobaczewski, A.R. Peaker and K.B. Nielsen, *Journal of Applied Physics*, **96**, 4689 (2004).

30. (a) M. Chabala, R. Levi-Setti and Y.L. Wang, *Applied Surface Science*, **32**, 10 (1988); (b) G. Slodzian, in *Proceedings of SIMS VI*, John Wiley & Sons, Ltd, Chichester, UK, p. 3 (1988).

31. J.L. Hunter, R.W. Linton and D.P. Griffis, *Journal of Vacuum Science and Technology*, **A9**, 1622 (1991).

32. D.G. Welkie and R.L. Gerlach, *Journal of Vacuum Science and Technology*, **20**, 1064 (1982).

33. J. Walton and N. Fairley, *Surface and Interface Analysis*, **38**, 1230 (2006).

34. M. Fartmann, C. Kriegeskotte, S. Dambach, A. Wittig, W. Sauerwein and H.F. Arlinghaus, *Applied Surface Science*, **231**, 428 (2004).

35. F. G. Rudenauer, *Analytica Chimica Acta*, **297**, 197 (1994).

36. K.H. Gray, S. Gould, R.M. Leasure, I.H. Musselman, J.J. Lee, T.J. Meyer and R.W. Linton, *Journal of Vacuum Science and Technology*, **A10**, 2679 (1992).

37. A. Scandurra, A. Licciardello, A. Torrisi, A. Lamantia and O. Puglisi, *Journal of Materials Research*, **7**, 2395 (1992).

38. S. Chandra, *Applied Surface Science*, **231**, 467 (2004).

39. G. Gillen, A. Fahey, M. Wagner and C. Mahoney, *Applied Surface Science*, **252**, 6537 (2006).

40. M.G. Dowsett, A. Adriaens, M. Soares, H. Wouters, V.V.N. Palitsin, R. Gibbons and R.J.H. Morris, *Nuclear Instruments and Methods in Physics Research*, **B239**, 51 (2005).

41. C.W. Magee, *Journal of the Electrochemical Society*, **126**, 660 (1979).

42. K. Wittmaack, *Applied Physics Letters*, **50**, 815 (1987).

43. P. Williams, *Surface Science*, **90**, 588 (1979).

44. R.T. Lareau and P. Williams, in *Proceedings of SIMS V*, p. 149, Springer-Verlag, Berlin (1986).

45. R.L. Hervig and P. Williams, in *Proceedings of SIMS V*, p. 152, Springer-Verlag, Berlin (1986).

46. K. Wittmaack, *International Journal of Mass Spectrometry and Ion Physics*, **17**, 39 (1975).

47. R. Hill and P.W.M. Blenkinsopp, *Applied Surface Science*, **231**, 936 (2004).

48. R. Hill, P.W.M. Blenkinsopp, A. Barber and C. Everest, *Applied Surface Science*, **252**, 7304 (2006).

49. G. Gillen, J. Batteas, C.A. Michaels, P. Chi, J. Small, E. Windsor, A. Fahey, J. Verkouteren and K.J. Kim, *Applied Surface Science*, **252**, 6521 (2006).

50. A.G. Sostarecz, S. Sun, C. Szakal, A. Wucher and N. Winograd, *Applied Surface Science*, **231**, 179 (2004).

51. N. Davies, D.E. Weibel, P. Blenkinsopp, N. Lockyer, R. Hill and J.C. Vickerman, *Applied Surface Science*, **203**, 223 (2003).

52. M.S. Wagner, *Analytical Chemistry*, **76**, 1264 (2004).

53. E.A. Jones, J.S. Fletcher, C.E. Thompson, D.A. Jackson, N.P. Lockyer and J.C. Vickerman, *Applied Surface Science*, **252**, 6844 (2006).

54. D.E. Weibel, N. Lockyer and J.C. Vickerman, *Applied Surface Science*, **231**, 146 (2004).

55. P. Sigmund and A. GrasMarti, *Nuclear Instruments and Methods*, **182–183**, 25 (1981).

56. B.V. King and I.S.T. Tsong, *Journal of Vacuum Science and Technology*, **A2**, 1443 (1984).

57. K. Wittmaack, *Surface Science*, **112**, 168 (1981).

58. R. Kelly, *Journal of Vacuum Science and Technology*, **21**, 778 (1982).

59. W. Reuter and K. Wittmaack, *Applied Surface Science*, **5**, 221 (1980).

60. W. Wach and K. Wittmaack, *Journal of Applied Physics*, **52**, 3341 (1981).

61. K. Wittmaack and W. Wach, *Nuclear Instruments and Methods*, **191**, 327 (1981).

62. R. Treichler, H. Cerva, W. Hosler and R. v. Criegern, in *Proceedings of SIMS VII*, John Wiley & Sons, Inc., New York, NY, pp. 259 and 262 (1990).

63. R. v. Criegern, Presented at SIMS V (Abstract only) (1986).

64. S.D. Littlewood and J.A. Kilner, *Journal of Applied Physics*, **63**, 2173 (1988).

65. H.U. Jager, J.A. Kilner, R.J. Chater, P.L.F. Hemment, R.F. Peart and K.J. Reeson, *Thin Solid Films*, **162**, 333 (1988).

66. P.D. Augustus, G.D.T. Spiller, M.G. Dowsett, P. Kightley, G.R. Thomas, R. Webb and E.A. Clark, in *Proceedings of SIMS VI*, John Wiley & Sons, Ltd, Chichester, UK, p. 485 (1988).

67. M.G. Dowsett, D.M. James, I.W. Drummond, M.M. El Gomati, T.A. El Bakush, F.J. Street and R.D. Barlow, in *Proceedings of SIMS VIII*, John Wiley and Sons, Ltd, Chichester, UK, p. 359 (1992).

68. M.G. Dowsett, *Fresenius Journal of Analytical Chemistry*, **341**, 224 (1991).

69. W. Vandervorst and J. Remmerie, in *Proceedings of SIMS V*, Springer-Verlag, Berlin, p. 288 (1986).

70. J.A. Kilner, G.P. Beyer and R.J. Chater, *Nuclear Instruments and Methods in Physics Research*, **B84**, 176 (1994).

71. O. Vancauwenberghe, N. Herbots and O.C. Hellman, *Journal of Vacuum Science and Technology*, **A10**, 713 (1992).

72. M.G. Dowsett, S.B. Patel and G.A. Cooke, in *Proceedings of SIMS XII*, Elsevier, Amsterdam, p. 85 (2000).

73. G.P. Beyer, S.B. Patel and J.A. Kilner, *Nuclear Instruments and Methods in Physics Research*, **B85**, 370 (1994).

74. M.G. Dowsett, N.S. Smith, R. Bridgeland, D. Richards, A.C. Lovejoy and P. Pedrick, in *Proceedings of SIMS X*, John Wiley & Sons, Ltd, Chichester, UK, p. 367 (1996).

75. S. Sun, C. Szakal, T. Roll, P. Mazarov, A. Wucher and N. Winograd, *Surface and Interface Analysis*, **36**, 1367 (2004).

76. M.G. Dowsett, J.H. Kelly, G. Rowlands, T.J. Ormsby, B. Guzman, P. Augustus and R. Beanland, *Applied Surface Science*, **203–204**, 273 (2003).

77. D.P. Chu and M.G. Dowsett, *Physics Review*, **B56**, 15167 (1997).

78. D.S. McPhail and M.G. Dowsett, *Vacuum TAIP*, **36**, 997 (1986).

79. M. Meuris, P. De Bisschop, J.F. Leclair and W. Vandervorst, *Surface and Interface Analysis*, **14**, 497 (1989).

80. J.H. Kelly, M.G. Dowsett, P. Augustus and R. Beanland, *Applied Surface Science*, **203–204**, 260 (2003).

81. F.A. Stevie, P.M. Kahora, D.S. Simons and P. Chi, *Journal of Vacuum Science and Technology*, **6**, 76 (1988).

82. W.T. Bulle-Lieuwma and P.C. Zalm, *Surface and Interface Analysis*, **10**, 210 (1987).

83. E.H. Cirlin, J.J. Vajo, R.E. Doty and T.C. Hasenberg, *Journal of Vacuum Science and Technology*, **A9**, 1395 (1991).

84. E.H. Cirlin, J.J. Vajo and T.C. Hasenberg, *Journal of Vacuum Science and Technology*, **B12**, 269 (1994).

85. Z.X. Jiang, P.F.A. Alkemade, E. Algra and S. Radelaar, *Surface and Interface Analysis*, **25**, 285 (1997).

86. A. Juhel, F. Laugier, D. Delille, C. Wyon, L.F.T. Kwakman and A. Hopstaken, *Applied Surface Science*, **252**, 7211 (2006).

87. A.R. Chanbasha and A.T.S. Wee, *Applied Surface Science*, **252**, 7243 (2006).

88. D. Krecar, M. Rosner, M. Draxler, P. Bauer and H. Hutter, *Applied Surface Science*, **252**, 123 (2005).

89. S.E. Harton, F.A. Stevie and H. Ade, *Journal of Vacuum Science and Technology*, **A24**, 362 (2006).

90. S. Hofmann, *Applied Surface Science*, **241**, 113 (2005).

91. S Hofmann, *Philosophical Transactions of the Royal Society of London Series A – Mathematical Physical and Engineering Sciences*, **362**, 55 (2004).

92. M.G. Dowsett, *Applied Surface Science*, **203**, 5 (2003).

93. A.M. Huber, G. Morillot, N.T. Linh. J.L. Debrun and M. Valladon, *Nuclear Instruments and Methods*, **149**, 543 (1978).

94. Y. Homma and Y. Ishii, *Journal of Vacuum Science and Technology*, **3**, 356 (1985).

95. K. Wittmaack, *Journal of Applied Physics*, **A38**, 235 (1985).

96. J.B. Clegg, in *Proceedings of SIMS V*, Springer-Verlag, Berlin, p. 112 (1985).

97. V.R. Deline, *Nuclear Instruments and Methods in Physics Research*, **218**, 316 (1983).

98. H.N. Migeon, C. Le Pipec and J.J. Le Goux, in *Proceedings of SIMS V*, Springer-Verlag, Berlin, p. 155 (1985).

99. K. Wittmaack and J.B. Clegg, *Applied Physics Letters*, **37**, 285 (1980).

100. J.A. McHugh, in *SIMS*, National Bureau of Standards, Special Publication 427, J. Heinrich and D.E. Newbury (Eds), National Bureau of Standards, Washington, DC, p. 179 (1975).

101. R. von Criegern, I. Weitzel, H. Zeininger and R. Langegieseler, *Surface and Interface Analysis*, **15**, 415 (1990).

102. K. Wangemann and R. Langegieseler, *Fresenius Journal of Analytical Chemistry*, **341**, 49 (1991).

103. J.M. McKinley, F.A. Stevie T. Neil, J.J Lee, L. Wu, D. Sieloff and C. Granger, *Journal of Vacuum Science and Technology*, **B18**, 514 (2000).

104. E. Napolitani, A. Carnera, V. Privitera and F. Priolo, *Materials Science in Semiconductor Processing*, **4**, 55 (2001).

105. K. Wittmaack, *Surface and Interface Analysis*, **21**, 323 (1994).

106. K. Wittmaack and N. Menzel, *Applied Physics Letters*, **50**, 815 (1987).

107. M.G. Dowsett, S.H. Al-Harthi, T.J. Ormsby, B. Guzmán, F.S. Gard, T.C. Noaks, P. Bailey and C.F. McConville, *Physics Review*, **B65**, 113412 (2002).

108. D. Giubertoni, M. Bersani, M. Barozzi, S. Pederzoli, E. Iacob, J.A. van den Berg and M. Werner, *Applied Surface Science*, **252**, 7214 (2006).

109. K. Wittmaack, *Journal of Vacuum Science and Technology*, **A15**, 2557 (1997).

110. C.H. Heiss and F. J. Stadermann, *Advances in Space Research*, **19**, 257 (1977).

111. Y. Zou, L.W. Wang and N.K. Huang, *Thin Solid Films*, **515**, 5524 (2007).

112. B. Guzmán de la Mata, A. Sanz-Hervás, M.G. Dowsett, M. Schwitters and D. Twitchen, *Diamond and Related Materials*, **16**, 809 (2007).

113. D. Álvarez, S. Schomann, B. Goebel, D. Manger, T. Schlosser, S. Slesazeck, J. Hartwich, J. Kretz, P. Eyben, M. Fouchier and W. Vandervorst, *Journal of Vacuum Science and Technology*, **B22**, 377 (2004).

114. I. Garcia, I. Rey-Stolle, B. Galiana and C. Algora, *Journal of Crystal Growth*, **298**, 794 (2007).

115. K. Wittmaack, *Applied Physics Letters*, **29**, 552 (1989).

116. M.G. Dowsett, E.A. Clark, M.H. Lewis and D.J. Godfrey, in *Proceedings of SIMS VI*, John Wiley & Sons, Ltd, Chichester, UK, p. 725 (1987).

117. H. Gnaser, *Journal of Vacuum Science and Technology*, **A12**, 542 (1994).

118. G.D.T. Spiller and J.R. Davis, in *Proceedings of SIMS V*, Springer-Verlag, Berlin, p. 334 (1986).

119. M.G. Dowsett, E.H.C. Parker, R.M. King and P.J. Mole, *Journal of Applied Physics*, **54**, 6340 (1983).

120. M.G. Dowsett, E.H.C. Parker and D.S. McPhail, in *Proceedings of SIMS V*, Springer-Verlag, Berlin, p. 340 (1986).

121. M.G. Dowsett, R. Morris, P.-F. Chou, S.F. Corcoranm, H. Kheyrandish, G.A. Cooke, J.L. Maul and S.B. Patel, *Applied Surface Science*, **203–204**, 500 (2003).

122. Z.X. Jiang, K. Kim, J. Lerma, A. Corbett, D. Sieloff, M. Kotte, R. Gregory and S. Schauer, *Applied Surface Science*, **252**, 7262 (2006).

123. P.R. Boudewijn, M.R. Leys and F. Roozeboom, *Surface and Interface Analysis*, **9**, 303 (1986).

124. E.A. Clark, *Vacuum*, **36**, 861 (1986).

125. S W Downey, A B Emerson and R F Kopf, in *Proceedings of the 1st International Conference on Measurement and Characterization of Ultra Shallow Doping Profiles in Semiconductors*, eds. C Osburn and G McGuire Microelectronics Centre of North Carolina, Research Triangle Park, NC, p. 172 (1991).

126. M.L. Yu and W. Reuter, *Journal of Vacuum Science and Technology*, **17**, 36 (1980).

127. V.R. Deline, C.A. Evans Jr and P. Williams, *Applied Physics Letters*, **33**, 832 (1978). (see also K. Wittmaack, *Journal of Applied Physics*, **52**, 527 (1981) and the following).

128. C.H. Meyer, M. Maier and D. Bimberg, *Journal of Applied Physics*, **54**, 2672 (1983).

129. A.A. Galuska and G.H. Morrison, *International Journal of Mass Spectrometry and Ion Processes*, **61**, 59 (1984).

130. J. Bellingham, M.G. Dowsett, E. Collart and D. Kirkwood, *Applied Surface Science*, **203**, 851 (2003).

131. P. Williams and J.E. Baker, *Applied Physics Letters*, **36**, 842 (1980).

132. N.J. Montgomery, J.L. Macmanus Driscoll, D.S. McPhail, B. Moeckly and K. Char, *Journal of Alloys and Compounds*, **251**, 355 (1997).

133. H.J. Mathieu and D. Leonard, *High Temperature Materials and Processes*, **17**, 29 (1998).

134. B.V. King, M.J. Pellin, J.F. Moore, I.V. Veryovkin, M.R. Savina and C.E. Pripa, *Applied Surface Science*, **203**, 244 (2003).

135. D. Lipinsky, R. Jede, O. Ganschow and A. Benninghoven, *Journal of Vacuum Science and Technology*, **A3**, 2007 (1985).

136. H. Gnaser, J. Fleischauer and W.O. Hofer, *Journal of Applied Physics*, **A37**, 211 (1985).

137. H. Oeschner, W. Ruhe and E. Stumpe, *Surface Science*, **85**, 289 (1979).

138. H. Oeschner, G. Baumann, P. Beckmann, M. Kopnarski, D.A. Reed, S.M. Baumann, S.D. Wilson and C.A. Evans, in *Proceedings of SIMS V*, Springer-Verlag, Berlin, p. 371 (1985).

139. M.J. Pellin, C.E. Young, W.F. Calaway and D.M. Gruen, *Surface Science*, **144**, 619 (1984).

140. D.L. Donohue, W.H. Christie, D.E. Goeringer and H.S. McKown, *Analytical Chemistry*, **57**, 1193 (1985).

141. G. Blaise, *Scanning Electron Microscopy*, **1**, 31 (1985).

142. H.S. Fox, M.G. Dowsett and R.F. Houghton, in *Proceedings of SIMS VI*, John Wiley & Sons, Ltd, Chichester, UK, p. 445 (1988).

143. M. Beebe, J. Bennett, J. Barnett, A. Berlin, and T. Yoshinaka, *Applied Surface Science*, **231**, 716 (2004).

144. S.F. Corcoran and S.B. Felch, *Journal of Vacuum Science and Technology*, **B10**, 342 (1992).

145. J.B. Clegg, *Surface and Interface Analysis*, **10**, 332 (1987).

146. R. Valizadeh J.A. Vandenberg, R. Badheka, A. Albayati, D.G. Armour and D. Sykes, *Nuclear Instruments and Methods in Physics Research*, **64**, 609 (1992).

147. C.M. Ng, A.T.S. Wee, C.H.A. Huan, C.S. Ho, N. Yakolevand and A. See, *Surface and Interface Analysis*, **33**, 735 (2002).

148. S. Miwa, *Applied Surface Science*, **231**, 658 (2004).

149. E.E. Rees, D.S. McPhail, M.P. Ryan, J. Kelly and M.G. Dowsett, *Applied Surface Science*, **203**, 660 (2003).

150. S. Fearn, D.S. McPhail and V. Oakley, *Physics and Chemistry of Glasses*, **46**, 505 (2005).

151. S. Fearn, D.S. McPhail and V. Oakley, *Applied Surface Science*, **231**, 510 (2004).

152. A. Benninghoven, B. Hagenoff and H.W. Werner (Eds), *SIMS X: Proceedings of the 10th International Conference on Secondary Ion Mass Spectrometry*, John Wiley & Sons, Ltd, Chichester, UK (1997).

153. G. Gillen, R. Lareau, J. Bennett and F. Stevie (Eds), *SIMS XI: Proceedings of the 11th International Conference on Secondary Ion Mass Spectrometry*, John Wiley & Sons, Ltd, Chichester, UK (1998).

154. A. Benninghoven, P. Bertrand, H.N. Migeon and H.W. Werner (Eds), *SIMS XII: Proceedings of the 12th International Conference on Secondary Ion Mass Spectrometry*, Elsevier, Amsterdam (2000).

155. A Benninghoven, Y. Nihei, M. Kudo, Y. Homma, H Yurimoto and H. W. Werner, (Eds), *SIMS XIII: Proceedings of the Thirteenth International Conference on Secondary Ion Mass Spectrometry and Related Topics*, *Applied Surface Science*, **203–204** (2003).

156. A. Benninghoven, J.L. Hunter, B.W. Schueler, H.E. Smith and H.W. Werner, (Eds), *SIMS XIV: Proceedings of the Fourteenth International Conference on Secondary Ion Mass Spectrometry and Related Topics*, *Applied Surface Science*, **231** (2004).

157. J.C. Vickerman, I.S. Gilmore, M.G. Dowsett, A. Henderson and A. Benninghoven (Eds), *SIMS XV: Proceedings of the Fifteenth International Conference on Secondary Ion Mass Spectrometry*, *Applied Surface Science*, **252** (2006).

158. D.S. McPhail, *Journal of Materials Science*, **41** (40th Anniversary Issue), 873 (2006).

159. G.D.T. Spiller and T. Ambridge, in *Proceedings of SIMS V*, Springer-Verlag, Berlin, p. 127 (1986).

160. E.A. Clark, M.G. Dowsett, H.S. Fox and S.M. Newstead, in *Proceedings of SIMS VII*, John Wiley & Sons, Ltd, Chichester, UK, p. 627 (1990).

161. D.P. Leta and G.H. Morrison, *Analytical Chemistry*, **52**, 227 (1980).

162. J.N. Miller, *Nuclear Instruments and Methods in Physics Research*, **218**, 547 (1983).

163. K. Wittmaack, *Surface Science*, **112**, 168 (1981).

164. W. Wach and K. Wittmaack, *Nuclear Instruments and Methods in Physics Research*, **228**, 1 (1984).

165. H.E. Smith and G.E. Morrison, *Analytical Chemistry*, **57**, 2663 (1985).

166. R.L. Hervig and P. Williams, in *Proceedings of SIMS V*, Springer-Verlag, Berlin, p. 152 (1986).

167. D.S. Simons, P. Chi, R.G. Downing, J.R. Ehrstein and J.F. Knudsen, in *Proceedings of SIMS VI*, John Wiley & Sons, Ltd, Chichester, UK, p. 433 (1988).

168. F. Toujou, S. Yoshikawa, Y. Homma, A. Takano, H. Takenaka, M. Tomita, Z. Li, T. Hasegawa, K. Sasakawa, M. Schumacher, A. Merkulov, H.K. Kim, D.W. Moon, T. Hong and J.Y. Won, *Applied Surface Science*, **231**, 649 (2004).

169. J.B. Clegg and R.B. Beall, *Surface and Interface Analysis*, **14**, 307 (1989).

170. J.B. Clegg and I.G. Gale, *Surface and Interface Analysis*, **17**, 190 (1991).

171. M.G. Dowsett, G. Rowlands, P.N. Allen and R.D. Barlow, *Surface and Interface Analysis*, **21**, 310 (1994).

172. M.G. Dowsett and D.P. Chu, *Journal of Vacuum Science and Technology*, **B16**, 377 (1998).

173. M. Nojima, A. Maekawa, T. Yamamoto, B. Tomiyasu, T. Sakamoto, M. Owari and Y. Nhei, *Applied Surface Science*, **252**, 7293 (2006).

174. P.N. Allen, M.G. Dowsett and R. Collins, *Surface and Interface Analysis*, **20**, 696 (1993).

175. M.G. Dowsett and R. Collins, *Philosophical Transactions of the Royal Society*, **A354**, 271 (1996).

176. D.S. McPhail, M.G. Dowsett, R.A. Kubiak, S.M. Newstead, S. Biswas and S.D. Littlewood, in *Proceedings of SIMS VII*, Monterey, California, A.M. Benninghoven, C.A. Evans, K.D. McKeegan, H.A. Storms and H.W. Werner (Eds), John Wiley and Sons, p. 103 (1990) (ISBN 0 471 92738 4).

177. D.E. Sykes and R.T. Blunt, *Vacuum*, **36**, 1001 (1986).

178. M. Meuris, W. Vandervorst and H.E. Maes, *Surface and Interface Analysis*, **12**, 339 (1988).

179. D.C. Sing and M.J. Rendon, *Nuclear Instruments and Methods in Physics Research*, **237**, 318 (2005).

180. D.S. Simons, P. Chi, R.G. Downing, J.R. Ehrstein and J.F. Knudsen, in *Proceedings of SIMS VI*, John Wiley & Sons, Ltd, Chichester, UK, p. 433 (1988).

181. L. Capello, T.H. Metzger, M. Werner, J. A. van der Berg, M. Servidori, L. Ottaviano, C. Bongiorno, G. Mannino, T. Feudel and M. Herden, *Journal of Applied Physics*, **100**, 6340 (1983).

182. J.B. Clegg, I.G. Gale, G. Blackmore, M.G. Dowsett, D.S. McPhail, G.D.T. Spiller and D.E. Sykes, *Surface and Interface Analysis*, **10**, 338 (1987).

183. S. Mitra, M. van Rao, N. Papanicolaou, K.A. Jones, M. Derenge, O.W. Holland, R.D. Vispute and S.R. Wilson, *Journal of Applied Physics*, **95**, 69 (2004).

184. P.C. Zalm, K.T.F. Janssen, G.M. Fontijn and C.J. Vriezema, *Surface and Interface Analysis*, **14**, 2039 (1988).

185. J. Kempf, *Surface and Interface Analysis*, **4**, 116 (1982).

186. A. Merkulov, O. Merkulova, E. de Chambost and M. Schuhmacher, *Applied Surface Science*, **231–232**, 954 (2004).

187. M.G. Dowsett, R.D. Barlow, H.S. Fox, R.A.A. Kubiak and R. Collins, *Journal of Vacuum Science and Technology*, **B10**, 336 (1992).

188. K. Wittmaack, *Philosophical Transactions of the Royal Society*, **A354**, 2731 (1996).

189. B.G. de la Mata, M.G. Dowsett, A. Tajani and M. Schwitters, *Surface and Interface Analysis*, **38**, 422 (2006).

190. I.H. Wilson, N.J. Zheng, U. Knipping and I.S.T. Tsong, *Physical Review*, **B38**, 8444 (1988).

191. I.H. Wilson, N.J. Zheng, U. Knipping, and I.S.T. Tsong, *Applied Physics Letters*, **53**, 2039 (1988).

192. G. Carter, J.S. Colligon and M.J. Nobes, *Radiation Effects*, **31**, 65 (1977).

193. P. Sigmund, *Journal of Materials Science*, **8**, 1545 (1973).

194. G.K. Wehner and D.J. Hajicek, *Journal of Applied Physics*, **42**, 1145 (1971).

195. B. Fares, C. Dubois, B. Gautier, N. Baboux, J.C. Dupuy, F. Cayrel and G. Gaudin, *Applied Surface Science*, **252**, 6448 (2006).

196. G.A. Cooke, M.G. Dowsett, P.N. Allen, R. Collins and K. Miethe, *Journal of Vacuum Science and Technology*, **B14**, 132 (1996).

197. H. Gries, *Surface and Interface Analysis*, **7**, 29 (1985).

198. D.S. McPhail and M.G. Dowsett, in *Proceedings of SIMS VI*, John Wiley & Sons, Ltd, Chichester, UK, p. 269 (1988).

199. G. Horcher, A. Forchel, S. Bayer, H. Nickel, W. Schlapp and R. Losch, in *Proceedings of SIMS VII*, John Wiley & Sons, Ltd, Chichester, UK, p. 631 (1990).

200. D.K. Skinner, *Surface and Interface Analysis*, **14**, 567 (1989).

201. C.M. Hsu and D.S. McPhail, *Mikrochimica Acta*, **13**, 317 (1996).

202. C.M. Hsu, V.K.M. Sharma, M.J. Ashwin and D.S. McPhail, *Surface and Interface Analysis*, **23**, 665 (1995).

203. C.M. Hsu and D.S. McPhail, *Nuclear Instruments and Methods in Physics Research*, **B101**, 427 (1995).

204. S. Fearn and D.S. McPhail, *Applied Surface Science*, **252**, 893 (2005).

205. R. Srnanek, A. Satka, J. Liday, P. Vogrincic, J. Kovac, M. Zadrazil, L. Frank and M. El Gomati, in, and J.M. Rodenburg (Eds), Institute of Physics Conference Series, **153**, p. 453 (1997).

206. R. Srnanek, P. Gurnik, L. Harmatha and I. Gregora, *Applied Surface Science*, **183**, 86 (2001).

207. R. Srnanek, R. Kinder, B. Sciana, D. Radziewicz, D.S. McPhail, S.D. Littlewood and I. Novotny, *Applied Surface Science*, **177**, 139 (2001).

208. R. Srnanek, J. Geurts, M. Lentze, G. Irmer, D. Donoval, P. Brdecka, P. Kordos, A. Forster, B. Sciana, D. Radziewicz and M. Tlaczala, *Applied Surface Science*, **230**, 379 (2004).

209. T. Achtnich, G. Burri, M.A. Ply and M. Ilegems, *Applied Physics Letters*, **50**, 1730 (1987).

210. R.T. Lareau, in *Proceedings of SIMS VI*, John Wiley & Sons, Ltd, Chichester, UK, p. 437 (1988).

211. J.G.M. van Berkum, E.J.H. Collart, K. Weemers, D.J. Gravesteijn, K. Iltgen, A. Benninghoven and E. Niehuis, *Journal of Vacuum Science and Technology*, **B16**, 298 (1998).

212. D.G. Welkie and R.L. Gerlach, *Journal of Vacuum Science and Technology*, **20**, 1064 (1982).

213. M.G. Dowsett and G.A. Cooke, in *Proceedings of the 1st International Conference on Measurement and Characterization of Ultra Shallow Doping Profiles in Semiconductors*, C. Osburn and G. McGuire (Eds), Microelectronics Centre of North Carolina, Research Triangle Park, NC, p. 116 (1991).

214. M. Nojima, Y. Kanda, M. Toi, B. Tomiyasu, T. Sakamoto, M. Owari and Y. Nihei, *Bunseki Kagaku*, **52**, 179 (2003).

215. G.A. Cooke, P. Pearson, R. Gibbons and M.G. Dowsett, *Journal of Vacuum Science and Technology*, **B14**, 348 (1996).

216. G.A. Cooke, M.G. Dowsett, C. Hill, E.A. Clark, P. Pearson, I. Snowden and B. Lewis, in *Proceedings of SIMS VII*, John Wiley & Sons, Ltd, Chichester, UK, p. 667 (1990).

217. V.A. Ukraintsev, P.J. Chen, J.T. Gray, C.F. Machala, L.K. Magel and M.C. Chang, *Journal of Vacuum Science and Technology*, **B18**, 580 (2000).

218. M. Rosner, G. Pockl, H. Danninger and H. Hutter, *Analytical and Bioanalytical Chemistry*, **374**, 597 (2002).

219. H. Liebl, *Journal of Applied Physics*, **38**, 3227 (1967).

220. S.P. Thompson, M.G. Dowsett, J.L. Wilkes, N.A. Fairley, C.A. Corlett, J. Nuttall, D. Finbow, B.W. Griffiths, P. Blenkinsop, and S.J. Mullock, in *SIMS VIII: Proceedings of the 8th International Conference on Secondary Ion Mass Spectrometry*, A. Benninghoven, K.F. Janssen, J. Tumpner and H.W. Werner (Eds), John Wiley & Sons, Ltd, Chichester, UK, p. 183 (1992).

221. W. Compston, *Applied Surface Science*, **252**, 7089 (2006).

222. G. Slodzian, B. Daigne, F. Girard, F. Boust and F. Hillion, *Biology of the Cell*, **74**, 43 (1992).

223. K. Wittmaack, *Reviews in Scientific Instruments*, **47**, 157 (1976).

224. C. W. Magee, W.L. Harrington and R.E. Honig, *Reviews in Scientific Instruments*, **49**, 477 (1978).

225. B. Schueler, P. Sander and D.A. Reed, *Vacuum*, **41**, 1661 (1990).

226. M. Schuhmacher, B. Rasser and F. Desse, *Journal of Vacuum Science and Technology*, **B18**, 529 (2000).

227. M.G. Dowsett, S.P. Thompson and C.A. Corlett, in *SIMS VIII: Proceedings of the 8th International Conference on Secondary Ion Mass Spectrometry*, A. Benninghoven, K.F. Janssen, J. Tumpner and H.W. Werner (Eds), John Wiley & Sons, Ltd, Chichester, UK, p. 187 (1992).

228. W.O. Hofer, H. Liebl, G. Roos and G. Staudenmaier, *International Journal of Mass Spectrometry and Ion Physics*, **19**, 327 (1976).

229. K. Wittmaack, *Applied Physics*, **12**, 149 (1977).

230. H. Liebl, *Nuclear and Instrumental Methods*, **187**, 143 (1981).

231. M.G. Dowsett, J.L. Wilkes, N.A. Fairley, A.C. Lovejoy, P.D. Pedrick, S.J. Potter and S.P. Thompson, in *SIMS VIII: Proceedings of the 8th International Conference on Secondary Ion Mass Spectrometry*, A. Benninghoven, K.F. Janssen, J. Tumpner and H.W. Werner (Eds), John Wiley & Sons, Ltd, Chichester, UK, p. 191 (1992).

Problems

1. State what is meant by the following terms: SIMS depth profile, sputter rate, depth resolution, sensitivity, pre-equilibrium period, altered layer, differential

shift, dynamic range, mass interference, secondary ion yield, memory effect, ion implant, retained dose (in an ion implant).

2. Calculate the number of atoms of silicon in a cube 100 nm on side. Now work out the number of ions that will be detected if this volume of material is sputtered in a SIMS experiment, assuming useful ion yields of $10^{-3}, 10^{-4}, 10^{-5}$ and 10^{-6}. Repeat the calculations for a 10 nm cube. Comment on the significance of these calculations in the context of (a) SIMS depth profiling if a depth resolution of 1 nm is required and (b) imaging if a lateral resolution of 50 nm is required [silicon contains 5×10^{22} atoms/cm^3].

3. Describe the steps involved in converting the raw data from a SIMS depth profile into a quantitative plot of concentration as a function of depth. Now describe the procedures that must be followed for the quantitative analysis of a multilayer sample.

4. What is the difference between the sensitivity and the detection limit of a SIMS depth profile and what are the factors that can lead to this difference?

5. Discuss the various factors that limit the depth resolution in a SIMS depth profile.

6. Is there a fundamental limit to the depth resolution that can be achieved in a SIMS depth profile, and if so what is it?

7. Explain why an ultra-high vacuum is beneficial in SIMS depth profiling and why the quality of the vacuum can become even more important in an ultra-low energy SIMS depth profile.

8. The data points in SIMS depth profiles are rarely assigned error bars; discuss.

9. Explain the difference between optical and electronic gating. Which would give the ultimate performance in terms of dynamic range?

6 Low-Energy Ion Scattering and Rutherford Backscattering

EDMUND TAGLAUER

Max-Planck-Institut für Plasmaphysik, Garching bei München, Germany

6.1 Introduction

Energetic ions are unique probes for surface analysis. The impact of an ion with a few hundred electron volts or more of kinetic energy on a solid surface causes a series of collisional processes and electronic excitations; analysis of the energy spectra of backscattered ions shows that they provide detailed information about the atomic masses on the surface and about their geometric arrangement. Since the time for the scattering processes is very short compared to thermal vibrations or the lifetime of a collision cascade, this information can be generally considered as being representative of the instantaneous condition of the surface, unperturbed by the ion beam (this holds for an individual scattering process and does not exclude surface modifications due to high fluence bombardment).

There are two distinct parameter regimes (see Table 6.1) for ion scattering analysis of surfaces and near-surface layers, and accordingly two different techniques.

In low-energy ion scattering (LEIS) or ion scattering spectroscopy (ISS) [1], primary ion energies of 0.5–5 keV are used with noble gas ions (He^+, Ne^+, Ar^+) and also alkali ions (Li^+, Na^+, K^+). With this method, information is obtained from the topmost atomic layer, under certain circumstances also from the second or third layer [2].

In Rutherford Backscattering (RBS) [3] the primary ion energy ranges from about 100 keV (for H^+) to several MeV (for He^+ and heavier ions). The

Surface Analysis – The Principal Techniques 2nd Edition Edited by John Vickerman and Ian Gilmore
© 2009 John Wiley & Sons, Ltd

Table 6.1   Typical physical parameters in ion scattering

Method	Ions	Energy (eV)	de Broglie wavelength λ (Å)	Distance of closest approach r_0 (Å)
ISS	He^+, Ne^+ Li^+, Na^+	10^3	10^{-2}	0.5
RBS	H^+, D^+ He^+	10^6	10^{-4}	0.01
MEIS	H^+, He^+	10^5	10^{-3}	0.03

ion–target atom interaction can be described using the Coulomb potential from which the Rutherford scattering cross-section is derived, which allows absolute quantification of the results. Information in principle arises from a thickness of the order of 100 nm (10^{-5} cm), but analysis of surface layers is also possible by using channelling/blocking techniques. Scattering of H^+ with energies around 100 keV is sometimes referred to as MEIS (Medium Energy Ion Scattering), probably because it only needs a smaller type of accelerator, but physically it is within the RBS regime.

The physical principles are the same for both techniques (ISS and RBS): an ion beam is directed onto a solid surface, a part of the primary projectiles is backscattered from the sample and the energy distribution of these ions is measured (see Figure 6.1). Since the ion–target atom interaction can be described by two-body collisions, the energy spectra can be easily converted into mass spectra. The difference between ISS and RBS arises from the difference in the cross-sections and the influence of electronic excitations and charge exchange processes, which result in different information depths. In both cases, structural information is obtained from crystalline samples by varying the angles between beam and sample. Deduction of structural information from the data is straightforward, since both techniques are 'real space' methods which are based on fairly simple concepts. With ion scattering only the individual atoms of an element can be detected and no information on compounds or molecules can be gained.

In general, models are required for structure analysis. These models are usually based on results from diffraction techniques (X-ray crystallography or low-energy electron diffraction), which provide the symmetry of the unit cell but not, directly, the real atomic positions. A large number of such studies is published in the literature, demonstrating the useful and unique contributions of ion-scattering techniques to surface analysis.

In the following section the physical basis of ion scattering is explained in detail, for low- and high-energy techniques together, since they are based on the same principles. In the subsequent sections, experimental instrumentation, physical characteristics and typical results are described for RBS and ISS individually. The descriptions and the examples are selected

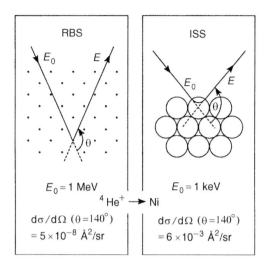

Figure 6.1 Schematic illustration of Rutherford Backscattering Spectroscopy (RBS) and Ion Scattering Spectroscopy (ISS)

with the intention to demonstrate and explain the potential and typical achievements of the methods. It is not useful in the present context to review the vast amount of published results for which the reader has to refer to the relevant literature.

6.2 Physical Basis

6.2.1 THE SCATTERING PROCESS

An important feature of analysis techniques based on energetic ions is that the scattering process can be considered as one or a sequence of classical two-body collisions. A simple estimate of the situation is possible using the parameters given in Table 6.1. Quantum effects are negligible for scattering angles larger than Bohr's critical angle θ_c [4], which is determined by the ratio of the de Broglie wavelength λ and the distance of closest approach r_0 (see also Figure 6.2); $\theta_c \approx \lambda/r_0$. This value is indeed very small compared to all practically used scattering angles. Also diffraction effects from periodic crystal lattices are negligible since $\lambda \ll d$; typical lattice constants d are of the order of a few Angstroms. Let us finally consider in brief the role of thermal vibrations of lattice atoms on the scattering process. Phonon energies are of the order of 0.03 eV and thus very small compared to the ion energies, i.e. phonon interaction cannot be detected in the ion energy spectra. Another way is to look at the collision times, which are about 10^{-15} s, or less for ISS and even shorter for RBS energies, whereas thermal vibration periods are about 10^{-12}–10^{-13} s. Therefore, the energetic ions virtually 'see'

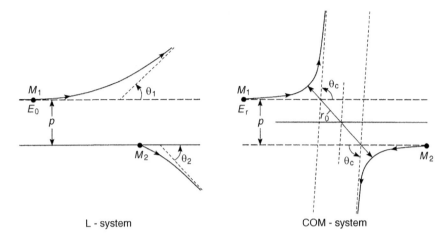

L - system   COM - system

Figure 6.2 Trajectories for the elastic collision between two masses M_1 and M_2 in the laboratory system (left) and the centre-of-mass system (right); p is the impact parameter and r_0 the distance of closest approach

a snap-shot of a rigid lattice with atoms thermally distributed around their ideal lattice positions. In suitable experiments ion scattering can well be sensitive to interatomic distances with an accuracy of 0.1 Å or less and therefore thermal displacements can be detected by ion scattering, as will be shown in Sections 6.3 and 6.4. It clearly follows from this discussion that the interaction between an ion and a target atom can be treated as a classical two-body collision and this is dealt with in the following.

6.2.2 COLLISION KINEMATICS

We consider the collision between two masses M_1 and M_2 which interact through a centrosymmetric potential $V(r)$. Figure 6.2 shows the trajectories in the laboratory and in the centre of mass (COM) system. The projectile mass M_1 has the initial energy E_0 and the target mass M_2 is initially at rest. From the conservation of energy and momentum the particle energies after the collision can be calculated [5] as a function of the scattering angles θ_1 and θ_2 in the laboratory system (see Figure 6.2). For the projectile we write:

$$E_1/E_0 = K, \qquad (6.1)$$

where K is the so-called kinematic factor:

$$K = \left(\frac{\cos\theta_1 \pm (A^2 - \sin^2\theta_1)^{1/2}}{1 + A} \right)^2 \qquad (6.2)$$

K only depends on the mass ratio $A = M_2/M_1$ and the scattering angle. The positive sign holds for $A > 1$ and both signs for $A < 1$. In this latter case,

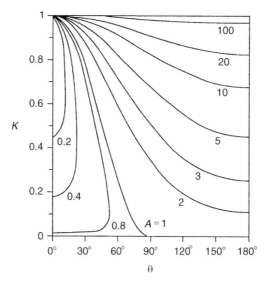

Figure 6.3 The kinematic factor K as a function of the laboratory scattering angle (see Equation (6.2)). The parameter is the mass ratio $A = M_2/M_1$

i.e. heavy projectile on a lighter target atom, the scattering angle is limited to $\theta_1 < \arcsin A$, while there are two scattering angles possible in that region. The function $K(\theta_1)$ is plotted in Figure 6.3. We also note that the kinematic factor, being based on the conservation laws, does not depend on the shape of the potential function.

Equation (6.2) becomes particularly simple for $\theta_1 = 90°$:

$$K(90°) = \frac{A-1}{A+1} = \frac{M_2 - M_1}{M_2 + M_1} \tag{6.2a}$$

and for $\theta_1 = 180°$:

$$K(180°) = \left(\frac{A-1}{A+1}\right)^2 \tag{6.2b}$$

The corresponding expression for the recoiling target atom is

$$\frac{E_2}{E_0} = \frac{4A}{(1+A)^2} \cos^2 \theta_2 \tag{6.3}$$

The applicability of Equations (6.2) and (6.3) for the identification of the scattering masses from ion energy spectra has been demonstrated in many cases [2, 3] and is extensively used in Sections 6.3 and 6.4. A schematic representation of energy spectra for scattering ^{4}He at an angle of 140° from a sample containing ^{108}Ag, ^{28}Si, and ^{16}O is shown in Figure 6.4. The arrows indicate the peak positions according to Equation (6.2). This figure shows the principal similarity of both techniques, ISS and RBS, and the specific differences which are discussed in the respective sections.

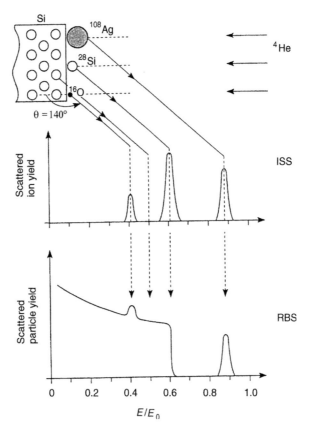

Figure 6.4 Schematic representation of energy spectra of He$^+$ ions scattered at an angle of 140° from a Si substance with Ag, Si and O on the surface. Upper spectrum; ISS ($E_0 \approx 1$ keV), lower spectrum: RBS ($E_0 \approx 1$ MeV)

Figure 6.4 also demonstrates that scattered-ion energy spectra are transformed into mass spectra by virtue of Equation (6.2). Consequently, the mass resolution also can be calculated from this equation:

$$\frac{M_2}{\Delta M_2} = \frac{E}{\Delta E} \frac{A + \sin^2 \theta_1 - \cos \theta_1 (A^2 - \sin^2 \theta_1)^{1/2}}{A^2 - \sin^2 \theta_1 + \cos \theta_1 (A^2 - \sin^2 \theta_1)^{1/2}} \tag{6.4}$$

which for the special case of $\theta_1 = 90°$ becomes

$$\frac{M_2}{\Delta M_2} = \frac{E}{\Delta E} \frac{2A}{A^2 - 1}$$

It can be deduced from Equation (6.4) and its representation in Figure 6.5 (assuming a constant relative energy resolution of the detector of $E/\Delta E = 100$), that mass resolution is best for large scattering angles and about equal ion and target atom masses. So the primary projectile mass has to be selected accordingly if mass resolution is important.

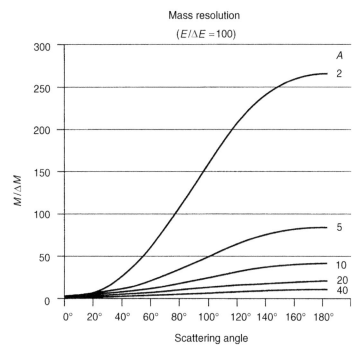

Figure 6.5 Mass resolution $M/\Delta M$ as a function of the scattering angle for a given analyser resolution $E/\Delta E$. The parameter is the mass ratio $A = M_2/M_1$ (see Equation 6.4)

6.2.3 INTERACTION POTENTIALS AND CROSS-SECTIONS

The collision kinematics treated in the previous section yields the positions of the peaks in the energy spectra. The peak intensities, i.e. the probability for scattering into a certain angular and energy interval, are usually given by a scattering cross-section which is determined by the interaction potential. This is briefly explained in the following.

We first need a relation between the scattering angle θ (in the COM system) and the impact parameter p (see Figure 6.2). It can be obtained by considering the conservation of angular momentum, which yields the so-called scattering integral:

$$\theta = \pi - 2 \int_{r_0}^{\infty} \frac{p\,dr}{r^2 \left(1 - \frac{p^2}{r^2} + \frac{V(r)}{E_r}\right)^{1/2}} \tag{6.5}$$

where $E_r = E_0 M_2/(M_1 + M_2)$ is the relative energy in the COM-system. Equation (6.5) gives the connection between the impact parameter p and the scattering angle θ (see Figure 6.2) which is required for calculating the differential scattering cross-section $d\sigma = 2\pi p dp$ and therefore also $d\sigma/d\Omega$ for scattering into a solid angle of $d\Omega$. The angles in the COM and laboratory system are connected through the mass ratio:

$$\tan\theta_1 = \sin\theta_c / [(M_1/M_2) + \cos\theta_c]$$
$$\theta_2 = 1/2(\pi - \theta_c) \tag{6.6}$$

An analytical solution of Equation (6.5) is only possible for certain simple potential functions $V(r)$, e.g. the important Coulomb potential:

$$V(r) = \frac{1}{4\pi\varepsilon_0}\frac{Z_1 Z_2 e^2}{r} \tag{6.7}$$

where Z_1 and Z_2 are the nuclear charges of projectile and target atom, respectively and e is the unit of electrical charge (in SI units). The Coulomb potential describes correctly the interaction in the RBS regime, i.e. the solution of Equation (6.5) with the potential of equation (6.7) yields the well known Rutherford scattering cross-section [6]:

$$(d\sigma/d\Omega)_c = \left(\frac{Z_1 Z_2 e^2}{4E_r \sin^2\theta_c/2}\right)^2 \tag{6.8}$$

in the COM-system, which also holds in the laboratory system if $M_1 \ll M_2$.
The general formula in the laboratory system is [7]:

$$\frac{d\sigma}{d\Omega} = \left(\frac{Z_1 Z_2 e^2}{2E_r \sin^2\theta}\right)^2 \cdot \frac{[(1 - \sin^2\theta/A^2)^{1/2} + \cos\theta]^2}{(1 - \sin^2\theta/A^2)^{1/2}} \tag{6.9}$$

As an example, the Coulomb scattering potential He–Ni is shown in Figure 6.6 and the Rutherford scattering cross-section as a function of Z_2 in Figure 6.7 (for $E_0 = 1$ MeV).

For the lower energies used in ISS, the screening of the nuclear charges by the electron cloud has to be taken into account and therefore frequently screened Coulomb potentials of the form

$$V(r) = \frac{Z_1 Z_2 e^2}{r}\Phi\left(\frac{r}{a}\right) \tag{6.10}$$

are used where a is the screening parameter in the screening function Φ and can be given according to Firsov [8] by:

$$a_F = \frac{0.8854 a_0}{\left(Z_1^{1/2} + Z_2^{1/2}\right)^{2/3}} \tag{6.11}$$

in which a_0 is the Bohr radius of 0.529 Å and the numerical factor is $(9\pi^2/128)^{1/3}$. A similar expression has been developed by Lindhard *et al.* [9]. For many cases, a value of the order of $0.8a_F$ has been found to give best agreement with experimental results.

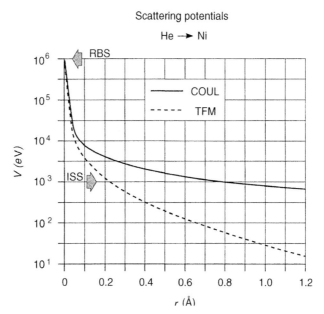

Figure 6.6 Interaction potentials as a function of atomic distance for He–Ni scattering: solid line, Coulomb potential (Equation (6.7)); dashed line, screened potential in the Thomas–Fermi–Molière approximation (Equations (6.10) and (6.12)). The operation regimes for RBS and ISS are indicated

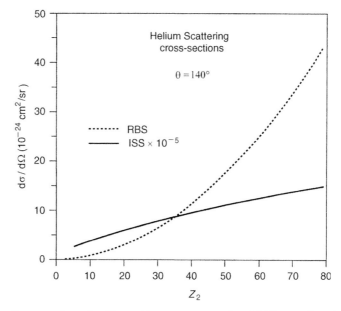

Figure 6.7 Cross-sections for He scattering at an angle of $140°$ as a function of target nuclear charge Z_2. Solid line, ISS ($E_0 = 1$ KeV); dashed line, RBS ($E_0 = 1$ MeV)

Several analytical expressions are given in the literature for the screening function $\Phi(r)$. The Molière approximation to the Thomas–Fermi function has become most widely used in ISS. It is given by a sum of three exponentials:

$$\Phi(x) = 0.35e^{-0.3x} + 0.55e^{-1.2x} + 0.10e^{-6x} \qquad (6.12)$$

with $x = r/a$.

For cross-section calculations, usually no distinction is made between charged and neutral projectiles, i.e. He^0 and He^+ or Ne^0 and Ne^+, because there is sufficient overlap in the electron clouds [8]. The interaction potential for He–Ni and the scattering cross-section for 1 keV He are also shown in Figures 6.6 and 6.7, respectively. From the different regimes in the internuclear distance, the relevance of the electronic screening becomes obvious.

Figure 6.7 shows that the scattering cross-section for ISS is several orders of magnitude larger than the RBS cross-section. While the latter increases with Z_2^2, the dependence on Z_2 is much weaker in the lower energy regime used for ISS.

6.2.4 SHADOW CONE

A very useful concept in structural surface analysis is the so called shadow cone [10, 11]. It is formed by the distribution of ion trajectories downstream of a scattering target atom (see Figure 6.8). The flux of primary ions, described by a beam of parallel trajectories, is deflected by the scattering atom such that a trajectory-free region is formed behind the scatterer. The envelope of trajectories forming this region is called the shadow cone. Obviously, scattering from another atom is not possible if it is located inside the shadow cone, but deviations from a static lattice position can lead to a temperature dependent scattering intensity and therefore to a determination of vibrational amplitudes.

In order to obtain structural information, i.e. to locate atomic positions, it is necessary to know the radius of the shadow cone as a function of distance d from the scattering atom and the intensity distribution across the shadow cone. These quantities can be gained analytically for the case of a Coulomb interaction potential and by using the momentum approximation (i.e. small scattering angles for which $\tan\theta \approx \theta$ and only momentum change but negligible energy loss in the scattering process). In that case the scattering angle θ is related to the impact parameter p according to:

$$\theta = \frac{Z_1 Z_2 e^2}{E_0 p} \qquad (6.13)$$

Hence the scattering angle is inversely proportional to the impact parameter, in this case.

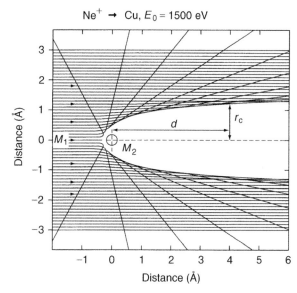

Figure 6.8 Trajectories of a parallel beam of projectiles of mass M_1 forming a shadow cone behind the scattering atom of mass M_2

Since we consider small-angle scattering, the distance R_s between the ion trajectory and a second atom located at a distance d from the scatter can be written as:

$$R_s = p + \theta d \tag{6.14}$$

The function $R_s(p)$ is plotted in Figure 6.9. It represents a situation corresponding to classical rainbow scattering since two primary impact parameters p can lead to the same radius (or secondary impact parameter) R_s. The function $R_s(p)$ has a minimum; the corresponding radius is the Coulomb shadow cone radius R_c given by:

$$R_c = 2(Z_1 Z_2 e^2 d/E_0)^{1/2} \tag{6.15}$$

In this approximation, the shadow cone has a square-root shape, R_c varies as $d^{1/2}$. From this treatment it can also be deduced that shadow cone formation leads to a flux peaking at the edge of the cone. The flux distribution at the position R_s is:

$$f(R_s) 2\pi R_s dR_s = f(p) 2\pi p dp \tag{6.16}$$

and if we normalize the primary flux $f(p)$ to one, we obtain:

$$f(R_s) = \frac{p}{R_s} \left| \frac{dR_s}{dp} \right|^{-1}. \tag{6.16a}$$

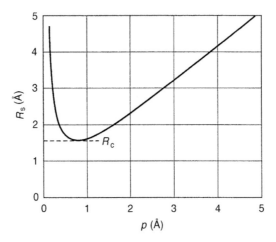

Figure 6.9 Distance R_s to the second atom as a function of the impact parameter to the first atom for 500 eV He scattered at a pair of Mo atoms with an interatomic distance of 3.15 Å. R_c is the corresponding shadow cone radius

This equation can be solved analytically by inserting Equation (6.14) giving the result:

$$f(R_s) = 0 \quad \text{for } R_s < R_c \quad \text{(inside the shadow cone)}$$
$$f(R_s) = \frac{1}{2}(1 - R_c^2/R_s^2)^{-1/2} + (1 - R_c^2/R_s^2) \quad \text{for} R_s > R_c \qquad (6.17)$$
$$\text{(outside the shadow cone)}$$

Equation (6.17) shows that we obtain a flux peak at the edge of the shadow cone represented by a square-root singularity. The distribution obtained from numerical differentiation of Equation (6.14) is plotted in Figure 6.10.

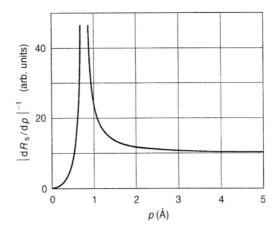

Figure 6.10 Intensity function across the shadow cone (see text)

For the low-energy case in which a screened Coulomb potential has be used, an analoguous expression for the shadow cone radius was calculated by Oen [12]:

$$R_s = R_c(1 + 0.12\alpha + 0.01\alpha^2) \qquad (6.18)$$

with $\alpha = 2R_c/a$ being between 0 and 4.5 and a similar expression for larger values of α. Values calculated using Equation (6.18) agree quite well with experiment if a screening length a between 0.8 a_F and 0.9 a_F is used [2] (see Equation 6.11).

6.2.5 COMPUTER SIMULATION

Many useful results can be obtained from ion scattering analyses by applying the collision kinematics described in Section 6.2.2. If multiple collision effects are important and if detailed structural analysis is aimed at, a more elaborate data analysis is necessary. For understanding of the relevant collision processes and for data interpretation it can then be very helpful to calculate the intensity distributions of the scattered ions with an appropriate model and according to the experimental parameters. Due to the complexity of the system and mathematical restrictions (see Section 6.2.1) this can generally not be done analytically and therefore numerical simulations are required. Various simulation programs have been developed for this purpose. Depending on the degree of sophistication, these programs can be run on a PC with fairly short computing times or they may require more powerful computer facilities. The general procedure is to propose a compositional and structural model of the investigated target and then vary the relevant parameters until satisfactory agreement with the experimental data is obtained. As criteria for the assessment of the agreement sometimes reliability factors (R-factors) are used [13]. A model-free evaluation of the data requires a more elaborate mathematical effort, as e.g. with Bayesian statistics [14] and is therefore not generally applied.

The simulation programs have to be based on sensible approximations that are justified by the physical conditions. Generally these calculations make use of the binary collision approximation based on the arguments given in Section 6.2.1. (The related collision physics is treated in various theory textbooks; a comprehensive description can e. g. be found in Eckstein [15].) The projectile trajectories are calculated from a single collision or a sequence of two-atom collisions. For scattering from polycrystalline or amorphous target material the spectrum is governed by one single collision with a large scattering angle and the corresponding energy loss. In the RBS regime additional electronic energy losses determine the scattered ion intensity distributions. For these situations the targets can be described by their atomic species and concentrations without crystalline structure;

adequate energy dependent electronic stopping must be included and multiple scattering if necessary.

In the RBS regime, i.e. for primary energies above about 100 keV, the collisions are determined by Coulomb interactions between projectile and target nuclei. Accordingly, the interaction potential is well known. Deviations from pure Coulomb interactions occur at low energies (ISS regime) due to electronic screening and with higher MeV energies due to the action of nuclear forces. Nuclear reaction analysis (NRA) has also been developed into a useful analytical technique, but it is not treated here because it is not really surface sensitive.

Among widely used computer programs for RBS analysis are e.g. RUMP [16] and the more recently developed simulation program SIMNRA [17] that comprises the simulation of RBS, NRA and ERDA (elastic recoil detection analysis). It is a versatile Microsoft Windows program with fully graphical user interface and allows the treatment of arbitrary multilayered targets. The targets are considered to be of amorphous structure, crystal effects are not included, i.e. channeling effects cannot be simulated. Surface roughness can also be included. The program moreover contains a large body of non-Rutherford and nuclear reaction cross-sections and includes appropriate possibilities for the treatment of energy loss and straggling. Typical computing times are in the range of several seconds. An example of a SIMNRA simulation is given in Figure 6.12 below.

For the ISS regime also a number of simulation programs have been developed. For basic surface composition analysis the application of Equation (6.2) and Equation (6.37) (see below) is sufficient for the determination of target atomic masses and surface concentrations. But in many cases structural information is required, i.e. the crystalline structure of the target surface is to be determined and thus the program must be able to calculate the scattered ion distribution from ordered atomic arrangements. A straightforward possibility for structural analysis is the shadow cone concept used with ICISS experiments as shown in Section 6.4.4. This is virtually a two-dimensional concept and inclusion of out-of plane scattering is difficult. Nevertheless it can be exploited to obtain useful information within the required accuracy.

The MARLOWE program that was initially conceived for calculating radiation damage in crystals [18] gives a fully three-dimensional treatment of the scattering process. It is a Monte-Carlo program, i.e. the incident particles hit a primary target atom according to a statistical algorithm, thus simulating the many different scattering processes occurring with an extended incident ion beam. Since a large part of these projectile trajectories does not end up in the limited detector acceptance solid angle, the calculations need extensive computer time. With the larger scattering cross-section in ISS (as compared to RBS) and particularly for forward scattering, this problem is reduced. Schemes have been developed to eliminate primary impact areas on the target and trajectories that do not contribute to the recorded scattered ion intensities. A trajectory resolving extension of MARLOWE has been developed and applied particularly for ISS applications [19] (see also Section 6.4).

The problem of calculating trajectories that are not recorded in the experiment is reduced in the 'hitting probability' concept [20]. Here the probability for a projectile to hit a target atom at a certain distance (impact parameter) after a collision with a preceding atom is calculated. The scattering yield is thus obtained after a series of two-atom collisions: multiple scattering effects are not well treated. It is therefore particularly useful for MEIS, for which it was originally developed, and also RBS. A similar program was also developed for ISS with respect to interpreting ICISS results [21]. It yields a continuous intensity distribution and is therefore a step beyond the simple shadow cone consideration.

An important issue in surface structure determination is the influence of thermal vibrations. The corresponding atomic displacements from the ideal lattice positions have to be included in the simulations in order to obtain realistic results. In most cases uncorrelated thermal vibrations of target atoms are considered, the vibrational amplitudes being given by the Debye model. This can be well done with the MARLOWE code and an example in Section 6.4 demonstrates its importance. In the hitting probability concept the relative thermal displacement of the pair of considered atoms is *a priori* contained in the probability calculation. Also correlated thermal vibrations have been considered [22].

The scattered ion yield in ISS is also determined by the probability that the projectile leaves the target surface as an ion, i.e. without undergoing neutralization. These electronic effects are generally not considered in the simulation programs. Their results are therefore particularly useful

Table 6.2 A Selection of frequently used computer programs for simulations of ion scattering

Program	Application	Dimension	Target	Features	Reference
SIMNRA	RBS, NRA, ERDA	3 D	Amorphous	Single atom including non-RBS	17
RUMP	RBS	3 D	Amorphous	Single atom	16
ICISS-SIM	ICISS	2 D	Crystalline	Hitting probability	21
FAN	ICISS	2D + 3 D	Crystalline	Backscattering simulation	2, 24
MARLOWE	ISS, DRS	3 D	Crystalline	Monte Carlo	8, 19
SARIC	ISS, DRS	3 D	Crystalline	BCA	25
TRIM	Backscattering	3D	Amorphous	Monte Carlo	15
VEGAS	MEIS	3D	Crystalline	Hitting probability	22

for charge-integrated results (time-of-flight experiments) or for experiments with known or large survival probability (e.g. with alkali ions). The extended MARLOWE code, yielding individual trajectory analysis, provides the possibility to include various neutralization models that have been applied successfully [23]. These models consider, e.g. the penetration depth in the sample, scattering angle or distance of closest approach.

Instructive examples of computer simulations of ion scattering for RBS, MEIS and ISS are given in the respective following sections. Table 6.2 gives a selection of some frequently used simulation programs together with the relevant references.

6.3 Rutherford Backscattering

6.3.1 ENERGY LOSS

An energetic ion penetrating into a solid loses its energy by a variety of collisional processes. At large impact parameters (of the order of lattice parameters, i.e. $\approx 1\text{Å}$) it transfers energy to valence electrons, about $10\,\text{eV}$ per collision with virtually no deflection. The cross-sections for such processes are high, about 10^{-16} cm^2. At smaller impact parameters, excitation of inner shell electrons can occur, which subsequently leads to de-excitation by X-ray emission, a process which is the basis for proton-induced X-ray analysis (PIXE) of surface layers. Only a small fraction of the primary ions come close enough to a target nucleus (impact parameters of the order of 10^{-12} cm) to undergo an elastic nuclear collision which is described by the kinematics given in the previous section. If such an ion is backscattered, its final energy is determined by the 'elastic' nuclear collision in a certain depth of the sample and the additional 'inelastic' energy loss to electrons on its way in and out of the target. The penetration of ions in matter and the corresponding energy loss processes have been the subject of a number of fundamental papers in this field [4, 10, 26]. Here, we are interested in the energy loss per unit length, $-\mathrm{d}E/\mathrm{d}x$ given in eV/Å and commonly called stopping power. The stopping cross-section S (eV/(atoms/cm^2)) relates this quantity to the atomic density N and is therefore more specific for a certain atomic species:

$$S = -\frac{\mathrm{d}E}{\mathrm{d}x}\frac{1}{N} \tag{6.19}$$

Extensive tables of stopping power data have been collected by Ziegler *et al.* [27] and they have become indispensable for practical analysis of RBS data. An example is given in Figure 6.11 which shows the stopping power for helium in nickel over a wide range of primary energies. Nuclear stopping (included in the dashed curve) is only significant at the lower energy part. The stopping power curve exhibits a broad maximum around 1 MeV and

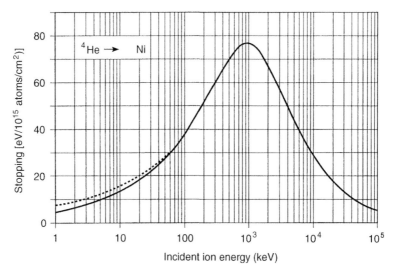

Figure 6.11 Stopping power for He ions in Ni as a function of incident energy. The solid line corresponds to electronic stopping while the dashed line also includes nuclear stopping (from [27])

that is the operational regime of RBS: here, the stopping power does not depend very much on the ion energy and hence can be assumed to be constant as a sufficient approximation in many cases; the stopping power there has its maximum value and therefore RBS its best depth resolution and in this energy range the nuclear interaction is exactly given by the Coulomb potential, giving RBS the advantage of an absolute analytical method. The shape of the stopping power curves is very similar for various ions and materials but the position of the maximum of course depends on the considered species. The high-energy part of the stopping power curve is theoretically well described by the famous Bethe–Bloch formula [27].

An RBS energy spectrum as, for instance, schematically shown in Figure 6.4 is not only determined by the two-body collision kinematics but also by the broad distribution of ions backscattered from deeper layers. One generally observes a surface 'edge' in the spectrum for each atomic species which is present in the target material, and an increasing scattered ion yield towards lower energies. A heavy adsorbate on the surface (Ag in the example) gives rise to an isolated peak at higher energy, obviously an analytically favourable situation. The peak of a lighter constituent (oxygen) generally sits on a broad background and is sometimes difficult to detect without special measures.

The energy spectra in RBS are therefore transformed into mass spectra via Equation (6.2) and into a depth distribution through the stopping power. The basic equation which relates the final energy E_1 of an ion to the scattering depth t in the approximation of a constant energy loss $(\mathrm{d}E/\mathrm{d}x)\,(E_0)$ on the way in and $(dE/dx)(E_1)$ on the way out, is for normal incidence:

$$E_1(t) = K\left[E_0 - t\left(\frac{dE}{dx}\right)(E_0)\right] - \frac{t}{|\cos\Phi|}\left(\frac{dE}{dx}\right)(E_1) \qquad (6.20)$$

A more complete analysis is based on the same concept but applies appropriate numerical techniques.

In the continuous distribution of the scattered ion spectrum, an energy interval ΔE_1 therefore corresponds to a layer of thickness Δt in the depth of the sample. In other words, an energy resolution ΔE_1 given by the apparatus, results in a depth resolution Δt that can be derived from Equation (6.20):

$$\Delta t = \Delta E_1 / \left[K\left(\frac{dE}{dx}\right)(E_0) + \frac{1}{|\cos\theta|}\left(\frac{dE}{dx}\right)(E_1)\right] \qquad (6.21)$$

It can be seen that the depth resolution strongly depends on the stopping power and it is best for most elements in the energy range of 1–2 MeV (for He) where the stopping power has its maximum. For the same reason, better depth resolution can be expected for heavier materials, higher Z elements having larger stopping power. If we consider a typical energy resolution of 15 keV for a solid state detector and backscattering geometry ($\theta \approx 180°$), we get a depth resolution of about 220 Å for He in nickel. This can be improved by setting the detector to a grazing exit angle and thus increasing the path length of the scattered particles in the material. Taking, in the example, a scattering angle of $\theta = 95°$ improves the depth resolution to about 40 Å.

In compound material the stopping is commonly calculated as the sum of the weighted elemental stopping cross-sections; this is called Bragg's rule [28]. If we consider, e.g. a compound of two constituents A and B with the relative abundances m and n, respectively ($m + n = 1$), then Bragg's rule yields:

$$S(AmBn) = mS(A) + nS(B) \qquad (6.22)$$

and the specific energy loss is:

$$\frac{dE}{dx}(AmBn) = N(AmBn)S(AmBn) \qquad (6.22a)$$

where $N(AmBn)$ is the atomic density of the compound material. It turns out that this simple rule implies an uncertainty of less than 10% in most practical cases [29].

Depth resolution can be optimized by detector resolution and grazing exit angles only for scattering from near-surface layers. In larger depths, energy 'straggling' occurs due to the statistical nature of the energy-loss process, i.e. if a number of particles has penetrated to a certain depth in the sample, their energies have a distribution of a certain width. The variance of this (Gaussian) distribution was calculated by Bohr [30] to be for normal incidence:

$$\Omega_B^2 = 4\pi Z_1^2 e^4 N Z_2 t(1 + 1/|\cos\theta|) \qquad (6.23)$$

We see that the mean-square value of the straggling in Bohr's treatment increases linearly with the nuclear charge Z_2 of the target material and with depth t and that it is independent of the ion energy. Bohr's calculation is only an approximation; improved values for straggling were obtained by Chu *et al.* [31]. For He scattering from Ni at a depth of 1000 Å and a scattering angle of 95°, Equation (6.23) yields $\Omega_B = 17$ keV. This is the value of the standard deviation; for comparison with detector resolution we have to take the full width at half maximum (FWHM), that is Ω_B has to be multiplied by a factor of $2\sqrt{(2\ln 2)} = 2.335$ which gives us an energy width of 40 keV, i.e. much larger than a typical detector resolution of 15 keV. Obviously in such a case the resolution of the system is determined by energy straggling.

We now give an estimate for the shape of the continuous energy spectrum determined by scattering from a thick target. The scattering yield from a slab of width Δt at a depth t can be written as:

$$Y(t)\Delta t = \frac{d\sigma}{d\Omega} N \Delta \Omega Q \Delta t \tag{6.24}$$

where $\Delta \Omega$ is the solid angle subtended by the detector (of 100% efficiency) and Q denontes the number of primary ions. We know from Equation (6.8) that the cross-section depends on the energy E_t in the depth t like:

$$\frac{d\sigma}{d\Omega} \approx E_t^{-2}$$

E_t can be estimated by assuming a constant ratio a of the energy loss on the inward and outward ion path:

$$a = \frac{\Delta E_{\text{out}}}{\Delta E_{\text{in}}} = \frac{KE_t - E_1}{E_0 - E_t} \tag{6.25}$$

For light projectiles we can assume $K \approx 1$ and then also $a \approx 1$ which yields:

$$E_t = \frac{1}{2}(E_0 + E_1) \tag{6.26}$$

Combining Equations (6.24) and (6.26), we see that the scattering yield $Y(E_1)$ in which E_1 varies with depth is

$$Y(E_1) \sim (E_0 + E_1)^{-2} \tag{6.27}$$

This is the form indicated in Figure 6.4 schematically and shown in the experimental results in the following sections.

6.3.2 APPARATUS

The principal components in RBS are those of a typical scattering experiment: (i) a source which provides energetic primary ions, (ii) a sample holder that allows us to position the target with the necessary degrees of freedom and the required precision and (iii) a detection system for measuring the energy distribution of the scattered particles. In most cases, particularly in fundamental surface research, manipulator and detector are mounted in an ultra-high vacuum (UHV) chamber.

The most widely used accelerator type for the energy region of interest for RBS is the van de Graaf type in which a high voltage is built up by a fast moving belt that carries electrons from a charging screen to a high-voltage terminal. This charging belt and the ion source at the terminal are housed in a gas-filled tank. Voltages up to about 2 MV, which are most useful for surface work, are readily accomplished, but much higher values (up to 30 MV) are achieved in large machines. The ion beam from the tank is directed through an evacuated beam line to the target chamber. For this purpose, a system of switching magnets, focusing quadrupole magnets, collimators, etc. has to be set up. For surface analysis work a beam current of about 100 nA on a beam spot of about 1 mm in diameter is typical and sufficient to give counting rates of 10 kHz or more. The high voltage requires tank dimensions of the order of metres and also the length of the beam lines is usually several metres. Therefore, major construction and financial investments are necessary for an accelerator and this limits the proliferation of instruments for that powerful method.

The UHV chamber and the manipulator are nowadays very much of a standard type used in surface science research, e.g. also in low-energy ion scattering (see Section 6.4). The manipulator usually has two rotational degrees of freedom, one around the main axis (defining the angle of incidence) and one around an axis perpendicular to the sample surface (defining the azimuthal position of the scattering plane). For channelling measurements, a precision for setting these angles of much better than $1°$, about $0.1°$, is usually required.

The manipulator can also provide heating (by electron bombardment) and cooling (liquid nitrogen) facilities if necessary. It is also very useful that it contains a calibrated scattering standard with a known concentration of a heavy element on a light support (e.g. Au on Si). It is then easy to determine the absolute aerial atomic density on a sample by comparison, without exact knowledge of the solid angle accepted by the detector.

An extremely useful instrument in the development of the RBS analysis technique turned out to be the silicon solid state particle detector. This is a fairly simple device which allows to detect particles and their kinetic energies, i.e. to record an energy spectrum without sweeping an energy window. Energetic particles which penetrate through the gold surface barrier, deposit their kinetic energy in the silicon by creating electron–hole pairs, losing 3.6 eV/pair. The reverse bias voltage separates the charges and creates a corresponding voltage pulse $\Delta V = Ee/3.6C$ (C being the capacity of the device), i.e. the signal is proportional to the energy of the incoming particle. The pulse height distribution, after appropriate amplification recorded by a multichannel analyzer or a computer, therefore represents the energy spectrum of the scattered particles. In this high-energy range, ions and neutrals are equally well detected with high probability. Internal fluctuations in the solid state detector limit the energy resolution to about 10 keV, in practical cases often 15 keV. This is then also a limitation of the mass resolution (see Section 6.2) and the depth resolution of the system.

Better resolution (at the cost of more complicated equipment and sequential recording) can be obtained by magnetic or electrostatic energy analyzers which have a constant relative energy resolution $\Delta E/E$. Typical values are of the order of 5×10^{-3}. For an electrostatic analyzer used with an accelerator in the 50 keV–400 keV range, a depth resolution of about 10 Å has been reported [32]. By this means MEIS becomes sensitive to surfaces and near surface layers and interfaces.

6.3.3 BEAM EFFECTS

An energetic primary ion beam impinging on a sample, modifies its surface by removing atoms from this surface. This effect, called sputtering [33] and for removal of adsorbed layers occasionally ion impact desorption, limits the sensitivity of ion-scattering methods. The questions of interest are: how many atoms are removed (sputtered) from the surface for obtaining a given scattered ion yield Q_D in the detector and what is the minimum atomic areal density which can be detected if we accept a fraction q of the layer to be sputtered away?

The scattered ion yield can be expressed as:

$$Q_D = N\frac{d\sigma}{d\Omega}\Omega Q \tag{6.28}$$

where Q is the number of primary ions, N the surface density of atoms of the considered species (atoms/cm^2), $d\sigma/d\Omega$ the differential Rutherford scattering cross-section (cm^2/sr), which in the following will be written as σ for simplicity, and Ω is the solid angle subtended by the detector (of unit efficiency). It follows from Equation (6.28) that we need an amount of primary ions $Q = Q_D/(N\sigma\Omega)$ in order to obtain a given signal Q_D. The number of atoms sputtered by these ions is QY, where Y (atoms/ion) is the sputtering yield that depends on the ion mass and energy and on the target material. Considering very thin layers (monolayers and below) it is convenient to define a total sputtering cross-section σ_s (cm^2/ion) [34] since the sputtering yield (the probability that a surface atom of the considered species is sputtered) depends on the coverage, in contrast to bulk elemental sputtering; σ_s can be connected to the sputtering yield via the monolayer density N_{ML}:

$$Y = \sigma_s N_{ML} \tag{6.29}$$

For a given Q_D the number of atoms removed by the beam with spot area a (cm^2) is then determined by the ratio of the sputtering and scattering cross-sections:

$$\Delta N = \frac{Q_D}{a\Omega}\frac{\sigma_s}{\sigma} \qquad (6.30)$$

If we then require that not more than a fraction q (e.g. 1%) of the initial coverage should be removed during the measurement, $\Delta N < qN$, we get an estimate for the minimum detectable coverage:

$$N_{min} = \frac{Q_D}{a\Omega}\frac{\sigma_s}{q\sigma} \qquad (6.31)$$

and also a condition for the maximum primary ion fluence by using Equation (6.28):

$$\frac{Q}{a} < \frac{q}{\sigma_s} \qquad (6.32)$$

Obviously, we get a good sensitivity for heavy elements since the Rutherford scattering cross-section σ varies with Z^2. Let us consider an example by using typical numbers for an Au layer on a light substrate [35]. We assume a minimum of 100 ion counts to give a statistically significant signal, i.e. $Q_D = 100$, a beam spot size of 10^{-2} cm^2, a solid angle of 4×10^{-2} sr (i.e. 1 cm^2 at a distance of 5 cm) and σ_s is about 10^{-18} cm^2 [34]. From these numbers we obtain a detection limit of 2.5×10^{12} Au atoms/cm^2 or about 1/400 of a monolayer for this special case; 1% therefore would be removed during the measurement which requires 10^{14} primary ions or slightly more than 10 µC. These estimates are not too far from experimental experience. An important general conclusion which can also be drawn from these considerations is that RBS can quite generally be regarded as a virtually non-destructive method for the analysis of surfaces and near-surface layers, another essential feature which makes this method so attractive.

6.3.4 QUANTITATIVE LAYER ANALYSIS

In Figure 6.4 we showed schematically the features of an RBS energy spectrum and, with the relations derived in the previous sections, we are able to understand in principle the quantitative interpretation of an actual RBS spectrum. As an example, Figure 6.12 shows an energy spectrum taken from a model catalyst consisting of a TiO$_2$ layer on Ti as support material and a Rh metal overlayer as the active component [23]. This is the situation of a heavy adsorbate on a light substrate and the amount of Rh can easily be determined by using Equations (6.24) or (6.28) to calculate N. In practice the solid angle $\Delta\Omega$ and the detection efficiency are taken into account by using a calibrated standard as mentioned above. For the present example we obtain a Rh coverage of 5×10^{14} Rh/cm^2 which is about one half monolayer on the nominal surface and demonstrates nicely the high sensitivity of RBS in such a case. In the continuous part we first see the Ti edge from the TiO$_2$ layer

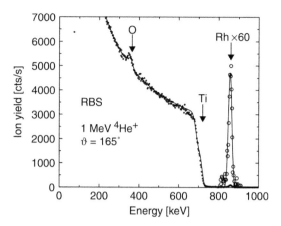

Figure 6.12 RBS spectrum for scattering of 1 MeV He$^+$ ions from TiO$_2$ with about half a monolayer coverage of Rh [36]. The open circles are experimental data while the solid line is a simulation calculation using the SIMNRA program

and a shoulder from the Ti at the Ti/TiO$_2$ interface. The oxygen (from the titania layer) sits on the continuous part of the spectrum which is related to scattering from Ti in the depth.

The decomposition of the spectrum of such a layered structure is demonstrated in more detail in the example given in Figure 6.13. It shows a spectrum taken with 2.5 MeV He^{2+} from a superconducting high-T_c film of YBa$_2$Cu$_3$O$_7$ on (100) SrTiO$_3$ [37]. The upper part shows the actual spectrum (noisy trace) and a numerical fit (solid line). The solid arrow lines indicate the edges corresponding to scattering from the respective elements at the sample surface (cf. Equation (6.1)) and the dashed arrow lines correspond to scattering from the substrate surface, i.e. they include the additional energy loss of the ions in the film (cf. Equation (6.20)). This is further demonstrated in the lower part of Figure 6.13 where the spectral contributions of the film components are separated and the scattered ion distribution for each of the elements in the film (Ba, Y and Cu) shows a leading edge for scattering from the surface and a trailing edge corresponding to the film-substrate interface.

The stoichiometry of the film can now be calculated from the height H_i of the spectra of the various elements i. Let us for simplicity consider normal incidence and a scattering angle close to 180° and assume the validity of the constant dE/dx approximation. Then a detector with an energy resolution ε counts backscattered particles from a layer with thickness Δt and from Equation (6.21) we find:

$$\varepsilon = [E]n\Delta t = [S]\Delta t \tag{6.33}$$

where $[S]$ is the so-called energy loss factor:

$$[S] = K\left(\frac{dE}{dx}\right)_{in} + \left(\frac{dE}{dx}\right)_{out} \tag{6.34a}$$

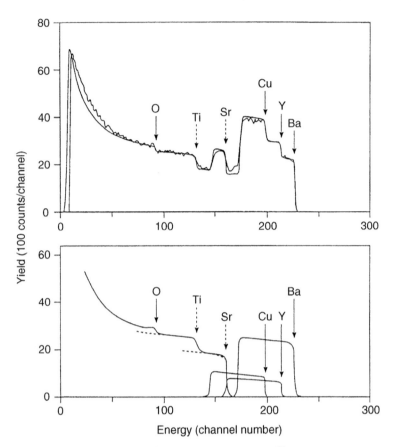

Figure 6.13　RBS spectrum for scattering of 2.5 MeV He^{2+} from a superconducting YBa$_2$Cu$_3$O$_7$ film. Upper part: experimental spectrum (noisy line) and numerical simulation (smooth line) (redrawn from [37]). Lower part: decomposition into the scattering contributions from the various constituents. Solid arrows correspond to the surface of the film, dashed arrows to the interface with the SrTiO$_3$ substrate

and [E] is the stopping cross-section factor:

$$[S] = [E]n \tag{6.34b}$$

both expressions in our approximation; n is the number of atoms per unit volume.

The number of counts from species i in a layer Δt at the surface is then:

$$H_i = \frac{d\sigma_i}{d\Omega}(E_0)\Delta\Omega Q \frac{\varepsilon}{[E]_i} \tag{6.35}$$

and the ratio for two species i and k is:

$$\frac{H_i}{H_k} = \frac{n_i\sigma_i[E]_k}{n_k\sigma_k[E]_i} \tag{6.36}$$

from which the stoichiometric ratios $n_i/n_k/$, etc. can be determined using the experimentally gained peak-height ratios. If we assume $[E]_k = E_i$ we obtain:

$$\frac{n_i}{n_k} = \frac{H_i}{H_k}\left(\frac{Z_k}{Z_i}\right)^2$$

as a crude estimate of the stochiometry.

Detailed numerical analysis, as e.g. shown in Figure 6.13, have to take the correct angular situation, the dependence of dE/dx on stoichiometry (Bragg's rule) and energy into account, but are based on the same principles.

6.3.5 STRUCTURE ANALYSIS

Analysis of crystalline surface structures can be achieved with RBS by exploiting collective scattering phenomena similar to the flux modifications discussed in connection with the shadow cone (Section 6.2.4). Structure analysis is based on the channelling effect [38–40] and the (in this respect) inverse process of blocking. Using these techniques, surface layer analysis can be performed, although in principles the depth resolution in RBS is of the order of 30–100 Å, as pointed out in the previous section.

Channelling occurs if a collimated ion beam impinges on a monocrystalline target along a low-index direction (i.e. close to a high-symmetry axis). In that case most primary particles have large impact parameters with the atoms of the first layer, i.e. they suffer only small angle deflections. This can be continued in the following deeper layers and the ions are then steered in the channels (axial channelling) or between crystal planes (planar channelling). The situation is schematically shown in Figure 6.14. It is obvious that the backscattering yield is reduced in such a case compared to the 'random' situation. In the idealized limit (rigid lattice, no beam divergence) only scattering from the top atomic layer would be possible. If the detector also defines a scattered ion direction along a high-symmetry axis, scattering from atoms in regular lattice positions is blocked. This double-alignment technique has been proven to be particularly useful for surface analysis [40, 41].

The analytical principle is then, that deviations from the idealized lattice structure change the backscattered flux intensity and angular distribution and so can be used for investigating effects such as thermal vibrations [42], lattice relaxation [43], lattice reconstruction [44], interstitial atom positions [45], adsorbate locations [46] and surface disorder [47].

The role of thermal vibrations has been studied in a number of investigations on silicon and metal single crystals [42]. It could be shown that the intensity of the surface peak scales with the ratio ρ/R_c where ρ is the root-mean-square amplitude of the thermal vibrations and R_c the shadow cone radius at the second atom. The surface peak intensity (in atoms/row) varies from 1 for $\rho/R_c < 0.3$ up to values of 4 atoms/row for $\rho/R_c > 1.5$.

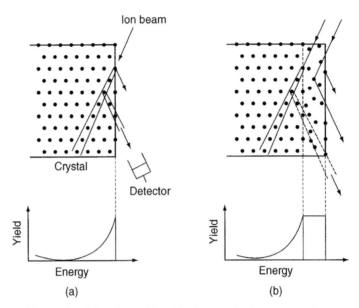

Figure 6.14 Schematic of the channeling-blocking technique: scattering geometry (ion incidence and detection along high-symmetry directions) and energy spectrum for: (a) a well ordered crystal and (b) a crystal with a disordered overlayer (after [47])

The channelling and blocking effects cause minima in the angular scan curves of the scattered ion yield. This is schematically shown in Figure 6.15 for a double alignment situation [41]. The figure also shows how the angular shift is related to the lattice spacing between the top surface layer and the bulk. Using this technique, surface relaxation and reconstruction has been identified on a number of metal (Ag, Ni, Pt, W) and Si single crystal surfaces.

An illustrative example for adsorbate position determination by channelling techniques, is a study of deuterium adsorption on Pd (100) [46]. Here transmission channelling through a thin (3000 Å) Pd crystal was used. The principle for site location and the experimental results are shown in Figure 6.16 giving data for 2 MeV He^+ scattering from Pd and D elastic recoil detection. Deuterium is obviously in a four-fold hollow site position, with different vertical displacements for two adsorbate phases, $\Delta z = 0.3$ Å for $p(1 \times 1)$ and $\Delta z = 0.45$ Å for the $c(2 \times 2)$ phase. This kind of application of the channelling technique is, in principle, very similar to the method used for analysing the positions of solute atoms in a bulk crystal, e.g. for discriminating between substitutional, interstitial or random solute atom positions [45].

Another very successful application of the channelling–blocking technique relates to the thermodynamics of ordered surfaces, in particular to the presently very actively studied field of surface melting. These studies again nicely demonstrate the power of the technique and an illustrative

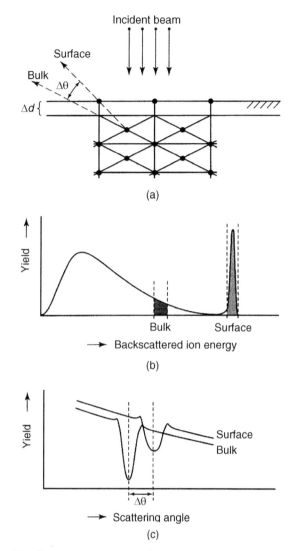

Figure 6.15 Double alignment experiment for measuring surface relaxation: (a) scattering geometry showing the different blocking directions for surface and bulk scattering, (b) corresponding energy spectra taken with an electrostatic analyzer and (c) angular intensity distributions showing the shift in the blocking minimum due to surface relaxation (from [41])

example is given in Figure 6.17 [47]. On a well ordered Pb (110) surface at 295 K the backscattered ion yield in double alignment geometry only shows the surface peak. By increasing the crystal temperature, the development of a disordered surface layer on an ordered bulk crystal can be seen. A typical 'random' spectrum is obtained only after heating to the bulk melting temperature of 600.7 K.

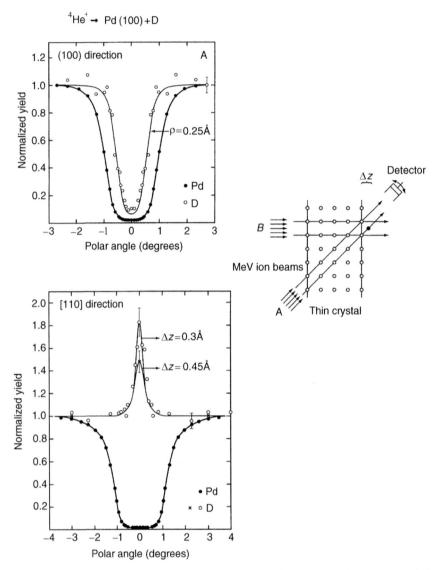

Figure 6.16   Transmission channelling experiment for determining the position of deuterium adsorbed on Pd(100) using 1.9 MeV ^{4}He$^+$. Pd scattering and D recoil intensities both show a minimum for incidence along the [100] direction (upper left panel), whereas in the [110] direction the D recoil intensity has maxima (lower left panel), demonstrating the D position in a four-fold hollow site, 0.3 Å (for p (1 × 1), open circles) and 0.45 Å (for C (1 × 1), crosses) above the surface (from [46])

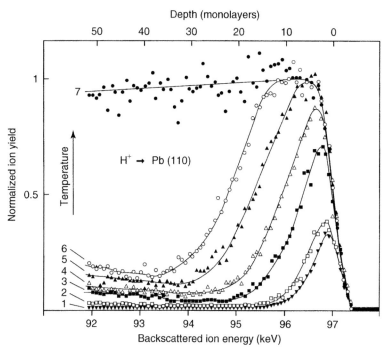

Figure 6.17 RBS spectra for 97.5 MeV protons scattered from a Pb(110) surface in the scattering geometry shown in Figure 14. The temperature variation (1:295 K, 2:452 K, 3:581 K, 4:597 K, 5:599.7 K, 6:600.5 K, 7:600.8 K) demonstrates the development of a disordered layer by surface melting. Spectrum 7 (above the bulk melting temperature of 600.7 K) corresponds to scattering from a 'random' solid (after [47])

6.3.6 MEDIUM-ENERGY ION SCATTERING (MEIS)

MEIS is the variety of RBS using ion energies in the range of about 100–400 keV, typically H^+ and He^+. It has been developed [40] as a technique to analyse composition and geometrical structure of crystalline surfaces and surface layers. The information arises from the formation of shadowing and blocking cones by the flux of ions into and out of the target surface (see Figure 6.15). The shadow cones are narrower than those for ISS energies (see Section 6.2.4) and therefore potentially higher precision in the determination of atomic positions can be obtained. The measured blocking patterns can be directly related to the positions of surface atoms. For the determination of atomic displacements an accuracy of 0.003 nm has been claimed [48]. By making use of the inelastic energy losses to target electrons also depth information can be included and thus subsurface compositional analysis is possible. This 'non-constructive' depth profiling is the same as in RBS, but with MEIS a shallower surface region is analysed and with the use of electrostatic energy analysers better depth resolution can be achieved, in principle a resolution of one atomic layer.

At the AMOLF Institute, NL, where the technique was initially applied, also the computer code VEGAS [20], using the hitting probability concept (see also Section 6.2.5) was developed for data interpretation. More recently a new facility dedicated to MEIS applications was established in Daresbury, UK [49]. It allows parallel detection of a range of scattering angles and energies and thus rapid accumulation of blocking patterns in double alignment. For the determination of best-fit structural parameters different reliability factors, 'R-factors' are evaluated, in analogy to procedures frequently used for data interpretation in low-energy electron diffraction, LEED [48]. A substantial number of successful MEIS applications can be found in the published literature. Figure 6.18 shows as an instructive example blocking patterns for the Ni (100)c(2×2)-O surface [50]. The obtained values for the oxygen nickel and outermost Ni-Ni spacings are consistent with previous LEED studies.

6.3.7 THE VALUE OF RBS AND COMPARISON TO RELATED TECHNIQUES

The outstanding strength of RBS is that it provides absolute quantitative analysis of elemental compositions. Surface coverages can be measured easily with an accuracy of 5% or better. For surface layers and thin films it provides quantitative depth profiles for a thickness of up to about 1 μm with depth resolution of the order of 30–100 Å. In many practical cases the high-energy ion beam used for the analysis causes relatively little damage to the sample. This is particularly so for metals and most semiconductors which are virtually unaffected by the energy transfer through electronic energy loss processes. Only the small nuclear energy loss (see Figure 6.11) contributes to permanent damage. The situation is different for insulating material such as oxides, alkali halides, polymers, etc. in which considerable radiation damage can be caused by electronic energy loss processes. Moreover, charging of the sample often disturbs the analysis in these cases.

Depending on the type of analysis, RBS requires high vacuum or ultra-high vacuum (for surface analysis) in the target chamber. In special cases analysis under atmospheric pressure has also been reported [51].

Using the channelling and blocking techniques, RBS can provide information with monolayer resolution and thus it can be very successfully used as a near-surface structural tool.

The high-energy ion beam used in RBS also provides the possibility to apply related analytical techniques in the same apparatus. Forward recoil detection or elastic recoil detection (ERD) is useful for light sample constituents such as hydrogen isotopes. It is a natural complement to the scattering process, the kinematics is similar (Section 6.2.2) and an example is given in the previous section.

Another related technique is the analysis of proton induced X-ray emission (PIXE) [52]. It is also a quantitative method, the element identification

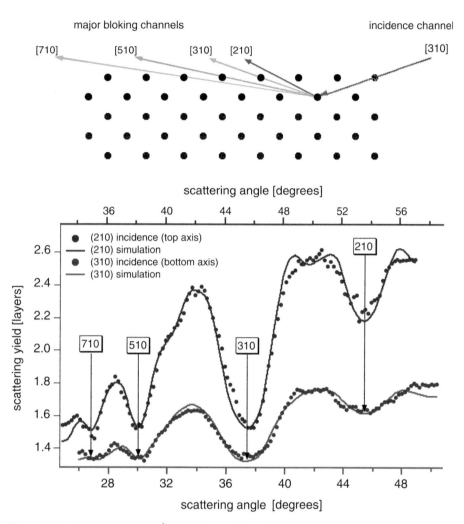

Figure 6.18 MEIS (100 keV H$^+$) blocking patterns for the Ni(100)c(2x2)-O phase, experimental data and best fit calculations with the VEGAS code. Note the shifted angular scales for the two different angles of incidence. Surface atom arrangement and scattering geometry for one direction are shown in the top panel [50]

is often more unambiguous than in the case of RBS which sometimes suffers from limited mass resolution. Compared to electron induced X-ray analysis (EIX), protons have the advantage of producing much less background due to bremsstrahlung. Some of these features can be seen in the example shown in Figure 6.19 [53] which shows the analysis of the identical surface area by RBS with 2 MeV He$^+$, PIXE with 1.5 MeV H$^+$, EIX with 15 keV electrons and Auger Electron Analysis (AES) [54] with 3 keV electrons. The

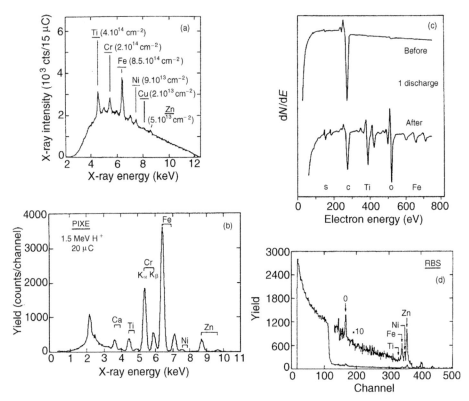

Figure 6.19 Analysis of a graphite surface exposed to one fusion plasma discharge with different techniques: (a) electron induced X-rays ($E_0 = 15$ keV, 175 μm Mylar filter); (b) proton induced X-rays; (c) AES ($E_0 = 3$ keV, 50 μA); (d) RBS with 2 MeV $^4He^+$ (from [53])

sample is a graphite foil whose surface was exposed to the flux of particles coming from one discharge in the fusion device ASDEX [53] to the vessel wall. It can be seen that the metal components are better separated by their core level electron energies showing up in the X-ray spectra, whereas light constituents (oxygen) are only accessible to RBS and AES. In this study RBS was again most useful due to its capability of absolute quantification and was used to calibrate the other methods.

6.4 Low-Energy Ion Scattering

6.4.1 NEUTRALIZATION

A fundamental property of low-energy ion scattering (ISS) that distinguishes this technique from RBS shows up in Figure 6.4: while ISS spectra exhibit a peak for each atomic species on the sample surface, RBS generally yields an edge in the spectrum, followed by a broad distribution towards lower energies.

This is a consequence of the fact that in ISS only particles backscattered from the top surface layer have a significant chance to survive the scattering process as ions and can thus be detected in an apparatus using an electrostatic analyser (see Section 6.4.2).

It is due to this selective property of the neutralization effect that ISS can be used as an extremely surface sensitive method in the sense that the scattering signal exclusively originates from the topmost atomic layer. This is most pronounced for noble gas ion scattering (He^+, Ne^+, Ar^+) in combination with an electrostatic analyser (modifications which occur by using alkali ions or neutral particle detection are discussed below). For these noble gas ions, the probability P for surviving as ions is of the order of about 5% for scattering from the first layer and at least an order of magnitude lower for scattering from deeper layers [2]. From this it immediately follows that the scattered ion yield is not only determined by the cross-section (cf. Equation (6.28) for the RBS case) but also to a large extent by the neutralization effect, expressed as the ion survival probability P.

The ion current I_i^+ arising from scattering from species i with a surface density N_i therefore can be written as:

$$I_i^+ = I_0^+ T N_i \frac{d\sigma_i}{d\Omega} \Delta\Omega P_i \qquad (6.37)$$

where I_0^+ is the primary ion current and T is a factor taking the transmission of the apparatus and the detector sensitivity into account. The cross-section $d\sigma/d\Omega$ can be calculated with sufficient accuracy as described in Section 6.2.3. The ion survival probability P, however, is generally not well known. As a rough estimate for scattering from metal surfaces, values of about 10% for 1 keV He^+ and 5% for 1 keV Ne^+ can be taken. For quantitative composition analysis, calibration with elemental standards has been used successfully, particularly for metal alloys [55, 56]. But it has also been observed [57, 58] that the survival probability can be trajectory dependent, i.e. it not only depends on the specific target atom i but also on the electron density encountered by the projectile on its way into and out of the target. This behaviour was studied in some detail, e.g. for oxygen adsorbed on Ni surfaces [57, 59]. It should, however, be noted that for close to normal incidence and large scattering angles these effects are virtually negligible. ISS can then be considered being without matrix effects, i.e. the scattered ion intensity from one surface atomic species is independent of its chemical environment.

The theoretical description of the various neutralization processes is based on Hagstrum's work [60] and is schematically shown in Figure 6.20. For noble gas ions with large ionization potentials (between 16 eV and 24 eV), Auger neutralization (AN) is the dominant process. The ion survival

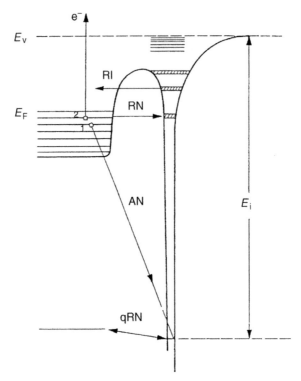

Figure 6.20 Energy levels for an ion (atom) close to a surface: E_V, vacuum level; E_F, Fermi energy; RI and RN, resonance ionization and neutralization, respectively; AN, Auger neutralization; qRN, quasi resonant neutralization; E_i, ionization energy of the atom

probability can be described within a model in which the electron transition rate depends exponentially on the distance of the ion from the surface. This results in an expression for the ion survival probability which depends on the ion velocity perpendicular to the surface, v_p:

$$P = e^{-v_0/v_p} \tag{6.38}$$

The parameter v_0 depends on the ion–target combination but is very generally of the order of 10^7 cm/s. Basically an expression like Equation (6.38) has to be taken into account for the incoming and the outgoing trajectory, but experimental results can often be described with sufficient accuracy by only considering the final velocity perpendicular to the surface. In addition to these neutralization processes due to charge exchange with the surface, a contribution from the close encounter during the large angle collision has been postulated [58, 61, 62]. For this to occur a small enough distance of closest approach, i.e. a primary projectile energy above a threshold, $E > E_{th}$, is required. These observations indicate that the final charge state is primarily determined on the outgoing trajectory. The incoming ion is effectively neutralized and reionization can take place during the violent collision with a target atom. The related energy loss, corresponding to the ionization energy of the projectile, was in fact observed for ions above a certain threshold of kinetic energy [63–65]. The effects of collision induced neutralization, CIN, and collision induced reionization, CIR, have recently been studied in detail for He^+ scattering from Cu [66]. A comprehensive discussion can be found in a recent review [67].

For alkali ions, resonance neutralization (RN) is most important. Their ionization potentials are close to the values of work functions of many materials, particularly metals, i.e. of the order of a few eV. Depending on the energetic position of the valence level involved, the ion yield can decrease or increase with increasing kinetic energy of the projectile. This behaviour has been reviewed and theoretically discussed in Los and Geerlings [68] and Brako and Newns [69]. Generally, ion yields are very much larger for alkali ions than for noble gas ions, of the order of 50 to 100%. If the work function of a surface is reduced, e.g. by alkali ion adsorption, resonant charge exchange with excitation levels of noble gas ions also becomes important and a dependence on the distance of closest approach is observed [62].

A special case of so called quasi resonant (qRN) charge exchange occurs if the ionized energy level of the projectile lies in energetically close vicinity to a target atom core level, as e.g. in the case of He 1s and Pb 5d levels. Then a Landau–Zener type of charge exchange occurs which results in an oscillatory dependence of the ion yield on the projectile velocity [65, 70].

From the analytical point of view the discussion given above can be summarized as follows: in general the ion escape probability for noble gas ions cannot be given *a priori* and therefore it poses a problem on quantitative analysis. However, if proper calibration can be provided, quantitative analysis is possible and a linear dependence on N_i, as suggested by Equation (6.37) has been established in a number of cases (see also Section 6.4.3). One way to circumvent these problems is to use alkali ions for which the values of P are close to one if the target work function is not too small. The other possibility consists of the detection of neutral (or ion plus neutral) scattered particles. This can indeed be done with substantial advantage, as discussed in the following sections. In both cases, however, the beneficial part of the neutralization effect, i.e. the exclusive surface sensitivity, is largely lost.

6.4.2 APPARATUS

Because of its extreme surface sensitivity, ISS requires vacuum conditions which allow the sample surface to be cleaned or prepared and maintained in a defined state for a sufficiently long period. Therefore the scattering chamber must be a UHV system with a base pressure for reactive gases (H_2, CO, H_2O) below 10^{-9} mbar. If, for example, a surface with a contamination below 10^{-2} monolayers should stay clean for an hour, a pressure below

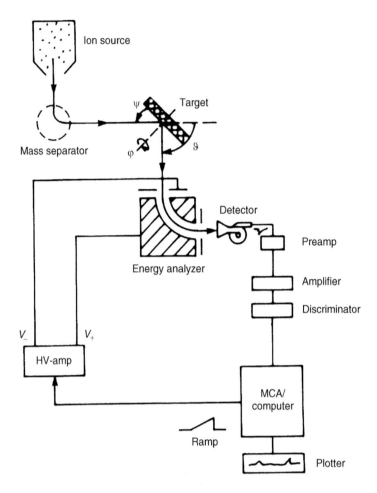

Figure 6.21   ISS arrangement with a $90°$ spherical sector electrostatic analyzer [76]

10^{-11} mbar is required for gases with a sticking coefficient around one. For reactive gases such values are not uncommon on metal surfaces [71]. In many systems the partial pressure of noble gases from the ion source reaches pressures between 10^{-7} and 10^{-6} mbar in the scattering chamber. These values are tolerable since the sticking probability of thermal noble gas atoms is virtually zero on all surfaces at room temperature [72].

    A typical ion-scattering apparatus using an electrostatic energy analyser is schematically shown in Figure 6.21. The essential components are the ion source, the target manipulator and the analyser and detector system. Electron impact ion sources are most convenient for noble gas ions. For energies around 1 keV they easily provide a constant ion current of 10–$30\,nA$ which is sufficient for surface analysis and causes limited surface erosion due to sputtering. For a beam spot diameter of 1–$2\,mm$ a total fluence of the order of $10^{13}\mathrm{He}^+/\mathrm{cm}^2$ is required to record an energy spectrum over the whole range of secondary energies. With a typical He^+ sputtering yield of about 10^{-1} atoms/ion [73] only about 10^{-3} of a monolayer is removed from the surface for one spectrum. For light adsorbates, however, much higher yields are possible [74] and the adsorbed layer must be restored after short bombardment intervals. Lower ion fluences are necessary for alkali ions and for neutral particle detection, as discussed below. If, on the other hand, erosion of the surface by ion bombardment is desirable for cleaning purposes or in order to obtain near-surface depth profiles, then higher current densities can generally be applied [74]. Mass separation is not absolutely necessary for electron impact noble gas ion sources, but with plasma ion sources and solid dispenser sources it is a necessity, since they produce a variety of ionic species and they also emit reactive neutral gas particles. Sources for alkali ions are commercially available. They contain appropriate minerals that release Li^+, Na^+ or K^+ ions at elevated temperatures.

For sample holding, various commercially available UHV-target manipulators exist and for special requirements, appropriately designed manipulators have been developed [75]. For structure investigations, generally two axes of rotation are necessary, one in the sample surface defining the incidence and exit angles, and one perpendicular to the surface for azimuthal variations. For the angular settings, generally an accuracy of $0.5-1.0°$ is sufficient. For composition analysis a fixed scattering geometry, preferably with large incidence and exit angles, can be used. The manipulators also provide facilities for sample heating by electron bombardment and in some instances also cooling by liquid nitrogen. The electrical and mechanical leads necessary for temperature control and measurement, together with the required rotational degrees of freedom call for intricate and precise constructions.

For the energy analysis of the scattered ions, electrostatic sector fields are very convenient and most commonly applied. Their relative energy resolution $\Delta E/E$ is given by the ratio of the aperture width to the radius of the central trajectory, i.e. $\Delta E/E = s/r$. For ISS purposes a resolution of $1-2\%$ is sufficient. It follows from this relation that the energy window of an electrostatic analyser increases in proportion to the energy detected and this has to be corrected for absolute measurements. The possibility to operate with constant pass energy is not often reported to be used with ISS. Spherical sector analysers (with a sector angle of $90°$, see Figure 6.21) can be mounted on a UHV manipulator system, such that the scattering angle is variable from $0°$ to large angles of $160°$ or more. This is very useful for the experimental determination of the direction, the energy and the width in angle and energy of the incident beam. Variation of the scattering angle is often useful for peak assignment and large scattering angles have become important for structure analysis (see Section 6.4.4). For the detection of charged particles, usually channeltron electron multipliers are used in the counting mode.

Cylindrical mirror analysers (CMA) have the advantage of large acceptance angles and consequently higher scattering signals (up to a factor of 30 compared to spherical sectors [76]). This high intensity and the relatively good mass resolution due to the large scattering angle (137°) make CMAs very useful for standard surface composition analysis. Since the scattering geometry is fixed, they are generally not suited for structure determination. For very high detection efficiency, special systems were developed with a spiral shaped position-sensitive detector for simultaneous angular and energy distribution recording [77]. An energy dispersive toroidal prism has also been successfully used to measure energy and angular distributions simultaneously in a multichannel mode [78].

A very successful alternative means for measuring the energy distributions of scattered or recoiling particles is the time-of-flight (TOF) method [79–85]. Since here the energy, or rather the velocity, is determined by the flight time, this method is applicable equally to charged and neutral particles. Typical elements of a TOF systems are shown in Figure 6.22 [84]. After mass separation, the primary ion beam is pulsed by a square-wave voltage applied to two orthogonal pairs of deflection plates. Bunches of primary particles thus impinge on the surface and, after scattering, they pass through a drift tube at ground potential until they hit the particle detector. The optional double deflection unit helps to switch electronically

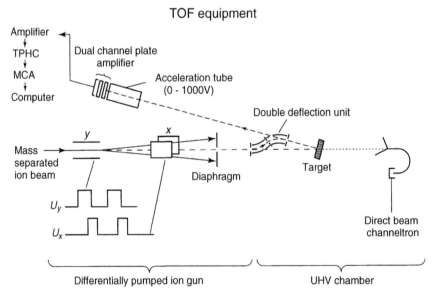

Figure 6.22　Schematic of a time-of flight (TOF) equipment for ion and neutral detection (from [84])

between two scattering angles, i.e. $165°$ and $180°$. The time distribution of the chopped-ion beam is measured after scattering and this distribution can be transformed into an energy spectrum through the relation

$$E = \frac{1}{2}M_1L^2/t^2 \tag{6.39}$$

where L is the length of the flight path. The counts per constant time increment $\Delta N(t)$ are converted into constant energy increments via the expression:

$$\Delta N(E) = (t^3/M_1L^2)\Delta N(t) \tag{6.40}$$

For distances of the order of 1 m, flight times of microseconds are obtained and the corresponding electronics have to chop the beam with rise times of some ten nanoseconds in order to obtain an energy resolution of about 1%.

By applying an appropriate potential to the drift tube, the scattered ions can be separated from the neutrals and charge fractions in corresponding parts of the energy spectrum can be determined. Particle detection can be achieved by channeltrons or open multipliers, just as in the case of electrostatic analysers.

If the particle energies are sufficiently high, secondary electron multipliers respond equally well to neutral particles as to ions (i.e. if the kinetic secondary electron emission is much higher than the potential emission with ions). Therefore the scattered particle energies should be above about 1 keV [86] and consequently TOF experiments are usually carried out with primary energies of 2 keV or more, up to 10 keV (thus reducing the surface specificity). Neutral particle spectra exhibit, in general, similar features as alkali ion spectra: quantification is easier in principle but large contributions from multiple scattering can make the interpretation of spectra more complicated and computer simulations are sometimes necessary for this purpose. Since the neutral particle fraction is usually more than 90% and due to the simultaneous recording of the entire spectrum (rather than scanning with an energy window), primary ion fluences for one spectrum are two orders of magnitude or more lower than for noble gas ISS with an electrostatic analyser, i.e. 10^{11} ions/cm^2 or less. Therefore, the surface damage is correspondingly low and thus this method is also non-destructive.

6.4.3 SURFACE COMPOSITION ANALYSIS

It is pointed out in the preceding sections that ISS has the capability of analysing the elemental composition of the outermost atomic layer of a solid surface. The method therefore appears well suited for routine surface composition analysis. This, however, is true only within certain limitations which are inherent to the method:

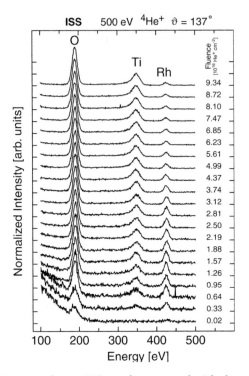

Figure 6.23   He$^+$-ISS spectra from a TiO$_2$ surface covered with about one half monolayer of Rh (compare Figure 6.12 for the RBS case) (from [36])

1. 'Technical' surfaces which are not prepared in a well defined way generally have a hydrogen-rich surface contamination layer (hydrocarbons, water) and the light H atoms do not contribute to the scattering signal under many common scattering conditions. Useful spectra are therefore only obtained after removing the contamination layer by sputtering with the analysing ion beam (for an example see Figure 6.23).

2. Mass resolution for heavier masses and consequently mass identification is restricted by collision kinematics (cf. Section 6.2.2) and therefore unambiguous determination of heavier masses without any pre-knowledge is not guaranteed.

The areas of research in which ISS has been shown to be particularly successful are surface composition analysis of adsorbates, analysis of catalyst and of metal alloy surfaces. The especially useful combination with structure determination is treated in Section 6.4.4.

Adsorbates.   According to Equation (6.37) the ion scattering signal should ideally increase linearly with the surface density N_i of the adsorbed species *i*. This has in fact been demonstrated for a number of systems such as S, O,

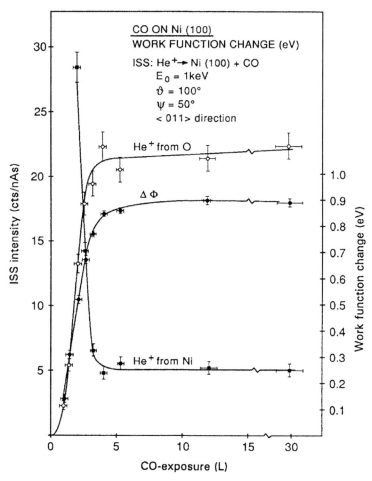

Figure 6.24 Helium ISS intensities and work function change for a Ni (100) surface as a function of CO exposure – see text (from [88])

CO and Pb adsorbed on Ni surfaces in which the surface coverage could be independently calibrated using other methods (neutron activation analysis, work function change, RBS, LEED) [87]. An example is given in Figure 6.24 for CO adsorption on Ni (100) [88]. The oxygen signal from CO increases linearly with coverage (and parallel to the work function change $\Delta\phi$) up to a saturation value of one monolayer (showing a c (2×2) structure). The linearity is maintained despite the large work function change of 0.9 eV which obviously does not influence the neutralization probability (P in Equation (6.37)). This linearity is lost if the work function is drastically decreased by adsorption of Cs [62].

Adsorbed species cover the substrate atoms, whose scattering signal should then decrease accordingly. This is also demonstrated in Figure 6.24 by the Ni intensity which shows a drastic linear decrease with CO adsorption.

The substrate signal I_s can therefore be expressed in analogy to equation (6.37) by:

$$I_s = I_0{}^+ T(N_s - \alpha N_i)(d\sigma_s/d\Omega)\Delta\Omega P_s \qquad (6.41)$$

The shadowing factor α indicates how many substrate atoms are excluded from scattering by one adsorbate molecule; an initial value of 4 is found for the example of Figure 6.24 and decreases with increasing coverage due to overlap of the shadows.

Shadowing can also yield information on the orientation of adsorbed molecules. For the case of CO on Ni shown in Figure 6.24, the energy spectra only exhibit an oxygen peak and no scattering from C can be observed. This gives a very direct and simple indication that CO is adsorbed on Ni in a vertical orientation, the O pointing away from the surface, a result which is also deduced from data of other techniques in a more indirect manner.

Information about the geometric surface structure of adsorbates, i.e. exact adsorbate positions and bond lengths are of major importance in adsorbate studies. These have been carried out with ion scattering techniques, e.g. for H, S, and O adsorption on a number of high-symmetry metal surface (Cu, Ni, W) and are further discussed in Section 6.4.4. For light adsorbates (H, D) direct recoil spectroscopy is a very useful variety of ion beam techniques.

The analysing ion beam causes sputtering or ion-induced desorption of adsorbed layers. This has to be taken into account in adsorbate structure investigations, but it can also be used for determining desorption cross-sections and for obtaining near-surface concentration profiles. There is a fundamental and practical interest in desorption cross-sections for the basic understanding of multicomponent material sputtering, for surface cleaning in UHV systems and in large vacuum vessels such as storage rings or fusion devices [88]. In the monolayer coverage region, the desorption cross-section σ_D can be easily determined in an ISS apparatus by simultaneously using the ion beam for monitoring the surface coverage and for the ion impact desorption. The signal from the adsorbed species I_i then decreases with time t or fluence it, where i is the primary current density:

$$I_i{}^+/I_0{}^+ = \exp(-it\sigma_D) \qquad (6.42)$$

In the above equation, σ_D can be connected with the sputtering yield Y through the relation $Y = \sigma_D N_{ML}$ where N_{ML} is the monolayer areal density. Using ion scattering these cross-sections have been determined for a number of ion-adsorbate–substrate combinations [89, 90]. Their values range from 10^{-16} cm^2 (e.g. 500 eV He-Ni-O) up to such large values as 7×10^{-15} cm^2 (500 eV Ne–Ni–CO).

Catalysts.   Supported catalysts as commonly used in heterogeneous catalysis are adsorption systems of a particular kind and represent a research field in themselves due to their enormous technical importance [91, 92]. These catalysts generally consist of a high surface area (about 100 m^2/g) support material and one or more finely dispersed adsorbed components ('active components' and promotors) that determine the efficiency, selectivity and stability of the catalyst. The support material usually consists of highly insulating metal oxides (Al_2O_3, TiO_2, SiO_2) and the active components can be metal oxides such as MoO_3, WO_3 or V_2O_5, at least in a precursor state. These insulating materials are hardly accessible to the electron spectroscopies generally used in surface analysis, because of charging effects on the sample. In the case of ISS, charging can be compensated by flooding the specimen with electrons from a filament. And since ISS monitors the composition of the outermost atomic layer – which is most important for the performance of the catalyst – with this technique useful contributions for understanding the composition and structure of supported catalysts can be made [93, 94].

An example of such studies is shown in Figure 6.23 above [23] which shows He$^+$-ISS spectra of Rh on TiO_2 as a model catalyst. These spectra are taken from the same sample as the RBS spectrum shown in Figure 6.12 and so a direct comparison of both techniques is possible on this basis. The ISS spectra show almost no scattering signal in the beginning due to surface contamination, as discussed above. With increasing He$^+$ ion fluence distinct O, Ti, and Rh peaks are observed. The sequence of spectra represents a composition depth profile and the increase in Ti intensity when the Rh adlayer is sputtered away can clearly be seen. By these means, the layering sequence and the spreading for a number of real and model catalysts have been studied [94]. Results from a comparative investigation of MoO_3 adsorption (from a solution) on Al_2O_3 are shown in Figure 6.25 [95]. The

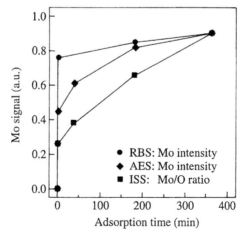

Figure 6.25 Normalized Mo intensities for RBS, AES and ISS from a Al_2O_3 surface impregnated with MoO_3 from a solution as a function of impregnation time (from [95])

increase in the amount of adsorbed Mo species with adsorption time is plotted from RBS, AES and ISS results. ISS, being most surface-sensitive, shows the flattest slope while RBS detects nearly 80% of the final amount already after 1 min of adsorption. AES is in between, according to its information depth of about five monolayers. This result means that, in the beginning, Mo is adsorbed in pores below the surface in less than 1 min; afterwards additional molybdate is adsorbed on the external surface in a timescale of hours.

In these cases, model catalysts were prepared by oxidizing metal sheets and the oxide layers were thin enough to allow the use of AES. If real catalysts are studied, wavers have to be pressed from powder material and for these samples, charging effects exclude electron spectroscopies. The surface of these wavers is very rough (on a μm scale) and poses the question of the influence of surface roughness on ISS results. It has been found [96] that surface roughness can decrease the ion-scattering signal by more than a factor of six compared to polished samples. If, however, intensity ratios are taken as a relative measure of surface coverage, the results from smooth and rough samples are very similar.

Alloys.   Low-energy ion scattering is also extremely valuable for the surface composition analysis of metallic alloys, again due to its 'monolayer' sensitivity. If surface segregation of one component occurs – and this is very generally the case for alloy systems – there is a discontinuity in the composition of the first and second atomic layer [97, 98]. Consequently this requires an analysis method which is able to discriminate between the first and the second layer. ISS has therefore been used for analysing a number of alloy systems and for obtaining data necessary for the fundamental understanding of surface segregation, often in combination with radiation enhanced diffusion and differential sputtering effects [99–102]. As an example, Figure 6.26 shows ion scattering spectra taken during the segregation of Al on the surface of a $Fe_3Al(110)$ single crystal surface [103]. At a temperature of 700 K, already after about 10 min, equilibrium is reached with a coverage of 95% Al in the top layer.

It has been proven in many cases that, with these metal surfaces, quantification is possible by using elemental standards. In a binary system with components A and B the surface concentration A can then be calculated to be:

$$X_A = 1 \left/ \left(1 + \frac{I_B}{I_A} \frac{S_A}{S_B} \frac{N_B}{N_A} \right) \right.$$

(6.43)

with $X_A + X_B = 1$. Here I refers to the scattering signal from the alloy and S to the signal from the elemental standard with surface density N, the subscripts denoting these values for the components A and B, respectively. If several data are taken from a surface with varying composition, then a calibration

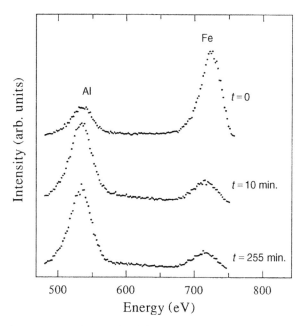

Figure 6.26 He^+ ISS spectra ($E_0 = 1$ keV) from a Fe_3Al (110) alloy surface held at 700 K, showing the surface segregation of Al with time [103]

without standards is also possible as long as all signals vary linearly with concentration and the sum of the concentration is unity, as stated above.

An extensive study was carried out for the $Cu_3Au(100)$ surface using 5 keV and 9.5 keV Ne^+ in a TOF technique which allowed a quantitative analysis of the top three crystal layers [99]. This alloy has an order–disorder transition (similar to the Fe_3Al alloy mentioned above in Figure 6.26) and upon heating above the transition temperature, the most significant composition changes were found in the second layer where the Au concentration increased from (ideally) zero to 0.25. Such layer-selective and mass-sensitive measurements are a unique feature of ISS. Analogous ion scattering results were obtained for Au_3Cu [104] and CuAu (100) [105]. The method of analysis of the latter by computer simulation using the MARLOWE code is indicated in Section 6.4.3.

Under the influence of ion bombardment, the surface composition of an alloy can be changed in addition to the thermally activated segregation by preferential sputtering and radiation enhanced diffusion. For a complete analysis then, data from the first and subsequent layers are necessary. They can either all be based on ISS results or on a combination with a method of larger information depth, e.g. AES. An example is given in Figure 6.27 which shows the surface composition of a series of PdPt alloy samples [102]. The combination of segregation and preferential sputtering of Pd leads to a moderate depletion in the first layer (as detected by ISS), but

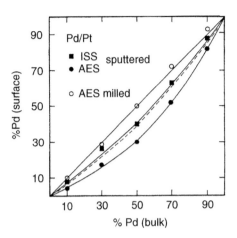

Figure 6.27   ISS and AES results for PdPt alloys with various compositions. The mechanically milled surface shows bulk composition: ISS concentrations are close to those calculated for mass conservation in steady state sputtering (dashed line) while AES indicates subsurface Pd depletion due to preferential sputtering [102]

AES averages also over the depleted sub-surface layers and therefore shows more deviation from the bulk composition. From these measurements, the segregation energies and diffusion coefficients can be determined as a function of alloy composition [106].

Multiple scattering and simulation.   Computer simulations are particularly useful if multiple scattering processes make a substantial contribution to the scattered ion intensities. This occurs for instance for scattering of heavier alkali ions (Na$^+$, K$^+$) that may be used to obtain good mass resolution and less neutralization effects. In addition, multiple scattering can be exploited to assess scattering from first and second layers quantitatively and thus to perform layer-selective composition analysis. Such studies are of great interest in the field of surface segregation as mentioned above. As an illustrative example we consider the scattering of 1.5 keV Na$^+$ from a Cu(100) surface. The example demonstrates the importance of using the adequate scattering potential (characterized by its screening length a, see Equation (6.10)) and the possible influence of thermal vibrations on the spectral distribution. Figure 6.28 (top) shows experimental and simulated spectra in two azimuthal directions. The intensity in the [001] direction originates mainly from scattering from the first atomic layer in the chosen geometry (45° incidence, 90° scattering angle). In the [1–11] direction second layer atoms of that surface are also exposed to the incoming ion beam and this results in a considerably higher scattered ion yield. This holds for the 'single' scattering peak at about 680 eV and the 'double' scattering peak at about 920 eV. The latter is dominated by two consecutive collisions with about half of the total scattering angle. The simulation with the Firsov

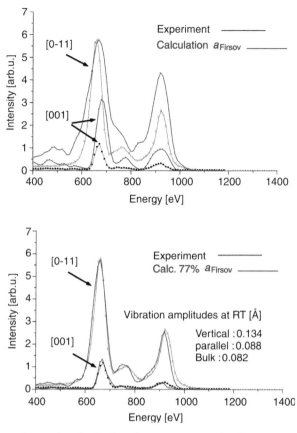

Figure 6.28   Experimental and calculated energy spectra for Na scattering from Cu(100) in two azimuthal directions. Top: calculated with Firsov screening lengths and isotropic thermal vibrations with bulk Debye temperature. Bottom: using calibrated screening lengths and unisotropc surface vibrations [23]

screening length and isotropic vibration amplitudes derived with the bulk Debye temperature reproduces the large difference of intensities in both directions, but the agreement between experiment and simulation is not satisfactory. By proper calibration it was shown that a screening length $a = 0.77a_{Firsov}$ is appropriate and anisotropic surface vibrations have to be taken into account. Using the Debye temperature values calculated by Jackson [107] for in-plane and perpendicular vibrations in the first layer and the bulk value for the second and deeper layers, good agreement could be obtained between experiment and simulation, see Figure 6.28 (bottom). Such results can be carried further to study layer selective segregation profiles, as e.g. shown for the CuAu system [105].

Data on quantitative analysis and technical surfaces.   In a large number of studies and applications it has been established that ISS is a reliable tool for quantitative surface composition analysis. This implies the fact that the method can be calibrated with appropriate reference standards and that matrix effects are very exceptional. These statements are supported by a recent review [67] that contains a huge compilation of investigated material for which the absence of matrix effects has been reported. These materials comprise elemental metals and metal alloys, compounds such as oxides and fluorides, organic polymers and even liquids. This impressive collection demonstrates the level of maturity ISS has reached as an analytical method.

6.4.4 STRUCTURE ANALYSIS

Principles. One of the outstanding strengths of low-energy ion scattering is the possibility of obtaining mass-selective structural information in real space about the topmost surface layers of a crystal. This means that the distances between neighbouring atoms in selected crystallographic directions can be measured. In order to determine the geometry of a surface, these distances are used in connection with structure models that are generally deduced from LEED (low-energy electron diffraction) results. Diffraction studies yield data in the reciprocal space from which the symmetry of the surface lattice is deduced but the exact position of surface atoms, particularly also for multicomponent crystals, requires substantial computational effort [108].

The specific feature of ISS used for structure determination is the shadow cone (see Section 6.2.4) with the peaked ion flux at the cone edge (cf. Figures 6.8 and 6.10). It was applied most successfully by Aono *et al.* [109] in the so called ICISS (impact collision ion-scattering spectroscopy) technique. Its principle can be explained by Figure 6.29 [109]: A parallel flux of ions is incident along a chain of surface atoms with constant distance d. For small angles of incidence Ψ each chain atom is in the shadow of the preceding one and therefore cannot contribute to scattering. By increasing the angle

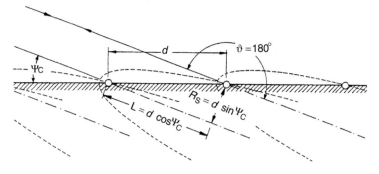

Figure 6.29 Scattering geometry illustrating the ICISS technique (after Aono *et al* [109])

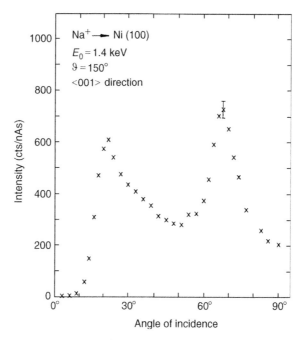

Figure 6.30 ICISS distribution: Na^+ scattering of a Ni(100) surface as a function of the angle of incidence Ψ. The first peak (at $\Psi \sim 20°$) appears when the edge of the shadow cone hits neighbouring atoms in the surface [100] direction; the second peak (at $\Psi \sim 70°$) is due to scattering from the second-layer neighbour atoms in the [101] direction [88]. For recording the scattering intensity the energy analyser is set to Na scattering from Ni at the selected scattering angle θ (Equation (6.1)), thus providing mass selective structure information

of incidence, a critical angle Ψ_c is reached at which the edge of the shadow cone (with its high ion flux) exactly hits the neighbouring atom. This results in a high backscattered intensity that, upon further increasing Ψ, falls to the average value of the primary flux. Consequently a scattered ion flux distribution is expected which represents the primary flux distribution shown in Figure 6.10. An example of an ICISS distribution is shown in Figure 6.30 [88]. The critical angle of Ψ_c is related to the shadow cone radius R_s at a distance L from the scattering centre through the atomic distance d as indicated in Figure 6.29. If the form of the shadow cone $R_s(L)$ is known (cf. Equation (6.18)) the interatomic spacing d can be determined by measuring Ψ_c:

$$d = R_s/\sin \Psi_c \tag{6.44}$$

The shadow cone radius $R_s(L)$ can be calibrated by ICISS measurements at a surface of known structure, e.g. a non-reconstructed surface which shows bulk termination (many clean low-index transition metal surfaces). This possibility of self-calibration is of course a very useful feature of ICISS and helps to avoid ambiguities arising from insufficiently well known interaction potentials. However, numerical calculations are generally useful and sometimes necessary for data interpretation. Shadow cones can be calculated using Equation (6.18). By numerically solving this equation for Ψ_c, the experimentally accessible quantity, a formula for a TFM potential was obtained [110]:

$$\ln \Psi_c = 4.6239 + \ln(d/a)(-0.0403 \ln B - 0.6730) \\ + \ln B(-0.0158 \ln B + 0.4647) \tag{6.45}$$

with $B = Z_1 Z_2 e^2 / (E_0 a 4\pi \varepsilon_0)$. For a given projectile–target combination it reduces to $\Psi_c \approx d^{-\gamma}$ with γ between 0.7 and 0.8.

In some cases, two-dimensional numerical codes were able to reproduce experimental results quite well [21, 103, 111–113] and thus improved data interpretation. In detailed investigations, particularly for cases of lower symmetry, three-dimensional codes such as the MARLOWE program [18] are useful. This has been applied for determining the position of H adsorbed on Ru(001) [113] or for studies of the Au(110) reconstruction [115] and CuAu alloys [105].

Ideally, the backscattered flux is measured at a scattering angle of $\theta = 180°$ which corresponds to an impact parameter $p = 0$ and therefore relates Ψ_c directly to the mean lattice position of the scattering atoms. Experimental designs with very nearly [114] or exact 180° [84] scattering are reported in the literature; see also the set-up shown in Figure 6.21. However, it turns out that scattering angles $\theta > 145°$ are generally large enough to allow a determination of d with an accuracy of about 0.1 Å. Limitation also arises from the thermal motion of surface atoms, resulting in a broadening of the ICISS intensity compared to the theoretical singularity at $\Psi = \Psi_c$ (Equation (6.17)).

The ICISS method can, of course, not only be used with noble gas ions but also with alkali ions and with neutral particle TOF techniques. The two latter varieties have been named ALICISS and NICISS, respectively [84]. Since they are not strongly influenced by neutralization, they also provide information about atomic positions in the second or deeper layers; see Figure 6.30.

The ICISS techniques have so far been very successfully applied to structure determinations in connection with surface reconstruction (clean surfaces and adsorbate induced reconstructions), for locating the position of adsorbed species, for determining the positions of the components on ordered alloy surfaces (in connection with surface segregation) and for determining thermal vibrational amplitudes of surface atoms (several reviews can be found in the literature [2]). A closely related technique is direct recoil spectroscopy (DRS) in which directly recoiling surface atoms can be identified by their energy (Equation (6.3)). This technique is particularly useful for light adsorbates (e.g. hydrogen isotopes) which are difficult to detect by scattering. Examples are presented in the following section.

Surface reconstruction. As an illustrative example of structure determination, we consider an ALICISS study of the reconstruction of the Ni (110) surface upon oxygen adsorption [116, 117]. The oxygen covered surface shows a (2×1) superstructure in the LEED pattern. The questions to be answered are: (i) Does the superstructure represent the oxygen overlayer or has a rearrangement of Ni surface atoms taken place? (ii) What is the reconstructed surface structure in the latter case? (iii) Where are the oxygen atoms positioned? The situation is depicted in Figure 6.31 [116].

The upper part shows the clean surface and the corresponding ICISS spectrum in the [112] direction taken with 2 keV Na^+ ions at the energy of the binary Na–Ni collision, $E = 0.216 \, E_0$. The slopes 1 and 2 represent the shadow cone enhanced scattering from first- and second-layer atoms, respectively, as indicated in the inset. The spectrum changes drastically for the reconstructed surface showing three distinct peaks. The following conclusions were drawn from this result, answering some of the questions posed above. The shift in the leading peak with a slope at $10°$ must be due to scattering from first-layer Ni atoms and thus demonstrates that the (2×1)

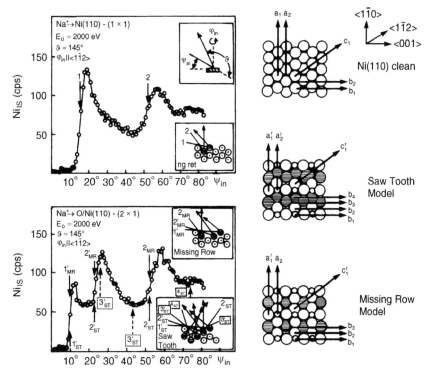

Figure 6.31 Na^+–ICISS distribution from a clean Ni(110) surface (upper left panel) and after oxygen induced (2×1) reconstruction (lower left panel). The surface structure models and scattering geometries are plotted on the right side. From [116]

superstructure is a result of rearrangement of Ni atoms (and not only due to the oxygen overlayer). The absence of the previously detected first-layer peak proves that the reconstruction is complete. Among the reconstruction models derived from the LEED pattern, the saw tooth (ST) structure and the missing row (MR) structure shown in Figure 6.31 remained as possible candidates. The ALICISS result unambiguously excludes the ST model, since the intensity increases expected from this model were not found, whereas all the expected MR features can be detected in the intensity distribution. The MR structure in which every second ⟨001⟩ row is missing was also confirmed by measurements in other azimuthal directions, and meanwhile confirmed by scanning tunnelling microscopy (with an added row growth mechanism) [118].

The ALICISS distributions can only be taken for Na^+–Ni scattering; no backscattering from oxygen is possible. The indicated long bridge O-position (in the [001] direction) cannot be directly demonstrated here, but it is compatible with the ALICISS results and it is directly deduced from recoil spectroscopy, as shown in the following section.

Direct recoil spectroscopy (DRS). DRS is a technique which is closely related to ion-scattering techniques (in the high-energy RBS regime it is usually called 'elastic recoil detection', ERD or ERDA). In this method, atoms or ions are detected which are removed from the surface by one single collision with an incoming projectile. They can therefore be identified by their kinetic energy according to Equation (6.3). An example is given in Figure 6.32 which shows a Ne^+–Ru scattering peak (at 1200 eV) and a Ne^+

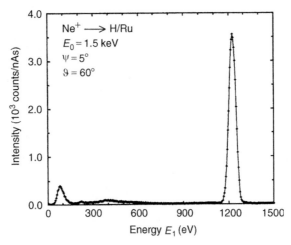

Figure 6.32 Ion energy spectrum from a hydrogen covered Ru(001) surface (at 138 K) showing a Ne^+–H^+ recoil peak at 68 eV and a Ne^+–Ru scattering peak at 1200 eV (from [113])

$-H^+$ recoil peak (at $68\,eV$) from a H covered Ru(001) surface [93]. In a TOF experiment, the corresponding flight time for a path length L is given by:

$$t = L(M_1 + M_2)\cos\theta_2/(8M_1E_0)^{1/2}. \tag{6.46}$$

Directly recoiling particles are therefore different from those secondary particles that originate from a collision cascade in a sputtering process and have a broad energy distribution around one to two eV. They are used in secondary ion or neutral mass spectroscopy (SIMS and SNMS; see Surman *et al.* [119] and Chapter 4). According to the collision kinematics (Section 6.2.2) DRS is a forward-scattering technique. It can be easily combined with ISS or ICISS studies if small and large scattering angles are available in the same apparatus (for examples, see Niehus and Comsa [84] and Schulz *et al.* [113]). Recoil cross-sections are generally of the same order as scattering cross-sections. They are largest towards $\theta_2 = 90°$, but then the recoil energy approaches zero. A useful energy range is obtained between $30°$ and $60°$.

The shadow-cone concept can also be applied with DRS and therefore, in principle, analogous measurements are possible as with the scattering techniques, i.e. elemental surface analysis, surface structure analysis and

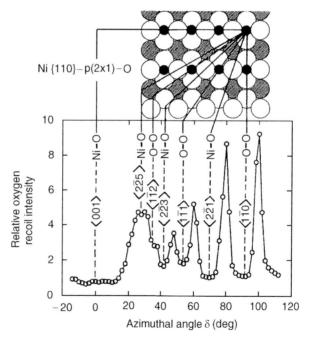

Figure 6.33 Oxygen recoil intensity distribution as a function of the azimuthal angle of the scattering plane for a Ni(110) surface with an oxygen induced (2×1) reconstruction (compare Figure 6.31). The blocking minima demonstrate the 'long bridge' position of the adsorbed oxygen (from [120])

analysis of charge exchange processes (by comparing recoiling ions and neutrals).

As an example of structural sensitivity, we consider the oxygen recoil intensity distribution from a Ni (110)–(2 × 1) reconstructed surface, the same surface as discussed in the previous section. The azimuthal intensity distribution due to 4 keV Ar^+ bombardment is shown in Figure 6.33 together with the structural model [120]. Oxygen recoils are shadowed in certain crystallographic directions either by surface Ni atoms or by other O adsorbates, which results in a distinctly structured intensity distribution. The correlation with the structure model reveals a perfect correspondence with the missing-row reconstruction, with the oxygen in the long-bridge position along the ⟨001⟩ rows. A large amount of so-called TOF-SARS (Time-of-Flight Scattering and Recoil Spectroscopy) studies have been carried out by the group at the University of Houston, USA, using an elaborate two-dimensional position sensitive detector [121].

The possibility of investigating hydrogen adsorbates by scattering is very limited; ^4He for example, can be scattered by H only into angles below 15°. Here DRS can be applied with advantage as demonstrated by a study determining the position of H on a Ru (001) surface [113]. A very instructive example for structure determination with light adsorbates is a study of deuterium adsorption on W(100) using a combination of ISS and DRS [122].The variations of D^+ and H^+ recoil intensities during scanning the azimuthal angle of the scattering plane are shown in Figure 6.34.

Without admitting D_2 gas to the scattering chamber, H^+ signals originate from adsorption from the residual gas. Upon D_2 admission the H^+ signal is suppressed and a strong D^+ signal is observed, exhibiting distinct maxima

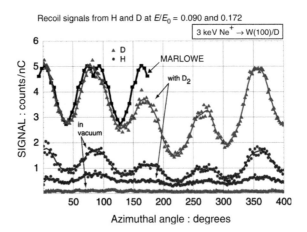

Figure 6.34 Azimuthal variation in the intensities of H^+ and D^+ from W(100) recoiled at 45° by 3 keV He^+ in vacuum and during exposure to 20 µPa D_2. MARLOWE simulations for the latter case are also shown (from [122])

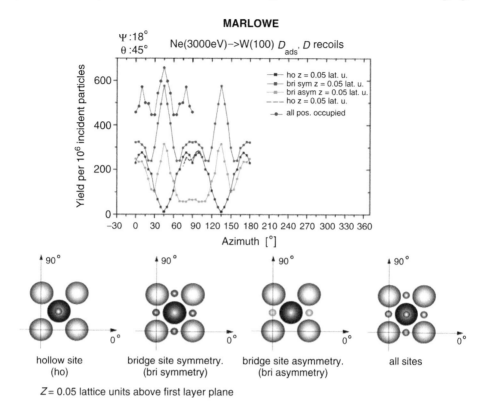

Figure 6.35 MARLOWE simulations of the recoil intensities shown in Figure 6.34 for four different high symmetry D adsorption sites. Agreement with experiment is only obtained for the hollow site [122]. Investigations of this kind nicely demonstrate the usefulness of DRS for structure studies of light adsorbates

and minima with variation of the azimuthal angle. To determine the position of the adsorbed D possible structure models are considered and analysed with MARLOWE simulations (Figure 6.35). Good agreement is obtained when the adsorbed D atoms are located in fourfold hollow sites.

6.4.5 CONCLUSIONS

Low-energy ion scattering is one of the many existing surface analytical techniques, all of which have their particular advantages and limitations. The outstanding features of ISS are the possibilities of obtaining mass-selective signals exclusively from the topmost atomic surface layer and to determine interatomic distances on the surface and surface structures by using straightforward simple concepts. Limitations associated with absolute quantification (due to uncertainties in neutralization and interaction potentials) can largely be overcome by appropriate calibration. ISS has been proven to be extremely useful for surface composition analysis e.g., of catalysts and alloy surfaces and for structure analysis of a large number of metal, semi-conductor and metal oxide surfaces and also adsorbates on these surfaces. In comparison with other common surface spectroscopies, AES is perhaps more generally applied (except for detection of hydrogen isotopes and for insulating material). It has an information depth of several atomic layers and is therefore less surface sensitive. Therefore, in many cases they cannot replace each other but rather should be used in a complementary way. SIMS certainly yields definite mass identification and often much higher sensitivity, but quantification problems can be much larger.

Considering structural analysis, the relation of ISS techniques to LEED and scanning tunnelling microscopy (STM) is important. LEED provides the bulk of structural information from surfaces, i.e. the basic crystallographic structure. ISS can nicely complement these results due to its mass sensitivity and its capability to determine atomic positions unambiguously. Scanning Tunneling Microscopy (STM), another real-space method [123], yields direct images of atomic arrangements on a microscopic scale (atomic resolution on an area of the order of 50 Å lateral dimension), without definite mass identification. This has, however, recently been achieved in some cases in connection with elaborate quantum mechanical calculations of tunneling probabilities [101]. Ion scattering averages over a comparatively large area (linear dimension of 1–2 mm). Again complementing measurements appear to be most useful for verifying structure models, positions of various atomic species and investigations of kinetic processes. Regarding this last point, sample temperature variations and time dependent measurements (with a resolution of about 1 s) are fairly straightforward with ion scattering, and often more difficult with STM so far, but rapid progress is presently being made in that field.

Acknowledgement

Thanks are due to Christian Linsmeier for helpful discussions and his assistence with preparing figures.

📖 References

1. D.P. Smith, *J. Appl. Phys.* **18**, 340 (1967).

2. For reviews, see for example: (a) E. Taglauer, in *Methods of Surface Characterization*, Volume 2, A.W. Czanderna and D.M. Hercules (Eds), Plenum Press, New York, NY, p. 363 (1991); (b) H. Niehus, W. Heiland and E. Taglauer, *Surf. Sci. Rep.* **17**, 213 (1993); (c) H. Niehus, in *Practical Surface Analysis* 2ⁿᵈ edition, Vol. 2. *Ion and Neutral Spectroscopy*,

D. Briggs and M.P. Seah (eds), John Wiley, Chichester, UK, p. 507 (1992); (d) J.A. van den Berg and D.G. Armour, *Vacuum* **31**, 259 (1981).

3. For comprehensive descriptions, see for example: (a) W.-K. Chu, J.W. Mayer and M.-A. Nicolet, *Backscattering Spectrometry*, Academic Press, New York, NY (1978); (b) J. Tesmer and M. Nastasi (Eds), *Handbook of Modern Ion Beam Materials Analysis*, Materials Research Society, Pittsburgh, PA, 1995.

4. N. Bohr, *Mat.-Fys. Medd. Kgl. Dan. Vid. Selsk.* **18**, 8 (1948).

5. See, for example H. Goldstein, *Classical Mechanics*, Addison-Wesley, Reading, MA (1965).

6. E. Rutherford, *Phil. Mag.* **21**, 669 (1911).

7. C.G. Darwin, *Phil. Mag.* **28**, 499 (1914).

8. O.B. Firsov, *JETP* **6**, 534 (1958).

9. J. Lindhard, V. Nielsen and M. Scharff, *Mat.-Fys. Medd. Dan. Vid. Selsk.* **36**, 10 (1968).

10. J. Lindhard, *Mat. Fys. Medd. Dan. Vid. Selsk.* **34**, 14 (1965).

11. A.G.J. de Wit, R.P.N. Bronkers and J.M. Fluit, *Surf. Sci.* **82**, 177 (1979).

12. O.S. Oen, *Surf. Sci.* **131**, L407 (1983).

13. D.P. Woodruff, D. Brown, P.D. Quinn, T.C.Q. Noakes and P. Bailey, *Nucl. Instr. Meth. Phys. Res.* **B183**, 128 (2001).

14. M. Mayer, R. Fischer, S. Lindig, U. von Toussaint, R.W. Stark and V. Dose, *Nucl. Instr. Meth. Phys. Res.* **B228**, 349 (2005).

15. W. Eckstein, *Computer Simulation of Ion-Solid Interactions*, Springer-Verlag, Berlin (1991).

16. R. Doolittle, *Nucl. Instr. Meth. Phys. Res.* **B 9**, 344 (1985).

17. M. Mayer, SIMNRA User's Guide, Technical Report IPP 9/113, Max-Planck-Institut für Plasmaphysik, Garching, 1997 [http://www.rzg.mpg.de/~mam/].

18. M.T. Robinson and I.M. Torrens, *Phys. Rev.* **B9**, 5008 (1974).

19. R. Beikler and E. Taglauer, *Nucl. Instr. Meth. Phys. Res.* **B182**, 180 (2001).

20. P.M. Tromp and J.F. van der Veen, *Surf. Sci.* **133**, 159 (1983).

21. R.S. Daley, J.H. Huang and R.S. Williams, *Surf. Sci.* **215**, 281 (1989).

22. J.F. van der Veen, *Surf. Sci. Rep.* **5**, 199 (1985).

23. R. Beikler and E. Taglauer, *Nucl. Instr. Meth. Phys. Res.* **B193**, 455 (2002).

24. (a) H. Niehus and R. Spitzl, *Surf. Interface Anal.* **17**, 287 (1991); (b) R. Spitzl, H. Niehus and G Comsa, *Rev. Sci. Instr.* **61**, 760 (1990).

25. V. Bykov, C. Kim, M.M. Sung, K.J. Boyd, S.S. Todorov and J.W. Rabalais, *Nucl. Instr. Meth. Phys. Res.* **B114**, 371 (1996).

26. H.A. Bethe, *Ann. Phys.* **5**, 325 (1930).

27. J.F. Ziegler, *Helium Stopping Powers and Ranges in All Elemental Matter*, Pergamon Press, New York, NY (1977).

28. W.H. Bragg and R. Kleemann, *Phil. Mag.* **10**, 318 (1905).

29. J.S.-Y. Feng, W.-K. Chu and M.-A. Nicolet, *Phys. Rev.* **B10**, 3781 (1974).

30. N. Bohr, *Phil. Mag.* **30**, 581 (1915).

31. W.-K. Chu, in *Ion Beam Handbook for Materials Analysis*, J.W. Mayer and E. Rimini (Eds), Academic Press, New York, p. 2 (1977).

32. J.F. van der Veen, R.G. Smeenk, R.M. Tromp and F.W. Saris, *Surf. Sci.* **79**, 212 (1986).

33. R. Behrisch (Ed.), *Sputtering by Particle Bombardment I*, Springer-Verlag, Berlin (1982).

34. E. Taglauer, *Nucl. Fusion* (Special Issue), 43 (1984).

35. L.C. Feldman, in *Methods of Surface Characterization*, Volume 2, A.W. Czanderna and D.M. Hercules (Eds), Plenum Press, New York, NY, p. 311 (1991).

36. (a) Ch. Linsmeier, H. Knözinger and E. Taglauer, *Nucl. Instr. Meth. Phys. Res.* **B118**, 533 (1996); (b) Ch. Linsmeier, *private communication* (2007).

37. P. Berberich, W. Dietsche, H. Kinder, J. Tate, Ch. Thomsen, and B. Scherzer, in *Proceedings of the International Conference on High Temperature Superconducting Materials*, Interlaken, CH (1988).

38. L.C. Feldman, J.W. Mayer and S.T. Picraux, *Materials Analysis by Ion Channeling*, Academic Press, New York, NY (1982).

39. D.S. Gemmel, *Rev. Mod. Phys.* **46**, 129 (1974).

40. J.F. van der Veen, *Surf. Sci. Rep.* **5**, 199 (1985).

41. W.C. Turkenburg, W. Soszka, F.W. Saris, H.H. Kersten and B.G. Colenbrander, *Nucl. Instr. Meth.* **132**, 587 (1976).

42. L.C. Feldman, *Nucl. Instr. Meth.* **191**, 211 (1981).

43. J.A. Davies, D.P. Jackson, J.B. Mitchell, P.R. Norton and R.L. Tapping, *Phys. Lett.* **54A**, 239 (1975).

44. T.E. Jackman, K. Griffiths, J.A. Davies and P.R. Norton, *J. Chem. Phys.* **79**, 3529 (1983).

45. L.M. Howe, M.L. Swanson and J.A. Davies, *Meth. Exp. Phys.* **21**, 275 (1983).

46. F. Besenbacher, I. Stensgard and K. Mortensen, *Surf. Sci.* **191**, 288 (1987).

47. J.W.M. Frenken, P.M.J. Marée and J.F. van der Veen, *Phys. Rev.* **B34**, 7506 (1986).

48. P. Bailey, T.C.Q. Noakes, C.J. Baddeley, S.P. Tear and D.P. Woodruff, *Nucl. Instr. Meth. Phys. Res.* **B183**, 62 (2001).

49. [http://www.dl.ac.uk/MEIS/].

50. T.C.Q. Noakes, P. Bailey and D.P. Woodruff, *Nucl. Instr. Meth. Phys. Res.* **B136–138**, 1125 (1998).

51. B.L. Doyle, D.S. Walsh and S.R. Lee, *Nucl. Instr. Meth. Phys. Res.* **B54**, 244 (1991).

52. (a) S. Raman, in *Applied Atomic Collision Physics*, Volume 4, S. Datz (Ed.), Academic Press, Orlando, FL, p. 407 (1983); (b) S.A.E. Johansson and J.L. Campbell, *PIXE-A Novel Technique For Elemental Analysis*, John Wiley & Sons, Inc., New York, NY (1988).

53. E. Taglauer and G. Staudenmaier, *J. Vac. Sci. Technol.* **A5**, 1352 (1987).

54. See Appendix 1 of this volume.

55. M.J. Kelley, D.G. Swartzfager and V.S. Sundaram, *J. Vac. Sci. Technol.* **16**, 664 (1979).

56. P. Novacek, E. Taglauer and P. Varga, *Fresenius J. Anal. Chem.* **341**, 136 (1991).

57. D.J. Godfrey and D.P. Woodruff, *Surf. Sci.* **105**, 438 (1981).

58. G. Engelmann, E. Taglauer and D.P. Jackson, *Nucl. Instr. Meth. Phys. Res.* **B13**, 240 (1986).

59. W. Englert, E. Taglauer, W. Heiland and D.P. Jackson, *Phys. Scr.* **T6**, 38 (1983).

60. H.D. Hagstrum, in *Inelastic Ion-Surface Collisions*, N.H. Tolk, J.C. Tully, W. Heiland and C.W. White (Eds), Academic Press, New York, NY, p. 1 (1976).

61. D.J. O'Connor, Y.G. Shen, J.M. Wilson and R.J. MacDonald, *Surf. Sci.* **197**, 277 (1988).

62. M. Beckschulte and E. Taglauer, *Nucl. Instr. Meth. Phys. Res.* **B78**, 29 (1993).

63. M. Aono and R. Souda, *Nucl. Instr. Meth. Phys. Res.* **B27**, 55 (1987).

64. A.W. Czanderna and J.R. Pitts, *Surf Sci.* **175**, L737 (1986).

65. W. Heiland and E. Taglauer, *Nucl. Instr. Meth.* **132**, 535 (1976).

66. M. Draxler, J. Valdés, R. Beikler and P. Bauer, *Nucl. Instr. Meth. Phys. Res.* **B230**, 290 (2005).

67. H.H. Brongersma, M. Draxler, M. de Ridder and P. Bauer, *Surf. Sci. Rep.* **62**, 63 (2007).

68. J. Los and J.J.C. Geerlings, *Phys. Rep.* **190**, 133 (1990).

69. R. Brako and D.M. Newns, *Rep. Prog. Phys.* **52**, 655 (1989).

70. R.L. Ericksen and D.P. Smith, *Phys. Rev. Lett.* **34**, 297 (1975).

71. See, for example A. Zangwill, *Physics at Surfaces*, Cambridge University Press, New York, NY (1988).

72. H.J. Kreuzer and Z.W. Gortel, *Physisorption Kinetics*, Springer-Verlag, Berlin (1986).

73. H.H. Andersen and H.L. Bay, in *Sputtering by Particle Bombardment I*, R. Behrisch (Ed.), Springer-Verlag, Berlin, p. 145 (1982).

74. E. Taglauer, *Appl. Phys.* **A51**, 238 (1990).

75. (a) E. Taglauer, W. Melchior, F. Schuster and W. Heiland, *J. Phys.* **E8**, 768 (1975); (b) F. Huussen, J.W.M. Frenken and J.F. van der Veen, *Vacuum* **36**, 259 (1986); (c) H. Dürr, Th. Fauster and R. Schneider, *J. Vac. Sci. Technol.* **A8**, 145 (1990).

76. E. Taglauer, *Appl. Phys.* **A38**, 161 (1985).

77. P.A.J. Ackermans, P.F.H.M. van der Meulen, H. Ottevanger, F.E. van Straten and H.H. Brongersma, *Nucl. Intr. Meth. Phys. Res.* **B35**, 541 (1988).

78. H.A. Engelhardt, W. Bäck, D. Menzel and H. Liebl, *Rev. Sci. Instr.* **52**, 835 (1981).

79. Y.-S. Chen, G.L. Miller, D.A.H. Robinson, G.H. Wheatley and T.M. Buck, *Surf. Sci.* **62**, 133 (1967).

80. T.M. Buck, G.H. Wheatley, G.L. Miller, D.A.H. Robinson and Y.-S. Chen, *Nucl. Instr. Meth.* **149**, 591 (1978).

81. S.B. Luitjens, A.J. Algra, E.P.Th. Suurmeijer and A.L. Boers, *Appl. Phys.* **A21**, 205 (1980).

82. J.W. Rabalais, J.A. Schultz and R. Kumar, *Nucl. Instr. Meth. Phys. Res.* **218**, 719 (1983).

83. D. Rathmann, N. Exeler and B. Willerding, *J. Phys.* **E18**, 17 (1985).

84. H. Niehus and G. Comsa, *Nucl. Instr. Meth. Phys. Res.* **B15**, 122 (1986).

85. R. Aratari, *Nucl. Instr. Meth. Phys. Res.* **B34**, 493 (1988).

86. H. Verbeek, W. Eckstein and F.E.P. Matschke, *J. Phys.* **E10**, 944 (1977).

87. E. Taglauer and W. Heiland, *Appl. Phys. Lett.* **24**, 437 (1974).

88. M. Beckschulte, D. Mehl, and E. Taglauer, *Vacuum* **41**, 67 (1990).

89. E. Taglauer, W. Heiland, and J. Onsgaard, *Nucl. Instr. Meth.* **168**, 571 (1980).

90. A. Koma, *IPPJ-AM*, **22** (1982).

91. E. Taglauer, in *Handbook of Heterogeneous Catalysis*, Volume 2, G. Ertl, H. Knözinger and J. Weitkamp (Eds), VCH, Weinheim, Germany, p. 614 (1997).

92. J. Knözinger, in *Fundamental Aspects of Heterogeneous Catalysis Studied by Particle Beams*, H.H. Brongersma and R.A. van Santen (Eds), Plenum Press, New York, NY, p. 7 (1991).

93. H.H. Brongersma and G.C. van Leerdam, in *Fundamental Aspects of Heterogeneous Catalysis Studied by Particle Beams*, H.H. Brongersma and R.A. van Santen (Eds), Plenum Press, New York, NY, p. 283 (1991).

94. E. Taglauer, in *Fundamental Aspects of Heterogeneous Catalysis Studied by Particle Beams*, H.H. Brongersma and R.A. van Santen (Eds), Plenum Press, New York, NY, p. 301 (1991).

95. K. Josek, Ch. Linsmeier, H. Knözinger and E. Tanglauer, *Nucl. Instr. Meth. Phys. Res.* **B64**, 596 (1992).

96. R. Margraf, H. Knözinger and E. Tanglauer, *Surf. Sci.* **211–212**, 1083 (1989).

97. R. Kelly, *Surf. Interface Anal.* **7**, 1 (1985).

98. J. du Plessis, Surface Segregation, in *Solid State Phenomena*, Volume 11, Editor-in-Chief: G. E. Murch, Scientific and Technical Publishers, Vaduz, Liechtenstein, p. 5 (1990).

99. T.M. Buck, in *Chemistry and Physics of Solid Surfaces IV*, R. Vanselow and R. Howe (Eds), Springer-Verlag, Berlin, p. 435 (1982).

100. U. Bardi, *Rep. Prog. Phys.* **57**, 939 (1994).

101. M. Schmid and P. Varga, in *The Chemical Composition of Solid Surfaces*, Volume 10, D.P. Woodruff (Ed.), Elsevier, Amsterdam, p. 118 (2002).

102. J. du Plessis, G.N. van Wyk and E. Taglauer, *Surf. Sci.* **220**, 381 (1989).

103. D. Voges, E. Taglauer, H. Dosch and J. Peisl, *Surf. Sci.* **269–270**, 1142 (1992).

104. S. Schömann and E. Taglauer, *Surf. Rev. Lett.* **3**, 1832 (1996).

105. E. Taglauer and R. Beikler, *Vacuum*, **73**, 9 (2004).

106. J. du Plessis and E. Taglauer, *Nucl. Instr. Meth. Phys Res.*, **B78**, 212 (1993).

107. D. P. Jackson, *Surf. Sci.* **43**, 431 (1974).

108. (a) M.A. van Hove, W.H. Weinberg and C.-M. Chen, *Low-Energy Electron Diffraction*, Springer-Verlag, Berlin (1986); (b) C.A. Lucas, *Chapter 8 of this volume*.

109. M. Aono, C. Oshima, S. Zaima, S. Otani and Y. Ishizawa, *Jpn. J. Appl. Phys.* **20**, L829 (1981).

110. Th. Fauster, *Vacuum* **38**, 129 (1988).

111. R. Spitzl, H. Niehus and G. Comsa, *Rev. Sci. Instr.* **61**, 1275 (1990).

112. W.D. Roos, J. du Plessis, G. N. van Wyk, E. Taglauer and S. Wolf, *J. Vac. Sci. Technol.* **A14**, 1648 (1996).

113. J. Schulz, E. Taglauer, P. Feulner and D. Menzel, *Nucl. Instr. Meth. Phys. Res.* **B64**, 588 (1992).

114. I. Kamiya, M. Katayama, E. Nomura and M. Aono, *Surf. Sci.* **242**, 404 (1991).

115. H. Hemme and W. Heiland, *Nucl. Instr. Meth. Phys. Res.* **B9**, 41 (1985).

116. H. Niehus and G. Comsa, *Surf. Sci.* **151**, l171 (1985).

117. J.A. van den Berg, L.K. Verheij and D.G. Armour, *Surf. Sci.* **91**, 218 (1980).

118. L. Eierdal, F. Besenbacher, E. Laegsgaard and I. Stensgard, *Ultramicroscopy*, **42–44**, p. 505 (1992).

119. (a) D.J. Surman, J.A. van den Berg and J.C. Vickerman, *Surf. Interface Anal.* **4**, 160 (1982); (b) H.J. Mathieu, *Chapter 2 of this volume*.

120. H. Bu, M. Shi, K. Boyd and J.W. Rabalais, *J. Chem. Phys.* **95**, 2882 (1991).

121. J. Wayne Rabalais, *Principles and Applications of Ion Scattering Spectrometry*, John Wiley & Sons, Inc., Hoboken, NJ (2003).

122. R. Bastasz, J.W. Medlin, J.A. Whaley, R. Beikler and E. Taglauer, *Surf. Sci.* **571**, 31 (2004).

123. G. Binnig, H. Rohrer, Ch. Gerber and E. Weibel, *Phys. Rev. Lett.* **49**, 57 (1982).

Problems

1. Why can thermal vibrations be neglected in the treatment of collision kinematics (Section 6.2.2)? Compare relevant timescales. Estimate the energy broadening resulting from momentum transfer due to the thermal motion of target atoms.

2. Which energy and angular resolutions are required to separate the stainless steel components (Fe, Cr, Ni)? At which experimental parameters (projectile mass, scattering angle) can these requirements be met in ISS or RBS?

3. Estimate the Y:Ba:Cu concentration ratios from the spectrum of the high T_c film displayed in Figure 6.13. How close is it to the nominal 1:2:3 ratios?

4. What are the accessible angular and energy regions for the detection of hydrogen isotopes by backscattering or by direct recoil detection? Which projectile masses and energies should be favoured (assuming Rutherford cross-sections)?

5. Estimate the interatomic distances in a Ni crystal from the ICISS distribution shown in Figure 6.30, using the appropriate Na^+ shadow cone radius.

6. Estimate the amount of damage to a Ni target surface which will occur when taking spectra with 1 keV He^+ ISS and 1 MeV He^+ RBS where the sputtering yields are 0.15 and 0.01 atoms/ion, respectively. Compare these with the damage resulting from the ISS analysis of a CO adlayer with a desorption cross-section of 10^{-15} cm^2.

7. Simulate RBS spectra such as shown in Figures 6.12 and 6.13 by using the SIMNRA program (download from www.rzg.mpg.de/~mam/). This is a shareware with a free 30 days trial period.

Key Facts

p. 269, 270. Ion scattering analyses the atomic masses on surfaces, not compounds or molecules.

p. 271, 272. For ion energies of 1 keV and more: (a) classical scattering, no quantum effects; (b) no diffraction effects from ordered crystal lattices; (c) collision times much smaller than thermal vibration periods.

p. 272, 273. By applying collision kinematics, the energy spectra of backscattered ions can be converted into mass spectra of surface atoms.

p. 273. Energy spectra: peak positions determined by kinematics, peak intensities by interaction potentials.

p. 278. Spotting neighbouring atoms: shadow cone yields interatomic distances by directing its high flux edge onto a nearby atom.

p. 297. Depth profiling: the inelastic energy loss depends on the path length and thus conveys information at which depth the large angle deflection occurred.

p. 284. Solid state detectors are uniquely simple devices for simultaneous recording energy spectra of ions (and neutrals as well).

p. 269, 285, 285. RBS can be considered as a non-destructive analysis method.

p. 293. Crystalline surface structures can be studied by exploiting channelling and blocking effects.

p. 298. RBS yields not only relative but absolute P numbers of concentrations of elements in near- surface layers.

p. 302. Neutralization: (a) pros – makes ISS exclusively surface sensitive; (b) cons – poses problems for absolute quantification.

p. 308. ISS yields quantitative results with proper calibration.

p. 314. Exact structure analysis and multiple scattering effects frequently require analysis with computer simulations.

p. 316. ISS yields mass selective structure information in real space (cf. reciprocal space for diffraction methods).

p. 320. Hydrogen isotopes cannot be detected by scattering (ISS) but by recoil detection (DRS).

7 Vibrational Spectroscopy from Surfaces[†]

MARTYN E. PEMBLE

Tyndall National Institute, Cork, Ireland

PETER GARDNER

School of Chemical Engineering and Analytical Science,
University of Manchester, Manchester, UK

7.1 Introduction

There can be few techniques as versatile to the chemist as vibrational spectroscopy. This also holds true for the surface chemist. This chapter reviews some of the techniques employed by surface chemists to measure vibrational spectra from molecules adsorbed at a variety of surfaces and gives specific examples of their use. The techniques may be summarized as involving the interactions of photons or particles with surfaces, which result in energy transfer to or from the surface adsorbed species via vibrational excitation or de-excitation. The discrete energies transferred of course correspond to vibrational quanta, and analysis of these energies provides the means of determining the structure of the surface species. The basic theory of vibrational spectroscopy is not covered here, but rather the reader is referred to the numerous undergraduate texts that deal with this material.

[†]This chapter is dedicated to the memory of Dr Peter Hollins, who passed away on 28th March 2005.

Surface Analysis – The Principal Techniques 2nd Edition Edited by John Vickerman and Ian Gilmore
© 2009 John Wiley & Sons, Ltd

Let us begin by asking the question, 'Why use vibrational spectroscopy as a surface probe? To answer this we must consider the usual methodology adopted by the surface chemist, who often notes that there is no one technique that provides all of the information required in order to solve a particular problem in surface chemistry. Many of the techniques available to the surface chemist involve the use of particle beams and hence high-vacuum systems, such that it becomes difficult to examine real surfaces under operating conditions, e.g. pertaining to corrosion, catalysis, etc. Yet as will be seen, there is a form of surface vibrational spectroscopy, which is capable of dealing with almost any surface one cares to imagine, under almost any conditions!

For the research and development chemist interested in the fundamental chemical mechanisms occurring on surfaces, vibrational spectroscopy is one of the most important methods of studying surface chemical reactions. However, we must introduce a note of caution at this point in that, although vibrational spectroscopy may allow us to identify intermediate species forming on surfaces, they may not be involved in the rate-determining step of the reaction. Such species, although of interest, may thus provide little insight into the nature of true surface processes – they are so called 'spectator species'. Despite this, since any mechanistic study requires not only the measurement of reaction rates but also the identification of possible intermediates, any technique, which reveals intermediates, has a part to play.

This chapter begins with a description of the most widely used form of surface vibrational spectroscopy, infrared spectroscopy.

7.2 Infrared Spectroscopy from Surfaces

IR spectroscopy is attractive as a means of studying surface species because of its enormous versatility, being applicable to almost any surface, capable of operating under both high and low pressure conditions and having a relatively low cost compared to, say, a technique which requires high vacuum for operation. Most modern surface IR facilities, whichever sampling mode is used, will utilise a Fourier transform IR (FTIR) spectrometer. The essential feature of an FTIR spectrometer as compared to a dispersive instrument is that all of the light from the source falls onto the detector at any instant. Since this inherently leads to increased signal levels this automatically improves the signal-to-noise ratio at any point in the spectrum. Wavelength or frequency identification is not achieved using monchromators but rather through a careful frequency analysis (Fourier analysis) of the periodic signal at the detector produced by the Michelson interferometer or similar device. This device induces a periodically varying path length difference between two, usually equally intense, beams from the source produced by a simple optical device designed to split the beam intensity – a so-called beam splitter. The two beams, which contain all wavelengths emanating from the source, are recombined and detected. The intensity measured depends on the overall effects of the phase difference for each component wavelength. The phase difference of course varies with each component wavelength. The mathematical operation of converting or transforming a signal, which varies with path length to a spectrum in which intensity varies with wavelength, is known as Fourier transformation.

The advantages that FTIR methods brings are usually described in terms of the Jaquinot or multiplex advantage which arises since all channels are sampled at once; the Fellgett or throughput advantage, which arises because – as compared to a dispersive instrument – the signal level at the detector is always higher; and the Connes advantage, which represents enhanced photometric accuracy arising from the in-built electronic calibration resulting from a single wavelength interferrogram produced by the interaction of an alignment laser, usually a HeNe, with the beam splitter. For further information on the operation of an FTIR instrument the reader is directed to Banwell and McCash [1], Ferraro and Basilo [2] and Griffiths and de Hareth [3].

There are both routine and complex research applications of surface IR spectroscopy and this section aims to cover briefly, most of these methods.

7.2.1 TRANSMISSION IR SPECTROSCOPY

As the name suggests, this mode of sampling involves the passage of the IR beam through the sample, which must therefore be at least partially transmissive in the IR region of the spectrum – Figure 7.1.

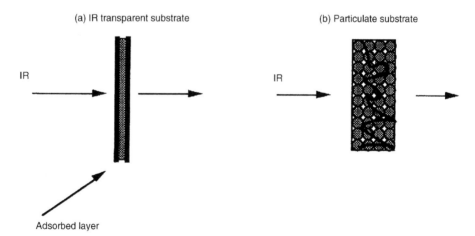

Figure 7.1 (a) An IR transmissive substrate with an adsorbate layer on both its exposed surfaces. (b) Depicting the case for a species adsorbed at the surface of a highly particulate medium

Figure 7.1(a) depicts the case of an IR transmissive substrate with an adsorbate layer on both its exposed surfaces while Figure 7.1(b) depicts the case for a species adsorbed at the surface of a highly particulate medium. In the case of sampling mode (b), samples are commonly prepared by pressing the particulate sample under high pressure, often with some additional support material such as KBr, such that it forms a self-supporting disc. This basic arrangement employing transmission techniques was used in the early pioneering experiments of Eischens *et al.*, who demonstrated for the first time, the utility of IR methods for the study of adsorbed species [4, 5].

Using transmission IR spectroscopy as an example, it is possible to see how IR measurements may be related to the amount of material present and its refractive index: If the incident IR beam has intensity I_0 and the transmitted beam has intensity I_t, then the relationship between these two quantities is:

$$I_0/I_t = \exp(-kcl) \tag{7.1}$$

where the dimensionless quotient I_0/I_t is known as the *transmittance*, or, when multiplied by 100, the *percentage transmittance* and k is the absorption coefficient, which corresponds to the imaginary part of the refractive index of the medium n, such that:

$$n = n + ik \tag{7.2}$$

The use of an imaginary function is a convenient way of representing an absorption process since the intensity of the transmitted radiation is proportional to the square of the refractive index, which therefore results in a 'negative reflectance' (absorption) for the imaginary term. Although Equation (7.1) is a common way of depicting the 'strength' of an infrared absorption, it also shows that transmittance is an exponential function of concentration. A linear form of this expression may be obtained by taking logarithms to base 10 and inverting such that we obtain:

$$\log_{10}(I_t/I_0) = \varepsilon cl \tag{7.3}$$

where $\varepsilon = k/\ln 10$ and is a constant known as the absorption cross-section. The term $\log_{10}(I_t/I_0)$ is known as the *absorbance* and is a linear function of concentration. Two things emerge from this simple treatment, which are worthy of note. Firstly, the relationship between the IR absorption cross-section and the refractive index of the material has been established. Secondly, in order to have a meaningful measurement, at least two intensity measurements are required which in practice means that one must record both a sample and reference spectrum.

Let us now examine a typical transmission IR spectrum of interest to the surface chemist. In order to record such a spectrum, the substrate may be any material that allows at least partial transmission of the IR beam. For example, this mode of IR spectroscopy is often used to study reactions that occur upon so-called 'supported' metal catalysts. The term supported, refers to the fact that the metal particles are chemically impregnated onto a support material, usually a high specific surface area oxide, which serves to increase the active working area of the catalysts and also prevents high temperature agglomeration of metal particles or 'sintering'. Such materials are usually black in colour and not at all the obvious choice of material for a transmission IR experiment. However, when pressed into a thin flat disc, it is found that up to 10% of the IR radiation falling onto the disc may be transmitted and this is more than sufficient to obtain an IR spectrum. The transmission process for these materials involves partial absorption by the supporting oxide but has also been shown by the group of Sheppard [6, 7] to involve a series of complex reflections from the metallic surfaces and for this reason the spectra obtained are often subject to the so called 'surface selection rule' which applies for species adsorbed on flat substrates where the spectrum is measured in reflection. This latter method, known as reflection–absorption IR spectroscopy or RAIRS, is described in detail in a later section. An example of a comparison between the application of RAIRS and transmission IR methods is shown in Figure 7.2.

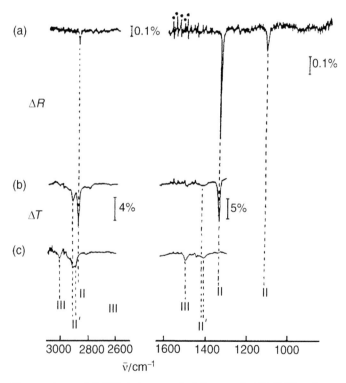

Figure 7.2 Comparison of RAIR data recorded from a flat Pt (111) substrate and data recorded in transmission from a Pt/SiO$_2$ catalyst sample, for the adsorption of ethene [7]

This figure demonstrates the similarity of the data obtained by the two methods of transmission and reflection. Note that the transmission spectra do not reveal any information in the spectral region below ca. $1300\,cm^{-1}$. This is due to the so-called oxide 'blackout' below $1300\,cm^{-1}$, which obscures this region of the spectrum. These data also reveal the wealth of information that may be obtained even from a 'simple' system such as this. Comparison of these data with data from other techniques such as electron energy loss spectroscopy and from IR spectra of inorganic cluster compounds enabled Sheppard and co-workers to ascertain that ethene undergoes a rearrangement on Pt surfaces near 300 K to produce the ethylidyne species, Figure 7.3.

Not too surprisingly from purely geometric considerations, this species is formed most readily on the (111) surfaces of face-centred cubic metals since these surfaces necessarily have the three-fold sites depicted above.

These examples serve to illustrate how transmission IR methods can be used to identify species formed at surfaces under static conditions, where species have formed as a result of the adsorption and reaction of molecules with surfaces and which remain on the surface until a deliberate perturbation, e.g. pumping, substrate heating, is introduced. However, transmission IR methods are actually much more versatile. Such methods have been used to study processes occurring almost in real time. As an example of such a study, here we highlight the work of Chabal *et al.* [8], who have used transmission IR methods to study the mechanisms of materials growth on silicon substrates. The particular growth processes concerned are among those currently of enormous importance to the semiconductor industry and are known as atomic layer deposition or ALD [9, 10]. In ALD, the aim is to grow ultra thin layers of materials from a combination of at least two precursor molecules. Such layers are designed for applications such as the high dielectric constant layers needed to replace SiO_2 as transistor structures are miniaturized still further, following the trend first described by Gordon Moore, founder of Intel, and which has now become known

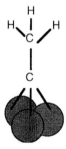

Figure 7.3 Schematic representation of the ethylidyne species formed via reaction of ethene over Pt surfaces

as Moore's law [11]. The substrate must of course be at least partially transparent in the spectral region in question. Substrates such as silicon and germanium satisfy this requirement, at least at low temperatures. At higher temperatures free carrier absorption reduced the effectiveness of the transmission approach. Fortunately, ALD is a relatively low temperature growth process and so is amenable to study using this approach. In order to maximize the signal-to-noise achievable in such an experiment it is necessary to maximize the signal level at the detector. With this in mind the angle of maximum transmission of the IR radiation by the substrate must be carefully selected. This angle in known as the Brewster angle and for silicon the Brewster angle is approximately 70 degrees to the surface normal.

To illustrate the type of data that it is possible to obtain consider we present transmission IR spectra recorded by the Chabal group [8], from a double-sided polished silicon substrate held at the Brewster angle, during the growth of HfO_2 from tetrakis-ethylmethylamino hafnium (TMEAH) and water (Figure 7.4). In their experiments, Ho *et al.* started with a substrate that had been cleaned using a procedure, which leaves the silicon surface in question, terminated with a single layer of adsorbed hydrogen atoms. The resulting Si–H vibrations are readily detectable using IR spectroscopy, as illustrated in Figure 7.4.

Figure 7.4 shows that as the number of half and full ALD cycles increases, so the intensity of the Si–H band at $2083\,cm^{-1}$ decreases while as the

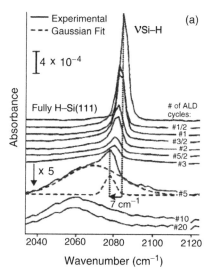

Figure 7.4 Transmission IR spectra recorded from both the front and back sides of a hydrogen passivated Si (111) substrate after various ALD half and full cycles (taken from Ho *et al.* [8])

reaction proceeds the band appears to split into two components, with the component at lower values of wavenumber being significantly broader. Ho *et al.* attribute the sharp band near 2083 cm^{-1} to the Si–H modes from patches of surface covered by unreacted hydrogen atoms, while the broader component at lower wavenumber is attributed to Si–H species surrounded by Si atoms that have reacted directly with either Hf or O atoms [8].

Spectra such as these are an incredibly powerful tool for the development of critical growth processes such as the one described above and are likely to have a very significant impact on the production of electronic devices at the 22 nm node and beyond.

7.2.2 PHOTOACOUSTIC SPECTROSCOPY

Some particulate materials are unsuitable for transmission IR studies because they either absorb too much radiation, or they divert the path of the radiation through scattering. It is still possible to record IR spectra from such materials using alternative sampling modes. For highly absorbing materials, the technique *photoacoustic IR* spectroscopy has been developed. This method relies upon the fact that when IR radiation is incident upon a highly absorbing material, the material effectively heats up. If the material is in contact with some inert gas, then the local heating caused by absorption of a specific wavelength of IR radiation gives rise to a minute thermal shock wave which propagates out into the gas. This shock wave is directly analogous to a sound wave and, as such, may be detected using a sensitive diaphragm, i.e. the sampling cell forms a microphone unit. This description makes the photoacoustic detector sound somewhat esoteric but this is not the case. Photoacoustic units are commonly available as off-the-shelf items. A schematic diagram of a photoacoustic detector is shown in Figure 7.5.

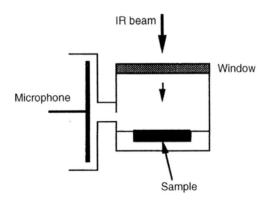

Figure 7.5   A schematic representation of an absorbance photoacoustic sampling system

Importantly, with the advent of the FTIR spectrometer PAS depth profiling has become a standard procedure allowing the surface/near surface region to be selectively probed. The depth d, from which a PAS signal can be observed, is given by $2\pi\mu$, where μ is the thermal diffusion length given by:

$$\mu = \left(\frac{\alpha}{\pi f}\right)^{1/2}$$

where α is the thermal diffusivity and f is the IR beam modulation frequency given by:

$$f = 2v\tilde{v}$$

In the above equation, v is the velocity of the moving mirror and $\tilde{v}$ is the wavenumber. Most FTIR spectrometers, working in rapid scan mode, allow the mirror speed and therefore the modulation frequency to be easily varied over a wide range thus giving the user control of the probe depth. Using values of f between 100 Hz and 1.0 kHz (typical in the wavenumber range of the FTIR spectrometer) the thermal diffusion length range for a typical organic sample is estimated to be 10 to 3.0 µm [12]. The disadvantage of using the rapid scan mode is that modulation frequency and therefore the penetration depth is wavenumber dependent meaning that different regions of the spectrum contain information from different depths below the surface. (The sampling depth varies by more than a factor of 3 over the range 4000 to 400 cm^{-1}.) This problem can be overcome by operating the spectrometer in step scan mode. In this mode, often referred to as phase modulation, the moving mirror is not continuously moved in one direction as the data is collected but is moved in a series of discrete steps. At each the step the mirrored is 'dithered' at a given modulation frequency. The advantage here is that the same modulation frequency applies to the whole spectrum so the wavenumber dependence of the thermal diffusion length is removed.

The particular arrangement shown in Figure 7.5 corresponds to absorbance sampling, although it is possible to obtain photoacoustic sampling accessories that will also work in diffuse reflectance or transmittance modes. In absorbance sampling, the photoacoustic signal arises from the sample heating produced by absorption of the IR radiation. Selective absorption such as this is obscured when using black highly absorbing samples and so transmittance sampling is employed, in which the radiation passing through the sample is absorbed into a gas, which then produces the photoacoustic signal at the microphone. It is thus possible to obtain both absorbance and transmittance spectra using this sampling mode. In absorbance mode, the photoacoustic signal is, as expected, directly proportional to the amount of substance present. It is widely believed that photoacoustic detection only provides an advantage over conventional methods when dealing with black samples. This is not the case, although it is true that for such materials it is hard to match the quality of a photoacoustic spectrum using another sampling technique.

7.2.3 REFLECTANCE METHODS

Many samples are effectively opaque to the IR radiation and thus cannot be studied in transmission mode. Reflectance methods are particularly useful here and have found wide application in both routine and research oriented surface analysis. In order to be able to optimize the efficiency of a reflectance experiment it is necessary to understand the equations governing the reflectance process.

The processes of reflection, refraction and absorption are all related. The purely directional changes that occur when electromagnetic radiation encounters an interface may be explained by considering the refractive indices of the interfacial system. Figure 7.6 depicts the processes of reflection and refraction at an interface.

From this figure we can establish a relationship between the angles of reflection and refraction that was first established by Snell and then later developed mathematically by Descart:

$$n_1/n_2 = \frac{\sin \theta_1}{\sin \theta_2} \tag{7.4}$$

where n_1 and n_2 are the refractive indices of the two media forming the interface. This simple ray diagram also defines the so-called plane of incidence, which is the plane containing the incident and reflected rays and the surface normal. To determine the intensity of the reflected and refracted rays, the treatment of Fresnel is employed. Fresnel, in the late 18th Century, considered the interaction of the electromagnetic wave with the surface between the two media as involving two extreme orientations or *polarizations* of the

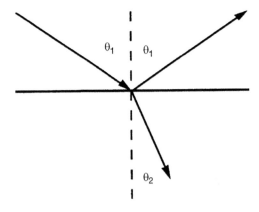

Figure 7.6　The relationship between reflection and refraction

electric vector – light polarized in the plane of incidence, p-polarization and light polarized perpendicular to the plane of incidence, s-polarization. For a more detailed description of these polarisation states, see Figure 7.11. Fresnel determined that the fraction of p-polarized monochromatic light reflected from the interface, R_p, is given by:

$$R_p = r_p r_p^* \tag{7.5}$$

$$\text{where } r_p = \frac{n_2 \cos \theta_1 - n_1 \cos \theta_2}{n_2 \cos \theta_1 + n_1 \cos \theta_2} \tag{7.6}$$

and r_p^* is the complex conjugate of r_p. The use of the complex conjugate may be attributed to the nature of the refractive index of the media, which as was shown in Section 7.2.1 consists of a real and imaginary part, the imaginary part being associated with absorption in the medium. Similarly for s-polarized light:

$$R_s = r_s r_s^* \tag{7.7}$$

$$\text{where } r_s = \frac{n_1 \cos \theta_1 - n_2 \cos \theta_2}{n_1 \cos \theta_1 + n_2 \cos \theta_2} \tag{7.8}$$

and again we note that the true form of the refractive index contains a complex part, such that for a vacuum solid interface, where $n_1 = 1.0$ and $n_2 = n + ik$, the reflectivity equations reduce to:

$$R_p = \frac{\cos^2 \theta_2 - 2n \cos \theta_1 \cos \theta_2 + (n^2 + k^2) \cos^2 \theta_1}{\cos^2 \theta_2 + 2n \cos \theta_1 \cos \theta_2 + (n^2 + k^2) \cos^2 \theta_1} \tag{7.9}$$

and

$$R_s = \frac{\cos^2 \theta_1 - 2n \cos \theta_1 \cos \theta_2 + (n^2 + k^2) \cos^2 \theta_2}{\cos^2 \theta_1 + 2n \cos \theta_1 \cos \theta_2 + (n^2 + k^2) \cos^2 \theta_2} \tag{7.10}$$

Equations (7.9) and (7.10) allow us to predict the variation in reflectivity as a function of angle of incidence for a particular system and thus determine the optimum angle of incidence to use in obtaining a specular reflection spectrum. Two extreme cases can be visualized as follows:

(i) *Near normal incidence.* This technique is commonly employed for the study of relatively thick films of material adsorbed onto highly reflecting substrates such as metallic surfaces. For films of sufficient thickness, this method is effectively a double-pass transmission experiment.

(ii) *Grazing incidence.* This technique is of most use when studying very thin films adsorbed at conducting surfaces. Known as reflection–absorption infrared spectroscopy (RAIRS), this technique is discussed in more detail in Section 7.2.3.

Attenuated Total (Internal) Reflection (ATR). Here the substrate to be analysed is pressed into intimate optical contact with a prism, which is transparent over the range of IR wavelengths to be studied. The IR radiation enters the prism and is incident on the surfaces of the prism at angles greater than the critical angle. If the geometry of the experiment is arranged correctly, then multiple internal reflection occurs (Figure 7.7).

From Figure 7.7 it may be seen that, by varying the angle of incidence, it is possible to vary the number of internal reflections within the ATR element. In practice, up to 100 internal reflections may be employed. The substrate surface is pressed against the ATR prism and at each reflection, the electric vector of the IR radiation samples the surface in contact with the prism via the so-called 'evanescent' wave, which extends beyond the boundary of the prism. To obtain internal reflectance, the angle of incidence must exceed the so-called 'critical' angle. This angle is a function of the real parts of the refractive indices of both the sample and the ATR prism:

$$\theta_c = \sin^{-1}(n_2/n_1) \qquad (7.11)$$

where n_2 is the refractive index of the sample and n_1 is the refractive index of the prism. The evanescent wave decays into the sample exponentially with distance from the surface of the prism over a distance on the order of microns. The depth of penetration of the evanescent wave d_{ev} is determined by:

$$d_{ev} = \lambda/\{2\pi n_1[\sin^2\theta - (n_2/n_1)^2]^{0.5}\} \qquad (7.12)$$

where λ is the wavelength of the IR radiation. Comparisons can be drawn between ATR spectra and transmission spectra by estimating the path length through the sample in the ATR experiment. This 'effective path length' may be taken as simply the penetration depth × the number of reflections.

An example of the application of this approach of current topical interest is the work of Gao *et al.* on the study of the growth of Al_2O_3 dielectric layers on H-terminated Si substrates using trimethyl aluminium and water as precursors [13]. The motivation behind such work is the same as that described earlier, for the transmission studies of atomic layer deposition

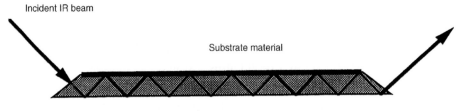

Figure 7.7   The path of the sampling IR beam as it passes through the wave-guide prism in contact with an opaque sample in the ATR experiment

processes. Why then should ATR be an option here, if simple transmission spectroscopy has been shown to be very effective?

The low temperatures used in ALD growth permit the use of transmission-type measurements using substrates such as silicon that would otherwise become opaque to the IR beam at higher temperatures. The use of ATR rather than simple transmission is justified in terms of the increased signal-to-noise that is obtained for an equivalent sampling time, thus pushing the measurement closer to true real-time analysis. Figure 7.8 depicts data obtained by Gao *et al.* As in the ALD application example presented in the section on transmission spectroscopy (page 339), Figure 7.8 focuses on the Si–H stretching region of the spectrum.

Before growth an intense sharp band is again seen at 2083 cm^{-1} together with a smaller, yet still sharp component at 2087 cm^{-1}. As described earlier, the 2083 cm^{-1} is ascribed to the Si–H stretching vibration arising on a high-quality H-terminated Si (111) surface. The component at 2087 cm^{-1} is ascribed to Si–H species present at step edges. After growth these bands effectively disappear and are replaced with a broader band centred at ca. 2060 cm^{-1}, which these authors ascribe to a 'softening' of the Si–H modes due to direct interactions between the H atoms and Al or O atoms.

It must be emphasized that due to the structure of the H-passivated Si (111) surface that all of these interactions occur within one molecular layer on the substrate surface. The quality of the spectra obtained by ATR and also by simple transmission (see earlier) is testament to the power of this approach.

Diffuse Reflectance. For highly scattering particulate samples such as white powders, photoacoustic detection is not efficient. An alternative

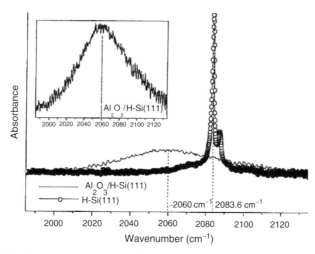

Figure 7.8 ATR IR spectrum recorded through a hydrogen passivated Si (111) substrate after various ALD full cycles (taken from Gao *et al.* [13])

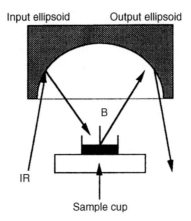

Figure 7.9   A schematic representation of the arrangement for diffuse reflectance sampling; B = beam blocker

method is available which collects the scattered light from the substrate surface and directs it onto the IR detector. This method of sampling is known as diffuse reflectance (Figure 7.9).

In this mode of analysis, only that radiation, which undergoes diffuse scattering, is considered to have penetrated into the surface of the particulate material, i.e. true specular scattering does not interact with surface species and is not absorbed. In order to maximise signal-to-noise performance it is necessary to prevent unwanted light from reaching the detector; this is partially achieved using a specular beam blocker, labeled B in the diagram. The form of the spectrum obtained in this manner does not directly coincide with that of a spectrum obtained in the conventional manner, since the spectral profile is dominated by the relationship between wavelength and scattering efficiency. This feature is eliminated by transforming the spectrum using a mathematical process known as a Kubelka–Munk transformation, which compensates for the wavelength: scattering power relationship. Specific instruments have this routine built into the permanent memory of the computational facility in addition to the normal Fourier routines. This is obviously a highly convenient method of obtaining a surface spectrum from a particulate material that involves almost no sample preparation! It is particularly useful for the study of solid catalysts, where almost any material may be studied and via the use of environmental chambers, gas pressures of up to 100 atmospheres may be introduced over the catalyst surface. For such purposes it is possible to obtain high pressure and temperature environmental chambers designed specifically for diffuse reflectance or transmission work. The resulting spectra, recorded under 'real' conditions, may reveal the presence of species not previously postulated to be intermediates in the catalytic process. An example of this is shown in Figure 7.10 in which

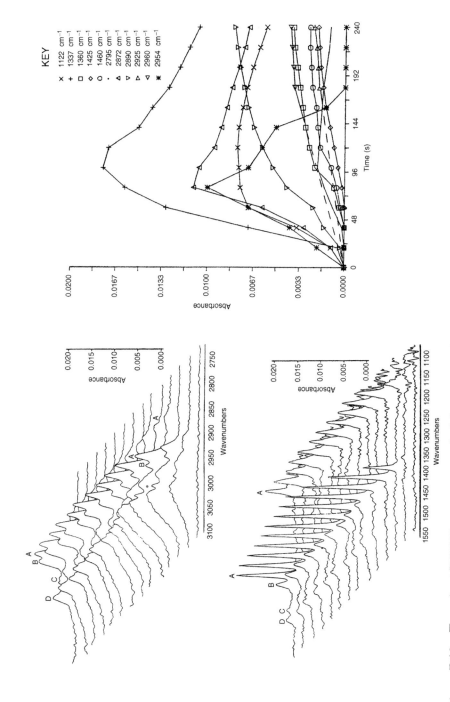

Figure 7.10 Transient DRIFT spectra recorded from an Ni/Al$_2$O$_3$ catalyst as a function of time following exposure to ethane (after Holmes *et al.* [14])

Holmes [14] and co-workers in Edinburgh have obtained spectra from a supported Ni catalyst under dynamic conditions.

This figure depicts the time-dependent evolution of an ethylidyne species similar to that detected on supported Pt catalysts (see Figure 7.2), which was not previously believed to be formed on Ni surfaces. It can be seen that this species is a true transient species, present for a limited time only during the reaction. The ability to record batches of spectra and to display them in this manner is also a good indication of the usefulness of the computer associated with a modern FTIR instrument.

Reflection Absorption IR Spectroscopy (RAIRS).   This technique has proved to be a particularly powerful research tool for the study of adsorbed layers on metal surfaces. For adsorbates on metallic or any conducting film it was Greenler [15] who first demonstrated that the absorption of IR radiation by the adsorbate overlayer is enhanced at high angles of incidence (near grazing) and involves only one polarization of the incident IR beam (Figure 7.11).

Figure 7.11 illustrates the incident and reflected electric vectors of the so called s and p components of the radiation where p refers to *parallel* polarized radiation and s to *perpendicular* polarized radiation with respect to the plane of incidence. Greenler [15] highlighted the fact that at the point of contact with the surface, the p-polarized radiation has net combined amplitude that is almost twice that of the incident radiation via the vector summation of E_p and $E_{p'}$. However, for the s-polarized radiation, the incident and emitted electric vectors E_s and $E_{s'}$ undergo a 180-degree phase-shift with respect to each other and so the net amplitude of the IR radiation parallel to the surface plane is zero (Figure 7.12).

Thus only radiation having a p-component may possess a finite interaction with the surface and hence the only active vibrations that may be observed in RAIRS must have a component of the dynamic dipole polarized in the direction normal to the surface plane. This is a statement of the so

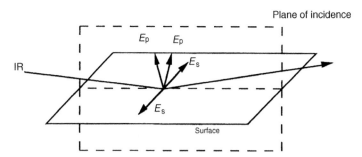

Figure 7.11   Schematic representation of the plane of incidence and the definition of s and p polarized radiation in the RAIRS experiment

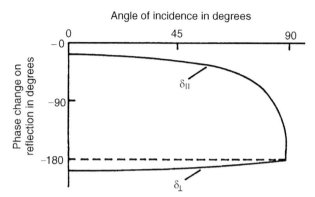

Figure 7.12   The phase shift for light reflected from a metal surface as calculated for light polarized both parallel to (p) and perpendicular to (s) the plane of incidence (after Greenler [15])

called 'surface selection rule' for reflection IR spectroscopy. The surface selection rule is discussed in greater detail in Section 7.4 where the group theory of analysis of surface vibrations is presented.

Obviously the same consideration would apply for all angles of incidence. However, it is possible to show via the use of Maxwell's equations, that the resultant amplitude of the p-polarized component of electromagnetic radiation that is oriented perpendicular to the surface plane reaches a maximum near grazing incidence, while the amplitude of the p-component oriented parallel to the surface plane is low and relatively structureless as a function of angle of incidence. Figure 7.12 shows that, as expected, the net amplitude of the s component upon reflection is zero. In order to determine the variation in *band intensity* expected over a range of angles of incidence, it is necessary to note that the spectral intensity will vary as a function of the *square* of the amplitude of the electric vector and also the variation in sampled surface area. It may be shown by trigonometry that the surface area sampled for a given incident beam diameter varies as a function of $1/\cos \theta$, i.e. sec θ. Thus the band intensity will vary as a function of E^2 sec θ. This function was first calculated for a series of model overlayers with varying refractive indices by Greenler [15]. A more recent summary of the behaviour expected has been produced by Chesters [16] (Figure 7.13).

From this figure it may be seen that the anticipated variation in band intensity reaches a maximum near grazing incidence. We are now in a position to state the optimum conditions for the measurement of RAIR spectra from species adsorbed at metallic surfaces:

Maximum spectral intensity will be observed when using p-polarized radiation oriented perpendicularly to the surface plane at grazing incidence.

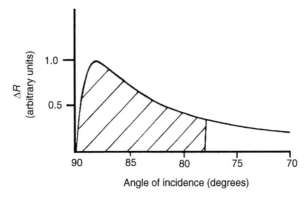

Figure 7.13   Schematic representation of the variation in band intensity with angle of incidence (after Chesters [16])

The RAIRS Experiment.   At the beginning of the section on IR methods, it was stated that most modern instruments utilize FTIR techniques. This is also true for the RAIRS experiment, which is therefore termed FT-RAIRS by some, although some of the highest quality RAIRS data have been obtained using purpose-built dispersive instruments designed to maximize sensitivity in one small spectral region such as that containing the C—O stretching vibration for molecularly adsorbed CO on certain metal surfaces. The particular methods employed to enhance signal-to-noise levels here are discussed in Section 7.2.4. However, a typical layout for the FT-RAIRS experiment is shown in Figure 7.14.

The resolution of IR spectroscopy is such that, for the analysis of surface species, the bandwidths are determined entirely by the heterogeneity of the surface and the nature of the surface–molecule interactions rather than any experimental artifacts (unlike electron energy loss spectroscopy – see Section 7.3). A typical RAIRS spectrum from a strong IR absorber, CO, adsorbed at a copper single crystal surface is shown in Figure 7.15 along with analogous data obtained using the technique of electron energy loss spectroscopy (see Section 7.3).

The RAIR data presented in Figure 7.15 reveal that at least two types of adsorbed CO are present on the saturated Cu (111) surface as indicated by two discrete bands in the C—O stretching region. It would be tempting to apply a semi-quantitative analysis to the band intensities depicted in Figure 7.15 and suggest that the more intense band corresponds to the majority species. However, in their seminal work, Hollins and Prichard demonstrated that for ordered overlayers such as this, where the dipole moments of the adsorbed species are also arranged in an ordered array, that the phenomenon known as dipole coupling may give rise to variations in band intensity such that intensities alone are not a reliable guide to quantity [17, 18]. Although the theory of dipole coupling is beyond the

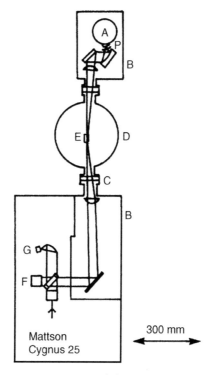

Figure 7.14 A schematic representation of the FT-RAIRS experiment as described by Chesters [16]

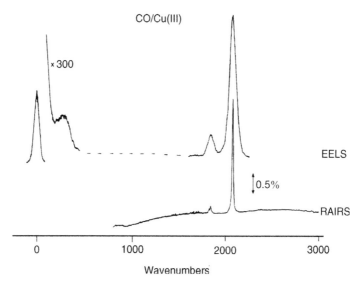

Figure 7.15 RAIRS spectrum of a monolayer of CO adsorbed on a Cu (111) surface at 95 K in comparison with analogous data obtained using electron energy loss spectroscopy (after Chesters *et al.* [16])

scope of this text, the effects of dipole coupling may be isolated via the use of dilute mixtures of isotopically substituted molecules which, as a result of the frequency shift arising from the variation of masses, do not take part in the collective dipole coupling process. Figure 7.15 also reveals the high signal-to-noise levels achievable using FT-RAIRS. The advent of FTIR techniques has opened up the use of RAIRS beyond the study of molecules, which have high IR absorption coefficients, to species such as simple organics. Although not a routine method by any means, it is now relatively easy to record RAIR spectra from sub-monolayer amounts of such weak absorbers on metallic surfaces. Figure 7.16 illustrates this by presenting data for cyclohexane adsorption at a Cu (111) single crystal surface.

Figure 7.16 demonstrates well that hydrocarbon species may be detected at the sub-monolayer level when adsorbed on metal surfaces using RAIRS. The particular spectral region described by Figure 7.16 depicts the C—H

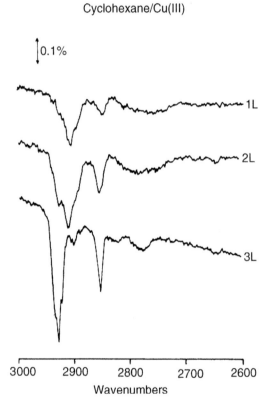

Figure 7.16   RAIRS spectrum of cyclohexane adsorbed on a Cu (111) surface at 95 K as a function of exposure in Langmuir (where I Langmuir $= 1 \times 10^{-6}$ Torr.s) (after Chesters *et al.* [19])

stretching region. The weak broad features near $2600 \, \text{cm}^{-1}$ are particularly worthy of note. These features arise due to the close proximity of C—H bonds to surface metal atoms, resulting in the formation of partial hydrogen bonds to the surface and generating so called 'softened' C—H modes. Obviously such states may well play a part in the mechanism of dehydrogenation of alkanes, which occur over such surfaces.

Most laboratory based RAIRS systems are restricted in the spectral range usually described as the mid-IR, i.e. $400 \, \text{cm}^{-1}$ to ca. $4000 \, \text{cm}^{-1}$. This restriction arises primarily due to the cut-off of the HgCdTe detector, typically used on such systems, but also due to the low intensity of thermal sources at lower wavenumber. However, the low wavenumber region can be accessed using synchrotron radiation as a source in combination with liquid He cooled detectors [20]. This region of the spectrum ($100–600 \, \text{cm}^{-1}$) is important because it is where the metal–adsorbate stretching modes of small molecules such as CO and NO can be found. In addition, it is also a region that contains many oxide and halide vibrations as illustrated in Figure 7.17. Here the formation of $MgCl_2$ on an Au surface in the production of a model Zeigler–Natta catalyst is monitored by Far IR RAIRS [21].

Even with a synchrotron, however, this region is still difficult to work in due to low intensity signals and the possibility of 'inverse' absorption band appearing in the spectra. In the case of small molecules adsorbed on metal surfaces, dipole forbidden modes, e.g. frustrated translational or rotational modes may be observed, as illustrated in Figure 7.18, which depicts the spectrum obtained in this region following the adsorption of CO on Cu (100). Here the Cu—C stretching vibration can clearly be seen at $340 \, \text{cm}^{-1}$ while the frustrated C—O rotation is observed as an *inverse* absorption band at $275 \, \text{cm}^{-1}$ [22].

In the case of adsorbates on thin oxide layered surfaces, optical interference effects can give rise to both positive and negative directions as in the case of $SnCl_4$ adsorption on a thin SnO_2 layer supported on tungsten [23, 24].

Polarization modulation RAIRS. Although RAIRS has become associated with the use of high-vacuum surface science techniques, several studies have shown that it is also possible to obtain RAIRS data under high overpressures of reactants. This is of course a tremendous advantage over many high-vacuum surface analysis methods. To understand how the gas-phase and surface-phase spectra may be decoupled we must consider the result shown in Figure 7.9. This figure demonstrates that radiation polarized parallel to the plane of incidence will not interact with surface species. This radiation will, however, interact with gas phase species. Thus if the s-polarized light is sensitive to gas phase species and the p-polarized light is sensitive to gas-phase and surface-phase species, then by utilizing a polarizer and gating the response of the detector accordingly, it is possible

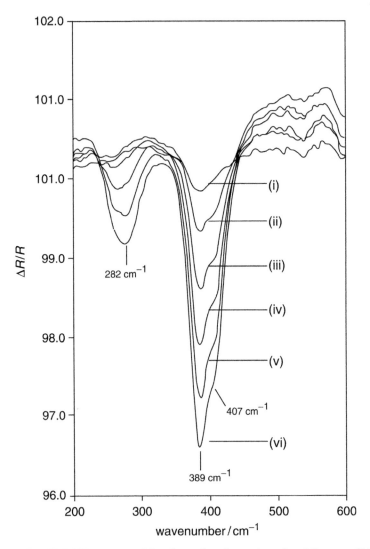

Figure 7.17 Far-IR RAIR spectra arising from the adsorption of multilayers of $MgCl_2$ on a gold substrate. Coverages of $MgCl_2$: (i) 2; (ii) 4; (iii) 6; (iv) 8; (v) 10; (vi) 12 ML (taken from Pilling *et al.* [21], Copyright 2005, Elsevier)

to extract a surface spectrum. This is known as polarization-modulation or PM-RAIRS.

   This particular technique used to be restricted to a few dedicated surface spectroscopists but technological developments that have occurred over the last ten years or so means that PM-RAIRS modules are now offered as off-the-shelf accessories by leading FTIR manufacturers. A schematic diagram of a typical experimental set up is shown in Figure 7.19.

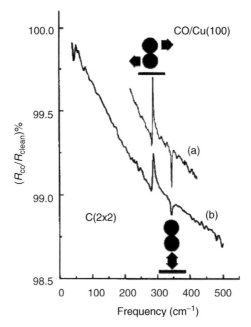

Figure 7.18 Far-IR RAIR spectra of CO adsorbed on a Cu(100) surface showing both conventional and so-called inverse absorption bands. Reflectance changes induced by a 0.5 monolayer coverage of CO/Cu(100) showing the vibrational modes and background changes at low frequencies. Curve (a), which has been shifted upwards for clarity, was taken at a resolution of 1 cm^{-1}, while curve (b) was taken at a resolution of 6 cm^{-1} (taken from Hirschmugl *et al.* [22])

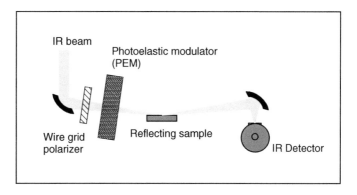

Figure 7.19 A schematic representation of the PM-RAIRS experiment

The arrangement depicted in Figure 7.19 is essentially the same as that for the conventional RAIRS experiment except that the IR beam passed through a polarizer and then through a photoelastic modulator (PEM) that alters the state of polarization between s and p at a modulation frequency of 37 kHz. The differential reflectance spectrum is given by:

$$\frac{\Delta R}{R} = \frac{R_p - R_s}{R_p + R_s}$$

The resulting signal at the detector can be demodulated using a lock-in-amplifier, but the need to separate the modulation of the IR signal resulting from the moving mirror and that resulting from the PEM means that slow mirror speeds need to be used. Alternatively real-time sampling electronics developed in the 1990s can be used in conjunction with conventional rapid-scan instruments, thus reducing data collection times [25]. A third possibility is to do away with the sampling electronics completely and demodulate the signal using digital signal processing software. At present this requires the use of a step scan instrument for data collection [26].

The driving force behind PM-RAIRS development has generally come from the surface science/catalysis community who are interested in extending the RAIRS method to the study of catalytic reactions over low area surfaces at elevated pressures. Typically single crystal samples are prepared using standard surface characterization methods under UHV conditions and then transferred to a 'high (atmospheric) pressure' cell for the PM-RAIRS measurement.

Figure 7.20 shows the adsorption of CO onto a Pd (111) single crystal surface at 600 mbar pressure and various temperatures. The surface absorption

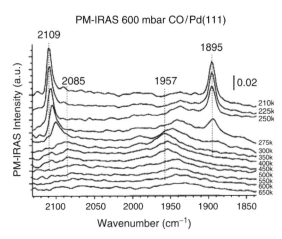

Figure 7.20 *In-situ* polarization modulation RAIR spectra obtained via for the adsorption of CO on (Pd (111) at 'high' CO overpressures (600 mbar) at 650K, followed by cooling to 210 K (taken from Ozensoy *et al.* [27])

bands are clearly observed but the gas phase bands are completely absent [27]. An advantage of the PM-RAIRS method is that it does not require the collection of a separate background spectrum of the clean surface. This is particularly useful where spectra of the metal surface before and after adsorption are difficult to obtain. The technique has thus been used extensively in the study of self-assembled monolayers (SAMs) where spectra of the monolayers, usually adsorbed on gold surfaces, can be recorded under ambient pressure conditions, without being referenced to a separate clean gold coated slide. A recent example of such a system is shown in Figure 7.21 where, as part of the development of a biological immunosensor, antigen immobilization on a mixed SAM surface is monitored using PM-RAIRS [28]. In this series of spectra one can clearly see the characteristic amide I and amide II vibrations at 1660 and 1550 cm^{-1} of the labeled antigen increasing, as the surface concentration of the antigen molecule increases as a result of binding to the appropriate immunoglobin antibody prepared as a mixed SAM layer on the substrate.

Importantly the PM-RAIRS method is also suitable for the study of surfaces under liquid environments provided that the liquid film above the surface is sufficiently thin to allow transmission of the incident and reflected beam (Figure 7.22).

The ability to probe surfaces through an aqueous layer opens up the possibility of probing the liquid solid interface of biologically important systems.

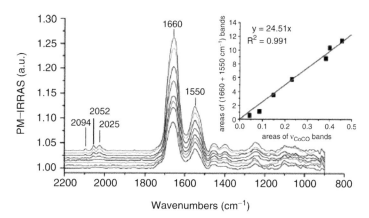

Figure 7.21   PM-RAIRS data for the immobilization of a labeled (using cobalt carbonyl) antigen on a gold surface prepared with the appropriate antibody. PM-RAIRS response of the rIgG$_{PrA}$ sensing surface to increasing concentrations of anti-rIgG labeled by cobalt–carbonyl probes in PBS containing goat serum (0.15 vol%). Inset: correlation between the ν_{MCO} and the amide I + II bands area. Relative standard deviations were estimated to be equal to 0.1 a.u. on each IR area value (taken from Briand *et al.* [28])

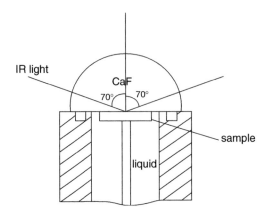

Figure 7.22 A schematic representation of the arrangement used to study surfaces immersed in liquids using PM-RAIRS (taken from Méthivier *et al.* [29])

Subtractively Normalized Interfacial FTIR Spectroscopy (SNIFTIRS). In an electrochemical environment alternative modulation methods can be employed to obtain a surface infrared spectrum. Bewick and co-workers at Southampton were able to obtain RAIR spectra from electrode surfaces immersed in aqueous solution by using thin layer cells coupled with electrochemical modulation methods [30].

The use of such methods has enabled not only traditional surface chemists to use RAIRS as a surface probe, but also electrochemists. Bewick and co-workers at Southampton were able to obtain RAIR spectra from electrode surfaces immersed in aqueous solution by using thin layer cells coupled with electrochemical modulation methods [30]. In these experiments, two electrode potentials are selected – one at which the adsorbate is adsorbed at the electrode surface and the other where there is no adsorption. By recording spectra and rapidly switching the electrode potential, the difference spectrum may be obtained. This method is known either as EMIRS (electrochemically modulated infrared spectroscopy) or the more flexible SNIFTIRS (subtraction and normalisation of interferrograms Fourier transform infrared spectroscopy), which is capable of compensating for high background absorptions such as those due to water. These techniques work well but do require that at the 'potential of no adsorption' there are no other processes occurring which may alter the state of the surface. For many electrochemical systems of interest this is not the case. An example of this concerns the formation of a corrosion layer at a nickel electrode surface in aqueous hydroxide media. This system is believed to involve the formation of a variety of hydroxides and oxy-hydroxides over a wide potential range and as such it is not possible to define a 'potential of no adsorption'. In order to be able to investigate this system the group at Southampton developed a method by which sets of interferrograms could be recorded at short time

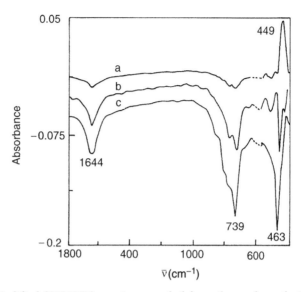

Figure 7.23 Modified SNIFTIRS spectra recorded from the surface of a Ni electrode in aqueous hydroxide media showing via the use of the time-dependent difference spectra the long-term evolution of a layer of α-Ni(OH)$_2$ (after Beden and Bewick [31])

interval separations, while the nickel electrode was also subjected to continuous potential cycling in order to stabilize the surface towards large changes over the timescale of the experiment. An example of the data arising from this 'modified SNIFTIRS' approach is presented in Figure 7.23.

Figure 7.23 shows that, despite the fact that the sample was immersed in aqueous solution, the modified SNIFTIRS approach allows the detection of the a-Ni species via the O—H stretching mode near 1644 cm^{-1}, amongst other bands. Thus there can be no doubt that these are very powerful methods for studying electrode surface processes *in situ*. A later section describes the application of the complementary technique of Raman scattering to the *in-situ* study of electrode surfaces.

Infrared Microscopy from Surfaces. FTIR microscope attachments are commonly available accessories for most instruments. Using these, which operate as beam condensing devices, it is possible to analyse by transmission or refection methods, samples as small as 10–20 microns. Below this the S/N falls dramatically since (a) the aperture cuts out most of the IR light from the source and (b) as the size of the aperture approaches that of the wavelength of light passing through it severe losses are encountered due to diffraction. Using a synchrotron as an IR source it is possible to obtain spectra down to the diffraction limit of $\sim \lambda/2$ thus at 2000 cm^{-1} a spatial resolution of 2.5 microns can be obtained [32, 33]. Using a synchrotron, high resolution maps can be obtained but long data collection times are required if large areas are to be mapped. This field however has been transformed due to the development of IR focal plane array (FPA) detectors. Using an FPA, a large area is illuminated and the light is detected on an array, typically 64×64 MCT detector elements. In this way an infrared image of the sample is obtained in a single scan with each of the 4096 'pixels' containing the full infrared spectrum. Typically each detector element is detecting signal from an area of 6.25^2 microns and thus an area of 400×400 microns can be mapped simultaneously.

While suitable for certain surface chemical applications, this method is better thought of as a bulk analysis method. For specific surface chemical applications, small area reflection accessories may be obtained, which permit the IR 'mapping' of a surface. Figure 7.24 shows a surface absorbance map

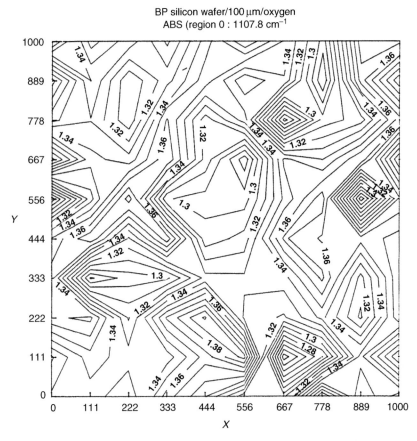

Figure 7.24 Absorbance map recorded from a Si wafer covered with oxide using a novel microscope attachment. The data refer to the detection of the 1108 cm^{-1} band of the SiO_2 overlayer present on the wafer (courtesy of Spectra Tech Europe Ltd and BP Research, Sunbury, UK)

obtained from an area of $100\mu \times 100\mu$ on a silicon wafer covered with a layer of oxide. The absorbance data obtained corresponds to the $1108\,\mathrm{cm}^{-1}$ Si–O mode of the oxide film.

Such sampling methods have obvious application in, say, the semiconductor industry as part of the quality control of large area wafers.

7.3 Electron Energy Loss Spectroscopy (EELS)

Also referred to as HREELS which stands for high resolution electron energy loss spectroscopy, the development of this method as a surface probe by Ibach and co-workers [16] in the early 1970s effectively revolutionized surface science. Up to this time, IR methods such as RAIRS were the only viable means of recording surface IR spectra and due to limitations with equipment and detectors, these early IR experiments were limited to the study of molecules with large dynamic dipole moments such as CO, NO, etc. In contrast, it was soon demonstrated that EELS was sensitive to sub-monolayer amounts of adsorbates possessing relatively weak dynamic dipoles, which could not be studied using IR methods at the time [34–37].

The EELS experiment was developed from gas-phase electron scattering experiments, which probed electronic states within molecules. For the analysis of surface vibrations, the technique utilizes the interaction of very low energy electrons (1–10 eV) with the surface electric fields produced by adsorbate molecules and substrate atoms (Figure 7.25).

Two types of scattering of electrons may be considered, 'elastic' and 'inelastic' (a third mechanism in which the incoming electron is trapped for a finite time within the surface forming a so called 'negative ion resonance' is not treated here). Electron scattering is directly analogous to neutron diffraction. This analogy with a diffraction experiment is a useful one, since

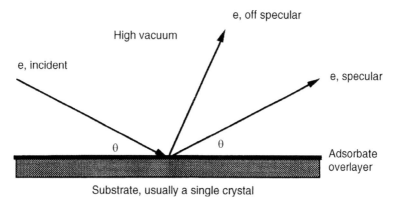

Figure 7.25 A schematic representation of the electron energy loss spectroscopy experiment

it illustrates why the best energy resolution in EELS, as measured by the full width at half maximum height (FWHM) of the elastically scattered electron peak, is obtained when using a single crystal substrate. Both elastic and inelastic scattering are broadened in energy terms when the scattering potential of the surface is poorly defined, e.g. a polycrystalline sample. Since the resolution of the experiment is determined largely by the efficiency of the energy selection system and the sample surface, typical values obtained are of the order of 30–40 cm^{-1}, which is very poor in comparison to IR or laser Raman methods but more than adequate to record good quality vibrational spectra from single crystal surfaces. Some modern EEL (HREEL) spectrometers are able to obtain resolution of <8 cm^{-1} from well-prepared surfaces.

7.3.1 INELASTIC OR 'IMPACT' SCATTERING

Inelastic or 'impact' electron scattering can be described in the following way: impact scattered electrons have effectively 'forgotten' their initial angle and plane of incidence and have been scattered by short-range interaction with surface atomic potentials, which are, of course, modulated at vibrational frequencies. If a net surface dipole exists, these electrons may lose energy into the corresponding vibrational mode of the surface–adsorbate complex. The efficiency of the scattering process, which is known as the scattering cross-section, depends upon the net momentum of the electrons *at the point of interaction to*gether with the 'magnitude' (amplitude and direction) of the dipole moment accompanying the surface vibration. Rather than deal with electric field strength as in the case of photons, for electrons we consider electron momentum sometimes known as wave vector for which we use the symbol k. Using these terms, the impact scattering experiment may be described as shown in Figure 7.26.

The wave vector in brackets represents the electron wave under conditions of *time-reversal*, illustrating that both energy and momentum of the electron must be conserved in the complete scattering event, the negative

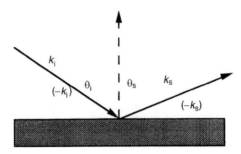

Figure 7.26 A schematic representation of the impact scattering process indicating the change in electron wave vector that is observed upon 'reflection' from the surface

signs arising from the imposition of a fixed coordinate frame. At the point of interaction, the net wave vector is given by the difference between the incident and scattered wave vectors, $\mathbf{k}_s - \mathbf{k}_i$. The scalar product of this function with a function describing the time-dependent electric field of the adsorbate vibration is the inelastic scattering cross-section. It may be seen that for the situation where $\theta_i = \theta_s$, i.e. specular scattering, the net wave vector will be oriented perpendicular to the surface. Since under specular scattering conditions the quantity $\mathbf{k}_s - \mathbf{k}_i$ has no component parallel to the surface plane, the cross-section for scattering from those vibrational modes oriented parallel to the surface must necessarily be zero. This then represents a statement of the surface selection rule that applies for impact scattering in the specular position. This may also be seen from a more general treatment of vibrational excitation.

Quantum mechanically the probability (cross-section) for excitation of a particular vibration in state $\mathbf{\Psi}_0$ to state $\mathbf{\Psi}_1$ is given by the overlap integral denoted as $<\mathbf{\Psi}_0|V|\mathbf{\Psi}_1>$, where V is the electron–vibration interaction potential. The symmetry of this function may be understood by considering a plan-view projection of Figure 7.27.

This figure reveals that, providing there is little change in the trajectory for the electron upon interaction with the surface, i.e. $k_i \approx k_s$, the shape of the electron potential as seen by the adsorbate is independent of the direction of propagation, i.e. it does not matter whether one considers the scattering event as occurring in the forward or reverse (time reversal) directions. The electron–vibration interaction potential V is said to be an even function. Since the ground vibrational state $\mathbf{\Psi}_0$ must by definition be even with respect to the symmetry operations for the point group of the

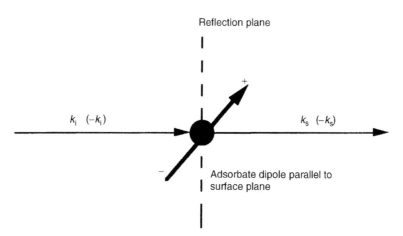

Figure 7.27 Schematic representation of the symmetry factors involved in the consideration of whether a vibrational mode will be observed via the impact scattering mechanism

adsorbed complex, the only allowed excited states must be those which are also even with respect to this point group. For a vibration which has its dynamic dipole oriented parallel to the surface as shown in Figure 7.27 it is clear that such a dipole transforms as odd with respect to the symmetry plane that represents time-reversal inversion and therefore will give rise to a zero scattering cross-section. The only orientation of the adsorbate dynamic dipole that does not violate this rule is where the dipole is parallel to the plane of reflection, but since this direction is orthogonal to both k_s and k_i, there can be no interaction anyway.

Thus it may be stated that, for impact scattering on specular, where the condition $k_i \approx k_s$ holds, a vibration which transforms as a vector parallel to the surface plane will not be observed in the energy loss spectrum. Obviously this rule may only be strictly observed where $k_i = k_s$ but in practice it is found that the cross-sections for excitation of parallel modes fall rapidly as one approaches the true specular position.

To determine whether a vibrational mode, having a component of its dynamic dipole parallel to the surface plane, will give rise to an impact scattered loss feature, the orientation of the dynamic dipole with respect to the plane of incidence of the electron beam must be considered (Figure 7.28).

Figure 7.28 depicts a particular orientation of an adsorbate dipole towards symmetry elements determined by the plane of incidence of the electron beam, which is usually chosen, in the case of single crystal substrates, to be coincident with a direction of symmetry of the surface. The symmetry

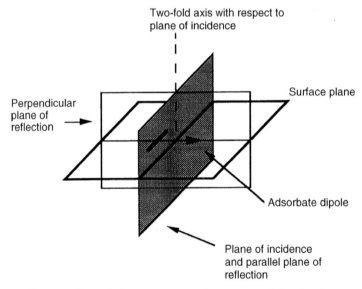

Figure 7.28 Construction of the symmetry elements used in the determination of whether a particular experimental geometry will permit vibrational modes oriented parallel to the surface plane to be observed via the impact scattering mechanism

elements in question are a two-fold axis of rotation and planes of reflection parallel to and perpendicular to the plane of incidence. The following rules may now be stated as follows:

1. If a vibration transforms as antisymmetric with respect to the two-fold axis oriented perpendicularly to the surface plane, then impact scattering is only observed in the *off-specular* directions.

2. If a vibration transforms as antisymmetric with respect to a plane of reflection oriented perpendicular to both the surface plane and the plane of incidence then impact scattering is only observed in the *off-specular directions confined to the plane of incidence*.

3. If a vibration transforms as antisymmetric with respect to a plane of reflection oriented perpendicular to the surface plane but parallel to the plane of incidence then impact scattering is *disallowed* both on and off-specular.

Thus unlike in RAIRS, it is possible to detect vibrational modes that produce dynamic dipoles even for the case when those dipoles are oriented parallel to the surface.

7.3.2 ELASTIC OR 'DIPOLE' SCATTERING

Elastic or 'dipole' scattering occurs when the change in momentum vector $\mathbf{k}$ of the electron upon reflection is minimal, which in practice means that small energy losses may occur such that $k_S \approx k_i$ (equal amplitudes) but the net *direction* of the momentum vector remains unchanged, i.e. $\mathbf{k_S} \approx \mathbf{k_i}$. The scattered electrons are therefore grouped in a small angular lobe of perhaps 2–3 degrees about the true specular position. Most of these electrons interact with surface dynamic dipoles via a long-range electrostatic interaction such that Coulombic energy transfer occurs while the electron is still some 100–200 Å above the surface. This 'dipole' scattering may be thought of as an entirely different scattering mechanism to that which occurs under the influence of short-range interactions. The electron in vacuum sees not only the surface dipole but also the response of the metallic conduction electrons to this dipole. This response is known as the *image dipole* and may be depicted for the two extreme orientations as shown in Figure 7.29.

From Figure 7.29 it may be seen that for an adsorbate dipole of magnitude p oriented parallel to the *surface normal*, the presence of the image dipole created by a redistribution of surface conduction electrons in response to the adsorbate dipole results in a net surface dipole of magnitude $\approx 2\mathbf{p}$, i.e. approximately twice that of the adsorbate dipole alone. However, for adsorbate dipoles oriented parallel to the *surface plane*, the image dipole acts to negate the surface dipole such that the net dipole is effectively zero. Thus

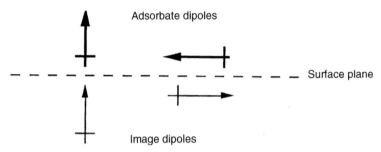

Figure 7.29 Schematic representation of the two extremes of dipole orientation with respect to the surface plane and the corresponding orientation of the image dipole formed via the response of the conduction electrons in the substrate

a surface selection rule exists that is effectively identical to that described for the RAIRS experiment in that for long-range, dipolar scattering, energy is only lost to those surface vibrations which have a component of their dynamic dipole oriented parallel to the surface normal. For this reason, a RAIRS spectrum and a dipolar EELS spectrum are entirely analogous in terms of band positions, although relative intensities may vary due to differences in scattering and absorption factors. The key features of dipole scattering, namely the narrow angular spread about the specular position and the long-range nature of the interaction are a direct consequence of an electrostatic model of the interaction of the surface dipole with the incident and outgoing electron. This treatment is beyond the scope of this text and the reader is referred to the work of Ibach and Mills [38] for a full description.

Considering both dipole and impact scattering, it may be seen that for surfaces having well defined symmetry elements such as single crystals, where the plane of incidence of the electrons may be well-defined, a comparison of dipolar and impact EELS data and application of the impact scattering selection rules often allows for the complete structural and orientational characterization of the adsorbed species. Examples of such analyses are given in a later section.

However, it is possible to take such analyses one stage further and arrive at a typical spectroscopic intensity *pattern* for an adsorbed species, which is attributable to a particular mode of bonding. For ethene adsorbed on a variety of single crystal metal surfaces this exercise has been carried out by Sheppard and co-workers [39] who have classified spectra as type I, with some di-σ character and type II, mainly π-bonding (Figure 7.30).

The assignment of the type I spectrum to that of a true di-σ species was made with the aid of data for the cluster compound $(C_2H_4)Os_2(CO)_8$ which is known to have this structure [39, 40]. Similarly, the assignment of the type II spectrum to that of a π-bonded species was made with the aid of IR data for Zeise's salt [41], $K^+ [(C_2H_4)PtCl_3)]^- H_2O$, which includes data for the C$=$C stretching mode, which does not possess a dipole moment.

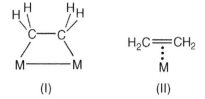

Figure 7.30 The classification of adsorbed molecular ethene species according to Sheppard and co-workers [39]

This apparent anomaly also occurs in the spectra, both IR and EELS of π-bonded ethene and may be explained by considering the bonding of the molecule–ligand to the surface–metal atom as via donation from the π-electron cloud, which periodically distorts at the frequency of the C=C vibration. Thus as the C=C bond vibrates, a similar periodic movement of electrons occurs in and out of the surface. This oscillation is of course entirely equivalent to a dipole oriented perpendicular to the surface plane, and thus is observed as either an absorption band (IR) or an energy loss feature (EELS).

7.3.3 THE EELS (HREELS) EXPERIMENT

If we consider the experimental requirements for EELS or HREELS, low energy electrons, single crystal substrates (usually), then it is clear that this is a high vacuum technique. In practice the working pressure limit for most EELS instruments is 10^{-6} mbar due to a gas-phase scattering processes. Figure 7.31 is a schematic of the essential features of an EEL spectrometer:

Usually at least one unit, the monochromator or analyzer, has some ability to rotate about an axis perpendicular to the plane of the experiment while the sample may also rotate. In this way it is possible to vary the angle of incidence over a wide angular range and still maintain the ability to detect at angles away from the specular position. The monochromators and analysers are electrostatic selection units based upon either 127-degree cylindrical capacitors or hemispherical capacitors. The potentials applied to these capacitors are such that only electrons having specific pass energies may reach first the sample surface and second the detector. In practice, the 127-degree systems produce the best experimental resolution although the hemispherical systems produce a higher beam current. To avoid the influence of 'stray' magnetic fields, the unit is normally encased in magnetic shielding. The working surfaces of the spectrometer are coated with an inert material such as graphite in order to minimize changes in spectrometer work function that may occur upon exposure to various gases.

The design of an EEL spectrometer is such that the energy spread of the incident low-energy electron beam is minimized while maintaining a

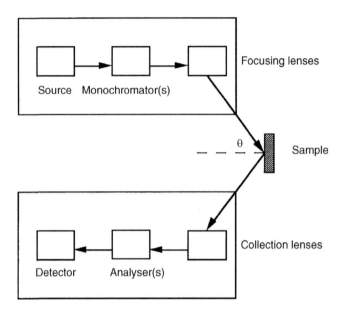

Figure 7.31   A schematic representation of the EELS experiment

reasonable electron flux. This primary consideration means that the area of surface sampled by the beam is of secondary importance. In some instruments, which utilize 127-degree sector capacitors, the resulting beam is ribbon shaped, having dimensions on the order of 5 mm × 0.5 mm. Other instruments, which utilize hemispherical sector capacitors, produce a spot focus beam of diameter of the order of 1 mm. Thus the spatial resolution is poor in comparison with other methods, although the applications of EELS are such that it should not be thought of along with other more 'routine' techniques.

7.4 The Group Theory of Surface Vibrations

7.4.1 GENERAL APPROACH

In earlier sections it was noted that for particular kinds of experiment only certain modes would be expected to be observed. This point was also made during discussion of the surface selection rule. In this section we will look at the way in which vibrations are classified using molecular symmetry in an attempt to be able to predict the form of various surface spectra.

It is well known that the link between spectroscopy and molecular structure is symmetry. Molecules may be classified into symmetry types or 'point groups' by invoking rules based upon the relationship between various types of symmetry operations. The mathematical process of allowing each symmetry element to operate on a system in such a way as to produce a related system is known as Group Theory, a detailed discussion of which is beyond the scope of this text. The interested reader is referred to the text by Bishop [42]. However, even the non-mathematician may easily see that it should be possible to write down the process of symmetry operation in the form of a series of equations. Accepting this, it is then possible to note that for each point group there will be a series of such equations that will apply. Sets of equations created in this manner may be expressed in matrix notation such that the process of symmetry operation may be expressed in terms of the product of two matrices. Fortunately, the experimentalist need not be too concerned with the detailed mathematics of this process since the coefficients of the matrices for each point group are readily expressed in terms of 'Character Tables' which not only list all of the operations, but also all of the unique combinations of operation which may be created. In the language of group theory these are the possible 'permutations', which may be derived, with each possible set of permutations describing a particular mode of motion. These modes include the so-called normal modes that we are used to studying in conventional vibrational spectroscopy, but also include translational and rotational motion. This factor must be accounted for when approaching the problem of predicting the form of a surface vibrational spectrum.

All of the above discussion is somewhat academic without illustration by suitable examples and so at this point we will turn to a particular example to provide some clarity.

7.4.2 GROUP THEORY ANALYSIS OF ETHYNE ADSORBED AT A FLAT, FEATURELESS SURFACE

For the purpose of this exercise let us view the surface in question as a flat, featureless plane. We realize at the outset that this is a gross assumption but it serves as a useful starting point for this analysis. We will also assume that the ethyne molecule adsorbs in a manner in which the molecular plane is parallel to the surface plane. The first step in the analysis is to determine the point group of the 'system', molecule plus surface. At its simplest, this may be achieved by simply listing the symmetry elements that the system possesses and then comparing the list with those tabulated for each point group. When the list matches that for a particular point group, this is the point group of the system. For gas-phase ethyne, there is an infinite axis of rotation, which corresponds to the molecular axis. This is labeled C_∞. The molecule also possesses mirror planes of reflection in both the horizontal and vertical directions. These are labeled σ_h and σ_v, respectively, while the presence of orthogonal planes of this nature tell us that the point group is likely to be of symmetry higher than C, in this case a D group. Overall, the symmetry of the ethyne molecule corresponds to the point group $D_{\infty h}$.

Figure 7.32   Ethyne adsorbed at a flat, featureless surface

However, this is not the point group of the adsorbed ethyne molecule. We may note here a generalization, which applies to all adsorbed molecules, namely, that 'the symmetry of the adsorbed species is either equal to or lower than that of the free molecule'. The inclusion of the surface can never increase the symmetry of the system. Let us now return to the adsorbed molecule (Figure 7.32).

If we list the symmetry elements of this system we may immediately note that inclusion of the surface plane removes the C_∞ element. In fact the highest axis of rotation is now the one which is normal to the surface and which passes thorough the C—C bond, a C_2 axis. The inclusion of the surface also removes the horizontal reflection plane since this would 'flip' the surface above the molecule! Thus the only elements of symmetry are the C_2 axis and reflection planes parallel to this axis and in the plane of the paper. We add to these three elements the so-called identity element, usually given the symbol I or E, which is the allowed operation of 'leaving the system alone'. Thus we have four symmetry elements E, C_2 and 2σ. At this stage we need to be able to distinguish the two mirror planes and also to be able to interpret the meaning of each operation on the motion of the molecule. For this reason we introduce a coordinate system in which the z-direction is always normal to the surface. The molecular plane thus becomes the $x - y$ plane and we may distinguish the two mirror planes as σ_{xz} and σ_{yz}. An examination of the character tables then attributes these four elements to the point group C_{2v}. This character table is shown in Table 7.1.

Looking at this table it is apparent that some further explanation is required. The labels forming the first column represent the four possible permutations of each symmetry element, whereas the 1 or −1 in the rows represent the character of each operation and indicate either that operation of the element produces an indistinguishable system to that with which we started (1) or that the system changes under the operation (−1). With this

Table 7.1   The C_{2v} point group character table

Permutation	E	C_2	σ_{xz}	σ_{yz}
A_1	1	1	1	1
A_2	1	1	−1	−1
B_1	1	−1	1	−1
B_2	1	−1	−1	1

in mind it may be easily seen that the highest symmetry permutation is A_1, since here operation of all elements results in an unchanged system. As we proceed down the first column we move to lower symmetry. Note that because of the special relationship between symmetry elements that must exist, it is not possible to have a permutation in which the character of every element is -1. In other character tables the permutations may be listed in a similar fashion, although the nomenclature is readily extended to represent more subtle variations in character, which may arise.

At this stage it would be possible, for simple molecules, to determine the symmetry of the vibrations which may be identified by inspection but this is not generally the case. For this reason we will proceed to determine the full nature of the motions of this system from first principles. Since each atom in the molecule possesses three degrees of freedom, it is convenient to place each atom at the centre of a coordinate system identical to that used for the whole molecule. In this way we create x, y and z axes centred on each atom. We then allow each element to operate on the molecule, giving a nominal value of 1 for an axis that remains unchanged and -1 if the atom is not moved but the axis in inverted. The easiest example is clearly the E operation, since this leaves all axes unchanged by definition. We then write onto the table, the net character for all atoms. Thus in the column under the E element, for ethyne we would write 12, being 4×3 axes unchanged (Table 7.2).

For the C_2 operation, the situation is also quite clear. The axis of rotation lies through the centre of the C—C bond and thus the rotation moves all axes because all atoms move. Thus in the column under C_2 we place zeros. For the σ_{xz} operation, since the $x - z$ plane is the plane of the paper, operation of the element leaves two axes unchanged for each atom (x and z) while the y-axis is inverted. The total character per atom is therefore $+2 - 1 = 1$. Since there are four atoms, we place the number 4 in the column under the σ_{xz} operation. For the σ_{yz} operation the situation is the same as for the axis of rotation in that all atoms move. Thus we place zeros in the column under σ_{yz}. The total value accumulated across the rows, remembering that each row represents an allowed symmetry permutation, will yield the number of each possible permutation when divided by the character of the table,

Table 7.2 Inclusion of the character arising from the operation of the C_{2v} elements in the character table for adsorbed ethyne as depicted in Figure 7.32

Permutation	E	C_2	σ_{xz}	σ_{yz}
A_1	1(12)	1(0)	1(4)	1(0)
A_2	1(12)	1(0)	-1(4)	-1(0)
B_1	1(12)	-1(0)	1(4)	-1(0)
B_2	1(12)	-1(0)	-1(4)	1(0)

which is simply the number of elements, in this case, 4. Thus for the A_1 permutation, the sum across the row is $12 + 0 + 4 + 0 = 16$. In this case this means that there are 4 possible A_1 permutations (being $16/4$). For the A_2 permutation, summing across the row yields $12 - 4 = 8$, giving in turn $2A_2$ permutations. Similarly for the B_1 and B_2 permutations, we find 4 and 2 permutations, respectively. Thus the total 'representation', sometimes referred to as the reducible representation, is $4A_1 + 2A_2 + 2B_2 = 12$. The total is in accord with that expected for four atoms each with three degrees of freedom. We now return to the point made earlier when we noted that this representation also includes translations and rotations of the system. To obtain just the vibrational part we must subtract those permutations which correspond to translation and rotation about x, y and z. In order to see which permutations to subtract from the reducible representation we return to the character table. In most published tables translations are marked in an additional column on the right-hand side as T_x, T_y or T_z while rotations are labeled R_x, R_y or R_z. A more complete version of the C_{2v} character table would therefore appear as in Table 7.3.

Note that in addition Raman activity is usually expressed in terms of the components of molecular polarizability (see Section 7.5), usually written as ∞_{ij} terms, where i and j represent combinations of the Cartesian coordinates x, y and z. Returning now to our example, we note that for the translational motion we must subtract $A_1 + B_1 + B_2$, while for the rotations we must subtract $A_2 + B_1 + B_2$. Removing these six permutations leaves us with the irreducible representation of the vibrational motion, $3A_1 + A_2 + 2B_1$. At this point we note an interesting observation – namely that, even if the surface is considered flat and featureless, then the number of vibrations expected is not that predicted by considering the system as a linear molecule, i.e. using the $3N - 5$ rule where N is the number of atoms. This arises because the presence of even a model surface reduces the symmetry of the system. Thus we have $3N - 6 = 6$ vibrations to consider.

Now that we have performed this exercise we may increase the complexity of the system by invoking the presence of a real surface, and necessarily introduce the symmetry of the surface.

Table 7.3 A general version of the C_{2v} character table

Permutation	E	C_2	σ_{xz}	σ_{yz}	Translation
A_1	1	1	1	1	T_z
A_2	1	1	−1	−1	R_z
B_1	1	−1	1	−1	T_x, R_y
B_2	1	−1	−1	1	T_y, R_x

7.4.3 GROUP THEORY ANALYSIS OF ETHYNE ADSORBED AT A (100) SURFACE OF AN FCC METAL

In this example we place the ethyne molecule in a two-fold bridge site on an unreconstructed (100) surface of an fcc metal such as Pt or Cu, which before adsorption is said to have 4 mm symmetry. The '4' refers to a four-fold axis of rotation while the mm refers to the presence of two orthogonal mirror planes in the surface (Figure 7.33).

Adsorption not only reduces the symmetry attributed to the ethyne molecule, but also the symmetry of the surface. Placing the ethyne molecule in the two-fold site over two Pt atoms creates a species with overall symmetry C_{2v}, i.e. the same as before. In this case we would obtain more vibrations in our analysis simply because there are more atoms involved. However, the method would be almost identical to that employed in the previous section.

Let us now make the system more complex by choosing to adsorb at the three-fold site involving two atoms from the top layer and one atom in the second layer, such that the plane formed by these three atoms is inclined with respect to the surface plane. Now since the plane of adsorption is inclined with respect to our surface plane, there can be no C_2 axis of rotation. There will still be one mirror plane in the system, but the other mirror plane found for C_{2v} disappears along with the axis of rotation. From character tables we would attribute the symmetry of this species as C_S, again noting the reduction in overall symmetry. The table for this point group is simpler than that above simply because of the reduced number of elements (see Table 7.4).

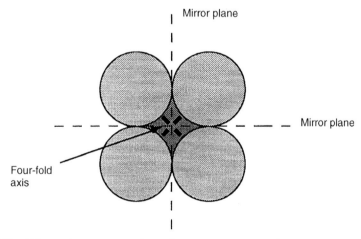

Figure 7.33 Schematic representation of an unreconstructed (100) surface for an fcc single crystal metal

Table 7.4 The C_s point group character table

Permutation	E	σ
A′	1	1
A″	1	−1

In our analysis we must now also include the three surface atoms, since these contribute to the symmetry of the species. Note that all of the permutations have the 'A' classification. This is to be expected since reducing the symmetry increases the number of types of motion, which effectively preserve this reduced symmetry. It then follows that a system with the lowest possible symmetry classification also has the highest number of A-type modes.

7.4.4 THE EXPECTED FORM OF THE RAIRS AND DIPOLAR EELS (HREELS) SPECTRA

Here we are invoking the surface selection rule described earlier, which states that a vibration must produce a component of its dynamic dipole perpendicular to the surface plane in order to be detectable. In our analysis so far we have not indicated any means by which we may select vibrations from our derived irreducible representation. To do this we return to the character table and study the form of our vibrations. Using the ethyne adsorbed at a flat featureless surface system as an example, we may examine the form of the vibrations and determine whether they satisfy the surface selection rule. Intuitively we may guess the form of the vibrations up to a point. There will of course be symmetric and asymmetric C—H stretching modes, both in the plane of the molecule. The symmetric stretch will not alter the C_{2v} symmetry and thus has a character of 1 under each operation. It is thus assigned to an A_1 mode. The asymmetric stretch must obviously have a character of 1 under E, but will have a character of −1 under C_2 and also −1 under σ_{yz}, because of the distortion to the overall symmetry of the system. It also follows that for the σ_{xz} the character remains 1. Thus from the table we assign this to a B_1 mode. The C—C stretching mode maintains the overall C_{2v} symmetry for all operations and hence must be A_1. The presence of the surface cannot be ignored and hence the vibration of the whole molecule against the surface must be considered. This is clearly an A_1 mode since the C_{2v} symmetry does not change simply if the distance between the molecule and the surface changes. Thus far we have found our three A_1 modes plus one B_1 mode. There are thus one B_1 and one A_2 modes to be found. Looking first at the B_1 mode we note that it must effectively remove the C_2 axis and the corresponding mirror plane. This is therefore an out-of-phase, out-of-plane wagging of the H atoms. The A_2 mode preserves the C_2 axis but removes both mirror planes. This is an in-plane twist of the molecule, which may be thought of in part as an out-of-phase in-plane wag of the H atoms. These then are the six modes that we wished to identify. In terms of our surface selection rule we must now examine which of these modes is likely to produce a changing dipole in the z-direction. Clearly the A_1 mode in which the whole molecule moves against the surface, meets this criterion. The C—C stretch, which is also A_1, does not at first sight satisfy this rule but we have implicitly placed a chemical bond between the molecule and the surface (implicitly because we have assigned this bond a vibration). As the C—C bond expands and contracts so the bond between the molecule and the surface must fluctuate in 'strength', at the same frequency. Thus this A_1 mode would also appear in our spectrum. No other modes have components of their dynamic dipoles perpendicular to the surface and thus of the six possible vibrations, only two appear in the RAIRS or dipolar EELS spectrum.

An interesting general conclusion emerges from this analysis:

Only modes with 'A' character (of the highest symmetry) may yield features in the RAIRS or dipolar EELS spectrum.

This combined with our prediction as to the influence of symmetry upon the number of A-type modes indicates that the RAIRS and dipolar EELS spectra associated with molecules adsorbed in such a manner as to produce low-symmetry systems will contain the largest number of features. Conversely, high-symmetry systems will yield the simpler spectra containing fewer features. Application of this type of analysis is clearly crucial in determining the nature of the adsorbate–substrate system.

7.5 Laser Raman Spectroscopy from Surfaces

Raman spectroscopy is often said to be complementary to infrared spectroscopy in that it is sensitive to those vibrational modes which are either not observed via IR or give rise to only weak IR absorption bands. Predicted by Smekal in 1921 [43] and first observed by Sir C.V. Raman in 1928 [44], the Raman effect relies upon the polarization of the electron cloud describing a chemical bond by the electric field of incident electromagnetic radiation, which induces a dipole moment, which in turn is time dependent due to the vibration of the atoms forming the bond. Thus it is the *polarizability* rather than the dipole moment, which is the important molecular parameter in determining Raman intensities. The Raman effect is a scattering phenomenon, and thus may be thought of as directly analogous to EELS in that one analyses the energy lost by, in this case, photons rather than electrons, in order to detect molecular vibrations. It is a weak effect, with perhaps only 1 in 10^{11} photons being inelastically scattered in a typical process

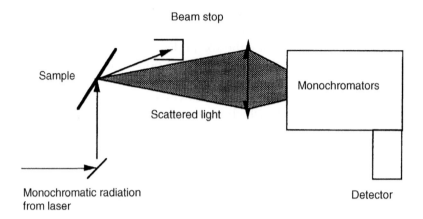

Figure 7.34   Schematic representation of the surface Raman experiment

and for this reason intense light sources are normally required. Early work utilized mercury-arc lamps but modern instruments employ lasers of some description. A schematic diagram of a typical Raman experiment is given in Figure 7.34.

From this figure it is clear that the Raman technique may be applied to the study of a wide range of materials, including most solid surfaces. The use of a laser as the excitation source also means that this technique is applicable to small areas of a sample surface. This ability to map a surface or to focus onto a small area of a surface is most effectively exploited in the Raman microprobe instrument. Modern instruments may employ holographic notch filters in preference to complex expensive monochromator systems in order to achieve the required elastically scattered light rejection capability.

7.5.1 THEORY OF RAMAN SCATTERING

The quantum mechanical description of the scattering process known as the Raman effect invokes a 'pseudo absorption' process in which the incident radiation is absorbed into a virtual electronic state of the molecule, followed by emission back to the first excited vibrational state – Case 1 in Figure 7.35.

The energy difference between the incident and emitted radiation is thus equal to one quantum of vibrational energy and the emitted photons are termed Stokes photons. An alternative situation is described in Case 2, where the molecular vibration is already described by $v = 1$ and, upon re-emission, photons of higher energy than the exciting energy result. These are known as the anti-Stokes photons. The virtual state is not generally a true electronic state of the molecule but a composite function involving all possible states, rotational, vibrational and electronic. Where the energy of the incident radiation does coincide with a true molecular state the

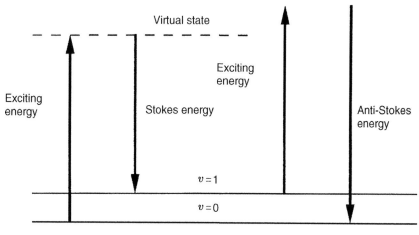

Figure 7.35 A description of the vibrational Raman effect based upon an energy level approach

cross-section for Raman scattering increases enormously. This situation is referred to as the resonance Raman effect, for obvious reasons, and may result in an increase in scattering levels of a factor of 10^6 as compared to conventional Raman scattering.

Classically the effect is described in terms of the induction of a dipole moment into the molecule of magnitude proportional to the polarizability, which is itself a time-dependent function, the time dependence arising from the distortion to the molecule that accompanies a vibration. The combination of the frequency of the incident radiation v_i and the superimposed vibrational frequency v_{vib} gives rise to inelastic scattering at frequencies $v_i + v_{vib}$ and $v_i - v_{vib}$, which are known as the anti-Stokes and Stokes components respectively. This classical description is conceptually easier to understand than the quantum mechanical description but cannot explain the relative intensities of the Stokes and anti-Stokes components, which are of course determined by Boltzmann (population) factors.

7.5.2 THE STUDY OF COLLECTIVE SURFACE VIBRATIONS (PHONONS) USING RAMAN SPECTROSCOPY

As with IR spectroscopy, there are a wide variety of applications of laser Raman spectroscopy to the study of surface processes. An application that illustrates some of the advantages of the technique is depicted in Figure 7.36.

Figure 7.36 illustrates immediately that via the use of multiple monochromators/notch filters and sophisticated detection systems, it is possible to

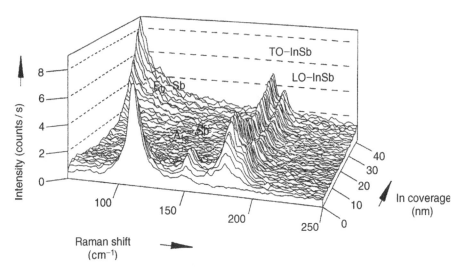

Figure 7.36 Raman spectra recorded from an Sb (111) surface as a function of varying in coverage (after Zahn [45])

detect extremely low frequency vibrations, i.e. at absolute frequencies close to that of the exciting laser frequency, which could only be accessed in the IR by using far-IR optics and sources. The particular vibrations in question are as a result of the collective oscillations of atoms in the surface layers. Such vibrations are known as *phonons*. The frequency of a surface phonon is a marked function of the structure of the surface layer and the extent of any surface symmetry. Thus it is noted that as the coverage of In increases, the phonon bands of InSb appear, indicating the reactive nature of the In overlayer with the Sb substrate. The width of the phonon bands is a measure of the degree of surface order.

The high level of sensitivity displayed in this example serves to illustrate that fact that Raman spectroscopy has many direct surface applications in the electronics industry where oriented wafers are used as substrates for the construction of devices based upon layer structures. However, the sensitivity obtained from these systems is somewhat atypical due to the careful choice of excitation wavelength. The materials under examination are direct bandgap semiconductors, capable of absorbing or emitting photons with reasonably high efficiency. The excitation wavelength is chosen so as to coincide with a particular electronic transition within either substrate or the deposited film. Under these conditions, the virtual levels displayed in Figure 7.35 become real levels and the process of Raman scattering becomes resonant. Resonance Raman systems similar to this are well known in other areas of chemistry and physics where the enhancement of the Raman scattering efficiency that occurs when the resonance condition is satisfied can reach up to 10^6 times that observed off-resonance.

7.5.3 RAMAN SPECTROSCOPY FROM METAL SURFACES

The most widely known application of Raman spectroscopy to metal surface chemistry involves a phenomenon known as surface enhanced Raman spectroscopy or SERS. This was first demonstrated by Fleischmann and co-workers [46, 47] in 1974 who were attempting to record vibrational spectra from silver electrode surfaces immersed in an aqueous solution containing KCl as a supporting electrolyte and pyridine as the specific adsorbate. The pyridine–silver system was chosen because pyridine has a relatively high Raman scattering cross-section, particularly for the ring 'breathing' modes and was known from other measurements to adsorb readily at silver electrode surfaces. In order to increase the sensitivity of the technique, the silver electrode was repeatedly oxidized and reduced *in situ*, resulting in a surface with a grey spongy appearance with a surface area about ten times that of a corresponding flat, polished surface. The resulting Raman spectra were intense, exhibiting signal to noise ratios in excess of 100:1 and were highly sensitive to changes in electrode potential (Figure 7.37).

Since only the first one or two molecular layers effectively feel the electrode potential, this was a remarkable display of sensitivity. It was soon realized that the spectra were too intense to be accounted for in the normal way and that the Raman scattering cross-section for the adsorbed pyridine species was between 10^4 and 10^6 times greater than that for liquid pyridine! [48, 49]. Much subsequent study of this and related systems revealed that the surface enhancement could only be observed from silver, copper and gold surfaces although the effect was not restricted to electrochemical systems and similar observations were made from metal surfaces under high vacuum and even from colloidal metal particles in solution. A large number of mechanisms were postulated in order to explain the enhancement process including resonance phenomena similar to the resonance Raman process discussed in the previous section, but involving novel surface complexes and surface plasmons activated by the electrochemical oxidation–reduction cycle. The 'conventional' resonance Raman process was discounted since molecules that exhibited resonance Raman scattering (i.e. those capable of absorbing the incident laser energy) gave rise to a further enhancement when adsorbed at a silver electrode. It is now felt that it is unlikely that any one mechanism will be sufficient to explain the phenomena observed from a large number of different surface chemical systems and that the limitation of the technique in terms of the 'active' metals precludes its use as a general method of surface analysis. However, for certain analytical applications of Raman spectroscopy, sensors made of silver, copper and gold have been employed in order to obtain increased sensitivity [50–52].

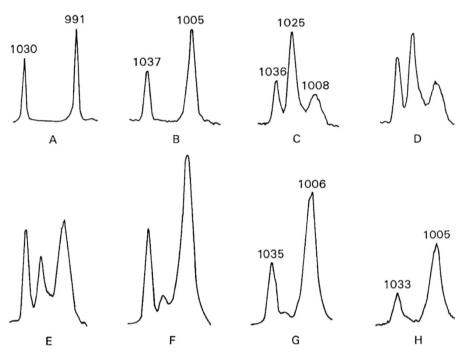

Figure 7.37   Raman spectra recorded from the surface of a polycrystalline Ag electrode immersed in an aqueous KCl–pyridine solution as a function of applied electrochemical potential, measured with respect to the potential of a saturated calomel reference electrode: (A) liquid pyridine; (B) 0.05 M aqueous pyridine: (C) silver electrode, 0 V (SCE); (D) −0.2 V (E); −0.4 V; (F) −0.6 V (G); −0.8 V; (H) −1.0 V (after Fleischmann *et al.* [46, 47])

7.5.4   SPATIAL RESOLUTION IN SURFACE RAMAN SPECTROSCOPY

As a photon-based technique, the resolution is typically limited to the best focus possible for a given exciting laser wavelength. This is not limited by the focusing optics but by self-diffraction of the light, which results in a beam 'waist' of fixed dimensions for a given wavelength that cannot be reduced. Using short wavelengths such as the 488 nm blue line available from a continuous-wave argon ion laser, it is possible to achieve a spot size of around 20 microns.

7.5.5   FOURIER TRANSFORM SURFACE RAMAN TECHNIQUES

The use of a visible laser source to stimulate Raman scattering from a sample, while convenient, is not without difficulties. Until recently, by far and above the major difficulty in this area was the stimulation of fluorescence from impurity species. Early studies of species adsorbed at supported metal catalyst surfaces were constrained by the appearance of intense fluorescence emission resulting from electronic excitation of impurity species such as vacuum greases. This fluorescence manifested itself as a broad spectral envelope on the Stokes side of the exciting laser line and was often too intense to enable the weak Raman bands to be observed. To overcome this, lasers were chosen such that the excitation energy was far below any possible electronic excitations, i.e. in the IR region of the spectrum. The difficulty associated with this was that spectral discrimination in the IR region was necessary and thus all of the advantages of working with visible radiation were lost. The advent of Fourier transform IR techniques overcame this difficulty. With some modification, interferometers were developed that could perform both conventional IR absorption measurements and IR-laser Raman experiments. This combination of techniques is proving particularly powerful for surface analysis.

7.6 Inelastic Neutron Scattering (INS)

7.6.1 INTRODUCTION TO INS

INS is a particularly powerful form of vibrational spectroscopy since both IR and Raman active modes, as well as those vibrations which are neither IR nor Raman active, may appear in an inelastic neutron scattering experiment. Like some of the other methods described in this chapter, INS is applicable to a wide range of samples, including surfaces and adsorbed species. It is not the intention to provide a comprehensive overview of this method here, but rather to concentrate on a particular surface chemical study and direct the interested reader to the review by Parker [53] where further references may be found.

The experiment is clearly not trivial since a supply of neutrons is required! Two methods are commonly employed to create a flux of neutrons – spallation and fission. The first of these, spallation, involves the shattering of nuclei using very high-energy protons. Generating high-energy protons is also non-trivial. Typically in the UK this is achieved using a proton accelerator or proton synchrotron such as the one based at the Rutherford Appleton Laboratories. The second process of nuclear fission requires a nuclear reactor to operate. Thus it may be seen that the generation of a neutron flux takes some considerable effort and expense. The expense is justified by the large number of experiments, which may then be performed, of which INS is only one type. The directional nature of the neutron flux coupled with the property of being able to penetrate most matter easily is used to great effect in INS, where the time-of-flight of the neutrons around the system is used to determine the energy transfer to the sample directly. Thus in terms of the other methods described in this section, the energy transfer is perhaps similar to that in Raman spectroscopy, while the mode of sampling is perhaps most similar to that of transmission IR methods.

7.6.2 THE INS SPECTRUM

There are no selection rules in INS. Thus, as has been stated earlier, modes that are IR-active, Raman-active or neither IR- nor Raman-active appear in the INS spectrum. However, an INS spectrum does not usually appear as a combination of IR and Raman spectra together with new bands, because of the nature of the neutron scattering cross-section – the fundamental property that determines band 'intensity'. This does not depend upon such factors as dynamic dipole or polarizability but rather on a more basic 'billiard ball' approach to momentum transfer. This approach predicts that momentum transfer occurs most efficiently between two particles of comparable mass. Thus the process is most efficient for modes involving hydrogen. In practice it is found that the inelastic cross-section for hydrogen is at least an order of magnitude larger than that of other elements, including deuterium, which makes possible the use of isotopic substitution for the immediate identification of hydrogenic modes. Thus a typical INS spectrum is dominated by bands arising from the motion of hydrogen atoms. While there are obviously many applications for a technique of this nature, it may also be exploited by the surface chemist. An example of which is presented in the next section.

7.6.3 INS SPECTRA OF HYDRODESESULFURIZATION CATALYSTS

Since crude oil contains a number of contaminants including sulfur, nitrogen and trace metals, part of the refining process involves hydrogenation to remove these contaminants as volatile hydrides. This process also serves to reduce the unsaturated fractions of the 'crude'. Of the contaminants, perhaps sulfur is the one where environmental factors place the most stringent limits on the levels of emission of substances such as SO_2, which may be formed if a sulfur-containing oil is combusted.

The present generation of hydrodesulfurization catalysts, as they are known, are unlikely to be able to reduce levels of sulfur in 'crude' sufficiently, given the particular economic factors that are involved and thus potential new catalysts are currently being sought. One possible new catalyst is the metal sulfide, Ru_2S. This material is capable of adsorbing hydrogen, which is then chemically activated towards the hydrodesulfurization process. The activity of the material has been directly correlated with the concentration of RuH species although until the application of INS, such metal–hydride species had not been formally identified by vibrational spectroscopy. Jobic *et al.* [54] have conclusively identified Ru–H species on partially desulfurized Ru_2S exposed to hydrogen (Figure 7.38).

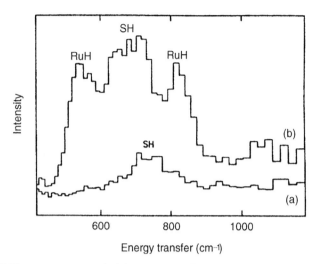

Figure 7.38 INS spectrum recorded from a partially desulfurised Ru_2S catalyst exposed to hydrogen, showing the features assigned to the formation of Ru—H species at energy transfer values of 540 and 821 cm^{-1} (after Jobic *et al.* [54])

These spectra also show bands assigned to overtones and combination modes – a feature which is typical of an INS spectrum. The cells used in INS are simply stainless steel or aluminium 'tubes', these metals being basically transparent to the neutron flux, and thus large amounts of material may be employed, e.g. up to 100 g. Under these circumstances it is fairly certain that the spectrum measured is indeed a true representation of the catalyst material. By comparison with other forms of surface vibrational spectroscopy described here, the following features are noteworthy:

(i) INS is extremely sensitive to hydrogenic modes, but as atomic number increases so its sensitivity decreases dramatically.

(ii) Neutron sources are comparatively weak, such that best results are obtained where there is a relatively large amount of sample, usually in the form of fine particulates.

(iii) Reliable neutron sources are not always readily available!

7.7 Sum-frequency Generation Methods

The reader will appreciate that many of the methods described here cannot be applied to surfaces in contact with liquids or significant overpressures of gases. For others, such as IR methods, gas-phase interferences require the use of complex normalization/background subtraction routines. In the mid-1980s researchers began to study the non-linear optical properties of surfaces, and one resulting technique, sum frequency generation (SFG), has been developed which in principle may be used to study certain substrate–adsorbate systems, irrespective of the medium over the surface. The technique is based upon the interaction of two photons at a surface, induced via wave mixing, such that, upon reflection from the surface, a single photon emerges. This process conserves energy (and also frequency), which is where the name sum-frequency generation originates. The full theory of non-linear optics required for a thorough understanding of this method is beyond the scope of this text. The reader is referred to the book by Shen for an in-depth review of the topic [55].

Three-wave mixing processes of this type rely upon the inherent dipole resulting either from the interface between two media, e.g. solid–liquid, solid–gas, liquid–liquid (for two immiscible liquids) or dipoles present in the bulk, found in ordered materials under circumstances where the unit cell lacks inversion symmetry. Thus if we consider a substrate such as a single crystal surface of metals like Ag, Cu, Ni, Pt etc., the non-linear optical process will not arise from photons that penetrate into the bulk since the bulk face-centred cubic structure possessed by these metals has inversion symmetry across the unit cell. However, the interface between the crystal surface and the medium in contact will be 'active' towards sum-frequency generation.

The simplest process to envisage is where both photons possess the same energy $E(\omega)$ (frequency ω), such that when the three-wave (two in, one out) process occurs, the outgoing photon has energy $2E(\omega)$ and frequency 2ω. This process is known as second harmonic generation.

The efficiency (intensity) of this process depends upon a variety of factors but where either $E(\omega)$ or $2E(\omega)$ correspond to allowed transitions of some description, e.g. electronic or vibrational transitions, then the process is said to become resonant, with the result that the efficiency of the process, and hence the intensity at 2ω is increased. Sum-frequency generation thus exploits this possibility by ensuring that one incident photon is generated from a source capable of producing radiation that may be tuned in energy across a suitable transition.

Thus as a form of surface vibrational spectroscopy, SFG operates as follows: one incident photon is usually produced by a tunable laser operating in the IR spectral region, producing radiation that may be absorbed directly by the species adsorbed at the surface. Another pump photon, which may be visible, or IR, is used to provide the intensity necessary to observe weak second-order effects such as this. If the pump photon has frequency ω_p and the tunable photon has frequency ω_t, then when ω_t corresponds to the frequency required to excite the surface vibration, there results a change in the intensity of the sum photon, frequency $\omega_p + \omega_t$. Due to the difficulty in providing laser systems with tunable output in the useful part of the IR spectral region this method is currently of limited application. However, as advances in laser technology are made it is likely to become more widely used.

An example of the application of SFG to a surface chemical problem is provided here. Bain and co-workers have studied the adsorption of surfactant molecules at the liquid–air interface [56, 57] (Figure 7.39).

Figure 7.39 shows a typical SFG spectrum in which the intensity of the sum-frequency component varies dramatically as the tunable laser sweeps over the frequency corresponding to the vibrations in the surfactant species. In this example, the observed resonances were interpreted in terms of a model in which, in the absence of the deuterated alcohol, the surfactant species were oriented such that the methylene groups were observed. While in the presence of the alcohol, the structure of the overlayer changed dramatically revealing sharp methyl resonances. The principal advantage of sum-frequency generation arises from its selectivity. For substrates in which the bulk unit cell possesses inversion symmetry bulk SFG, is disallowed. Furthermore, a gas or a liquid represents an isotropic medium to the incident laser beams. Isotropic media also behave as if

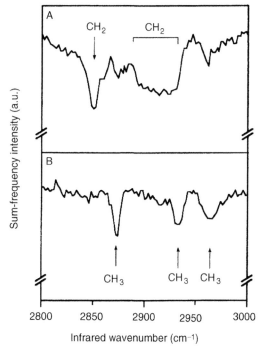

Figure 7.39 Sum-frequency generation spectrum recorded from a hydrophobic surface in contact with the anionic surfactant SDS (sodium dodecyl sulfate) and the analogous spectrum recorded in the presence of deuterated dodecanol (after Bain [57])

they possess inversion symmetry. Thus there can be no contribution from either a gas or a liquid in contact with the substrate. Therefore, the only SFG component that may be observed from such a system arises from the interface itself. In this respect given the correct choice of substrate, SFG is the ultimate pressure independent form of surface vibrational spectroscopy. The principle disadvantage of the method is that it is based upon an inherently weak phenomenon such that intense, usually pulsed laser sources are required in order to provide sufficient photons in a given time interval (typically 10 ns or less), in order to observe SFG. There is thus the possibility that surface damage may occur due to interaction of the intense laser pulses with the surface. A recent advance in SFG has come about due to further developments in laser technology and the application of basic physics: according to the Heisenberg uncertainty principal, $\Delta E \times \Delta t = h/4\pi$; thus as the duration of the pulse gets shorter it also becomes broader. At first sight this may seem to be a disadvantage but for very short pulses each pulse can be considered as a broad band of radiation. Broadband SFG is achieved by spatially overlapping a <100 fs IR beam with a longer pulse 3 ps visible beam. The sum frequency output is then dispersed over a CCD array. The wavenumber range of the experiment is determined by the centre frequency IR pulse and the linewidth, which in the case of 100 fs pulse is 330 cm^{-1} [58]. Thus this spectral range, which is sufficient to cover the whole C—H stretch or carbonyl region, can be collected in a single shot experiment, i.e. in 3 ps, although in reality several thousand shots may have to be co-added to obtain a reasonable signal-to-noise ratio [59].

📖 References

1. C.N. Banwell and E.M. McCash, *Fundamentals of Molecular Spectroscopy*, 4th Edition, McGraw-Hill, London, pp. 93–96, (1994).

2. J.R. Ferraro and L.J. Basilo, *Fourier Transform Infrared Spectroscopy: Application to Chemical Systems*, Volumes 1 and 2, Academic Press, New York, NY (1978 and 1979).

3. P.R. Griffiths and J.A. de Hareth, *Fourier Transform Infrared Spectroscopy*, John Wiley & Sons, Ltd, Chichester, UK (1986).

4. R.P. Eischens, W.A. Pliskin and S.A. Francis, *J. Chem. Phys.*, **22**, 1786 (1954).

5. R.P. Eischens, S.A. Francis and W.A. Pliskin, *J. Phys, Chem.*, **60**, 194 (1956).

6. C. Delacruz and N. Sheppard, *J. Chem. Soc., Chem Commun.*, **24**, 1854 (1987).

7. M.A. Chesters, C. Delacruz, P. Gardner, E.M. McCash, P. Pudney, G. Shahid and N. Sheppard, *J. Chem. Soc., Faraday Trans.*, **86**, 2757 (1990).

8. M.-T. Ho, Y. Wang, R.T. Brewer, L.S. Wielunski and Y.J. Chabal, *Appl. Phys. Lett.*, **87**, 133103 (2005).

9. J. Paivasaari, J. Niinisto, P. Myllymaki, C. Dezelah, C.H. Winter, M. Putkonen, M. Nieminen and L. Niinisto, *Rare Earth Oxide Thin Films: Growth, Characterization and Applications, Topics in Applied Physics*, Volume 106, pp. 15–32 (2007).

10. B.S. Lim, A. Rahtu and R.G. Gordon, *Nature Materials*, **2**, 749 (2003).

11. G.E. Moore, *Electronics*, **38**, 114–117 (1965).

12. T.J. Harvey, A. Henderson, E. Gazi, N.W. Clarke, M. Brown, E Correia Faria, R.D. Snook and P. Gardner, *Analyst*, **132**, 292 (2007).

13. K.Y Gao, F. Speck, K. Emtsev, Th. Seyller, L. Ley, M. Oswald and W. Hansch, Interface of atomic layer deposited Al_2O_3 on H-terminated silicon, *Phys. Stat. Solidi*, **203**, 2194–2199 (2006).

14. P.D. Holmes, G.S. McDougall, I.C. Wilcock and K.C. Waugh, *Catalysis Today*, **9**, 15 (1991).

15. (a) R.G. Greenler, *J. Chem. Phys.*, **44**, 310 (1966). (b) R.G. Greenler, *J. Vac. Sci. Technol.*, **12**, 1410 (1975).

16. M.A. Chesters, *J. Electr. Spectrosc. Related Phenom.*, **38**, 123 (1986).

17. P. Hollins, *Surf. Sci.*, **107**, 75 (1981).

18. P. Hollins and J. Pritchard, *J. Chem. Soc., Chem. Commun.*, **21**, 1225 (1982).

19. M.A. Chesters, S.F. Parker and R. Raval, *J. Electr. Spectrosc. Related Phenom.*, **39**, 155 (1986).

20. G.P. Williams, *Surf. Sci.*, **368**, 1 (1996).

21. M.J. Pilling, M.J. Cousins, A. Amiero Fonseca, K.C. Waugh, M. Surman and P. Gardner, *Surf. Sci.*, **587**, 78 (2005).

22. C.J. Hirschmugl, G.P. Williams, F.M. Hoffmann and Y.J. Chabal, *Phys. Rev. Lett.*, **65**, 480 (1990).

23. M.J. Pilling, S. Le Vent, P. Gardner, A. Awaluddin, P.L. Wincott, M.E. Pemble and M. Surman, *J. Chem. Phys.*, **117**, 6780 (2002).

24. P. Gardner, S. LeVent and M.J. Pilling, *Surf. Sci.*, **559**, 186 (2004).

25. M.J. Green, B.J. Barner and R.M. Corn, *Rev. Sci. Instrum.*, **62**, 1426 (1991).

26. J. Hilario, D. Drapcho, R. Curbelo and T.A. Keiderling, *Appl. Spectrosc.*, **55**, 1435 (2001).

27.].E. Ozensoy, D.C. Meier and D.W. Goodman, *J. Phys. Chem.*, **106**, 9367 (2002).

28. E. Briand, M. Salmain, C. Compère and C.-M. Pradier, Anti-rabbit immunoglobulin G detection in complex medium by PM-RAIRS and QCM: Influence of the antibody immobilisation method, *Biosens. Bioelectr.,,* **22**, 2884–2890 (2007).

29. C. Méthivier, B. Beccard and C.-M. Pradier, *Langmuir*, **19**, 8807 (2003).

30. A. Bewick and B.S. Pons, in *Advances in Infrared and Raman Spectroscopy*, Volume 12, R.J.H. Clark and R.E. Hester (Eds), Wiley, London, pp. 1–63 (1985).

31. B. Beden and A. Bewick, *Electrochim. Acta*, **33**, 1695 (1988).

32. N. Jamin, P. Dumas, J. Moncuit, W.-H. Fridman, J.-L. Teillaud, G.L. Carr and G.P. Williams, *Proc. Natl. Acad. Sci. USA*, **95**, 4837 (1998).

33. E. Gazi, J. Dwyer, N.P. Lockyer, J. Miyan, P. Gardner, C.A. Hart, M.D. Brown and N.W. Clarke, *Vibr. Spectrosc.*, **38**, 193 (2005).

34. H. Froitzheim, H. Ibach and S. Lehwald, *Rev. Sci. Instr.*, **46**, 1325 (1975).

35. H. Ibach, H. Hopster and B. Sexton, *Appl. Surf. Sci.*, **1**, 1 (1977).

36. P.A. Thiry, *J. Electr. Spectrosc. Related Phenom.*, **39**, 273 (1986).

37. M.A. Chesters, G.S. McDougall, M.E. Pemble and N. Sheppard, *Appl. Surf. Sci.*, **22/23**, 369 (1985).

38. H. Ibach and D.L. Mills, *Electron Energy Loss Spectroscopy and Surface Vibrations*, Academic Press, New York, NY (1982).

39. N. Sheppard, *J. Electr. Spectrosc. Related Phenom.*, **38**, 175 (1986).

40. B.J. Bandy, M.A. Chesters, D.I. James, G.S. McDougall, M.E. Pemble and N. Sheppard, *Phil. Trans. R. Soc. London, Ser. A*, **318**, 141 (1986).

41. M.J. Grogan and K. Nakamoto, *J. Am. Chem. Soc.*, **88**, 5454 (1966).

42. See, for example D.M. Bishop, *Group Theory and Chemistry*, Clarendon Press, Oxford (1973).

43. A. Smekal, *Naturwissenschaften*, **11**, 873 (1923).

44. C.V. Raman and K.S. Krishnan, *Nature*, **121**, 501 (1928).

45. D.R.T. Zahn, *Phys. Stat. Solidi, a*, **152**, 179 (1995).

46. M. Fleischmann, P.J. Hendra and A.J. McQuillan, *Chem. Phys. Lett.*, **26**, 163 (1974).

47. M. Fleischmann, P.J. Hendra, A.J. McQuillan, R.L. Paul and E.S. Reid, *J. Raman Spectrosc.*, **4**, 269 (1976).

48. R.P. Van Duyne and D.L. Jeanmaire, *J. Electroanal. Chem.*, **84**, 1 (1977).

49. M.G. Albrecht and J.A. Creighton, *J. Am. Chem. Soc.*, **99**, 5215 (1977).

50. W. Hill, B. Wehling, V. Fallourd and D. Klockow, *Spectrosc. Eur.*, **7**, 20 (1995).

51. K. Mullen and K. Carron, *Anal. Chem.*, **66**, 478 (1994).

52. K. Carron, L. Pietersen and M. Lewis, *Environ. Sci. Technol.*, **26**, 1950.

53. S.F. Parker, *Spectrosc. Eur.*, **6**, 14 (1994).

54. H. Jobic, G. Clugnet, M. Lacroix, S.B. Yuan, C. Mirodatos and M. Breysse, *J. Am. Chem. Soc.*, **115**, 3654 (1993).

55. Y.R. Shen, *The Principles of Non-Linear Optics*, John Wiley & Sons, Inc., New York, NY (1984).

56. D.C. Duffy, P.B. Davies and C.D. Bain, *J. Phys. Chem.*, **99**, 15241 (1995).

57. C.D. Bain, *Biosens. Bioelectr.*, **10**, 917 (1995).

58. E.L. Hommel, G. Ma and H.C. Allen, *Anal. Sci.*, **17**, 1 (2001).

59. V.L. Zhang, H. Arnolds and D.A. King, *Surf. Sci.*, **587**, 102 (2005).

Problems

1. Using group theory, show that for ethene adsorbed at a flat, featureless surface with the plane of the ethene molecule parallel to the surface plane, we predict a maximum of four possible bands in the RAIRS or dipolar EELS spectrum.

2. For the example of the formation of ethylidyne on Pt (111), predict the number of possible bands in the RAIRS or dipolar EELS spectrum.

3. Given that a surface vibration is observed at $3000\ cm^{-1}$ in a RAIRS experiment, calculate the energy loss in meV and the absolute energy in eV where it may appear in an EELS experiment involving the use of an electron beam of energy 5.0 eV.

4. The same band as in question 3 in found to have some Raman activity. Calculate the wavelength and absolute energy that this band would appear at in a Raman experiment using as excitation source an argon ion laser operating at 488.0 nm.

5. The molecule HCl is found to adsorb on a Pt (110) surface with the H—Cl bond parallel to the surface plane and aligned along the rows of atoms. Using the selection rules for impact scattering deduce the required orientation of the EELS experiment such that (a) H—Cl vibration is observed and (b) the H—Cl vibration is not observed.

6. In a RAIRS experiment, an infrared beam of diameter 10 mm strikes a pt (111) single crystal surface at an angle of incidence of 88°. Determine the area of the surface sampled in the experiment.

7. Calculate the real and imaginary parts of the bulk dielectric function of a simple metal in the infrared region of the spectrum given that the real imaginary parts of the refractive index of the metal in this spectral region are 3.0 and 30.0, respectively.

8. By means of a simple diagram, distinguish between a and p-polarized light in the context of a surface experiment and further distinguish between the p-perpendicular and p-parallel components.

9. Via the use of a microscopic dipole model, show how the so-called surface selection rule arises for the detection of the vibrations of a species adsorbed at a metal surface.

10. Determine the critical angle for total reflection from the boundary between vacuum and a hypothetical solid medium with refractive index $n = 2.7$ and $k = 0$ for infrared photons of energy equivalent to $2000\ cm^{-1}$, plus determine

the depth of penetration of the radiation into the solid at an angle of incidence of 60°.

11. Calculate the energy and frequency of the second harmonic generation (SHG) response produced via the interaction of a Nd;YAG laser operating at 1064 nm with a surface of an fcc metal.

12. Account for the fact that both SHG and SFG may be observed from the (001) surface of Pt and Si, but not from the (001) surface of GaAs.

13. An SFG experiment aims to measure the C—H vibrations in an adsorbed hydrocarbon species. Given that the pump laser is operating at a wavelength of 532 nm and that in a typical IR spectrum the C—H vibrations for the particular adsorbed species lie in the range 2800–3000 cm^{-1}, calculate the wavelength range that should be measured by the detector in the SFG experiment.

8 Surface Structure Determination by Interference Techniques

CHRISTOPHER A. LUCAS

Department of Physics, University of Liverpool, Liverpool, UK

8.1 Introduction

In the earlier chapters of this book, a range of techniques have been introduced which enable the chemical composition of a surface to be determined. Just as important as this is the way in which the atoms making up the surface are arranged relative to one another, as the geometry of the surface determines the way in which new molecules adsorb at the surface, and hence influences the reactivity of the surface. The problem of determining surface structure may be divided into two parts. First, it is necessary to know the *symmetry* of the surface atomic arrangement, i.e. where a surface has some *long range order* we would like to know the size and shape of the repeat unit on the surface (the surface unit cell). Secondly, we would like to know the precise details of the atomic positions themselves, i.e. the number of neighbours surrounding a particular atomic site, and the distances and direction vectors to each surrounding atom (this is known as the *short range order* of the surface).

Not all solid surfaces possess long range order; this will only normally be a property of the surfaces of single crystal materials. If we wanted to investigate the *long range order* in the *bulk* of a single crystal material,

Surface Analysis – The Principal Techniques 2nd Edition Edited by John Vickerman and Ian Gilmore
© 2009 John Wiley & Sons, Ltd

we would use the diffraction pattern obtained when a beam of X-rays or neutrons is incident on the sample. By carefully measuring the intensities of the diffracted beams, and comparing them to those calculated for various models of the bulk structure, we could also obtain the details of the atomic coordination, in other words, information about short range order in the bulk of the crystal. The general theory of diffraction from three-dimensional crystals is described in Section 8.1.1. These diffraction techniques can be adapted to give information about the *surface*, rather than the bulk of the material, giving analogous information to bulk diffraction techniques, i.e. details of symmetry *and* atomic coordination. This application of the diffraction theory to surfaces is described in Section 8.1.2 and the use of electron diffraction to probe the *long range order* at surfaces is described in Section 8.2. The application of X-ray diffraction to surfaces, surface X-ray diffraction (SXRD), has been stimulated by the development of dedicated synchrotron radiation sources and this is described in Section 8.3.3.

There are many important classes of material, such as polycrystalline and amorphous materials, glasses and gels, where no long range order exists, and diffraction techniques cannot be used. Similarly there are cases where locally ordered surface structures are not present in a coherent manner across a crystal surface. In these cases a technique is required which gives information about the *short range order* of the surface or bulk of a material. Again many of the techniques which have become available have resulted from the increased availability of synchrotron radiation. This is particularly true of the X-ray standing wave (XSW) technique (Section 8.3.4), which uses the X-ray standing wavefield set up by diffraction in the bulk of a crystal to determine local adsorption sites, and other techniques based on the creation of an electron wave at a particular atomic site (by absorption of a photon), i.e. surface extended X-ray absorption fine structure (EXAFS) or SEXAFS – see Section 8.3.2. These latter methods exploit the coherent interference of components of the photoelectron wavefield directly emitted from an atom core level and elastically scattered from the surrounding atoms. Photoelectron diffraction (Section 8.4) exploits the same coherent interference phenomena but, rather than measuring the total photoabsorption cross section, as in EXAFS, the photoelectrons emitted in a particular direction from a sample are detected themselves, i.e. both the emission direction and photoelectron energy can be determined.

8.1.1 BASIC THEORY OF DIFFRACTION – THREE DIMENSIONS

Before considering the particular case of surfaces, it is useful to review the important concepts underlying the treatment of diffraction from three-dimensional crystals, as most of the general concepts can be readily applied to the treatment of diffraction from surfaces. The most widely used technique for studying bulk crystal structure is X-ray diffraction. X-rays are scattered by the electrons distributed around atom cores in the solid. As the X-rays are uncharged, this interaction is rather weak, with the result that the X-rays penetrate deeply into the solid and act as a bulk probe (although we shall see in Section 8.3 that it is also possible to use X-rays to probe surfaces, so called surface X-ray surface diffraction). Diffraction effects will only be observed from crystalline solids. A crystal is distinguished from a glass or an amorphous material by possessing long range order. The crystal structure itself may be regarded as being made up of two parts, the *lattice* and the *basis*. The crystal lattice is a three-dimensional array of points which repeats itself periodically in all three dimensions, so providing the framework for the crystal structure. In fact, for this infinite repetition to occur, only 14 unique lattices are possible in three dimensions, known as the 14 Bravais Lattices [1]. The basis is the number and arrangement of atoms that are associated with each lattice point. When the basis is superimposed on the crystal lattice, the full crystal structure results:

$$lattice + basis = crystal\,structure$$

The *unit cell* for a particular structure is usually the simplest possible unit of the structure which contains everything which is unique about that structure, so that if the unit cell is repeated infinitely in space in three dimensions, the macroscopic crystal structure results. The unit cell lengths (the cell parameters) are usually denoted a, b and c (and the corresponding vectors in these directions $\boldsymbol{a}, \boldsymbol{b}$ and $\boldsymbol{c}$), and the cell angles (which need not be 90°) α, β and γ. In the simplest possible case of a cubic unit cell (e.g. NaCl), $a = b = c$ and $\alpha = \beta = \gamma = 90°$.

Any plane of atoms within a crystal structure must be defined by three coordinates and can be labelled by its *Miller Indices*, (h, k, l). The Miller indices of a plane are obtained by calculating the intercepts of the plane with the $\boldsymbol{a}, \boldsymbol{b}$ and $\boldsymbol{c}$ axes, as a fraction of a, b and c, then taking the reciprocals of these numbers, and expressing the result in the form of the smallest integer numbers. An example is shown in Figure 8.1. A bar above a figure is used where the intercept with the $\boldsymbol{a}, \boldsymbol{b}$ or $\boldsymbol{c}$ axis is negative – for example, the six faces of a cube of length a are denoted (100), (010), (001), $(\bar{1}00), (0\bar{1}0)$ and $(00\bar{1})$. A group of such equivalent planes is known as a family of planes, in this case the {100} family of planes. Directions in a crystal structure can also be denoted by the Miller indices and are written as $[hkl]$. An important and useful property of Miller indices is that the $[hkl]$ direction in a crystal is perpendicular to the (hkl) plane.

The Miller indices of a plane bear a reciprocal relationship to the real intercepts of the plane with the axes; in the same way the wavelength of the X-ray beam (a real distance) bears a reciprocal relationship to the wavevector, $\boldsymbol{k}$, of the beam, whose magnitude is defined as:

$$k = 2\pi/\lambda \qquad (8.1)$$

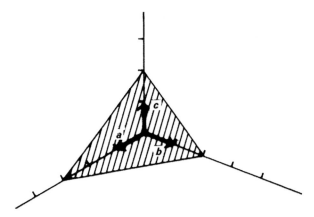

Figure 8.1 An example of the derivation of Miller indices for a plane. The plane has intercepts with the axes at $3a, 2b, 2c$. The reciprocals of these numbers are $1/3, 1/2, 1/2$. The smallest whole integers having the same ratio are 2, 3, 3, so the Miller indices for this plane are (233)

The wavevector is an important quantity, as it is a measure of the momentum of the incident and diffracted beams[1] – the change in wavevector of a beam on scattering from a plane of atoms will determine the direction of any emergent diffracted beams. Thus, the diffraction pattern obtained is a reflection of changes in wavevector (or the wavevector transfer). It is thus very useful when treating diffraction to work not in 'real' space, but in reciprocal space, which bears an inverse relationship to real space. Instead of the real crystal lattice, where diffraction is occurring, we can create a reciprocal lattice, where the distances between points are inversely proportional to the corresponding distances in the real lattice but are a direct measure of k (hence the alternative name of 'k-space'). A diffracted beam of X-rays will emerge from the crystal whenever constructive interference occurs between X-rays reflected from successive planes of atoms in the real lattice. For an X-ray beam incident as an angle θ to a series of atomic planes separated by a spacing d, it is straightforward to show that the condition for constructive interference is given by $n\lambda = 2d\sin\theta$, known as Bragg's law. We shall return to this shortly. The advantage of the reciprocal lattice is that the condition for constructive interference can easily be determined by applying the law of conservation of momentum in the reciprocal lattice, where distances are directly proportional to momentum. The geometrical construction employed to do this is known as the *Ewald sphere construction*. This is shown in Figure 8.2 for a two-dimensional cubic

---

[1]Remember that the inverse relationship between the wavelength and the momentum of a photon comes from the de Broglie equation, $\lambda = h/p$, where p is momentum and h is Planck's constant – hence $k = p/\hbar$.

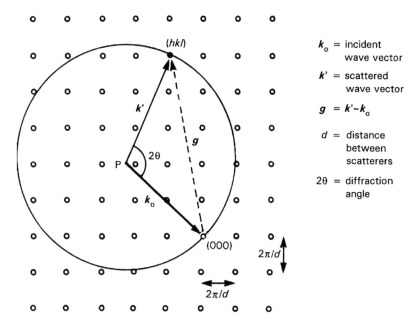

Figure 8.2 The Ewald sphere construction for a cubic lattice where (*hkl*) are the Miller indices of the point on the Ewald sphere. Reproduced from reference [3], Copyright 1994, Oxford University Press

lattice. The lattice shown here is the reciprocal lattice, where the distance between adjacent lattice points is $2\pi/d$, with d the distance between points in the real lattice. For a three-dimensional crystal, of course, the reciprocal lattice is three-dimensional but it is convenient to represent the Ewald sphere construction in only two dimensions. A vector, k_0, is drawn to scale on this diagram, with its tip pointing towards the origin of reciprocal space, (000), representing the wavevector of the incident X-ray beam. A circle is then drawn with radius $|k_0|$, centred at the origin of the vector, point P. This is the Ewald sphere. Its significance is that it maps the magnitude of k_0 onto the reciprocal lattice – by conservation of momentum, no diffraction events can occur outside this sphere. If any of the reciprocal lattice points are intersected by the Ewald sphere, then the condition for elastic scattering is satisfied (i.e. there is a change in momentum of the beam, but not its energy), with the scattered beam having a wavevector k'. By conservation of momentum:

$$k_0 = k' + g \tag{8.2}$$

where the change in momentum on scattering is represented by the reciprocal lattice vector, g. Note that for elastic scattering, $|k_0| = |k'|$, i.e. there is a change in *direction* of the beam, but not in the *magnitude* of the momentum of the beam, and this is represented by the radius of the Ewald sphere.

In Figure 8.2, the diffraction angle is shown as 2θ, in accordance with convention[2]. It can be seen that:

$$\sin\theta = \frac{|k_0|}{|g|/2} \tag{8.3}$$

Pythagoras' theorem (in three dimensions) gives the result:

$$|g| = (h^2 + k^2 + l^2)^{1/2} 2\pi/d \tag{8.4}$$

and knowing that $|k_0| = 2\pi/\lambda$, hence:

$$\sin\theta = (h^2 + l^2 + k^2)^{1/2}\lambda/2d \tag{8.5}$$

or the familiar Bragg condition:

$$n\lambda = 2d\sin\theta \tag{8.6}$$

where $n = (h^2 + k^2 + l^2)^{1/2}$ is known as the 'order of diffraction'.

The Bragg condition implies that diffraction will occur for all possible values of $(h^2 + k^2 + l^2)$. However, in many crystal systems this is not the case, in some cases the diffracted beams from one sub-set of planes may be exactly cancelled by diffracted beams from another sub-set having the same amplitude but exactly opposite phase. This leads to the absence of a diffracted beam at some θ values where one is expected, i.e. to *systematic absences*. These patterns of systematic absences are particularly useful in determining the symmetry of the lattice. By relating the interplanar spacing, d, in the Bragg equation to the unit cell lengths, a, b and c, the angles at which diffraction occurs may be used to calculate the unit cell size.

Information about the crystal lattice size and symmetry can thus be obtained by an analysis of the positions of the diffracted beams. However, this does not give information about the arrangement of atoms within the lattice, i.e. the crystal basis. The exact atomic positions can only be obtained from an analysis of the intensities, rather than the positions of the diffracted beams. Each atom within a lattice scatters X-rays to an extent which is dependent on the charge distribution around the atom (i.e. on the number of electrons surrounding the atomic core), so that each atom i has an atomic scattering factor, f_i associated with it (f_i is known as the atomic form factor [1, 9]). The scattering produced by the crystal as a whole will depend on the number of atoms in each unit cell, and on their positions relative to one another (as scattering from one atom, i, in the basis may interfere constructively or destructively with scattering from another atom j). This is summarized in the *structure factor*, F_{hkl}, which expresses the scattering amplitude expected for each diffracted beam and is obtained by summing

---

[2]Note that in the treatment of diffraction in two dimensions (Sections 8.1.2 and 8.2.2), the diffraction angle is often denoted as θ, not 2θ; this arises in particular from the special case where a beam is incident along the surface normal (Section 8.2.2).

the individual atomic scattering factors over all the atoms of the unit cell, taking into account their phase differences:

$$F_{hkl} = \sum_j f_j \exp[i2\pi(hu_j + kv_j + lw_j]$$ (8.7)

In this expression, the atomic scattering factors are summed over the atoms j in the unit cell, the exponential term representing the correction for their phase differences, u_j, v_j and w_j are the fractional coordinates of the jth atom in the unit cell, expressed as fractions of the unit cell lattice parameters, a, b and c. In many cases the exponential term is complex. Unfortunately, the intensity of a wave is the square of its amplitude, so that the intensity, I_{hkl}, of a diffraction spot with Miller indices (hkl) is related to F_{hkl} through:

$$I_{hkl} \propto |F_{hkl}|^2$$ (8.8)

This means that a measurement of the intensity of a diffracted beam tells us only the magnitude of the structure factor, F_{hkl}, and the complex information (the phase) is lost. This difficulty is known as the *missing phase problem*. It creates some complexity in the determination of atomic positions from diffraction data but, in general, the problem can be overcome. A number of methods have been developed to allow atomic positions to be extracted from the experimental data. Typically, the methods rely on making an initial guess at the atomic positions, consistent with the symmetry of the lattice determined from the positions of the diffracted beams. Results from other experiments, such as spectroscopic data, and general chemical knowledge of the expected coordination of particular atoms/ions may feed into this initial guess. The intensities of the expected diffracted beams are then calculated for this arrangement, as a function of a change in the diffraction conditions, usually a change in the angle of incidence, or azimuthal rotation of the sample around the surface normal. This calculation is compared with a detailed set of experimental measurements of a large set of diffracted beams, and the guessed structure is refined as a result. A new set of intensities is calculated, and the procedure is repeated until satisfactory agreement is obtained between theory and experiment. Alternatively recent methods have been devised that use the fact that the scattering power of the atoms, f_j, is really a complex quantity that depends on the energy of the incident X-ray beam. These direct methods are beyond the scope of this article but interested readers are referred to Als-Nielsen and McMorrow [9] for more details. Such methods are being increasingly used in many of the surface analysis techniques described in this chapter.

The interaction of the X-ray beam with the solid is quite weak and this means that, in the calculation of diffracted beam intensities, it may be assumed that each scattered beam undergoes only one scattering event before emerging from the solid. The theory of X-ray diffraction is said to be *kinematic*, in other words, multiple scattering effects are ignored, and this considerably simplifies the theoretical treatment. This is not the case when dealing with an electron beam, as electrons interact strongly with a solid and multiple scattering effects must be considered. This is described in Section 8.2.

8.1.2 EXTENSION TO SURFACES – TWO DIMENSIONS

In measurements of bulk atomic structure, X-ray and neutron diffraction are the most commonly used methods. X-rays are essentially scattered by the charge distribution around atoms; this scattering is very weak, so that X-rays penetrate materials very deeply, and bulk structure can be probed. Neutrons are even more weakly scattered by solids. These properties of X-rays and neutrons, while useful for bulk measurements, mean that they are not the ideal probes for surface sensitive measurements. (This does not mean that they cannot be used in such experiments – see Section 8.3.) Electrons, on the other hand, as charged particles, interact very strongly with matter; we have already seen in previous chapters that the mean free path length of low energy electrons (say < 500 eV) is very short, of the order of a few tens of Å. In addition, the wavelength of such electrons is of the order of an angstrom, slightly smaller than a typical interatomic spacing, and hence suitable for diffraction experiments. These properties, combined with the ease of producing a monochromatic electron beam, make electrons the primary tools for routine surface diffraction measurements.

The two most important 'surface diffraction' techniques in common use are Low Energy Electron Diffraction (LEED) and Reflection High Energy Electron Diffraction (RHEED). These are discussed in Section 8.2. First, however, we will discuss the main ways in which the theoretical treatment of diffraction is adapted to apply to surfaces. Firstly, if a very surface sensitive probe, such as a beam of low energy electrons, is being used to make the measurement, it is reasonable to suppose that only diffraction from the topmost layer of ordered atoms is observed, i.e. a *two-dimensional surface unit mesh*, rather than the three-dimensional unit cell observed in a bulk measurement. The reduction in the number of dimensions means that instead of the 14 Bravais lattices which are possible in three dimensions, there are only five possible unit meshes, or *surface nets*, which may be repeated infinitely in two dimensions to build up the planar, periodic net of a surface [1, 3]. In this net, every lattice point can be reached from the origin by translation vectors:

$$T = ma_s + nb_s \qquad (8.9)$$

where m and n are integers, and the vectors a_s and b_s define the unit mesh, with the subscript s denoting the surface. The five surface nets are illustrated in Figure 8.3 [3]. As in the case of three dimensions, generation of the complete surface structure requires the basis atoms to be attached to the

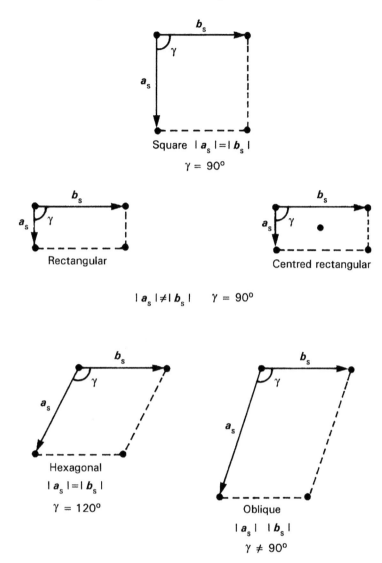

Figure 8.3 The five surface nets. Reproduced from [3], Copyright 1994, Oxford University Press

unit mesh consistent with certain symmetry restrictions. These restrictions mean that only 17 two-dimensional space groups are possible [1].

Notation for Surface Structures.   An ideal surface may be identified easily by reference to the bulk crystal plane of termination, e.g. Pt (100), Ni (110), NaCl (100). However, it is very common for the atoms in the topmost surface layer to rearrange themselves into a new net which is not a simple termination of the bulk lattice; this is known as *surface reconstruction*, and occurs essentially because the minimum energy configuration for the atoms at a newly created surface, which have a reduced number of nearest neighbours, may not be the same as for the same atoms in the bulk of the material. A new form of notation is required which describes the orientation of the new net of the reconstructed surface on the underlying bulk structure and which can also be used to describe the orientation of adsorbed overlayers on any surface. If the periodicity and orientation of the surface net is the same as the underlying bulk lattice, the surface is designated (1×1), i.e. unreconstructed. In this case the (1×1) surface may undergo relaxation, i.e. the top surface atomic layer or layers may undergo expansion or contraction of the bulk interlayer spacing. This phenomenon

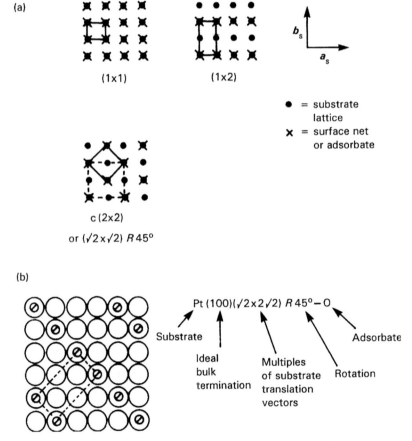

Figure 8.4 Surface net nomenclature: (a) some examples of surface nets, with their corresponding notation; (b) an example of full notation for a real surface (Pt(100) exposed to oxygen)

is different from *surface reconstruction* and is known as *surface relaxation*. It is often the case, however, that the translation vectors of the surface net differ from those of the underlying lattice, so that

$$a_s = Ma, b_s = Nb \qquad (8.10)$$

where *a* and *b* are the translation vectors of the ideal, unreconstructed surface. The nomenclature for this structure is $(M \times N)$. If, in addition, the surface net is rotated with respect to the underlying lattice by an angle ϕ degrees, the notation becomes $(M \times N)R\phi°$. If the surface net is best described using a centred, rather than a primitive net (i.e. one with a surface lattice point at the centre), this is indicated as $c(M \times N)$. If the overlayer consists of an adsorbate, rather than simply reconstructed substrate atoms, this is also usually indicated. These points are illustrated in Figure 8.4. The general nomenclature is known as Wood's notation. A listing of the known surface structures for clean surfaces and adsorbate systems is given in Somorjai [4].

The Ewald Sphere Construction in Two Dimensions.   In Section 8.1.1, some basic ideas about reciprocal space were discussed, and the Ewald sphere construction for three-dimensional diffraction was introduced. This construction is also helpful when considering two-dimensional diffraction from surfaces and is shown in Figure 8.5. It can be seen that instead of showing reciprocal lattice points (as in the case of diffraction in three dimensions (Figure 8.2), the diagram shows reciprocal lattice *rods*. These rods arise because the surface forms a completely two-dimensional net and, hence, the periodic repeat distance normal to the surface is infinite. As shown earlier, the distance between adjacent points in a reciprocal lattice is inversely proportional to the corresponding distance in the 'real' lattice. This means that the reciprocal lattice 'points' along the surface normal are infinitely dense, forming rods (Figure 8.5). The diffraction condition is satisfied for every beam that emerges in a direction along which the sphere intersects a reciprocal rod. By using a construction similar to that used in Figure 8.2 (for three dimensions), it can be shown that this corresponds to a Bragg relationship (Section 8.2.2), although this condition is often expressed slightly differently to the condition used in three dimensions. The diffracted beams, which each produce a spot in a LEED pattern, are indexed according to the reciprocal lattice vector which produces the diffraction. Due to the loss of periodicity in one dimension, only two Miller indices, *h* and *k* are needed to label a reciprocal lattice rod.

One important consequence of the loss of periodicity normal to the surface is that the conditions for observation of a diffraction pattern are relaxed relative to those for bulk diffraction. This is because observation of a diffraction pattern from the bulk rests on the constructive interference of the outgoing waves in the surface normal direction, which can no longer

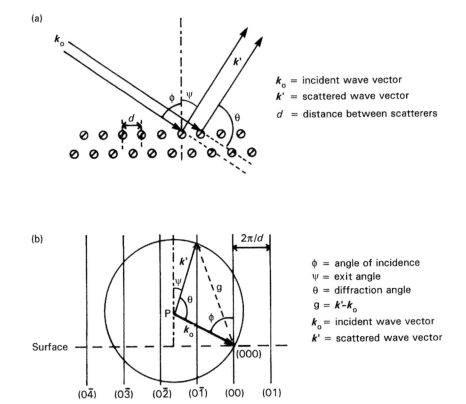

Figure 8.5 A schematic diagram of a diffraction process occurring at a surface in real space (a), with the corresponding Ewald sphere construction in reciprocal space shown in (b) (see text)

occur when the structure is not periodic in this direction. As a result the diffracted beams may occur at all energies and not just at certain discrete energy values, provided that the corresponding rod lies within the Ewald sphere.

8.2 Electron Diffraction Techniques

8.2.1 GENERAL INTRODUCTION

An electron beam of energy around 150 eV has a wavelength of around 1 Å, making it suitable for diffraction experiments. However, this energy is roughly at the minimum in the universal pathlength curve, giving these electrons optimum surface sensitivity. The elastic backscattering of low energy electrons incident normally on a crystal surface forms the basis of the technique of Low Energy Electron Diffraction (LEED). An alternative is to use high energy electrons incident at a grazing angle on the crystal surface. In this case the penetration depth of the electron beam into the surface is also very small, as the component of the incident electron momentum normal to the surface is very small. This forms the basis of Reflection High Energy Electron Diffraction (RHEED).

8.2.2 LOW ENERGY ELECTRON DIFFRACTION

Introduction. The first experiments which showed that electrons could be diffracted by crystalline solids, in the same way as X-rays, were conducted almost simultaneously by Davisson and Germer and by Thomson and Reid, in the late 1920s. The latter observed diffraction of a beam of electrons transmitted through a thin metal foil. However, it was Davisson and Germer who might be said to have performed the first LEED experiment, when they observed diffraction effects in the electrons backscattered from a single crystal of nickel [16]. At the time, these experiments provided evidence of the wave properties of electrons, and the wavelength of the electrons was found to be consistent with the value of h/mv (Planck's constant divided by the momentum of the electron) predicted by the then new theories of wave mechanics. Despite these initial experiments, however, the technique was not developed further until 1960, when work by Germer and co-workers led to the development of the modern LEED display system [17]. The development of UHV technology has subsequently seen LEED become one of the most widely and routinely used techniques for surface structure determination.

Theoretical Considerations. The simple production of a LEED photograph, without an analysis of the intensities of the individual spots, is by far the most widespread use of LEED. In most surface science laboratories around the world, LEED is often routinely used to check the cleanliness and order of surfaces being prepared for other experiments. LEED is sensitive to surface contamination and surface roughness and, hence, the appearance of a LEED pattern with bright, sharp spots is widely regarded by surface scientists as evidence of a completely clean, ordered surface. This is *not necessarily the case* as surfaces which show sharp LEED patterns can sometimes look like the surface of the moon in imaging experiments such as scanning tunnelling microscopy (STM) and atomic force microscopy (AFM) – see Chapter 9. The pattern of LEED spots can, however, be used to obtain information about surface symmetry or surface reconstruction, or about imperfections in the surface, such as steps or islands. It can also be used to determine whether any molecules on the surface are adsorbed in an ordered or random way. If an overlayer is ordered, its surface unit mesh size can be determined, and if the layer is adsorbed commensurately with the substrate, its orientation relative to the underlying substrate may be determined.

It was shown in Section 8.1.2 that in LEED, diffracted beams may occur at all energies, provided the corresponding rod lies within the Ewald sphere. Changing the incident beam energy will change the radius of the sphere, $|k_0|$, and so the number and directions of the scattered beams will vary (Figure 8.5). The LEED pattern, therefore, contracts towards the specularly reflected beam as the incident electron beam energy is increased. Usually, the incident electron beam is normal to the surface, in which case a symmetrical LEED pattern is obtained, which converges towards the (0,0) specular beam (i.e. the centre of the pattern) as the beam energy is increased. (In practice, of course, no (0,0) spot is observed, as the centre of the fluorescent screen is occupied by the electron gun.) The energy of the incident electron beam, E, is given by:

$$E = (\hbar^2/2m)k^2 \tag{8.11}$$

where $k = 2\pi/\lambda$. Thus the incident wavevector k_0 increases with increasing electron beam energy, so that the size of the Ewald sphere increases, cutting more and more rods. This means that, as the beam energy increases, more and more spots will appear on the LEED screen, and the spacing between spots will progressively decrease, i.e. the pattern converges towards the centre of the screen.

It should be remembered that the LEED pattern is a representation of reciprocal space – the distance between adjacent points in the LEED pattern (reflecting the reciprocal lattice) is inversely proportional to the distance between points in the corresponding direction of the real surface unit mesh (the direct lattice). Figure 8.6 shows a Ewald sphere construction for an electron beam at normal incidence, i.e. along the (0,0) direction. (Comparing this with Figure 8.5, ϕ is now zero.) Note that each Ewald sphere construction can show only the diffraction conditions which are satisfied in one azimuth, rather than all the diffracted beams which are emerging in all directions from the crystal. This is because only that part of reciprocal space in the plane of the paper can be drawn. In Figure 8.6 we have chosen to keep the Miller Index h constant (and equal to zero), and vary the index k, corresponding in real space to probing diffraction from a row of atoms along the b-direction of the surface unit mesh. The angle of diffraction between the incident wavevector and the backscattered beam having (h, k) equal to (0,2) is shown as θ. From the construction in the diagram, it can be seen that, for the (0,2) beam:

$$\sin\theta = \sin\theta' = \frac{2(2\pi/b)}{|k_0|} \tag{8.12}$$

or, as $|k'| = |k_0| = 2\pi/\lambda$, where λ is the wavelength of the electron beam:

$$\sin\theta = 2\lambda/b \tag{8.13}$$

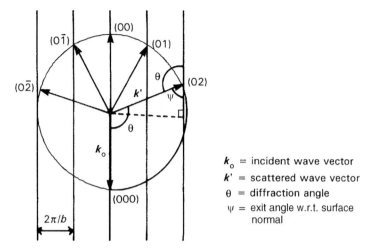

Figure 8.6 The Ewald sphere construction for an electron beam incident normal to the surface (i.e. along the (0,0) direction, compared with Figure 8.5, ϕ is now zero.) Here only diffracted beams where the Miller index, $h = 0$, are shown

for this reflection, or more generally for this azimuth:

$$b \sin \theta = k\lambda \tag{8.14}$$

The integer k is sometimes known as the *order of diffraction*. The rows giving rise to the (0,2) beam will, by definition of Miller indices, be spaced a distance $b/2$ apart, (and will be parallel to a). Thus we arrive at the condition:

$$\lambda = d_{02} \sin \theta \tag{8.15}$$

Using a Ewald sphere for the perpendicular direction ((0,0)...(h,0)) the corresponding condition for diffraction from atoms along the a axis is:

$$a \sin \theta = h\lambda \tag{8.16}$$

We have thus identified the conditions which must be satisfied in order to observe diffraction from two perpendicular sets of rows of atoms. In order to observe diffraction from our two-dimensional surface, both these conditions must be satisfied simultaneously. In the case of a simple cubic lattice ($a = b$), this gives the more general result:

$$a \sin \theta = (h^2 + k^2)^{1/2}\lambda \tag{8.17}$$

where $(h^2 + k^2)^{1/2}$ is the order of diffraction, sometimes written as n:

$$a \sin \theta = n\lambda \tag{8.18}$$

equivalent to the Bragg condition for diffraction in three dimensions, but expressed slightly.[3]

In the case above of a simple cubic mesh, the mesh side 'a' may be easily obtained from a plot of $\sin\theta$ versus λ for any spot (h,k). Obviously, the sample–screen distance and the size of the screen must be known in order to calculate the diffraction angle. More generally, the observed LEED pattern will reflect the symmetry of the surface under study, and the surface unit mesh size may generally be easily obtained. Some examples of the possible types of two-dimensional lattice and their corresponding LEED patterns are shown in Figure 8.7. In the case of a surface reconstruction of an ideal surface, or overlayer adsorption, new spots will usually appear in the LEED pattern. Figure 8.8 shows some examples of overlayer structures due to adsorbates or reconstructions, with their corresponding LEED patterns. Note the reciprocal relationship between the real overlayer structure on the surface and the spot density in the LEED pattern. It is important to remember that the surface atom arrangement can have *at most* the symmetry indicated by the LEED pattern; the true symmetry could be lower than that indicated by the LEED pattern. An example is shown in Figure 8.8, where the LEED pattern expected from a surface covered by domains of a (1×2) and a (2×1) overlayer is shown. The LEED pattern which results is a composite of the individual patterns of both domains, as the LEED beam is physically larger than the individual domain size. The resultant pattern has four-fold symmetry, even though neither of the domains has this property.

From the above discussion, it would seem likely that there is a finite distance over which LEED can detect features such as disorder occurring. In fact, because of the energy spread of the incident beam and its angular divergence, the electrons have a limited *coherence length* at the surface, typically ≈ 50–100 Å, depending on beam energy. Surface features which occur at larger scales than this will not be detected. For example, it will generally be difficult to distinguish an adsorbate forming large islands at the surface from one forming a uniform overlayer. Features such as surface steps give rise to distinct features in LEED patterns. In the case of a regular array of steps, the effect is to increase the repeat unit of the surface mesh in the direction perpendicular to the steps, as illustrated in Figure 8.9. This gives rise to a corresponding decrease in spot spacing in the corresponding direction in the LEED pattern as illustrated. Irregular steps give rise to streaking of spots in the direction of disorder (Figure 8.9). Facetted surfaces

---

[3]NB: in this treatment, θ is the angle between the incident wavevector, and the wavevector of the backscattered beam; some treatments use 2θ, and the corresponding angle in the three-dimensional treatment is usually denoted as 2θ. From Figure 8.5(a), the general pathlength difference between successive diffracted beams is $d(\sin\theta + \sin\chi)$. Usually in LEED, $\theta = 0°$, and so the condition for constructive interference becomes $\eta\lambda = d\sin\chi = d\sin\theta$, as θ is $180° - \chi$ when $\theta = 0°$.

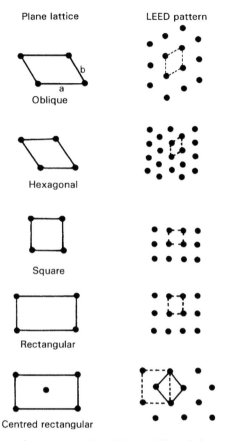

Figure 8.7 The five plane lattice types (see Figure 8.3) and their corresponding LEED patterns. Note the reciprocal relationship. In high-symmetry cases, systematic absences can occur – for example, compare the 'centred' rectangular pattern with the 'simple' rectangular pattern

may be distinguished in LEED by changing the beam energy. We have seen that for a perfect surface, the LEED pattern converges to the (0,0) specular beam as the beam energy is increased. In the case of a facetted surface, there are different (0,0) beams corresponding to different facets, and this should be apparent on increasing the beam energy.

The positions of the spots appearing in the LEED pattern may be used to determine the size and symmetry of the surface unit mesh. This is analogous to the situation in X-ray diffraction in three dimensions, where the positions of diffracted beams may be used to determine the size and symmetry of the unit cell of the bulk lattice. In X-ray diffraction, despite the complications due to the complex nature of the phase factor, it is generally possible to determine precise coordinates for the atoms inside each unit cell (forming the basis) from the intensities of the diffracted beams. It might be

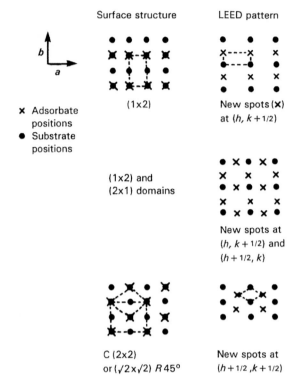

Figure 8.8 Some possible overlayer structures caused by absorbates or reconstructions, with their corresponding LEED patterns

anticipated, therefore, that it is possible to determine the positions of the atoms within each surface unit mesh by measuring the intensities of the diffracted beams in the LEED pattern. The intensities of particular LEED spots as the diffraction conditions are changed can be measured, for example as a function of beam energy (known as $I(V)$ curves), or as a function of azimuthal rotation ($I(\psi)$). Unfortunately, the theoretical interpretation of such data is extremely difficult. Paradoxically, this arises because of the extremely strong interaction between electrons and atoms, precisely the property which makes LEED surface sensitive. As the cross-section for elastic scattering (as well as inelastic scattering) of low energy electrons by atoms is high, it is possible for diffracted beams to be elastically scattered several times at the surface, and still emerge with a measurable intensity. This multiple scattering complicates the analysis of the intensities of the resultant beams. In contrast, in X-ray diffraction the interaction between X-rays and the charge distribution around atoms is very weak, and as a result, the probability of such multiple scattering effects may be regarded as negligible; each photon is backscattered after a single encounter with an atom. This type of scattering is known as *kinematic* and forms the basis

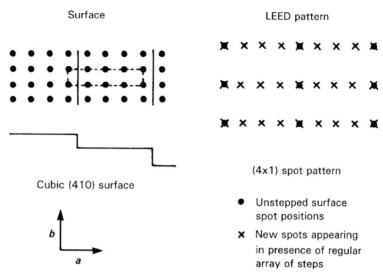

Surface   LEED pattern

(4x1) spot pattern

Cubic (410) surface

● Unstepped surface
spot positions

✗ New spots appearing
in presence of regular
array of steps

Figure 8.9 The effect on a LEED pattern of steps on a surface, where the distance between steps is less than the coherence length of the technique. In the example, the surface repeat unit is enlarged by four times in the horizontal direction, leading to the appearance of a LEED pattern which is roughly (4×1) (but not exactly, as the real repeat length is from step edge to step edge, which is $(17)^{1/2}$). The presence of irregular steps on a surface will lead to streaking of the LEED spots in the direction of the disorder

of the theoretical treatment used to interpret X-ray diffraction data (see Section 8.3.3). In LEED, however, multiple scattering effects cannot be neglected, and the intensity of a particular spot may only be obtained by adding together all the waves scattered into a particular direction from many different scattering sequences, taking into account their amplitude and phase differences. This is known as a dynamical theory, and is essential in the treatment of LEED.

Much of the development of a dynamical theory of LEED is due to Pendry [7, 8, 18]. The theoretical difficulty associated with computing the multiple scattering intensity is eased slightly by the fact that, although the cross-section for elastic scattering is high in LEED, the corresponding cross-section for inelastic scattering is high also. This means that the mean-free-pathlength of electrons in this energy range within the solid is of the order of a few tens of Å. There is thus a limit on the number of elastic scattering events which can occur before inelastic scattering destroys the coherence of the diffracted beam and reasonable agreement between theory and experiment may be achieved using a limited number of multiply scattered beams. The normal procedure in this type of analysis is to determine the variation in intensity of the LEED beam intensities, as a function of some change in the diffraction parameters, such as azimuthal rotation, or variation in beam energy. A purely kinematic theory of LEED would predict that there would be no variation in spot intensity as a function of azimuthal rotation of the sample, as this does not vary the angle of incidence ϕ (Figure 8.5). In fact, such plots tend to show strong intensity minima at certain angles where strong multiple scattering can occur in some direction other than that being measured. Similarly, a kinematic theory would predict that an $I(V)$ curve would only show strong maxima whenever the incident electron wavelength satisfies the diffraction conditions discussed in Sections 8.1.1 and 8.1.2 (known as 'Bragg peaks'). Usually, a number of secondary peaks, caused by multiple scattering, are observed in such curves. An example is shown in Section 8.2.2.

An iterative procedure has been developed by Pendry to determine the geometrical arrangement of surface atoms within the surface unit mesh from an experimentally determined set of $I(V)$ curves. The starting point for the calculation is an initial guess at the arrangement of atoms on the surface, which is chosen to be consistent with the symmetry of the LEED pattern. The intensity of a number of the diffracted beams expected for this arrangement is then calculated as a function of electron beam energy. This is done by solving the Schrödinger equation for the electron wavefunction in the first few atomic layers of the solid. The resulting calculated $I(V)$ curves are compared to the experimental result, and the guessed atomic arrangement is adjusted, and a new set of curves is calculated. The process is repeated until satisfactory agreement is obtained. One major drawback of this type of treatment is that the amount of computer time necessary to solve a particular structure scales exponentially with the size of the problem. For example, for a structure involving just three atoms (say CO on a Cu atom), nine coordinates in space are involved. If 10 trials are required for each coordinate to obtain a good fit to the data, then 10^9 trials must be run for the system as a whole! One treatment aimed at cutting down the computational effort involved has been developed by Rous and Pendry [19] and is known as 'Tensor-LEED'. The procedure involves guessing an initial trial structure as close as possible to the expected structure. Perturbation theory is then used to move the *individual* atoms about by small amounts, until a good fit is obtained. For the example above, treating the three atoms independently of one another, 10^3 trials are needed for each atom, and so only 3×1000 trials must be run for the whole system. This means that if the contribution to the LEED data from each atom can be picked out independently (easiest for heavy atoms which scatter strongly), the computer time required now scales linearly with the complexity of the problem. In its original form, Tensor-LEED was restricted to structural models within $0.1\,\text{Å}$ of the guessed reference structure (as first-order perturbation theory was assumed valid, so this limited the size of the perturbation possible). This model was developed further by Oed, Rous and Pendry, into a second-order Tensor-LEED approximation, which appears to be valid for displacements up to ≈ 0.2 Å perpendicular and 0.4 Å parallel to the surface [20].

In general, in the analysis of LEED data, the agreement between experiment and theory is not always good, and there may sometimes be more than one computed structure which fits equally well with the data. This can lead to arbitrary and subjective assignments. An attempt to overcome this is the use of reliability factors (R-factors), which provide objective criteria for the quantitative evaluation of the closeness of curve-fitting. There are various ways of calculating R-factors, but in general they are designed to emphasize features of the LEED data which are very sensitive to structural details, such as peak positions and shapes. Usually, a fit to LEED data will have an R-factor associated with it; the lower the value, the better the fit. Using the types of procedure described above, atom positions may now be obtained optimally to an accuracy of ± 0.01 Å, very close to the accuracy of bulk diffraction techniques.

Experimental Details.　A typical experimental arrangement used in a LEED experiment is shown in Figure 8.10. An electron beam of variable energy is produced by an electron gun, and is incident on the sample. The electrons are then backscattered from the sample surface onto a system of grids surrounding the electron gun. The backscattered electrons are of two types; elastically scattered electrons forming a set of diffracted beams which create the LEED pattern, and inelastically scattered electrons, which may make up 99 % of the total flux, but which are not required. After reaching the first grid, G1, which is earthed, the elastically scattered electrons are accelerated towards the fluorescent screen, S, which carries a high positive potential (of the order of 5 kV). This provides the electrons in the diffracted beams with enough energy to excite fluorescence in the screen, so that a pattern of bright LEED spots is seen. The grids G2 and G3 are held at an adjustable negative

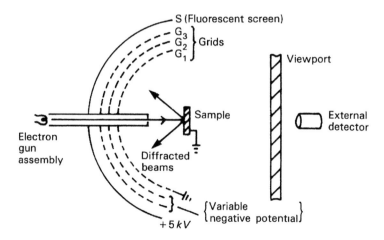

Figure 8.10　Schematic diagram of conventional RFA-type LEED optics

potential, and are used to reject the majority of the electron flux, which is made up of inelastically scattered electrons and which otherwise contribute to a bright, diffuse background across the whole of the LEED screen. The potential on these grids is adjusted to minimize the diffuse background to the LEED pattern.

The LEED pattern which is observed may be recorded using a still or video camera mounted onto a chamber window placed directly opposite the LEED screen. This of course has the disadvantage that parts of the pattern will be obscured by the experimental arrangement around the sample and, possibly, by any sources or detectors used for other techniques and mounted on the same UHV chamber. This arrangement is known as 'front view LEED' and has remained very popular as the screen and grid arrangement is essentially a retarding field analyser (RFA), and may also be used for Auger spectroscopy. The problem of obscuring of the LEED pattern can be alleviated by the use of 'reverse' or 'rear-view LEED', where the pattern may be viewed from a window placed on the rear side of the screen system (Figure 8.11).

In many laboratories, the type of LEED system described above is used almost entirely for measuring spot positions on the screen. Equally important, however, is a measure of the intensities of individual spots and their widths (particularly in applications where surface phase transformations are being measured). Spot intensities can be measured using a conventional system, but accurate analysis of spot profiles can be difficult. For this purpose, the spot profile analysis LEED system (SPA-LEED) was developed [21]. A schematic representation of this system is shown in Figure 8.12. An electron beam passes through a series of deflection plates, which provide two octopole fields, and strikes the crystal sample. The diffracted beam is detected by a channeltron. The angle between the electron gun and the channeltron is fixed; a spot profile is obtained by scanning the potential on the electrostatic deflection plates. As shown in the diagram, this has the

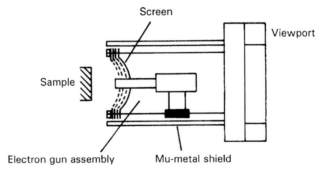

Figure 8.11 Schematic diagram of a reverse view LEED/Auger system. Reproduced with permission from Omicron

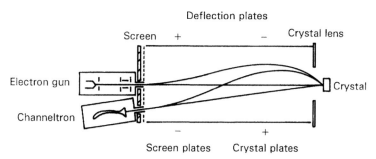

Figure 8.12 Schematic diagram of the SPA-LEED system, showing the beam paths with and without deflection voltages applied. Reproduced by permission of Elsevier Science from Scheithauer *et al.* [21]

effect of changing the angle of incidence of the beam. At constant beam energy, an arc is described through reciprocal space. In this mode, a small current of the order of 0.1–50 nA is used, to scan individual spots. As no mechanical movement is involved, and the deflection plate voltages are under computer control, spot profiles can be recorded with speed and accuracy. An overview of the diffraction pattern may be obtained using larger gun currents of up to 10 µA; the resulting pattern is then observed on a glass phosphor screen, which is viewed from the electron gun side (Figure 8.12). In this mode, the system has higher spatial resolution than RFA systems, as the screen–sample distance is greater.

Applications of LEED.   LEED remains one of the most widely used tools for surface structure determination. Here we describe a few examples of the many applications of LEED which may be found in the literature. Other representative examples may be found by consulting the reference list [22–24]. The first example given here concerns the surface reconstruction of a metal surface, Be (1120) [25]. Figure 8.13 shows the LEED patterns obtained from this surface at three different beam energies, and a temperature of −160 °C. In Figure 8.13(a), recorded with a beam energy of 94 eV, the {1, n} spots are evident. As the beam energy is increased (Figure 8.13(b)), these spots disappear, and the intensity is split into pairs of spots at the third-order positions {2/3, n} and {4/3, n}, indicating a (1×3) reconstruction of the surface. At higher beam energy still (Figure 8.13(c)) the original integer-order, unsplit spots are regained. These observations may be explained by a surface reconstruction in which every third row of atoms on the surface is missing. Two possible ways in which this might occur were considered (Figure 8.14), corresponding to a simple missing row (Figure 8.14(b)) and a situation where the removed row sits on top of the bridge sites between the remaining two rows (a facetted surface, Figure 8.14(c)). A simple kinematic analysis was initially applied to the data. For a simple stepped surface (Figure 8.14(b)) this predicts that when the scattering from the up-steps and

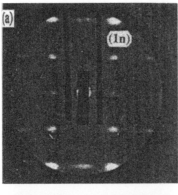

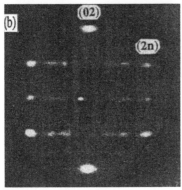

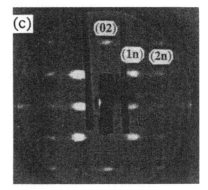

Figure 8.13   Photographs of LEED patterns obtained at (a) 94 eV, (b) 123 eV and (c) 171 eV from the Be (1120) surface. The orientation of the surface corresponds to Figure 8.14 below. Reproduced by permission of Elsevier Science from Hannon *et al.* [25]

down-steps is in phase, integral order diffraction spots will appear unsplit, whilst when the scattering is completely out-of-phase, the spot will be split in a manner related to the terrace width. The model therefore predicts that the split-spot intensity will oscillate completely out-of-phase with the integer-order spot intensity, as is observed. For this model, a value of the step height of 1.62 Å, 40 % larger than the interplanar spacing in the bulk, can be obtained. For the facetted surface (Figure 8.14(c)) the dependence of spot intensity on beam energy is more complex, and in particular the intensity of integral order spots does not vanish. The kinematic analysis therefore supports the missing row model (Figure 8.14(b)). However, as this treatment ignores multiple scattering effects, the analysis cannot be taken as definitive proof. Total energy calculations showed that in fact the facetted surface is likely to be more stable than the missing row structure, so a full dynamical calculation is necessary to resolve this problem.

The next example is one where a full multiple-scattering calculation has been carried out, and concerns the geometry of overlayers of Sb on

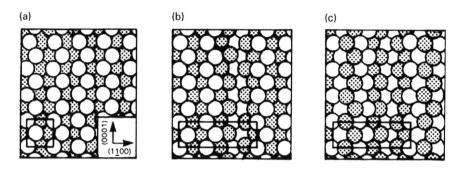

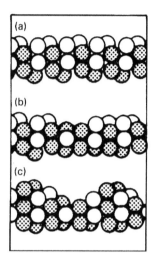

Figure 8.14 Top and side views of: (a) the bulk termination of Be (11̄20); (b) proposed surface structure based on the removal of every third surface chain; (c) the facetted surface produced by moving every third surface chain to the bridge site of the remaining two surface chains. Some atoms are shown hatched, merely to distinguish them. Reproduced by permission of Elsevier Science from Hannon *et al.* [25]

GaAs(110) [26]. Here the calculation was undertaken to distinguish between two alternative models proposed for the surface structure. Sb forms a simple overlayer on GaAs(110), denoted GaAs(110)-p(1×1)-Sb. The two proposed structures for this overlayer are shown in Figure 8.15. In the experiments, normal incidence LEED $I(V)$ curves were obtained at 2 eV intervals in the range ≈ 50 eV–300 eV. Figure 8.16 shows some of the resulting $I(V)$ curves, for the most intense diffracted beams, compared with the best fit to the data which can be obtained from the multiple scattering calculation. A number of geometries were tested, including those shown in Figure 8.15. Very satisfactory R-values (of around 0.2) were obtained for the model shown in Figure 8.15(a), but for the model in Figure 8.15(b), and for a disordered

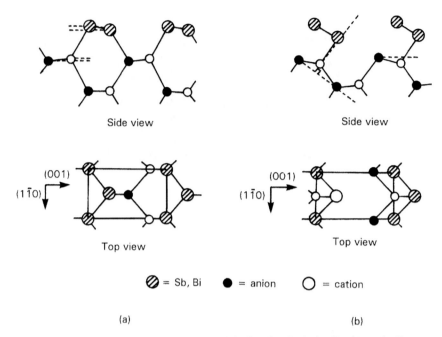

Figure 8.15 (a) and (b): two proposed models for the GaAs(110)-p(1 × 1)–Sb system. Reproduced by permission of the American Physical Society from Ford *et al.* [26]

model, *R*-values were unsatisfactory (>0.3). It was therefore concluded that the geometry shown in Figure 8.15(a) is the most probable atomic geometry for Sb/GaAs (110).

The two examples given so far relate to ordered surface structures or adsorbate systems. In many cases, however, adsorption at a surface does not occur in an ordered way and clusters of adatoms may be present in different binding sites and there may be no long range order. This gives rise to a diffuse intensity distribution in the LEED pattern. Rather than an isotropic background the diffuse LEED (DLEED) may be anisotropic and then this contains information regarding the local structure in adsorbed clusters. Modern LEED equipment is capable of measuring this diffuse background as shown in Figure 8.17 and enables the determination of local structure in the case of disordered adsorption. The W(100) surface exhibits a substantial reconstruction in c(2×2) symmetry when clean. The reconstruction is lifted when, for example, oxygen is adsorbed to a coverage of ∼20 % of a monolayer. As the c(2×2) superstructure spots disappear, strong diffuse intensities appear as shown in the upper left of Figure 8.17 together with the three-dimensional presentation for one quadrant in the middle. The relatively large coverage is likely to cause correlations between different adsorption clusters and so the *Y*-function (a function related to the Pendry *R*-factor – see Heinz [6] and Heinz *et al.* [27] for further details)

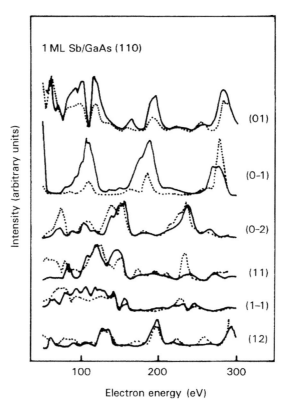

Figure 8.16 *I(V)* curves for the most intense diffracted beams for the GaAs(110)-p(1×1)−Sb system (solid lines), compared to the best fit to the data which can be obtained from multiple scattering calculations. A good fit is only obtained for the model shown in Figure 8.15(a). Reproduced by permission of the American Physical Society from Ford *et al.* [26]

has to be used for the retrieval of the local structure. It is displayed in the upper right panel of Figure 8.17 again for one quadrant and is apparently rather unstructured. Below, the calculated best-fit Y function is displayed which compares very well to the experiment corresponding to a Pendry *R*-factor, *R* = 0.05. This is achieved by oxygen adsorbed in the hollow site and variation of both the adsorption height and a local reconstruction of the substrate according to the atomic model given in the lower left of Figure 8.17. The *R*-factor map, also displayed in the figure, shows that oxygen is adsorbed at a height of 0.59 Å and induces a diagonal local reconstruction of tungsten atoms with an amplitude 0.15 × $\sqrt{2}$ Å = 0.21 Å.

Finally we note that the combination of LEED and STM measurements offers the possibility of solving complex surface structures which may be impossible by using only an imaging- or diffraction-based technique. The power of this experimental combination has been recently demonstrated

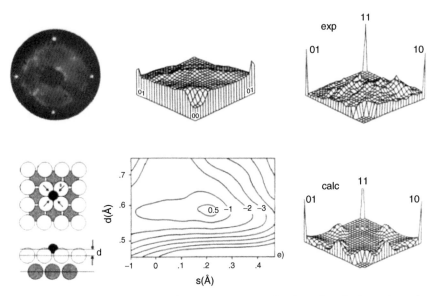

Figure 8.17   Retrieval of the local adsorption structure for disordered adsorption of oxygen on W(100). Top (from left): diffuse pattern at 41 eV, measured intensity distribution and Y-function for one quadrant. Bottom (from left): local adsorption model, R-factor map as a function of oxygen adsorption height and substrate atom displacements (in the x- and y-directions), best fit Y-function. Reproduced by permission of the Institute of Physics from reference [6]

[28] for a number of surface reconstruction and adsorbate systems and it is anticipated to be a growth area in structural studies of surfaces.

8.2.3 REFLECTION HIGH ENERGY ELECTRON DIFFRACTION (RHEED)

Introduction.   An alternative electron diffraction technique for the determination of surface structure is Reflection High Energy Electron Diffraction (RHEED). Here, a relatively high energy electron beam (5–100 keV, electron mean free pathlength 20–100 Å) is used, but the electron beam is directed towards the sample at a very grazing angle of incidence. Surface sensitivity is obtained because the component of the incident electron momentum normal to the surface is very small (even though the electron mean free pathlength is longer than for the low energy electrons used in LEED), and the penetration of the electron beam is small. The high energy electrons are scattered through small angles, and sample only the first one or two atomic layers of the target material under these conditions. RHEED has been developed during the last forty or so years, alongside LEED, and in parallel with the development of UHV technology. There are some disadvantages to the technique when compared with LEED, which have resulted in RHEED being less widely and generally used than LEED. There are, however, specific applications where the RHEED technique offers unique advantages and this is discussed in Section 8.2.3. The grazing incidence arrangement means that RHEED is more sensitive to surface roughness on an atomic scale than LEED (where roughly normal angles of incidence are usually used); this property is exploited in the applications of the technique.

Theoretical Considerations.   The major difference in appearance between RHEED and LEED patterns is caused by the higher energy (small wavelength, large $|\mathbf{k}_o|$, Equations (8.1) and (8.11) of the incident beam in RHEED. The Ewald sphere (Figure 8.5) is now very large compared to the reciprocal lattice vectors. This means that the Ewald sphere cuts the (0 0) rod almost along its length (Figure 8.18). In the resulting RHEED pattern, the (0 0) rod will give rise to a long streak, rather than a spot. Other reciprocal lattice rods which intersect the Ewald sphere will also give rise to streaks, but in practice the sphere is so large that these are very few. In order to explore the arrangement of reciprocal lattice rods in three dimensions, it is necessary to change the angle of incidence, ϕ, so that additional diffraction

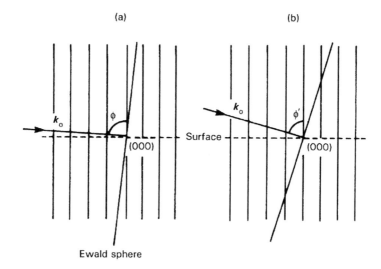

$\mathbf{k}_o$ = incident wave vector

ϕ, ϕ' = angles of incidence

Figure 8.18   The Ewald sphere construction of RHEED, at two slightly different angles of incidence, ϕ (a) and ϕ' (b). Because the sphere is very large, the sector shown is almost linear, and the angle of incidence relative to the surface normal must be reduced (as in (b)) to cut reciprocal rods away from (0, 0)

conditions may be satisfied, as the Ewald sphere cuts other reciprocal lattice rods (Figure 8.18). For most general purposes, this requirement to change the diffraction geometry is something of a disadvantage when compared with LEED, where a large number of reciprocal lattice rods may be probed simultaneously, allowing a surface unit mesh size and arrangement to be quickly and easily obtained. A change in ϕ is usually obtained by rocking the sample about an axis in its surface, although this has the disadvantage that it changes the component of the incident electron beam normal to the surface, and thus changes the surface sensitivity of the technique during an experiment. As an alternative, the sample may simply be rotated about its surface normal, in which case, well-defined streak patterns will be produced from a single crystal surface when the incident electron beam lies along high symmetry directions of the surface.

The RHEED pattern may be used to obtain the size of the surface unit mesh. Taking s as the separation between the streaks in the pattern, it can be seen from Figure 8.19 that the diffraction angle, θ, is given by

$$\tan\theta = s/L \tag{8.19}$$

where L is the distance between the sample and the screen. As θ is very small in RHEED, it may be difficult to measure s. For this reason, RHEED cameras are constructed with L as long as possible, as this increases the streak separation on the screen (Equation (8.19)). Using the diffraction condition obtained in Section 8.2.2 (Equation (8.17)) for a square surface unit mesh, and remembering that in RHEED, θ is very small, so that $\sin\theta \approx \tan\theta$, gives:

$$a = (h^2 + k^2)^{1/2}\lambda(L/s) \tag{8.20}$$

allowing a to be determined.

Due to the finite energy spread of the incident beam, and its angular divergence, the RHEED beam has a limited coherence length at the surface, analogous to that in LEED. This is longer than the corresponding length in LEED, and is typically of the order of 2000 Å. The use of grazing incidence in the experiment means that RHEED is extremely sensitive to surface roughness on scales smaller than the coherence length. It thus can be very difficult to obtain a surface which is sufficiently flat to give a clear pattern of streaks. In some cases, the high energy RHEED beam may pass through any protuberances on the surface, giving rise to bulk diffraction and hence spots, rather than streaks, are observed in the RHEED pattern.

As in the case of LEED, information may be obtained about the arrangement of atoms within the surface unit mesh from an analysis of the variation in intensity of the diffraction features as a function of change in the diffraction conditions. The variables in this case are the angle of incidence, ϕ, or the azimuthal angle, ψ (Figure 8.19). Intensity data is usually in the

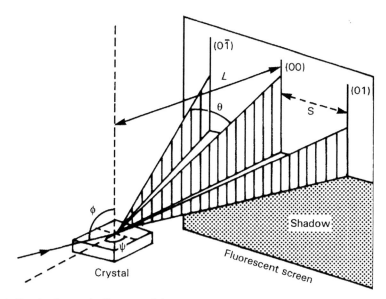

Figure 8.19 A schematic diagram of the experimental geometry of a RHEED experiment: ϕ is the angle of incidence (which is close to 90°), ψ is the azimuthal angle and θ is the diffraction angle. Reproduced from reference [3], Copyright 1994, Oxford University Press

form of 'rocking curves', $I(\phi)$, as the sample is rocked about an axis in its surface, changing the angle of incidence, or rotation diagrams, $I(\psi)$, produced as the sample is rotated around its surface normal. The cross-section for elastic scattering at the high beam energies used in RHEED is smaller than in LEED, which leads to less multiple scattering in the diffraction process. Unfortunately, however, the cross-section for inelastic scattering is also smaller than in LEED, which leads to a longer electron mean free pathlength. This means that diffracted beams may travel a longer distance through the solid before losing their coherence by inelastic scattering. The net effect is that any accurate description of the intensities of the diffracted beams must include multiple scattering effects and, as for LEED, a *dynamical* theoretical treatment is essential.

Experimental Details. The experimental geometry employed in the technique is illustrated in Figure 8.19. A high energy (5–100 keV) fine parallel beam of electrons is incident on the surface at a very grazing angle of incidence (near $\phi = 90°$). A RHEED pattern may then be collected in a similar manner to LEED, for example with a retarding field analyser (RFA), so that electrons which have lost energy of more than a few electronvolts are removed. The elastically diffracted electrons may then be detected by a fluorescent screen and photomultiplier. Alternatively, the intensity of specific diffracted beams may be monitored, for example by using an optical fibre with a well-collimated entrance aperture and a photomultiplier. Scanning High Energy Electron Diffraction (SHEED) attachments are available, which enable the diffraction pattern to be scanned in a raster way in order to build up a map of intensity of the RHEED spots along various directions in the diffraction pattern.

Applications of RHEED.   RHEED is extremely sensitive to surface roughness and so has been extensively used in the study of thin surface coatings, surface disorder and processes such as surface passivation and hardening of metals. One of the most important applications has been its use to monitor the layer-by-layer growth, necessary for the production of semiconductors (e.g. GaAs) by molecular beam epitaxy (MBE). During MBE deposition, gated sources or effusion (Knudsen) cells of the elements are used to lay down successive alternating single atomic layers of the components (e.g. Ga and As in the case of GaAs) in a UHV environment. The ability to deposit material in this layer-by-layer (Frank–Van der Merwe or FV) manner is essential to the process, but can be a difficult process to control.

RHEED has been shown to be a particularly reliable monitor of FV growth. In this experiment, the intensity of the specular diffracted beam is simply monitored as a function of time during the layer growth process [29, 30]. The intensity exhibits very regular oscillations as a function of time (Figure 8.20). In the example shown (the growth of GaAs (001)) control of the Ga beam in the presence of a continuous flow of As is crucial, as the rate determining step is the sticking of As in the MBE growth process. The period of the oscillations corresponds exactly to the growth rate of one layer of GaAs in the [001] direction (i.e. one layer of Ga plus one layer of As). The maxima in the reflectivity correspond to atomically smooth surfaces, i.e. before deposition of a layer, and when deposition is complete ($\theta \approx 0$ and $\theta \approx 1$ in Figure 8.20). The reflectivity minima correspond to the most disordered surfaces, i.e. $\theta \approx 0.5$. As can be seen in Figure 8.20, the intensity of the oscillations progressively decreases as the overall surface roughness gradually increases as more and more layers are deposited. The RHEED technique for monitoring MBE growth is simple, practical and accurate. It is particularly well suited to the geometry of the MBE process – as the Knudsen effusion cells producing the molecular beams of the elements (e.g. Ga and As) generally occupy positions fairly directly in front of the substrate, LEED cannot be used to monitor the growth. The grazing geometry of RHEED, however, means that it is one of very few techniques which can be used inside the MBE machine without interfering with the growth process.

In addition to semiconductor growth processes RHEED has also been used to obtain real-time information on the top monolayers of a surface during surface phase transitions [31] and, recently, during the growth of liquids onto solid substrates [32]. Vapour condensation is a process that

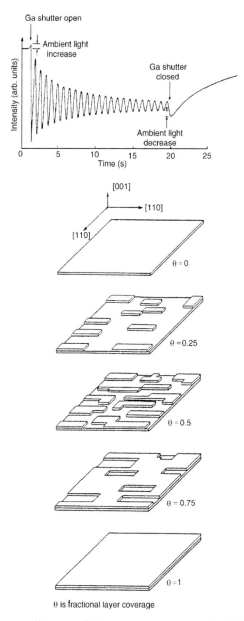

Figure 8.20 Intensity oscillations of the specular beam in the RHEED pattern from a GaAs(001)-(2×4) reconstructed surface, during semiconductor growth by MBE. Intensity maxima correspond to atomically smooth surfaces ($\theta \approx 0$ and $\theta \approx 1$), while minima correspond to completely disordered surfaces ($\theta \approx 0.5$). The period of the oscillations then exactly corresponds to the growth rate of a single Ga + As layer. Inflections at the beginning and end of growth result from ambient light change as the effusion cell shutters are opened and closed. Reproduced with permission of Elsevier Science from Dobson *et al.* [30]

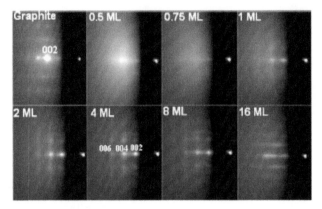

Figure 8.21   Real-time RHEED patterns taken during Bi deposition at room temperature. The graphite spot intensity decays and that of Bi starts to appear after a deposition of ~0.5 ML. The spot intensities of Bi increase with the deposited thickness up to ~8 ML. Elongated RHEED streaks at 16 ML indicate coalescence and formation of asymmetric shape crystallites. Reproduced with permission of the American Physical Society from Zayed and Elasayed-Ali [32]

has a direct impact on many surface phenomena such as the wetting transition and surface reconstruction. Using RHEED it was possible to follow the condensation of liquid bismuth onto a graphite (002) surface at temperatures far below the bulk melting point of bismuth. Figure 8.21 shows real-time RHEED patterns taken during Bi deposition at room temperature. The graphite spot intensity becomes dim at a coverage of 0.5 ML and a diffuse background appears. Indexing of the spots that appear at higher coverages indicates that they are characteristic of a rhombohedral structure of Bi and the fact that the intensity of the graphite spot decreased continuously indicated that the Bi is growing in islands. By performing these experiments at different substrate temperatures it was possible to show that the condensation of Bi followed two regimes: at low temperature (<415 K) corresponding to solid film deposition and at higher temperatures (>415 K) to liquid phase condensation. The morphology of the Bi films formed in the different growth regimes was found to be dependent on the degree of liquid supercooling.

8.3 X-ray Techniques

8.3.1 GENERAL INTRODUCTION

In Section 8.2 the diffraction of electrons from surfaces was described and it was demonstrated that this is an inherently surface sensitive process due to the strong interaction of electrons with a material. The information contained in the electron diffraction data concerned mostly the long range order of the crystalline surface, although DLEED does allow some disorder to be probed. In contrast to electrons, the interaction of X-rays with materials is relatively weak although this can be an advantage in analysing diffraction data as the kinematic approximation (single scattering) can be used. In this section the use of X-rays to probe surface structure is described. The application of X-rays in surface studies has been stimulated by the recent developments in the production of synchrotron radiation (briefly described in Section 8.3.1) as new sources provide intense beams of X-ray radiation that can be tuned across a broad energy range. The brilliance of these X-ray beams means that, despite the weak interaction with matter and, hence, the small contribution to the total scattering made by the surface atoms, it is still easily measurable. Surface X-ray diffraction and X-ray standing wave (XSW) techniques are described in Sections 8.3.3 and 8.3.4, respectively. The tuneability of the X-ray synchrotron radiation has also led to the development of the techniques Extended X-ray Absorption Fine Structure (EXAFS), its surface equivalent (SEXAFS) and near edge XAFS (NEXAFS) which are capable of probing both long range and short range order (Section 8.3.2).

Synchrotron Radiation. Synchrotron radiation is a generic term describing the radiation from charged particles travelling at relativistic speeds in curved paths caused by the application of applied magnetic fields. These days synchrotron radiation is produced in dedicated storage rings where electrons (or positrons) are kept circulating at constant energy. The radiation produced is emitted tangentially to the electron orbit confined to a narrow cone with an opening angle of $1/\gamma$ around the instantaneous velocity. The characteristic features of the radiation depend on two key parameters:

1. The cyclic frequency ω_0 of the orbiting electron.

2. γ, the electron energy in units of the rest mass energy, $\gamma = E_e/mc^2$.

Typically this gives $\gamma^{-1} \sim 1$ milliradian (mrad). In addition to the radiation obtained from the bending magnets that keep the electron beam travelling in the circular orbit, radiation is also produced from insertion devices such as wigglers or undulators situated in the straight sections of the storage ring. These devices use alternating magnetic fields to force the electron beam to follow an oscillating path. In a wiggler the radiation from different wiggles adds incoherently whereas in an undulator the electron produces coherent addition of the radiation from each oscillation. The emitted spectrum is very broad, ranging from the far infrared to the hard X-ray region. The brilliance allows comparison of the quality of X-ray beams from different sources and is defined as:

$$\text{Brilliance} = \frac{\text{Photons/second}}{(\text{mrad})^2(mm^2\text{source area})(0.1\,\%\text{ bandwidth})} \tag{8.21}$$

where the photon energy range contributing to the measured intensity is defined as a fixed energy bandwidth.

There are now many sites worldwide where synchrotron radiation research is carried out (a current listing of these sites can be found at http://www.esrf.eu/UsersAndScience/Links/Synchrotrons). At present the third generation of sources is operational and construction of the next generation of sources based on the free-electron laser is currently underway. A schematic of the key components of a typical experimental beamline is shown in Figure 8.22. The details depend on the particular application or technique that is being used but many of the components shown in Figure 8.22 are found on any given beamline. One of the key components is the monochromator which is used to select a particular wavelength with a bandwidth defined by the monochromator crystal. Typically this monochromator can be adjusted so that the energy of the X-ray radiation incident on the sample can be tuned across a defined range. The broad range of frequencies available means that synchrotron radiation may be used for a wide variety of different techniques, ranging from photoemission to X-ray diffraction, in addition to XAS, EXAFS and XSW. Figure 8.23 shows a plan of

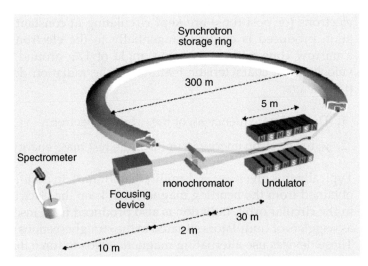

Figure 8.22 A schematic of a typical X-ray beamline at a synchrotron radiation source. Radiation from an insertion device passes through a number of optical elements, such as a monochromator, focusing devices, etc., so that a beam of radiation with the desired properties is delivered to the sample. Typical distances are indicated. Reproduced with permission of John Wiley & Sons, Ltd from Als-Nielsen and McMorrow [9]

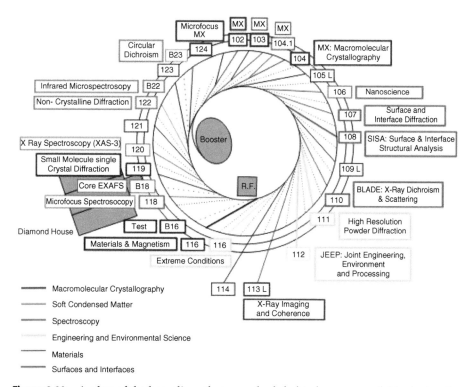

Figure 8.23 A plan of the beamlines that are scheduled to become available during the first five years of operation at the Diamond Light Source at the Rutherford Appleton Laboratory, Oxfordshire, UK. © Diamond Light Source Ltd, 2008

the beamlines that are due to become available at the Diamond Light Source at the Rutherford Appleton Laboratory, Oxfordshire, UK, which became operational at the beginning of 2007. The schematic indicates the beamlines that will become available during the first five years of operation (more details can be found at http://www.diamond.ac.uk/) and illustrates the importance of synchrotron radiation in the study of surface and interfaces.

8.3.2 X-RAY ADSORPTION SPECTROSCOPY

Introduction. The parameter used to describe the absorption of photons by a medium is the absorption coefficient α, defined according to Beer's law:

$$I = I_o \exp(-\alpha \ell) \tag{8.22}$$

i.e. a photon beam having intensity I_o at $l = 0$ is attenuated by absorption on travelling through a homogeneous medium; after travelling a pathlength l, its intensity is reduced to a value I. The extent of absorption and thus the absorption coefficient, α, depend very strongly on the photon energy; the photon beam may be transmitted right through a medium unless it is of a particular energy to cause an excitation within the medium.

In the UV/visible energy range, photons are generally sufficiently energetic to cause excitation of the outermost valence electrons of a species in solution or an atom in a solid. Examples are the d–d and charge–transfer transitions of transition metal ions in solution, which generally give rise to the colours of these solutions. In metallic solids, valence electrons may be excited into the empty density of states above the Fermi level. In semiconductors and insulating solids, photon absorption will begin when the UV/visible photons are sufficiently energetic to cause excitation of electrons from the filled valence levels across the forbidden band gap to the unfilled conduction levels. If higher energy UV photons are supplied, these electrons may be excited into an unbound continuum state above the vacuum level, i.e. ionized (Figure 8.24).

In contrast, X-ray photons are sufficiently energetic to excite core electrons, either into empty valence states, or at higher energies into the ionization continuum (Figure 8.24). The general form of the X-ray Absorption Spectrum (XAS) for an atom (such as, say, Xe) will consist of a series of core level excitation thresholds (each of which occurs close to

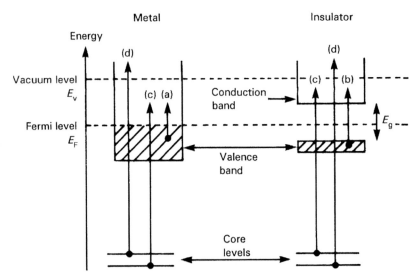

Figure 8.24   Optical absorption processes in solids for metals and insulators or semiconductors where E_g is the forbidden band gap of the semiconductor. (a) Low-energy transition from the filled to empty band states of a metal. (b) Interband transition from valence to conduction band in an insulator or semiconductor, typically observed in UV/visible absorption. (c) Higher energy transitions from filled core levels to empty bound states below the vacuum level. (d) Transitions to continuum states above the vacuum level, E_V corresponding to ionization

the ionization potential of the core level), corresponding to excitation from $n = 1, n = 2, n = 3 \ldots$ levels (i.e. K-, L- and M-edges) of the atom. In fact, although the K-edge is a single edge, higher edges are a collection of edges close together. This is because while the K-edge corresponds to excitation of electrons having $n = 1$ and orbital angular momentum $l = 0$, the L-edge corresponds to $n = 2$ ($l = 0, 1$) and states 2s, $2p_{1/2}$ and $2p_{3/2}$. The L-edge thus consists of three separate components known as L_1, L_2 and L_3. Similarly, the M-edge gives rise to M_1, M_2, M_3, M_4 and M_5 components, corresponding to states 3s, $3p_{1/2}$, $3p_{3/2}$, $3d_{3/2}$ and $3d_{5/2}$. The probability, P_{if} of an optical transition occurring between an initial state $|\psi_i\rangle$ and a final state $|\psi_f\rangle$ is given by:

$$P_{if} \propto |\langle \psi_i | \mu | \psi_f \rangle|^2 \tag{8.23}$$

where μ is the dipole operator. The contribution of each optical transition to the total optical absorption coefficient is proportional to the corresponding transition probability P_{if}. The study of the absorption coefficient as a function of photon energy, $\alpha(h\nu)$, provides important information about the initial and final states involved in the transitions. The magnitude of α is clearly strongly dependent on the overlap between the initial core level wavefunction and the final state wavefunction. After crossing the core level threshold (at the ionization energy), α decreases, with decreasing overlap of the initial state wavefunction with the photoelectron wavefunction. However, absorption into the continuum will generally continue far above the absorption threshold. The general form of an XAS spectrum for a single absorption threshold is shown in Figure 8.25.

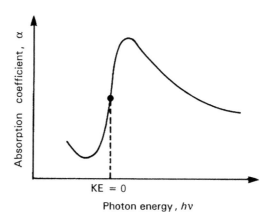

Figure 8.25   The general form of an X-ray absorption spectrum of an isolated atom for a single core-level threshold. In the treatment of EXAFS data (see Section 8.3.2, Equation (8.28)), it is necessary to know the photon energy, $h\nu$, at which the kinetic energy of the photoelectron created is zero. This is usually chosen to be halfway up the absorption edge

Figure 8.25 shows only an idealized form of an X-ray absorption edge. In general, in the case of species which are not isolated atoms, two types of fine structure are superimposed on the X-ray absorption edge. The first of these, which gives structure up to around 50 eV above the absorption edge is known as Near Edge X-ray Absorption Fine Structure (NEXAFS), or alternatively as X-ray Absorption Near Edge Structure (XANES) (Figure 8.26). This near-edge structure is determined by the details of the final density of states, the transition probability, and resonance and many-body effects and, as a result, the analysis of NEXAFS may be very complex.

In addition to the near-edge structure, the optical absorption coefficients of molecules and condensed media show fine structure and this extends from about 50 eV above the absorption threshold for several hundred eV. This second type of fine structure is known as Extended X-ray Absorption Fine Structure, or EXAFS (Figure 8.26). This structure is due to interference effects in the wavefunction of the excited electron. After absorption of a photon, this wavefunction propagates away from the atomic core where excitation occurred and is partially backscattered by the surrounding atoms in the medium. Interference between the outgoing wave and the backscattered wave produces the extended oscillations observed above the absorption edge (Figure 8.26). As the structure arises due to the presence

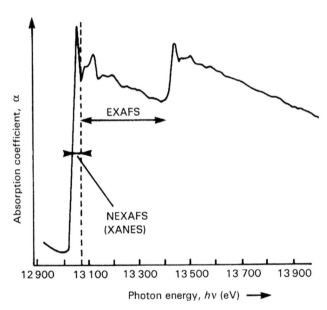

Figure 8.26 A typical X-ray absorption spectrum of a condensed medium. The sample is $BaPb_{1-x}Bi_xO_3$, and the figure shows the L_{III} absorption edge of Pb, with the NEXAFS and EXAFS regions indicated. Absorption above $\sim 13\,400$ eV is due to the L_{III} edge of Bi, which is close in energy to the Pb edge

of atoms around the core which absorbs a photon, the effect is not observed in the case of isolated atoms. In the case of molecules and condensed media, it is possible to extract information about local coordination number and coordination distances from the amplitude and period of the EXAFS oscillations. At ~50 eV beyond the absorption threshold, the final excited electron state can be regarded as a nearly free electron wavefunction, i.e. a nearly spherical wave centred on the emitting atom (like the ripples caused by throwing a stone into a pond) (Figure 8.27). The backscattering of this wave, and the resulting interference effects which give rise to the EXAFS can be described to a first approximation using atomic quantities which are independent of the chemical environment of the atom, and may be calculated or determined experimentally. The interpretation of EXAFS is thus much easier than the interpretation of NEXAFS data.

EXAFS. From the preceding sections of this chapter, you will be familiar with the idea that the mean free pathlength of low energy (say 50–500 eV) electrons in solids is extremely small – of the order of a few atomic spacings in some cases. In addition, the amplitude of the outgoing spherical electron wavefunction is inversely proportional to the radius. This means that EXAFS arises from backscattering only from atoms very close to the emitting atom, with the nearest neighbours playing the most important role. By careful

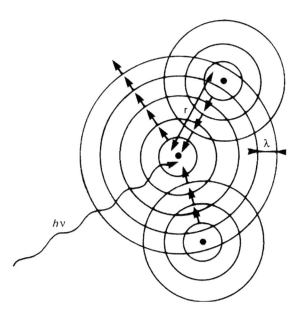

Figure 8.27 In the EXAFS process, the excited electron wavefunction (wavelength λ) propagates away from the atom at which photon absorption occurs, and is backscattered from surrounding atoms. The interatomic distance is r

analysis, some information may be obtained about the second nearest neighbour coordination shell, or sometimes about more distant shells. EXAFS is thus a very important probe of local coordination and interatomic distances. This is in contrast to the diffraction techniques where information is collected simultaneously on a large number of atoms of the system, probing the long range order. In contrast, in EXAFS, if an absorption edge belonging to a particular atomic species of interest is selected, it is possible to probe the structure specifically around that chemical species. There are many applications where such information is very important, such as metal atom coordination in supported catalysts or in biomolecules such as haemoglobin. As the EXAFS oscillations are produced only by atoms close to the emitting atom, EXAFS studies are not limited to systems having long range order. Systems which have only a well defined coordination around a central atom may thus be studied. These include polycrystalline and amorphous materials, glasses, gels and solutions: examples include amorphous semiconductors, supported catalysts and biological systems where single crystal samples may be very difficult and expensive to grow.

It has been shown in a very qualitative way how the EXAFS modulations arise. Whether the interference between the outgoing and backscattered waves is constructive or destructive will depend on the phase shift between the two waves, in turn determined by the difference in pathlengths. Assuming, for the time being, that there is no phase shift due to the backscattering process itself, and that backscattering occurs only from the shell of nearest neighbours itself, then constructive interference will occur when the pathlength of the backscattered wave, $2r$ (Figure 8.27), is equal to a whole number of wavelengths:

$$n\lambda = 2r \qquad\qquad (8.24)$$

where λ is the wavelength of the excited electron, and n is an integer. If, on the other hand:

$$n\lambda/2 = 2r \qquad\qquad (8.25)$$

destructive interference will occur. For other values, some intermediate interference will occur. This interference modulates the amplitude of the final state wavefunction, which in turn implies modulation of the transition probability P_{if} (Equation (8.23)) and thus the absorption coefficient α. Normally, an EXAFS spectrum will be recorded as a plot of the variation of α with photon energy, $\alpha(h\nu)$. Thus, it is necessary to relate the photon energy to the electron wavelength, λ. The wavelength λ can be related to the electron kinetic energy, E, by noting that:

$$E = p^2/2m \qquad\qquad (8.26)$$

where p is the momentum, and m the mass of the electron, and remembering the de Broglie equation:

$$p = h/\lambda \tag{8.27}$$

where h is Planck's constant. The excited electron only possesses kinetic energy if it is created above the ionization threshold and at the ionization threshold itself, E is zero.
Thus:

$$E = h\nu - h\nu_{E_0} \tag{8.28}$$

where $h\nu_{E_0}$ is the photon energy at which the electron kinetic energy is zero. This is usually chosen as half-way up the absorption edge (Figure 8.25); however, there are limits on the validity of this assumption, and in practice, E_0 is used as an adjustable parameter. Combining Equations (8.26–8.28) gives the following result:

$$\lambda = h[2m(h\nu - h\nu)_{E_0}]^{-1/2} \tag{8.29}$$

In other words, the excited electron wavelength decreases as the photon energy $h\nu$ increases above the absorption threshold. Combining with Equation (8.24) gives the condition for constructive interference:

$$(2r/h)[2m(h\nu - h\nu_{E_0})]^{1/2} = n, \text{an integer} \tag{8.30}$$

This means that if an EXAFS spectrum is acquired as $\alpha(h\nu)$, the period of the EXAFS oscillations increases with $h\nu$ (an example is seen in Figure 8.28. The analysis of the EXAFS is made easier if the absorption coefficient, α, is plotted not as a function of $h\nu$, but as a function of the electron wavevector, $k = 2\pi/\lambda$. We then obtain the following:

$$k = (1/\bar{h})[2m(h\nu - h\nu_{E_0})]^{1/2} \tag{8.31}$$

$$\text{or } k = 0.5123(\bar{h}\nu - h\nu_{E_0})^{1/2} \tag{8.32}$$

where k is in Å^{-1}, and $h\nu$ is in eV. To a first approximation, the oscillations in a plot of $\alpha(k)$ have a constant, rather than an increasing period (Figure 8.28 [33]). It follows that the amplitude of the interference oscillations will depend on the number of nearest neighbours, with higher coordination numbers characterized by larger amplitudes. In addition, examination of Equation (8.30) suggests that shorter bonds (smaller r) will give rise to larger spaced oscillations (Figure 8.29 [34]).

In order to be able to work out absolute values of the bond distance and coordination number, the physics of the absorption/scattering process must be examined in more detail. Mathematically, the EXAFS oscillations are often expressed as the 'EXAFS modulation', χ, corresponding to the difference between the EXAFS-modulated part and the unmodulated part of

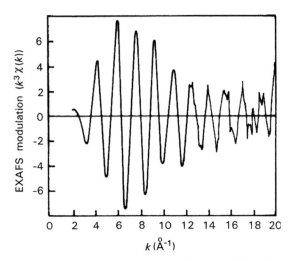

Figure 8.28 The EXAFS modulation expressed as a function of k, rather than $h\nu$, showing roughly regularly-spaced oscillations. The sample is a light-exposed As_2S_3 film. Data are taken at the As K-edge, and the primary modulation is caused by backscattering from S nearest neighbours [33]. The quantity plotted on the y- axis is $k^3\chi(k)$, and is a measure of the EXAFS modulation

the absorption coefficient, normalized to the background due to absorption by core levels at lower IP, namely:

$$\chi(h\nu) = (\alpha - \alpha^* - \alpha_0)/\alpha_0 \qquad (8.33)$$

Here $\chi(h\nu)$ is the EXAFS modulation per absorbing atom corresponding to ionization of a particular core level and α^* is the background absorption coefficient representing background due to absorption by core levels with ionization potentials at lower photon energy; α_0 is the unmodulated part of the absorption coefficient, known as the 'atomic' absorption coefficient (referring to the fact that there is no EXAFS for isolated atoms).

Assuming that the scattering can be treated within a plane wave approximation, then the EXAFS modulation for core level absorption can be written as a function of wavevector in an expression of the following type:

$$\chi(k) = \sum_i A_i(k)\sin[2kr_i + \Phi_s^i + \Phi_d^i(k)] \qquad (8.34)$$

where it is no longer assumed that backscattering occurs only from the shell of nearest neighbours: r_i is the interatomic distance to the ith shell of neighbouring atoms, and contributions to $\chi(k)$ from shells other than that nearest to the emitting atom ($i = 1$, Figure 8.30) are included by summing over a range of i values, although the major contribution is from the nearest neighbours of the emitting atom.

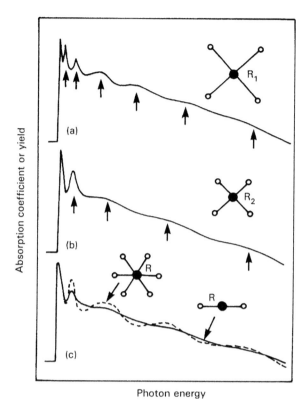

Figure 8.29 The effect of coordination number and nearest-neighbour distance on the form of the EXAFS spectrum. Shorter bonds give rise to larger spaced oscillations (compare R_1(a) and R_2(b)), whilst an increase in coordination number gives a larger amplitude (compare the traces for a two-coordinate and a six-coordinate species in (c)) [34]. Shorter bonds are characterized by larger-spaced EXAFS oscillations, and the EXAFS amplitude increases with the number of nearest-neighbour atoms

Equation (8.34) has two basic components: the sine term determines the frequency of the oscillations, while the prefactor A determines their amplitude. The phase shift difference between outgoing and backscattered waves is:

$$\Delta\Phi = 2\pi \cdot 2r_i/\lambda = 2kr_i \qquad (8.35)$$

(Figure 8.27), which appears in the argument of the sine term. The basic EXAFS modulation then corresponds to a sine function for each value of r_i. This phase shift is adjusted by further angles, Φ_s and Φ_d. Φ_s reflects the fact that the initial assumption that there was no phase shift due to the backscattering process itself was not strictly correct; it is the sum of the phase shifts induced for the outgoing and backscattered electron wavefunctions

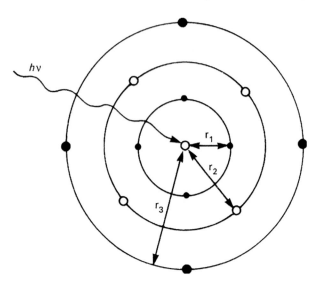

Figure 8.30 Summation over successive shells ($i = 1$, 2, 3) of neighbouring atoms in the EXAFS equation (Equation (8.34)). Nearest neighbours correspond to $i = 1$ and are predominantly responsible for the EXAFS

by the core atomic potentials of the emitting atom and the backscattering atoms. The appearance of this second angle, which is itself a function of k means that $\chi(k)$ does not in fact show oscillations of exactly constant period. A further correction to the argument of the sine function, $\Phi_d^i(k)$, may be necessary to account for deviations of the actual positions of the atoms from their ideal positions, which may arise due either to thermal motion or to structural disorder. In the case of thermal vibrations, the phase shift correction $\Phi_d^i(k)$ is zero, although as we will see, thermal motions modulate the amplitude term.

The amplitude term, $A_i(k)$ consists of a number of components:

$$A_i(k) = (\pi m_0 / h^2 k)(N_i^* / r_i^2) F_i(k) \exp(-2\sigma_i^2 k^2) \exp(-2r_i / \lambda(k)) \qquad (8.36)$$

$F_i(k)$, the atomic backscattering amplitude, is an important term, as it describes the amplitude of the EXAFS oscillations. It is dependent on the atomic number, Z, of the backscattering species, so its variation with k may be used to recognize the species around the emitting atom. $A_i(k)$ also contains two exponential terms. The first of these, $\exp(-2\sigma_i^2 k^2)$, is the Debye–Waller factor, and describes the amplitude correction due to thermal displacement of the backscatterer relative to the emitting atom. σ_i is the mean square deviation from the ideal distance, r_i. A more complex correction may be necessary in cases where deviations from ideal distances arise due to structural disorder. The second exponential term, the attenuation factor, describes the fact that the excited photoelectron has only a short mean-free-pathlength, λ in the solid. This causes an $\exp(-2r_i/\lambda(k))$ attenuation in the intensity of the electron wave over the path $2r_i$ (in accordance with Beer's law). N_i^* is the effective coordination number, given by:

$$N_i^* = 3 \sum_i^{N_i^{j=1}} \cos^2 \theta_j \qquad (8.37)$$

The summation extends over all j neighbours in the ith shell. The angle θ_j is the angle between the electric vector of the light, and the vector connecting the absorber and the backscatterer. Electron emission is strongest in the electric vector direction. This means that if the electric vector is pointing along the axis between the emitter and the backscatterer, the backscattering amplitude is at a maximum. This may be used to great effect in studies of ordered materials, such as single crystals, or their surfaces. However, the effect is averaged out in media having no long range order, such as liquids, gels and amorphous samples. In this case, N_i^* is simply equal to N_i, the true coordination number.

The basic method for detecting EXAFS is to perform an optical absorption measurement. In principle, the absorption coefficient can then be measured (via Equation (8.22)) by measuring the intensities of the incident and emergent beams in parallel, if the thickness of the sample is known. X-ray absorption measurements themselves are not new; as early as the 1930s measurements were being carried out using a continuum source of X-rays, a dispersive spectrometer and a film detector. However, the technique progressed very little until the advent of synchrotron radiation sources. The most usual method of measurement of the intensities of the incident and transmitted light uses gas-filled ion chambers (Figure 8.31). The incident synchrotron radiation ionizes the gas in ion chamber 1 before passing through the sample (the ion count provides a measure of I_0). The transmitted intensity, I, is measured by ion chamber 2. It is very important that I_0 and I are measured simultaneously, allowing for cancellation of sudden changes

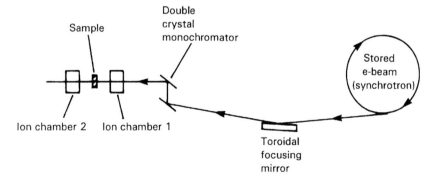

Figure 8.31 Schematic experimental arrangement for transmission EXAFS

in intensity of the incident photon beam. Even using a synchrotron radiation source, EXAFS data may take some time to accumulate; in general, the data is best acquired in the form of several short scans of the entire spectral range, rather than a single slow scan, as this minimizes the effects of a sudden step loss in the beam current in the storage ring.

In a transmission EXAFS experiment, the absorption coefficient of the entire system is measured, as α at a given photon energy is affected by all the absorption processes whose thresholds occur at lower photon energies. This can make it very difficult to measure the EXAFS of a dilute atomic species, as the EXAFS for that species is a very small modulation of the total absorption coefficient. An important alternative is fluorescence detection. This relies on detection of the X-ray fluorescence which occurs when the core hole created in the initial X-ray absorption process is filled by electron decay from an upper, filled level (Figure 8.32. The energy of the photon emitted is dependent on $(E_v - E_c)$, and is thus characteristic of the element in question. It is thus possible, using an energy dispersive X-ray detection system, to single out the absorption coefficient of the species of interest by selecting one of the characteristic emission frequencies for that species. For small values of the absorption coefficient, α_i, for the species, the X-ray fluorescence is proportional to α_i. Using the enhanced sensitivity which fluorescence detection gives, it is possible to detect isolated impurity atoms with concentrations of 10^{19} cm^{-3} or less.

EXAFS does not rely on long range order within a sample and has thus been one of the work-horses of materials science, being applied to a wide range of applications (see, for example, Hasnain [35]). These include determination of metal ion coordination in biomolecules, such as ferritin and

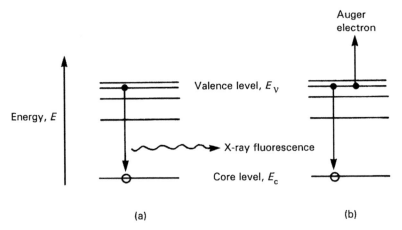

(a)   (b)

Figure 8.32   (a) The core hole created in XAS may be filled by an electron from a higher level, with the excess energy emitted as X-ray photons. (b) Alternatively, the excess energy may be dissipated by the emission of a second, Auger electron

haemoglobin, studies of glasses and gels, determination of metal cluster size in supported metal catalysts, studies of impurity sites in semiconductors and in naturally occurring minerals, and studies of local coordination in complex materials, such as high temperature superconductors. The purpose of this chapter is to discuss surface structure determination, however, and one way in which the surface sensitivity of EXAFS may be improved, without moving into a UHV environment as with SEXAFS (discussed below), is simply to use a glancing angle of incidence. The EXAFS oscillations may then be picked up using either fluorescence detection, or, if the angle of incidence is small enough, in the externally reflected beam (rather than the transmitted beam, as with conventional EXAFS). By varying the angle of incidence, the depth of the surface sampled may be varied from nanometres to millimetres. This technique is known as glancing angle XAFS.

Extension to Surfaces – Surface EXAFS SEXAFS).   SEXAFS is the surface-sensitive modification of EXAFS, and may be used for the determination of local surface geometry and coordination, for example of molecules chemisorbed at a surface. As will be shown, the technique can be made surface specific, rather than simply surface sensitive (like glancing angle XAFS) and, as such, provides detailed information on the nature of adsorption sites in the topmost plane of the surface. The development of SEXAFS was driven by the availability of synchrotron radiation and an interesting account of the development phase can be found by Citrin [36].

The surface of a material represents a very small part of the solid (for a single crystal, around 1 ppm for a 1 mm thick crystal). The main problem to be overcome in obtaining SEXAFS is thus one of obtaining sufficient sensitivity to the surface species. The usual transmission technique is not suitable, as the signal from the bulk of the crystal dominates. A number of alternative techniques are available which measure the absorption coefficient indirectly, and which give enhanced sensitivity in SEXAFS. All the methods require the use of ultra-high vacuum instrumentation. Several methods rely on the decay of the core hole created in the absorption event. These include fluorescence detection (as used in EXAFS), Auger yield and total electron yield detection. The core hole created in the initial absorption process may decay with the emission of a photon (X-ray fluorescence) or the hole may decay via the ionization of another Auger electron (as described in Chapter 4, see also Figure 8.32(b)). The relative proportions of the two decay pathways depend on the atomic number of the species in question, with Auger electron production being favoured for the lighter elements. The energy of the released Auger electron is dependent on the energy difference $(E_v - E_c)$, and thus Auger emission occurs at characteristic energies for each element. The yield of Auger electrons is proportional to the number of core holes produced, and hence to the absorption coefficient of the species, $\alpha_i(h\nu)$.

As discussed in previous chapters, and in Section 8.3.2, the mean free pathlength of low energy electrons in solids is of the order of a few atomic spacings. Thus, the Auger electrons being detected come mainly from the top 10 Å of a solid which gives the method its surface sensitivity. The Auger electrons are detected in an electron energy analyser similar to those used in XPS experiments described in Chapters 3 and 4. A related detection technique is total electron yield detection, in which *all* electrons emitted from the sample are detected. In the main, these are electrons which were emitted as Auger electrons, but which have lost varying amounts of energy in inelastic scattering processes in the solid. The total electron yield is thus also proportional to the absorption coefficient. One disadvantage of the technique compared with Auger detection is that it is not element specific, as all absorption processes contribute to the low energy tail of 'secondary' electron emission. The technique is surface sensitive for the same reasons as Auger detection. However, most of the electrons collected have very low energies and have a longer mean free pathlength in the solid ($\approx$ 50 Å) than the primary Auger electrons. In addition, the mean free pathlength changes rapidly at low energies, so that the surface sensitivity of detection can be varied by changing the energy range of the detection window. A further advantage of total electron yield detection is that it requires only a biased metal plate to collect the electrons, coupled to a picoammeter.

Although both Auger yield and total electron detection are *surface sensitive*, neither is specific *only* to surfaces. The surface signal may be very difficult to disentangle from the bulk signal. Complete surface specificity is only obtained by ensuring that the adsorbate species studied is present only at the surface. The signal in SEXAFS experiments may be very low, necessitating long data accumulation times (several hours). As discussed in earlier chapters, surface-sensitive techniques of this type require the use of ultra-high vacuum techniques, but even with pressures in the 10^{-10}–10^{-11} mbar range, it may be difficult to avoid unwanted surface contamination for the duration of data accumulation.

One further technique which may be used is ion yield SEXAFS. Figure 8.32(b) shows that following emission of an Auger electron, the emitting atom is left in a very unstable doubly positively charged state. This state may be produced via either inter- or intra-atomic process, and the positive ions produced at the surface may spontaneously desorb, and can be detected using a time-of-flight mass spectrometer. Due to the high probability of re-neutralization, and the very low escape depth for ions, only ions produced on the surface can escape into the detector, giving the technique inherent surface specificity. However, the low yield for the desorption process means that it cannot be used to study dilute surface species, and competition with fluorescence decay for heavier elements means that the ion yield method is only applicable to the lighter elements.

Although in principle, SEXAFS probes local structure, and so can be used to study amorphous and polycrystalline materials, in practice these types of surface give rise to a large amount of random electron scatter, so a high photon flux is necessary to carry out these experiments. In the case of a single crystal material, we saw from Equations (8.36) and (8.37) that the backscattering amplitude from a particular atom is at a maximum if the electric vector of the light is pointing along the axis between the emitter and the backscatterer. Thus by rotating the crystal sample to vary the angle between the electric vector and the surface, and observing the effect on the SEXAFS, the geometry of, say, an adsorption site can often be determined. This approach has been quite widely adopted, for example, in the determination of the coordination site of sulfur on Ni (110)-c(2×2)-S [37], and the coordination site of O on Ni(111)p(2 × 2)-O [35]. Figure 8.33 shows the SEXAFS data of Haase *et al.* [38] for the latter system, taken at the oxygen K-edge, and recorded in partial electron yield mode, together with the corresponding Fourier transforms. Both normal incidence ($\theta = 90°$) and grazing incidence ($\theta = 20°$) have been used. The plots of $\chi(k)$ versus k show nearly sinusoidal forms, and the corresponding Fourier transforms are dominated by a single peak, corresponding to the nearest-neighbour O–Ni distance of around 1.85 ± 1.03 Å. In the normal incidence data, there is a small second peak (labelled R_2), which can be tentatively assigned to the next-nearest neighbour O–Ni distance, corresponding to around 3.1 Å. Having determined the O–Ni nearest-neighbour distance, the amplitudes of the SEXAFS oscillations expected in both polarizations for different coordination sites with this nearest-neighbour distance may be calculated. The ratio of the amplitudes in the polarizations is then compared with the experimentally determined value. No satisfactory agreement can be obtained for oxygen in atop or bridged sites, and the most convincing agreement by far is obtained when oxygen is in a three-fold site. Further information about whether the surface is reconstructed around the three-fold site can be obtained from an analysis of the coordination shells outside the first. Due to the polarization dependence, Ni atoms in the second layer of the sample do not contribute to the normal incidence SEXAFS. However, the grazing incidence data are affected by the separation between the first and second Ni layers and so should be capable of distinguishing between *fcc* and *hcp* three-fold coordinated sites. Best agreement is obtained for the *fcc* site, shown in Figure 8.34.

SEXAFS can be particularly useful when combined with other surface structure techniques. For example, Cheng and co-workers used a combination of SEXAFS and XSW (see Section 8.3.4) to investigate the adsorption and incorporation of Co^{2+} ions at the surface of calcite [39]. Figure 8.35 shows the polarization-dependent SEXAFS results for Co incorporated at the calcite (10$\bar{1}$4) surface after adsorption from a dilute water solution.

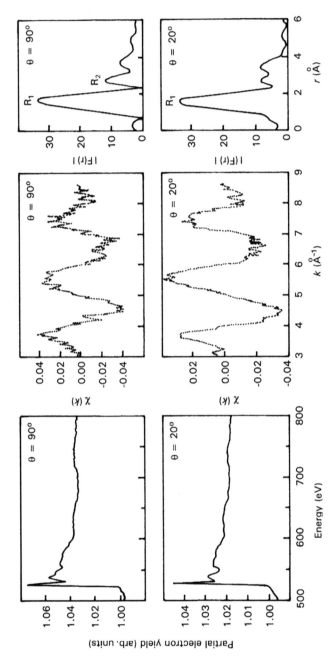

Figure 8.33 Oxygen K-edge SEXAFS data for Ni(111)p(2×2)–O, for both normal inci-dence ($\theta = 90^0$) and grazing incidence ($\theta = 20^0$). Raw data as function of photon energy is shown (left), together with $\chi(k)$ versus k plots (centre) and finally the Fourier trans-forms of the data (right). Reproduced with permission of Elsevier Science from Haase *et al.* [38]

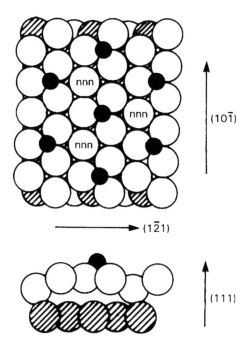

Figure 8.34 Top and side views of the Ni(111)p(2×2)-O system, with oxygen atoms (the small filled circles) in three-fold coordinated *fcc* sites, suggested by the SEXAFS data of Figure 8.33. Next-nearest-neighbours are marked 'nnn'. Reproduced from Haase *et al.* [38], Copyright 1992, Elsevier Science

The open circles correspond to background-subtracted, k^2-weighted data from the in-plane and the normal polarization measurements (data from a powder reference sample $CoCO_3$ are also shown). The fits to the data give the first shell bonding information of the coherently incorporated Co^{2+} and its nearest neighbour O atoms of the surrounding CO_3^{2-} groups. XSW measurements on the same surface were used to determine the Co lattice sites (see Section 8.3.4) and the combination of the results yield a model for the local structure of the Co^{2+} atom as shown schematically in Figure 8.36. Similarly to the combined use of imaging techniques and LEED described in Section 8.2.2, by combining two methods for structural analysis (utilizing the relative strengths of each experimental technique) it is possible to build a detailed model of the surface structure.

NEXAFS (XANES) in Surface Structure Determination. In Section 8.3.2, it was noted that NEXAFS may be used to yield structural information, but that the analysis of NEXAFS data can be complex. In the EXAFS part of the spectrum, the interaction between the photoelectron wave and the backscatterer is relatively weak, so single scattering predominates. However, in the NEXAFS region, the interaction is much stronger, and

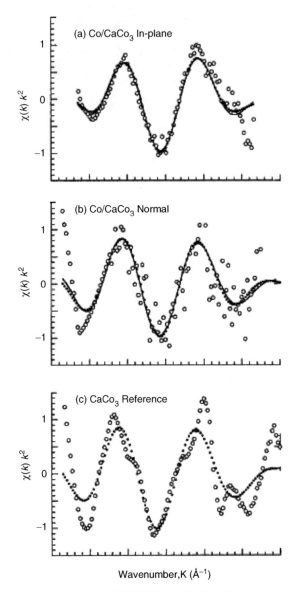

Figure 8.35 Polarization-dependent SEXAFS results for Co incorporated at the calcite (10$\bar{1}$4) surface. Raw data after background subtraction (open circles), the Co-O first shell component (dotted lines) and the best fit to the first shell component (solid lines) for (a) the in-plane, (b) the out-of-plane polarization measurements and (c) the CoCO$_3$ standard compound. Reproduced with permission from the American Physical Society from Cheng *et al.* [39]

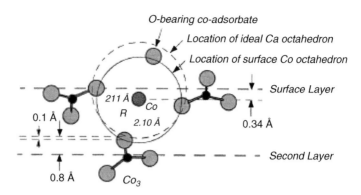

Figure 8.36 Side view of a model describing the local structure of the Co^{2+} ion incorporated at the calcite ($10\bar{1}4$) surface. Reproduced with permission from the American Physical Society from Cheng *et al.* [39]

multiple scattering of the outgoing photoelectron must be considered. As the photoelectron wavefunction is strongly dependent on the form of the potential around the absorber, in principle NEXAFS can be used to investigate the oxidation state of the absorber and its coordination geometry. Investigation of the oxidation state is relatively straightforward, as the position of the absorption edge changes as the charge on the absorber changes (i.e. there is a chemical shift in the edge position). For small molecules adsorbed on surfaces, the NEXAFS at an adsorbate core level threshold is dominated by resonances within the adsorbed molecule, which means that the technique may sometimes be very powerful in this type of study. However, for atomic adsorbates, particularly those on metallic substrates, long range multiple scattering dominates, so that modelling the spectra then requires intensive multiple scattering calculations. As an example, in a study of Ni(110)-c(2 × 2)-S [40], five shells of Ni atoms surrounding a surface S atom (corresponding to 42 atoms) needed to be included in the calculation in order to obtain good agreement with experiment, indicating significant multiple scattering within this cluster. In the corresponding SEXAFS measurement from the same system, only one shell of Ni atoms (four atoms) contributed significantly.

The polarization dependence of NEXAFS can be used in the same way as that in SEXAFS, for example to determine the orientation of a small molecule on a surface. A good example of this is the adsorption of CO on Ni(100) [41]. In this study, NEXAFS at the C K-edge is observed, using C-KVV Auger emission detection. The NEXAFS in this case is dominated by intermolecular excitations, within the CO molecule. When the photons are incident at $10°$ to the (100) plane, two peaks are observed (labelled A and B in Figure 8.37). As the angle of incidence is increased to $90°$, peak B disappears. The two peaks are due to transitions from the C 1s level to the $2\pi^*$ (A) and σ^* (B) molecular orbitals of CO. (Peak B is much broader than peak A. This is because the $2\pi^*$ state is pulled below the vacuum level on ionization of the C 1s electron, so peak A is a bound state resonance. In contrast, peak B is what is known as a continuum or shape resonance, and arises because an electron excited into the σ^* orbital remains above the vacuum level and will escape from the solid. The σ^* wavefunction has a large amplitude near the C atom, but tails off rapidly to a free-electron

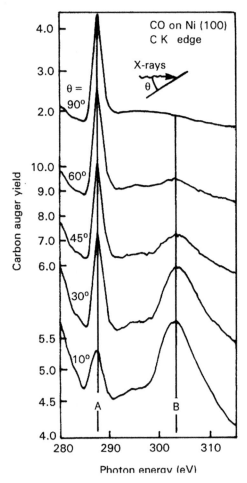

Figure 8.37　NEXAFS data above the C K-edge for CO on Ni(100) at 180 K, as a function of incidence angle. Reproduced with permission from the American Physical Society from Stöhr and Jaeger [41]

wavefunction.) Using the dipole selection rule, it is easy to show that the C 1s $(\sigma) \rightarrow 2\pi^*$ transition is only allowed if the electric vector of the light is perpendicular to the CO bond axis, whilst the C 1s $(\sigma) \rightarrow \sigma^*$ transition is allowed if the vector is parallel to the CO bond axis. This means that the intensity ratio between peaks A and B can be used to determine the orientation of the CO molecule on the surface, by comparing calculated ratios for different CO orientations with the experimentally obtained ones. This produces the result that the CO molecule is perpendicular to the Ni(100) plane, in agreement with photoemission results.

NEXAFS has recently been widely used in the study of organic molecules adsorbed onto metal surfaces [42] and semiconductor surfaces [43, 44]. This is because NEXAFS is an ideal method to determine molecular orientation, interface bonding and electronic structure and, also, by the fact that the energy resolution that has become available by the use of undulator beam-lines at the third generation of synchrotron sources has revealed new fine structure allowing detailed analysis of the electronic and chemical state. This increased energy resolution has also opened up the possibility of studying vibrational coupling to electronic excitations in large organic molecules, well established in optical and electron spectroscopies but, due to poor structural order, little studied by NEXAFS in the solid state. Such effects are nicely illustrated in the study by Scholl and co-workers [45] who reported rich fine structures in NEXAFS spectra of large organic molecules (NCTDA) adsorbed onto Ag(111). Figure 8.38 shows the C K-edge NEXAFS spectra that were obtained for the NCDTA multilayers on Ag(111). In the extended energy scale shown on the left-hand side of Figure 8.38 three prominent sharp resonances are observed and the dependence of these π^*-resonances on the polarization of the incident X-ray beam can be used to determine the molecular orientation. In the expanded energy scale on the right-hand side of Figure 8.38 a rich fine structure with numerous well-resolved peaks and shoulder is observed. These fine structures were interpreted as coupling of the electronic transitions to vibronic excitations in the core-excited final state. The results on this model system illustrate that NEXAFS can be a subtle probe for organic solids.

8.3.3 SURFACE X-RAY DIFFRACTION (SXRD)

Introduction. The initial negative assessment of X-rays as a probe of surface structure was based on the weak interaction between X-rays and solids, with the result that X-rays penetrate very deeply into solids. However, this weak interaction also means that multiple scattering effects can be ignored and X-ray diffraction can be treated by a kinematic theory. From this point of view, a technique which uses X-rays, rather than electrons, to probe surface structure should have considerable advantages over LEED and RHEED, as the theoretical description of the scattering will be much simpler, and

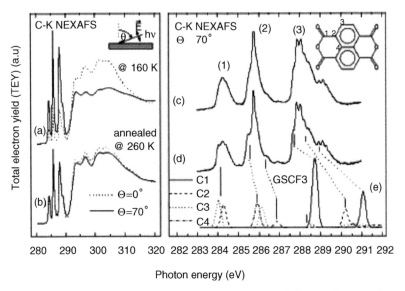

Figure 8.38 Left: C K-NEXAFS spectra for (a) NTCDA multilayers deposited at 160 K and (b) annealed at 260 K for grazing ($\theta = 70°$) and normal ($\theta = 0°$) X-ray incidence. Right: (c)–(d) π^* region of the C K-NEXAFS spectra ($\theta = 70°$) and (e) calculated NEXAFS resonances. The electronic transitions involving the atoms C1–C4 (see inset) are assigned by guidelines. Reproduced with permission of the American Physical Society from Scholl *et al.* [45]

the arrangement of atoms within the surface unit mesh may be obtained in a relatively straightforward way. In fact SXRD has become a major tool in surface structure determination due, partly, to this straightforward interpretation, but also due to the development of synchrotron radiation sources that has allowed satisfactory signal levels to be obtained even from sub-monolayer quantities of material. Furthermore, the fact that X-ray radiation is penetrating means that the surface can be probed in a buried (non-UHV) environment, such as a buried solid–solid interface, the interface between a solid and a liquid and the solid/high-pressure gas interface. This means that SXRD can be used in a broad area of science, for example ranging from *in-situ* metal organic chemical vapour deposition (MOCVD) growth processes to electrochemistry and crystal growth, but also including standard UHV surface science. For reviews the reader is referred to the general texts and review articles listed in the references.

Theoretical Considerations.   The index of refraction of solids at X-ray wavelengths is only very slightly less than unity and so, from Snell's Law, at very grazing angles of incidence the incoming X-ray beam undergoes total external reflection (Figure 8.39). For example, for a typical X-ray wavelength of around 1.5 Å the critical angle from the surface, below which total external

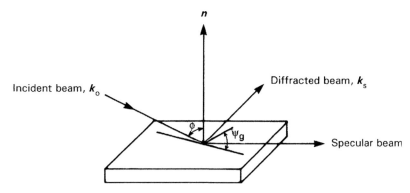

Figure 8.39 The grazing angle geometry used in surface X-ray diffraction. An incident beam of wavevector k_0 is incident at a grazing angle to the surface, such that the polar angle, ϕ, with the surface normal n is typically $\approx 89.5°$. After diffraction through an azimuthal angle ψ_B, the beam leaves the surface at a grazing angle similar to the angle of incidence

reflection occurs, is around $0.2°–0.6°$, depending on the material. Under these conditions there is a finite penetration depth of the X-ray beam into the solid, caused by absorption of X-rays by the sample and, for angles of incidence below the critical angle, the penetration depth is of the order of $10–50$ Å. This means that scattering is essentially from the reciprocal lattice vectors of the surface. The two-dimensional treatment, considered for LEED and RHEED in Section 8.2, is again appropriate, so that diffracted beams emerge wherever the Ewald sphere intersects the surface reciprocal lattice *rods* (see, e.g. Figure 8.5). The additional advantage, however, is that multiple scattering effects are negligible, so that the diffraction process may be treated by a *kinematic* rather than a dynamical theory. The two-dimensional X-ray diffraction experiment will, of course, still be subject to the phase problem but, using refinement techniques developed from conventional X-ray diffraction, atom positions can be estimated to within $\pm 0.01–0.03$ Å using Equations (8.7) and (8.8) to calculate structure factors for comparison with the measured values.

Of course in a general X-ray diffraction experiment the grazing incidence geometry is not always used and the X-rays penetrate to the bulk of the sample crystal. For a 3D crystal the summation which gives rise to the Laue equations (for an arbitrary crystal with N_1, N_2, N_3 unit cells, lattice vectors $\mathbf{a}$, $\mathbf{b}$, $\mathbf{c}$) for the scattered intensity at a wavevector $\mathbf{q}$ in the reciprocal lattice can be written as:

$$I(\mathbf{q}) \propto |F(\mathbf{q})|^2 \frac{\sin^2(\tfrac{1}{2}N_1\mathbf{q}\cdot\mathbf{a})}{\sin^2(\tfrac{1}{2}\mathbf{q}\cdot\mathbf{a})} \frac{\sin^2(\tfrac{1}{2}N_2\mathbf{q}\cdot\mathbf{b})}{\sin^2(\tfrac{1}{2}\mathbf{q}\cdot\mathbf{b})} \frac{\sin^2(\tfrac{1}{2}N_3\mathbf{q}\cdot\mathbf{c})}{\sin^2(\tfrac{1}{2}\mathbf{q}\cdot\mathbf{c})} \tag{8.38}$$

where $F(\mathbf{q})$ is the structure factor that determines the points in the reciprocal lattice at which diffraction peaks are observed. For scattering from an isolated 2D monolayer, this is equivalent to setting $N_3 = 1$ in Equation (8.38), when the **c** vector is along the surface normal and **a** and **b** lie in the 2D plane. In this case the diffracted intensity is independent of the component of the wavevector transfer $\mathbf{q} \cdot \mathbf{c}$ in the surface normal direction of the reciprocal lattice and intensity is observed for all values of the Miller index l, when the Laue conditions are met in the plane of the surface (integer values of h and k). This is schematically shown in Figure 8.40(d) where the scattering is sharp in both directions parallel to the surface but diffuse along the surface normal. If the 2D monolayer has the same periodicity as the underlying bulk crystal then the diffuse scattering is superimposed upon the bulk scattering as indicated in Figure 8.40(e) and the spots now indicate the bulk Bragg reflections. In the case of a real surface there is actually interference between the scattering from the surface atoms and the

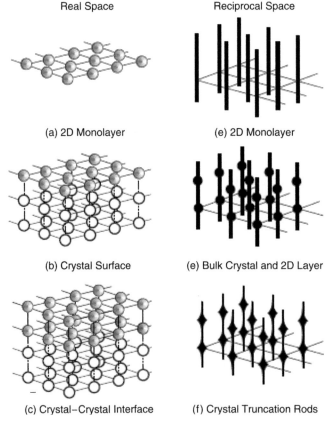

Real Space   Reciprocal Space

(a) 2D Monolayer   (e) 2D Monolayer

(b) Crystal Surface   (e) Bulk Crystal and 2D Layer

(c) Crystal–Crystal Interface   (f) Crystal Truncation Rods

Figure 8.40 Real space structures for (a) a 2D monolayer, (b) the surface of a crystal and (c) a crystal–crystal interface. The corresponding X-ray diffraction representations in reciprocal space are shown in (d), (e) and (f), respectively

bulk and the intensity becomes modulated along the l direction as shown schematically in Figure 8.40(f). The diffracted intensity in between the Bragg reflections forms the so-called crystal truncation rods (CTRs) as they are rods of scattering that arise from the truncation of the crystal lattice at the surface [10, 46]. When the in-plane Laue conditions are satisfied, the scattered intensity as a function of the surface normal wavevector transfer is then given by:

$$I(q) \propto |F(q)|^2 N_1^2 N_2^2 \frac{1}{2\sin^2(\frac{1}{2}q \cdot c)} \qquad (8.39)$$

This has sharp peaks at the integer l positions (the Bragg reflections) but the intensity between the Bragg reflections contains information regarding the surface structure and its registry with the underlying bulk lattice. The rich amount of structural information that can be obtained in a SXRD experiment can thus be understood from Figure 8.40. For a surface structure that has a different symmetry than the underlying bulk crystal, the rods (Figure 8.40(d)) are separated in reciprocal space and can be measured independently. Surface structures that are commensurate with the underlying bulk crystal can also be probed by detailed measurement and modelling of the CTRs (Figure 8.40(f)).

Experimental Details. As mentioned above, the fact that X-ray radiation is penetrating means that the surface under study can be buried or contained in a non-UHV (liquid, high pressure gas) environment. It is, however, also possible to study surfaces in the UHV environment and in this case the standard equipment for performing X-ray diffraction experiments, i.e. the diffractometer, is coupled to a UHV chamber so that the sample may be manipulated within the UHV environment. Access to the UHV chamber for the incident and scattered X-ray beams is provided by a Be window on the chamber. A photograph of the X-ray station on beamline BM32 at the ESRF, Grenoble is shown in Figure 8.41 and this instrument is representative of several other facilities at synchrotrons around the world (see also, for example, Ferrer and Comin [47]). The Surface under Ultra-high Vacuum (SUV) station on this beamline is devoted to studies of surfaces, interfaces and thin films under UHV conditions. It consists of a large UHV chamber mounted on a four-circle diffractometer. The diffractometer supports the UHV chamber with allowance for a rotation of the whole chamber defining the incidence angle of the X-ray beam with respect to the vertical sample surface. A goniometer contained inside the UHV chamber allows two perpendicular tilts and two perpendicular translations of the sample. The rotary motion of the sample on the diffractometer is obtained through the rotation of this goniometer via a differentially pumped rotary feedthrough. The UHV chamber has large Be windows that give access to large input

Figure 8.41   Photograph of the SUV station on beamline BM32 at the ESRF, Grenoble. The X-ray beam comes in through a small Be window (the right-hand side) and exits through a large Be exit window. Reproduced with permission of BM32-CRG-IF, Grenoble, France

and exit angles for the X-ray beam. The chamber is equipped with several evaporation sources for *in-situ* epitaxial deposition and also has RHEED, AES, Residual Gas Analysis (RGA) and ion sputtering capabilities.

The experimental procedures for performing SXRD experiments are well established. Typically the diffractometer is aligned so that the incident X-ray beam passes through the centre of rotation of the diffractometer axes. The sample is then aligned by finding the angular positions of two or more Bragg reflections and this allows the angular positions of the diffractometer to be related to the sample reciprocal lattice through specialist diffractometer software. The experiment can then be conducted in terms of the reciprocal lattice enabling the features shown schematically in Figure 8.40 to be measured. The rod-like scattering features, due to the surface (see Figure 8.40), are broadened in the surface plane reciprocal space direction by a combination of instrumental resolution, finite coherence of the structure itself and the mosaicity of the crystalline substrate. This finite width can be accounted for by setting the diffractometer to the desired (h, k, l) position and then rocking the sample about its azimuth. This 'rocking-scan' gives a background-subtracted integrated intensity that is representative of the scattered intensity at that (h, k, l) position. After correction for various instrumental effects (for details see, for example, Feidenhans'l [48]) the data are ready for modelling using Equations (8.7) and (8.8) and modified versions of Equation (8.39).

Applications of SXRD. The last twenty years has seen an explosion in the number of surface studies using SXRD that has been driven by the availability of beamlines at the synchrotron radiation sources. Examples include surface reconstruction of semiconductors and metals [49], oxides [50], surface phase transitions, surface roughening, surface melting and numerous studies of surface adsorption and growth processes [9, 10, 51]. As noted earlier, the penetrating nature of hard X-ray radiation means that surface studies are not restricted to the UHV environment and many of the UHV experiments have analogous experiments performed at solid–solid, solid–liquid and solid–high pressure gas interfaces [52].

To illustrate the methodology used in an SXRD experiment we use a relatively early example in which the adsorption of Sb onto Ge(111) was studied [53]. In this experiment azimuthal scans (made by rotating the crystal around the surface normal) were used to determine the in-plane structure of the Sb layer, and its registry with the Ge (111) substrate, whilst rod scans were used to determine out-of-plane displacements. Figure 8.42 shows one of the azimuthal scans around a fractional order peak (0.5,2,0.3), due to the adsorbate. A structure factor analysis of this data gives the in-plane registry of the Sb on the substrate. This showed that the Sb atoms are arranged on the surface in zigzag chains (Figure 8.42), with a nearest-neighbour distance very close to the bulk Sb—Sb bond length. The Sb atoms are slightly out of registry with the first layer of underlying Ge atoms, which implies some tilting of the downward Sb—Ge bonds. Rod-scans were used to investigate the out-of-plane structure. An example of a rod scan for the (1 0) rod is shown in Figure 8.42. Several different calculated rod scans were determined, for different models of the heights of the Sb atoms above the Ge surface. The best fit corresponds to the situation shown in Figure 8.43, where it is found that the Sb atoms are at two different heights, with half lying 0.2 Å above the other half. This buckling continues into the Ge substrate.

As the X-ray scattering power of an atom varies with the square of its atomic number, Z, many of the early studies of surface structures by SXRD were for high $-Z$ materials. Given that the surface scattering may be separated from the scattering from the bulk of the crystal, however, if the adsorbate adopts a symmetry that is different to the bulk termination of the lattice, the high intensity X-ray beams available at synchrotron sources make it feasible to study low-Z adsorbates in certain circumstances. This is nicely illustrated in studies of the adsorption of CO onto a Pt (111) surface despite the fact that this surface was studied under liquid, i.e. at the charged electrochemical interface [52, 54]. By saturating the electrolyte that covered the Pt (111) surface with CO it was possible to form an ordered CO structure

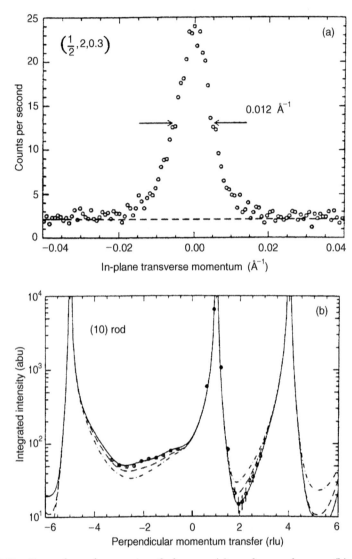

Figure 8.42 Examples of an azimuthal scan (a) and a rod scan (b), from the Ge(111)(2 × 1)-Sb surface, used to determine the surface structure shown in Figure 8.43 below. In (b), the open circles represent experimental data, while the dashed and full curves represent the fits to the data obtained for different models for the heights of the Sb atoms above the surface. The best agreement is found for the full curve, which corresponds to the model shown in Figure 8.43. Reproduced with permission of Elsevier Science from van Silhout *et al.* [53]

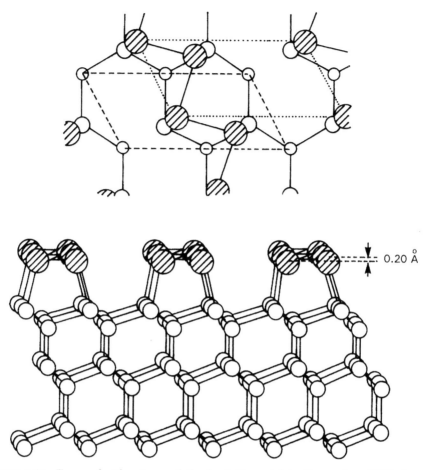

Figure 8.43 Top and side views of the best fit model to the surface XRD data of Figure 8.42, for the Ge(111)(2×1)–Sb surface. Half the Sb atoms lie slightly above the other half. Larger hatched atoms are Sb. Reproduced with permission of Elsevier Science from van Silhout *et al.* [53]

on the Pt (111) surface with a p(2×2) unit cell. Figure 8.44 shows a rocking scan through one of the p(2×2) reflections that is due only to scattering from the adsorbed CO monolayer. Inset to the rocking scan shows the intensity dependence along one of the rods of scattering due to the CO monolayer. The lack of any significant modulation implies that the adsorbed structure is essentially a 2D monolayer, as shown schematically in Figure 8.40(d). The kinematical nature of SXRD means that Fourier techniques can be used in the structure determination. In this case the measured structure factors can be used to calculate the Patterson function which relates to the electron distribution in the surface unit cell [10, 48]. The straightforward nature of the Patterson function (Figure 8.44) allows a structural determination to be

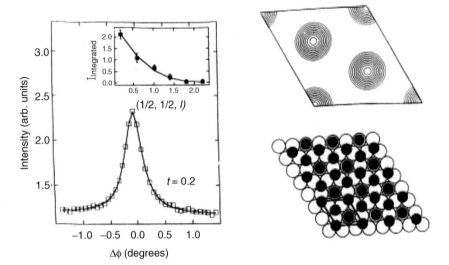

Figure 8.44 (left) A rocking scan through the (1/2, 1/2, 0.2) reciprocal lattice position which shows a peak due to the p(2×2) CO adlayer; (inset) the integrated intensity along the (1/2, 1/2, *l*) rod. (right) The Patterson function for the p(2×2) unit cell calculated from the measured in-plane structure factors. The structural model for the p(2×2) adlayer shows the open circles as surface Pt atoms and the solid circles as adsorbed CO molecules. Reproduced with permission of Elsevier Science from Lucas *et al.* [54]

made in this case showing that the unit cell consisted of three CO molecules, i.e. with a coverage of $\theta = 0.75$ per Pt surface atom (a structural model is shown in Figure 8.44). Due to the weak scattering power of CO compared to the underlying Pt substrate, the registry of the CO monolayer with the substrate could not be determined by measurement of the CTRs from the Pt surface. In this case the CTRs give information only about the relaxation at the Pt surface. In a latter study of a lower coverage CO phase (the $(\sqrt{19} \times \sqrt{19})R23.4°$-13 CO phase, $\theta = 0.68$) under similar conditions (i.e. at the charged electrochemical interface), it was shown that CO adsorption could lead to significant distortion of the underlying Pt surface and this distortion, extending several atomic layers into the Pt surface, was resolved by detailed measurement and modelling of the CTR data [55].

8.3.4 X-RAY STANDING WAVES (XSWs)

Introduction. So far we have seen that surface structural techniques can be broadly divided into those that probe long range order (LEED, SXRD) and those that probe short range order (SEXAFS, NEXAFS). The latter methods have the advantage that they also possess element specificity. Another technique that probes short range order but combines elements of X-ray diffraction is that of X-ray standing waves (XSWs). When an X-ray beam is incident on a surface, interference between coherent incident and diffracted beams will result in the generation of an X-ray standing wave above (and below) the surface. This means that the electric field intensity associated with the X-ray photons will vary in a periodic way above the surface. An example of an X-ray standing wave is shown in Figure 8.45. The XSW periodicity, D is related to θ, the angle of incidence (relative to the surface) by a Bragg relationship in the following way:

$$D = \lambda/(2\sin\theta) \tag{8.40}$$

Thus the period of the standing wave may be altered either by changing the energy of the incident radiation, and hence, λ, or by changing its angle of incidence at the surface. Atoms lying within the XSW may absorb X-ray radiation of a suitable frequency, with a number of possible results, including photoemission and X-ray fluorescence. As we saw in Section 8.3.2, the fluorescence yield, total electron yield or Auger emission intensity (yield denoted as Y_p) following X-ray absorption may be shown to be directly related to the probability of initial absorption. In the case of absorption from the XSW, this is in turn directly related to the intensity of the sinusoidally varying electric field above the surface. For a particular set of experimental conditions, this can be related to the distance of the atom from the surface. The basic XSW experiment therefore involves measuring the X-ray fluorescence or photoemission observed from an atom of interest as the phase of the standing wave is varied.

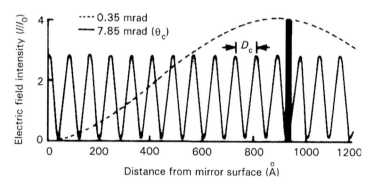

Figure 8.45 An example of an X-ray standing wave. The diagram shows the variation of the normalized electric field intensity generated during specular reflection of a 9.8 keV X-ray plane wave from the surface of a gold mirror. Dashed and solid lines correspond to the field profile with the beam incident at $0.02°$ (0.35 mrad) and $0.449°$ (7.85 mrad, the critical angle) to the gold surface. The shaded rectangle shows the position of Zn atoms 925 Å above the surface, when the gold film is coated with a lipid multilayer (Figure 8.48 below). Reproduced with permission of the Nature Publishing Group from Wang *et al.* [56]

Theoretical Considerations.   The theory of XSW was originally developed by Batterman [57] and requires a dynamical (multiple scattering) treatment of X-ray diffraction. Without reproducing the details of the calculations, it can be seen from Figure 8.45 that the XSW field is produced by the coherent superposition of the incoming plane wave (with electrical field vector $\mathbf{E_0}$) with the Bragg-reflected wave (with electrical field vector $\mathbf{E_H}$). The (normalized) spatial intensity variation, $I^{SW}(\mathbf{r})$, within and above the substrate is then given as:

$$I^{SW}(\mathbf{r}) = \frac{|\mathbf{E_0} + \mathbf{E_H}|^2}{|\mathbf{E_0}|^2} = 1 + \frac{|\mathbf{E_H}|^2}{|\mathbf{E_0}|^2} + 2P\frac{|\mathbf{E_H}|}{|\mathbf{E_0}|}\cos(\phi - 2\pi\mathbf{H}\cdot\mathbf{r}) \qquad (8.41)$$

where ϕ is the phase of the complex amplitude ratio, $\frac{E_H}{E_0} = \frac{|E_H|}{|E_0|}\exp(i\phi)$, $\mathbf{H}$ is the reciprocal lattice vector and P is the polarization of the X-ray beam ($P = 1$ for σ polarization and $P = \cos 2\theta_B$ for π polarization; θ_B is the Bragg angle). The energy dependence of ϕ and the reflectivity, $R = |E_H/E_0|^2$, can be calculated within the framework of dynamical diffraction theory [58]. As described above, the XSW technique measures the yield Y_p of a spectroscopic signal that is characteristic of the adsorbate. In the dipole approximation Y_p is proportional to the square of the electric field at the centre of the atoms and thus is proportional to $I^{SW}(\mathbf{r})$. If the adsorbed atoms are all on equivalent adsorption sites (the fully coherent case) then:

$$Y_p = 1 + R + 2P\sqrt{R}\cos(\phi - 2\pi\mathbf{H}\cdot\mathbf{r}) \qquad (8.42)$$

For an incoherent system with atoms on non-equivalent adsorption sites, a function describing the spread of the atom positions around the average lattice position has to be introduced such that the yield is modified by a normalized distribution function $n(\mathbf{r})$:

$$Y_p = \int Y_p(\mathbf{r})n(\mathbf{r})\,d\mathbf{r} = 1 + R + 2P\sqrt{R}\int n(\mathbf{r})\cos(\phi - 2\pi\mathbf{H}\cdot\mathbf{r})\,d\mathbf{r} \qquad (8.43)$$

Equation (8.43) can be rewritten in terms of a coherent position, $P_H = \mathbf{H}\cdot\mathbf{r}$, and a coherent fraction, f_H, such that:

$$Y_p = 1 + R + 2P\sqrt{R}f_H\cos(\phi - 2\pi P_H) \qquad (8.44)$$

Equation (8.44) is the working equation for most XSW experiments since P_H and f_H contain all of the structural information on the adsorbate system. The coherent position $0 \leqslant P_H \leqslant 1$ gives the adsorbate position relative to the diffraction planes associated with the vector $\mathbf{H}$ and the coherent fraction $0 \leqslant f_H \leqslant 1$ describes the degree of coherent order in the adsorbate system. It can be shown that the coherent fraction f_H is the first Fourier component of the distribution function $n(\mathbf{r})$ multiplied by a phase factor containing the coherent position P_H. By measuring XSW yields for several reciprocal lattice vectors, $\mathbf{H}$, it is thus possible to determine the Fourier representation of the adsorbate.

Experimental Details. A variety of photon energy ranges and geometries have been used in XSW experiments and the design of the experimental equipment is related to that used for SXRD (Figure 8.41), in that the technique involves X-ray diffraction. Most measurements have been carried out in the 'Bragg geometry'. This utilizes the XSW associated with a particular diffracted beam. Measurement of the X-ray fluorescence or photoelectron yield as the phase of the XSW is varied at this Bragg condition permits the location of impurity or adatoms along the direction of the reciprocal lattice vector associated with the diffracted beam. Some experiments using this geometry have used a highly collimated beam of hard X-rays as a source of photons, providing both surface and bulk information, sometimes about surfaces exposed to high pressure liquids or gases. The Bragg relationship (Equation (8.40)) indicates that by lowering the energy of the X-ray beam, and hence increasing its wavelength, the angle of incidence relative to the surface plane may be increased. In fact, by using soft X-rays (with energies say in the range 800 eV–5 keV), it is possible to make the incident beam normal to the lattice plane for most metal and semiconductor surfaces. This forms the basis of normal incidence standing X-ray wavefield absorption (NISXW), first applied by Woodruff *et al.* to Cl/Cu(111) [59, 60]. The method offers a number of advantages over the hard X-ray method. The width of a Bragg reflection becomes narrower as the angle of incidence is increased towards $90°$, and this allows the method to be applied to less perfectly crystalline surfaces, such as those made up of a mosaic of small crystallites of differing orientations. Precise collimation of the soft X-ray beam is in general not required, which means that synchrotron beamlines, used for example for SEXAFS measurements, can be used without modification, as both techniques require a UHV chamber and similar methods of detection involving, for example, electron yield or fluorescence yield measurement.

Applications of XSW. The largest group of systems studied by XSW is adsorbed layers on semiconductors as semiconductor materials have near-perfect crystalline order that permits the XSW to be used from any Bragg reflection [12]. The development of NIXSW, in which the problem of crystal mosaicity is overcome, has broadened the application to many more materials systems [13]. As described in Section 8.3.2, XSW measurements are particularly powerful when they are used in combination with other surface sensitive probes. In such a combined study SEXAFS and XSW measurements were used to study the local structure of Co^{2+} incorporated into the surface of calcite [39]. The SEXAFS results are shown in Section 8.3.2. Figure 8.46 shows the XSW results obtained from the same system. By combining XSW measurements from three mutually noncolinear diffraction vectors, an unambiguous three dimensional position of the impurity atom can be obtained. The fit to the data used Equation (8.44), where the fitting parameters are P_H giving the $\Delta d_H/d_H$ position and f_H giving the static and

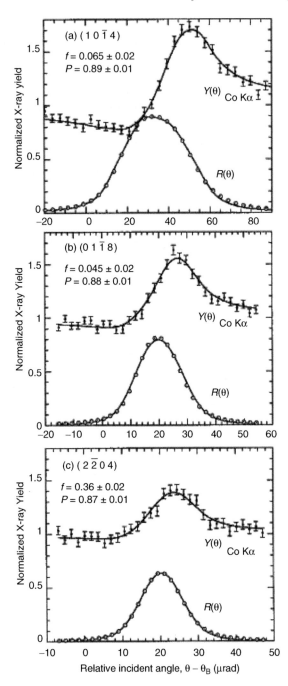

Figure 8.46 XSW triangulation results for Co^{2+} incorporated at the calcite surface, showing experimental data (dots) and theoretical fits (solid-line curves) of the normalized X-ray reflectivity, R, and CoK_α fluorescence yield Y, for (a) the ($10\bar{1}4$), (b) the ($01\bar{1}8$) and (c) the ($2\bar{2}04$) Bragg reflections of calcite. Reproduced with permission of the American Institute of Physics from Cheng *et al.* [39]

Table 8.1 XSW coherent fractions and positions for Co incorporated at the calcite surface and the Co positions derived from the coherent positions and according to the structural model. Reproduced with permission of the American Institute of Physics from Cheng *et al.* [39]

H	d_H(Å)	f_H	XSW measured P_H	h_H(Å)	Model h_H(Å)	(± 0.04)
$(10\bar{1}4)$	3.04	0.65 ± 0.02	0.89 ± 0.01	2.70 ± 0.03		
$(01\bar{1}8)$	1.91	0.45 ± 0.02	0.88 ± 0.01	1.68 ± 0.04	1.69	1.62
$(2\bar{2}04)$	1.93	0.36 ± 0.02	0.87 ± 0.02	1.67 ± 0.04	1.72	1.54

dynamic spread of the atomic distribution, are shown by the solid lines and give the parameters listed in Table 8.1. From the XSW-measured coherent positions, the projected height h_H of the mean Co^{2+} position relative to the H diffraction planes in the plane normal direction can be calculated with the equation $h_H = P_H \times d_H$. Accordingly, the projected heights of Co^{2+} in the $(10\bar{1}4)$, $(01\bar{1}8)$ and $(2\bar{2}04)$ directions are $h = 2.70 \pm 0.03$ Å, $h = 1.68 \pm 0.04$ Å and $h = 1.67 \pm 0.04$ Å, respectively. To locate the lattice sites of Co^{2+} the h_H values are compared with the corresponding h_H values for ideal Ca^{2+}, which are 0 Å for all three h values. This indicates that the Co^{2+} ion is displaced from the ideal Ca^{2+} site, or relaxed. The magnitude of the displacement is small in comparison to the Ca–O nearest-neighbour distance of 2.35 Å such that the Co^{2+} ion is still located within the volume of the ideal CaO_6 octahedron. A model showing the substitutional Co^{2+} structure according to the XSW data is shown in Figure 8.47. The combination of the XSW and SEXAFS (Figures 8.35 and 8.36) results, enables the detailed determination of the lattice sites and local bonding structures of an impurity atom incorporated into the calcite surface and illustrates the advantages of combining surface sensitive techniques.

XSW is one of few techniques which may be used to probe surface structure at rather large distances above a surface. The distance above a surface is limited by the distance over which the XSW remains coherent. Figures 8.45, 8.48 and 8.49 (below) illustrate the result of such a study on a lipid multilayer (Figure 8.48), where the metal atoms in a zinc arachidate bilayer were located at Å resolution at a distance of almost 1000 Å above the surface of a gold mirror. Figure 8.49 shows the Zn fluorescence yield and the reflectivity of the surface as θ is increased from zero to θ_c. From Figure 8.45, it can be seen that for the heavy atom layer at around 925 Å above the surface, 12 maxima in the XSW will have passed through the heavy atom end of the bilayer during this process, giving the oscillations in fluorescence intensity observed. This information is used to calculate the position of the heavy atom layer relative to the surface, using the fact that the fluorescence yield is proportional to the modulated part of the absorption coefficient, χ (as for SEXAFS); thus χ^2 is calculated for a number of possible

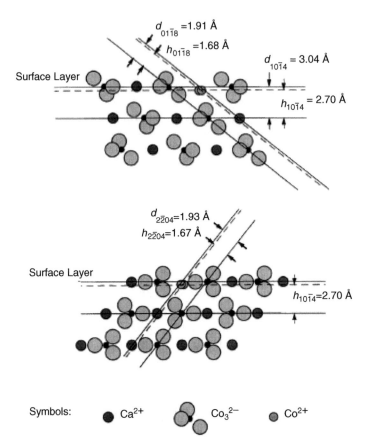

Figure 8.47 Side views of the calcite $(10\bar{1}4)$ surface showing the average height h_H of Co^{2+} with respect to three lattice planes according to XSW measurements and the position of the two crystallographically nonequivalent Co^{2+} ions according to the structural model. Reproduced with permission of the American Institute of Physics from Cheng *et al.* [39]

configurations of the heavy atom layer, and the best fit to the data over a range of θ values is taken as the solution to the problem.

The fact that XSW can probe these large distances above the surface, does not require long-range order (unlike say SXRD) and is element specific (and in some cases, chemical-state specific) means that it is an ideal probe for studying molecular adsorbates, using NIXSW to study molecular adsorbates on metal surfaces. In the last few years a number of molecular adsorption systems have been studied by XSW, for example alkane thiols $(CH_3(CH_2)_{n-1}SH)$ which have attracted great interest because of their

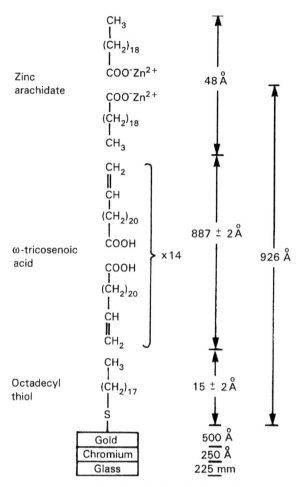

Figure 8.48 Schematic drawing of lipid multilayer LB films deposited on a gold surface. The Zn atoms at 925 Å from the surface gave rise to the XSW data shown in Figure 8.49. Reproduced from Wang *et al.* [56], Copyright 1991, Nature Publishing Group

ability to form self-assembled monolayers (SAMs) [61], amino acids, such as glycine (NH_2CH_2COOH) adsorbed onto Cu(100) [62], endohedral fullerene molecules, in which one or more atoms are trapped within the carbon cage modifying the physical properties of the deposited films [63] and planar organic molecules, such as end-capped quarterthiophene [64], which are of interest in the context of molecular electronics. With the anticipated development in synchrotron radiation sources, particularly in the low energy range, and the improvement in detection techniques, it is likely that XSW will remain as an important surface technique for a number of years.

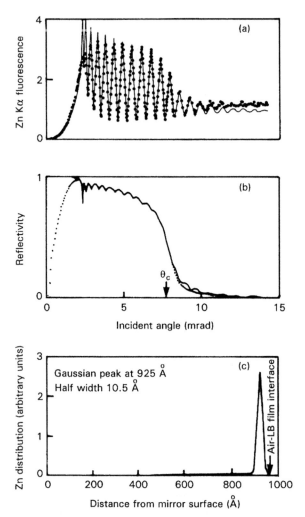

Figure 8.49 The experimental (dots) and theoretical (lines) angular dependence of the ZnKα fluorescence yield (a) and specular reflectivity (b) at 9.8 keV photon energy for the lipid multilayer shown in Figure 8.48. The resulting calculated zinc distribution above the surface is shown in (c). Reproduced from Wang *et al.* [56], Copyright 1991, Nature Publishing Group

8.4 Photoelectron Diffraction

8.4.1 INTRODUCTION

In Section 8.2 we saw how electron-based techniques (LEED, RHEED) could provide structural information on the long-range order at a surface, with the extension to DLEED also giving some short-range order information.

The X-ray based techniques (EXAFS, SXRD and XSW) also give structural information about both short- and long-range order. SXRD (Section 8.3.3) is similar to LEED although interpretation of the data is more straightforward and the penetrating nature of the X-ray radiation means that it can also be used to study a buried surface. XSW (Section 8.3.4) gives element-specific adsorbate site information relative to the underlying bulk lattice but does not give reliable adsorbate-substrate bondlength information. In Section 8.3.2 the basic principles of SEXAFS were introduced and we noted that information may be obtained about the local coordination of a particular surface species, by selecting the incident X-ray wavelength to correspond with an absorption edge of that species and by analysing the SEXAFS modulations at that edge. At the absorption edge, deep-lying core electrons are excited into empty valence states or at higher energies into the ionisation continuum (Figure 8.24). The SEXAFS modulations in the absorption signal above the absorption edge then arise due to interference between the outgoing excited electron wavefunction and the waves backscattered by the surrounding atoms in the medium. In practice in SEXAFS, the Auger or fluorescence yield is measured following deep core hole production, as these reflect the absorption coefficient (Section 8.3.2).

A related technique, which also involves the production of deep-lying core holes is X-ray Photoelectron Diffraction (XPD) and this technique has seen increasing use in recent years. The principles of photoelectron spectroscopy were introduced in Chapter 3. In Photoelectron Diffraction (PD), adsorbate-specific surface structural information (similar to that obtained from SEXAFS) may be gained by measuring the direct photoelectron current following creation of a core hole. In this case the scattering interferences produce variations in the measured photoemission intensity both as a function of emission direction and photoelectron energy. The structural information obtained is the relative position of the emitter to the near-neighbour atoms and, hence, both the distances to the near neighbours and their relative positions in space can be determined. Measurements of the angular variation of the photoemission intensity are effectively measurements of a photoelectron hologram with the directly emitted photoelectron field being the reference wave of the hologram. Energy scan photoelectron diffraction, in which the photoemission intensity is measured as a function of the photoelectron energy, is related to EXAFS in that EXAFS is essentially a spherical average of energy scan photoelectron diffraction.

8.4.2 THEORETICAL CONSIDERATIONS

As we saw in Section 8.3.2, the probability, P_{if} of a transition occurring between an initial state $|\psi_i\rangle$ and a final state $|\psi_f\rangle$ is given by Equation (8.23) where μ is the dipole operator. After creation of a photoelectron by X-ray absorption, the excited photoelectron wavefunction propagates away from

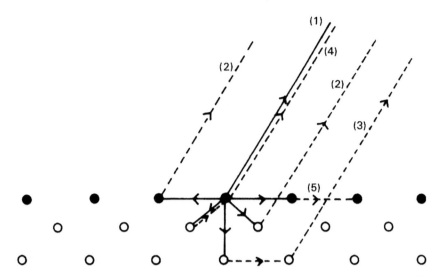

Figure 8.50 Photoelectron scattering events occurring close to a surface, following X-ray absorption. The photoelectron wavefunction propagates away from the absorbing atom (solid arrows). The direct beam (1) propagates freely away from the central absorption site. Some beams may propagate first to a neighbouring atom and undergo elastic scattering into the **k** direction (2). Other contributions to this **k** direction may arise from beams scattered by several atoms (3) or from beams backscattered to the emitting atom, (4) (these last give rise to EXAFS). However, most scattering is in the forward direction (e.g.(5))

the absorbing atom, with wavevector **k**. Some beams may leave the solid without undergoing a scattering event, but others may propagate first to a neighbouring atom, and undergo elastic scattering into the **k** direction (Figure 8.50). Other contributions may arise from beams scattered by several other atoms, or from beams backscattered to the emitting atom (the latter beams being those which give rise to EXAFS modulations). Analysis of the interference between the outgoing and scattered beams will provide bond length information.

There are two fundamentally different ways in which photoelectron diffraction can provide structural information and these arise from the angular dependence of the electron scattering cross section by atoms as a function of scattering angle and electron energy:

1. At higher electron energies the scattering factor is completely dominated by the zero angle forward scattering. An example of how this may be exploited occurs when a diatomic molecule is adsorbed on a surface in a well-defined orientation and the atom closest to the surface acts as the photoelectron emitter. For zero-angle forward scattering of the photoelectrons by the outer atom of the molecule there is no pathlength difference between the directly emitted and forward scattered

components of the wavefield and so the two components interfere constructively along the intermolecular axis. As the scattering angle increases a pathlength difference is introduced and, as the scattering angle increases, this eventually leads to destructive interference. Thus there is a peak in the angular dependence of the photoemission intensity aligned along the intermolecular axis and one can determine interatomic directions in adsorbed molecules by searching for these forward scattering, zero-order diffraction, peaks in the photoelectron angular distribution. This method requires the emitter atom to lie below the scatterer atom relative to the detector and so no information on the location of adsorbates relative to the substrate is provided. This is the basis of the traditional method of X-ray photoelectron diffraction (XPD) [14].

2. At lower photon energies the scattering cross-section is comparable over the full range of scattering angle with a tendency to peak in the 180° backscattering direction as well as the forward direction. In the backscattering geometry all of the scattering paths involve pathlength differences relative to the directly emitted component of the photoelectron wavefield and so the detected photoemission in any particular direction is modulated as the photoelectron energy is changed. Backscattering measurements are typically undertaken at low photoelectron energies (below ~ 500 eV) using a synchrotron source and can provide information on adsorbate–substrate registry and bondlengths via the acquisition of either angle-scan or energy-scan data. In the energy-scan mode this technique has been given the acronym PhD [15] although in original studies it was know as angle-resolved photoemission fine structure (ARPEFS) [65].

The first theoretical model of photoelectron diffraction was developed by Liebsch [66]. For the photoelectrons created in this process with an energy greater than ≈ 200 eV, some simplifications are introduced into the analysis of the scattering. This is because the cross-section for elastic scattering of high energy electrons is lower than that for low energy electrons (as we noted when comparing LEED with RHEED), and so *multiple* scattering of the photoelectron beams is unlikely. The intensities of diffracted beams may then be analysed simply in terms of the interference between the outgoing wave and waves which have undergone one scattering event (Figure 8.50). This scattering event will usually be at the nearest neighbouring shell of atoms, as in EXAFS, because of the relatively short mean-free-pathlength, λ, of the photoelectron wave due to inelastic scattering (Section 8.3.2). In photoelectron diffraction, therefore, the outgoing photoelectron signal is modulated by interference from scattered beams, and carries information about the local coordination of the emitting atom. Liebsch estimated the modulation to be of the order of 40 %.

The simplest expression for the scattered intensity assumes single scattering processes only, approximates the wave at the scatterer as a plane wave and considers emission from an initial s-state (for which the final state is a pure p wave). The intensity of the photoelectron wavefield with wavevector $\mathbf{k}$ is then given by:

$$I(\mathbf{k}) \propto \left| \cos\theta_k + \sum_j \frac{\cos\theta_r}{r_j} f_j(\theta_j, k) W(\theta_j, k) \exp(-L_j/\lambda(k)) \right.$$
$$\left. \times \exp(ikr_j(1 - \cos\theta_j) + \phi(\theta_j, k)) \right|^2 \qquad (8.45)$$

where the first term accounts for the directly emitted component of the emitted p-wave, the $\cos\theta_r$ terms accounts for the same effect at the scattering atoms whose positions are defined by r_j relative to the emitter and the summation j is over the scattering atoms with scattering factors f_j dependent on the scattering angle θ_j and the electron energy. W is a Debye–Waller factor accounting for atomic vibrations and the term $\exp(-L_j\lambda(k))$ describes the additional attenuation due to inelastic scattering experienced by the scattered component which has an additional pathlength, L_j, through the crystal relative to the directly emitted component. The phase of each scattered component is determined by the scattering phase shift, ϕ, and by the phase difference associated with the pathlength difference $kr_j(1 - \cos\theta_j)$. Equation (8.45) has been refined to account for the 'curved wave' character of the photoelectron wavefield at the scattering atom. Multiple scattering effects have also been considered using methodology similar to that used to describe multiple scattering in LEED. With these modifications it is then possible to use trial and error methods (similar to the R-factor analysis used in LEED) to determine structural information from the measured photoelectron diffraction [15].

Expanding equation (8.45) leads to a set of terms describing the interference of the directly emitted component with each of the scattered waves together with a large number of cross terms describing the interferences between different scattered waves. Typically these cross terms average to zero and an approximate expression for the modulation function as a function of the photoelectron wavevector can be obtained, namely:

$$\chi(\mathbf{k}) \propto \sum_j \frac{\cos\theta_r}{r_j} f_j(\theta_j, k) W(\theta_j, k) \exp[-L_j/\lambda(k)] \cos[kr_j(1 - \cos\theta_j) + \phi(\theta_j, k)] \qquad (8.46)$$

which can be related to the equivalent EXAFS Equations (8.34), (8.35) and (8.36). The harmonic cosine term depends on the structure not only through the emitter-scatterer distance, r_j, but also on the real-space orientation of the vector $\mathbf{r}_j$ relative to the detection through the scattering θ_j which appears in the term $kr_j(1 - \cos\theta_j)$. This means that the PD spectra recorded in different emission directions probe the structural environment of the emitter in a real-space directional fashion.

8.4.3 EXPERIMENTAL DETAILS

The PD experiment requires a photon source of the type generally used in a conventional photoemission experiment. This may be a lab-type fixed energy soft X-ray source, producing say MgKα or AlKα radiation (Chapter 3), or a synchrotron source which may be tuned through a range of X-ray wavelengths. The experiment is carried out in UHV, and the photoelectrons are conveniently collected using a hemispherical electron energy analyser (of the type discussed in Chapter 3). A typical experimental arrangement is shown in Figure 8.51 [67]. In order to collect the diffracted beams emerging in different directions from the sample, the analyser must be rotatable about two axes, in order to change both the exit angle relative to the surface normal, ϕ and the azimuthal angle ψ (Figure 8.52).

The intensities of the diffracted beams are measured as a function of a change in the diffraction conditions. This may be a change in the exit angle, $I(\phi)$, or a change in the azimuthal angle, $I(\psi)$, (as in the RHEED experiment, Section 8.2.3). This is known as 'angle resolved photoelectron diffraction' or APD. Plots of the azimuthal (ψ) dependence of the photoemission intensity display characteristic interference maxima and minima consistent with the symmetry of the surface. In practice, the location of an adsorbate relative to

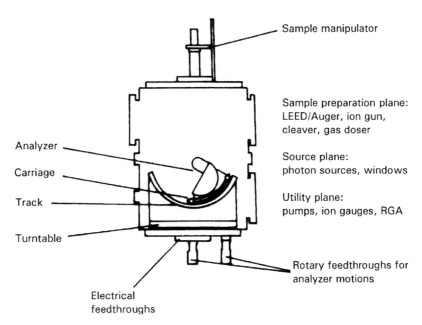

Figure 8.51   A typical experimental chamber for photoelectron diffraction, showing a rotatable hemispherical electron energy analyser mounted inside the vacuum chamber on a carriage allowing both azimuthal and in-plane rotations. Reproduced from Margoninski [67], Copyright 1986, Taylor & Francis

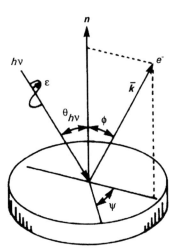

Figure 8.52 Schematic diagram of the experimental geometry used in photoelectron diffraction: ε is the polarization vector of the incident radiation, θ_{hv} is the polar angle of the incident radiation, k is the photoelectron wavevector and n is the surface normal. The polar angle relative to the surface normal, ϕ, of the emerging photoelectrons and their azimuthal angle, ψ, are varied in the experiment. Reproduced from Margoninski [67], Copyright 1986, Taylor & Francis

a substrate is determined by comparing measured XPD $I(\phi)$ and $I(\psi)$ scans to those calculated for different geometries. Use of a synchrotron source offers the opportunity to vary the incident radiation wavelength, through a change in its energy, $I(E)$. In this case, all angles are kept constant, as is the binding energy of the photoelectrons (i.e. only electrons coming from the same core levels are considered), while the energy of the incident radiation is varied. Thus the kinetic energies of the photoelectron are varied. In both the scanned-energy mode and angle-scan mode of PD limited data sets can lead to errors in structure determination. Similarly to the case of SXRD, structural determinations are also subject to having a correct starting model for structure refinement. Alternative methods based on direct methods have been developed, so called 'holographic inversion' of the data to provide a real space image of the structural environment of the emitter atom [68]. Such methods are beyond the scope of this chapter and the interested reader is referred to the recent review of PhD [15].

8.4.4 APPLICATIONS OF XPD AND PHD

The technique of XPD, i.e. high energy forward scattering, is nicely illustrated in the study of CO adsorption onto Ni(110) [69]. In these experiments a fixed X-ray source (1253.6 eV) and hemispherical analyser were used to collect photoelectrons in a polar range of $\theta = \pm 60°$ by rotating the sample about an axis parallel to the [001] direction. The angle $\theta = 0°$ corresponds to the surface normal. Three XPD scans for increasing CO exposure are shown in Figure 8.53 illustrating the change of CO orientation from perpendicular to tilted. Perpendicular CO is observed at low CO coverage while tilted CO develops at high coverage with simultaneous ordering into a p2mg structure with the tilt angle being determined as 21°. Note that the simple determination is not only based on the assumption of strong forward scattering but also on a small scattering phase shift as otherwise the interference in the direction of the molecular bond could be destructive. In a similar way XPD has been used to determine a number of adsorbate orientations, from diatomic molecules to much more complicated molecular species and

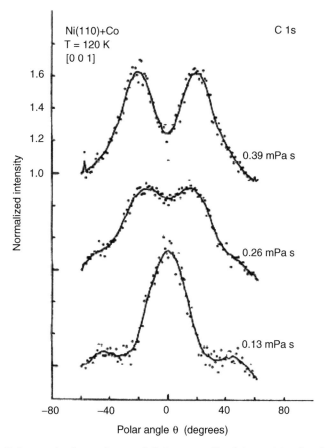

Figure 8.53 Polar angle dependence of C 1s normalized intensities for CO adsorbed at 120 K on Ni(110). The azimuth of measurement is parallel to the [001] direction. Reproduced with permission of Elsevier Science from Wesner *et al.* [69]

in the grazing-incidence geometry has also been used to obtain information on substrate–adsorbate registry [15].

One of the first successful studies to apply EXAFS data analysis techniques to XPD was the work of Shirley *et al.* on Ni(001)c(2 × 2)-S [70]. Here, two polar angles were used, 0° (NPD) and 45°, and the sample and analyser were also rotated in tandem with respect to the incident beam, to change the polarization vector of the light relative to the crystal. This is used to enhance the scattering from specific Ni atoms, making use of the fact that photoemitted intensity is peaked strongly in the direction of the electric vector of the light. Thus if the electric vector is polarized parallel to the surface normal, the photoemission current is directed into the sample and strong scattering from substrate atoms occurs. If the vector is polarized parallel to the sample surface, strong intra-layer scattering from the topmost sample layer occurs. This effect is analogous to the polarization dependencies observed in SEXAFS (Section 8.3.2). In this study, S 1s photoemission was measured, at kinetic energies up to around 500 eV (Figure 8.54). In the data a background function representing the atomic contribution to the photoemission current was subtracted from the experimental photoemission intensity, from which the normalized oscillatory part, $\chi(k)$, was obtained, and then Fourier transformed. The resulting transforms are shown in Figure 8.54. For the Ni/S surfaces, the first three peaks in the transformed data are related to scattering from the four nearest Ni neighbours around the S hollow site. The resulting Ni–S distance is 2.23 Å, in excellent agreement with SEXAFS

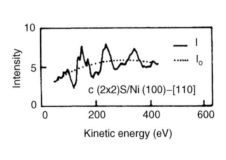

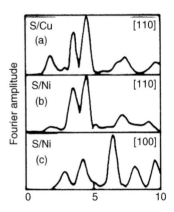

Figure 8.54   XPD data for Ni(001)-c(2×2)–S at the S K-edge, showing the application of EXAFS data analysis techniques. The left-hand panel shows the oscillations in the photoemission intensity, (with I_0 approximating to the atomic contribution to the photoemission current). The right-hand panel shows Fourier transforms of data recorded for different polarizations of the light relative to the sample, for the Ni system and for Cu(001)-p(2×2)–S. Reproduced from Barton *et al.* [70], Copyright 1984, American Institute of Physics

data. Interpretation of the data for this system was then used to determine the geometry of an unknown system, Cu(001)-p(2 × 2)–S. A similar four-fold hollow site was determined, with a Cu–S distance of 2.28 Å. The Ni(001)-c(2x2)-S system was also used as a demonstration of the holographic technique [68] as the structure was known from the previous study. In fact this system was used to test various methods for reconstruction of the atomic structure from XPD data [71].

The low energy (below ∼500 eV) backscattering method (PhD) has seen widespread use in recent years as it is a technique that requires a synchrotron radiation source but is well suited to providing quantitative information on adsorbate–substrate registry. Key to successful structure determination using this method is to sample the photoemission intensity over a sufficient range of both emission angle and photoelectron energy. Numerous adsorption systems relevant to heterogeneous catalysis have been studies using PhD, originally as the technique was developed by Shirley and co-workers [72–74] and, later, as the technique was further developed by Woodruff's group [75, 76]. As PhD gives local structural information on the individual constituent atoms of a molecule, it is possible to determine intermolecular bondlengths directly, as demonstrated in numerous studies of CO adsorption onto metals, oxides and semiconductors [15]. Studies have been performed of many molecular adsorbates on metals, ranging from SAMs, in which the results obtained by XSW were further supported by PhD measurements [77], to a variety of hydrocarbons (benzene, acetylene, ethylene, etc.) [15]. A particularly challenging system was that of the carbonate species CO_3 adsorbed onto Ag(110) which was studied using C 1s emission data [78]. Figure 8.55 shows a plan view of the best-fit structure which has extremely low symmetry with the consequence that the PhD spectra are only weakly modulated by the substrate scattering. Spectra were recorded at near normal incidence emission and at grazing emission and detailed simulation of the spectra was used to obtain the azimuthal orientation of the molecule and its location relative to the underlying substrate. In this study the use of other surface analysis techniques was vital in the final determination of the structure. In particular, STM studies showed that the (1 × 2) periodicity exhibited by the phase was due to 'added' Ag row atoms and not the CO_3 species and with the bondlengths obtained from the PhD spectra this meant that the molecule had to be located above the Ag atoms in the outermost unreconstructed layer. This was later confirmed in density functional theory calculations [79].

As pointed out in Section 8.4.1, photoelectron diffraction is element-specific which is a useful asset in surface studies. In addition, however, atoms of the same element can exhibit different core level shifts in the photoelectron binding energy, depending on the structural and electronic environment of the atom and this can be exploited to separate the photoemission from atoms in different environments and provide local

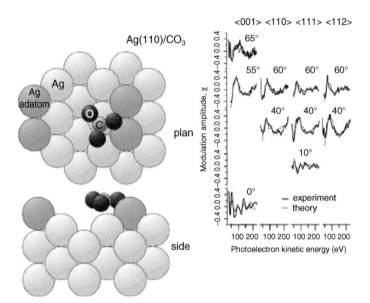

Figure 8.55 Plan and side views of the optimized local structure of the carbonate species CO_3 on Ag(110). Also shown is a comparison of the experimental C 1s PhD spectra recorded at different polar and azimuthal emission directions with theoretical simulations for the best-fit structure. Reproduced with permission of Elsevier Science from Kittel *et al.* [78]

structural information. There are a number of situations in which this additional resolution is important, for example, in distinguishing between atoms of the same element in different environments within a single adsorbed molecule [80] or for systems in which the adsorbate and substrate contain the same element. With the increasing energy resolution that is becoming available with the development of synchrotron radiation sources it is anticipated that such studies will become increasingly important in the future.

📖 References

Useful general texts and review articles for the surface analysis techniques are listed below:

General Texts

1. C. Kittel, *Introduction to Solid State Physics*, John Wiley & Sons, Inc., New York, NY (1996).

2. G. Margaritondo, *Introduction to Synchrotron Radiation*, Oxford University Press, New York, NY (1988).

3. M. Prutton, *Introduction to Surface Physics*, Oxford University Press, Oxford (1994).

4. G.A. Somorjai, *Introduction to Surface Chemistry and Catalysis*, John Wiley & Sons, Inc., New York, NY (1994).

5. A. Zangwill, *Physics at Surfaces*, Cambridge University Press, Cambridge (1988).

Electron Diffraction Techniques

6. K. Heinz, LEED and DLEED as modern tools for quantitative surface-structure determination *Rep. Prog. Phys.*, **58**, 637 (1995).

7. J.B. Pendry, *Low Energy Electron Diffraction*, John Wiley & Sons, Inc., New York, NY (1974).

8. J.B. Pendry, *LEED and the Crystallography of Surfaces, Tutorials on Selected Topics in Modern Surface Science*, T. Barr and D. Saldin (Eds), Chapter 2, Elsevier Science, Amsterdam (1993).

X-ray Techniques

9. J. Als-Nielsen and D. McMorrow, *Elements of Modern X-ray Physics*, John Wiley & Sons, Ltd, Chichester (2001).

10. I.K. Robinson and D.J. Tweet, Surface X-ray Diffraction, *Rep. Prog. Phys.*, **55**, 599 (1992).

11. D.P. Woodruff, Fine Structure in Ionization Cross Sections and Applications to Surface Science, *Rep. Prog. Phys.*, **49**, 683 (1986).

12. D.P. Woodruff, Surface Structure Determination using X-ray Standing Waves, *Rep. Prog. Phys.*, **68**, 743 (2005).

13. J. Zegenhagen, Surface Structure Determination with X-ray Standing Waves, *Surf. Sci. Rep.*, **18**, 199 (1993).

Photoelectron Diffraction

14. C. Westphal, The Study of the Local Atomic Structure by Means of X-ray Photoelectron Diffraction, *Surf. Sci. Rep.*, **50**, 1 (2003).

15. D.P. Woodruff, Adsorbate Structure determination using Photoelectron diffraction: Methods and Applications, *Surf. Sci. Rep.*, **62**, 1 (2007).

References Cited in the Text

16. C.J. Davisson and L.H. Germer, *Phys. Rev.* (2nd series), **30**, 705 (1927).

17. E.J. Scheiber, L.H. Germer and C.D. Hartman, *Rev. Sci. Inst.*, **31**, 112 (1960).

18. J.B. Pendry, *Low Energy Electron Diffraction*, John Wiley &Sons, Inc., New York, NY (1974).

19. P.J. Rous and J.B. Pendry, *Surf. Sci.*, **219**, 355 and 373 (1989).

20. W. Oed, P.J. Rous and J.B. Pendry, *Surf. Sci.*, **273**, 261 (1992).

21. U. Scheithauer, G. Meyer and M. Henzler, *Surf. Sci.*, **178**, 441 (1986).

22. J. Falta and M. Henzler, *Surf. Sci.*, **269/270**, 14 (1992).

23. M. Sidoumou, T. Angot and J. Suzanne, *Surf. Sci.*, **272**, 347 (1992).

24. P. Zschack, J.B. Cohen and Y.W. Chung, *Surf. Sci.*, **262**, 395 (1992).

25. J.B. Hannon, E.W. Plummer, R.M. Wentzcovitch and P.K. Lam, *Surf. Sci.*, **269/270**, 7 (1992).

26. W.K. Ford, T. Guo, D.L. Lessor and C.B. Duke, *Phys. Rev. B*, **42**, 8952 (1990).

27. K. Heinz, D.K. Saldin and J.B. Pendry, *Phys. Rev. Lett.*, **55**, 2312 (1988).

28. K. Heinz and L. Hammer, *J. Phys. Chem. B*, **108**, 14579 (2004).

29. J.H. Neave, B.A. Joyce, P.J. Dobson and N. Norton, *Appl. Phys. A*, **31**, 1 (1983).

30. P.J. Dobson, N.G. Norton, J.H. Neave and B.A. Joyce, *Vacuum*, **33**, 593 (1983).

31. Y. Shigeta and Y. Fukaya, *Appl. Surf. Sci.*, **237**, 21 (2004).

32. M. K. Zayed and H. E. Elasayed-Ali, *Phys. Rev. B*, **72**, 205426 (2005).

33. C.Y. Yang, J.M. Lee, M.A. Paesler and D.E. Sayers, *J. Phys. Colloq.*, **C8–47**, 387 (1986).

34. J. Stöhr, in Emission and Scattering Techniques, Proceedings of the NATO Advanced Studies Institute, Volume 73, P. Day (Ed.), p. 215, D. Reidel, Dordrecht, The Netherlands (1981).

35. S.S. Hasnain (Ed.), *X-ray Absorption Fine Structure*, Ellis Horwood, Chichester (1991).

36. P.H. Citrin, *Surf. Sci.*, **299–300**, 199 (1994).

37. D.R. Warburton, G. Thornton, D. Norman, C.H. Richardson, R. McGrath and F. Sette, *Surf. Sci.*, **189–190**, 495 (1987).

38. J. Haase, B. Hillert, L. Becker and M. Pedio, *Surf. Sci.*, **262**, 8 (1992).

39. L. Cheng, N.C. Sturchio and M.J. Bedzyk, *Phys. Rev. B*, **61**, 4877 (2000).

40. D. Norman, R.A. Tuck, H.B. Skinner, P.J. Wadsworth, T.M. Gardiner, I.W. Owen, C.H. Richardson and G. Thornton, *J. Phys. Colloq.*, **C8–47**, 529 (1986).

41. J. Stöhr and R. Jaeger, *Phys. Rev. B*, **26**, 4111 (1982).

42. M. Kiguchi, R. Arita, G. Yoshikawa, Y. Tanida, S. Ikeda, S. Entani, I. Nakai, H. Kondoh, T. Ohta, K. Saiki and H. Aoki, *Phys. Rev. B*, **72**, 075446 (2005).

43. N. Witkowski, F. Hennies, A. Pietzsch, S. Mattsson, A. Folisch, W. Wurth, M. Nagasono and M.N. Piancastelli, *Phys. Rev. B*, **68**, 115408 (2003).

44. F. Zheng, J.L. McChesney, X. Liu and F.J. Himpsel, *Phys. Rev. B*, **73**, 205315 (2006).

45. A. Scholl, Y. Zou, L. Kilian, D. Hubner, D. Gador, C. Jun, S.G. Urquhart, Th. Schmidt, R. Fink and E. Umbach, *Phys. Rev. Lett.*, **93**, 146406 (2004).

46. I.K. Robinson, *Phys. Rev. B*, **33**, 3830 (1986).

47. S. Ferrer and F. Comin, *Rev. Sci. Instrum.*, **66**, 1674 (1995).

48. R. Feidenhans'l, *Surf. Sci. Rep.*, **10**, 105 (1989).

49. S.J.G. Mochrie, *Curr. Opinion Solid State Mater. Sci.*, **3**, 460 (1998).

50. G. Renaud, *Surf. Sci. Rep.*, **32**, 5 (1998).

51. P.H. Fuoss and S. Brennan, *Annu. Rev. Mater. Sci.*, **20**, 360 (1990).

52. C. A. Lucas and N. M. Markovic, *In-situ* surface X-ray scattering and infra-red reflection adsorption spectroscopy of CO chemisorption at the electrochemical interface, in *In-Situ Spectroscopic Studies of Adsorption at the Electrode and Electrocatalysis*, S.G. Sun, P.A. Christensen and A. Wieckowski (Eds), Chapter 11, Elsevier, Amsterdam (2007).

53. R.G. van Silhout, M. Lohmeier, S. Zaima, J.F. van der Veen, P.B. Howes, C. Norris, J.M.C. Thornton and A.A. Williams, *Surf. Sci.*, **271**, 32 (1992).

54. C.A. Lucas, N.M. Markovic and P.N. Ross, *Surf. Sci.*, **425**, L381 (1999).

55. J. Wang, I.K. Robinson, B.M. Ocko and R. R. Adzic, *J. Phys. Chem. B*, **109**, 24 (2005).

56. J. Wang, M.J. Bedzyk, T.L. Penner and M. Caffrey, *Nature*, **354**, 377 (1991).

57. B.W. Batterman, *Phys. Rev.*, **133**, A759 (1964).

58. B.W. Batterman and H. Cole, *Rev. Mod. Phys.*, **36**, 181 (1964).

59. D.P. Woodruff, D.L. Seymour, C.F. McConville, C.E. Riley, M.D. Crapper and N.P. Prince, *Phys. Rev. Lett.*, **58**, 1460 (1987).

60. D.P. Woodruff, D.L. Seymour, C.F. McConville, C.E. Riley, M.D. Crapper and N.P. Prince, *Surf. Sci.*, **195**, 237 (1988).

61. G.J. Jackson, D.P. Woodruff, R.G. Jones, N.K. Singh, A.S.Y. Chan, B.C.C. Cowie and V. Formoso, *Phys. Rev. Lett.*, **84**, 119 (2000).

62. J.H. Kang, R.L. Toomes, M. Polcik, M. Kittel, J.-T. Hoeft, V. Efstathiou, D.P. Woodruff and A.M. Bradshaw, *J. Chem. Phys.*, **118**, 6059 (2003).

63. R.A.J. Wooley, K.H.G. Schulte, L. Wang, P.J. Moriarty, B.C.C. Cowie, H. Shirohara, M. Kanai and T.J.S. Dennis, *Nano Lett.*, **4**, 361 (2004).

64. L. Kilian, W. Weigand, E. Umbach, A. Langner, M. Sokolowski, H.L. Meyerheim, H. Maltor, B.C.C. Cowie, T. Lee and P. Bauerle, *Phys. Rev. B*, **66**, 075412 (2002).

65. S.D. Kevan, D.H. Rosenblatt, D.R. Denley, B.-C. Lu and D.A. Shirley, *Phys. Rev. Lett.*, **41**, 1565 (1978).

66. (a) A. Liebsch, *Phys. Rev. Lett.* **32**, 1203 (1974).(b) A. Liebsch, *Phys. Rev. B*, **13**, 544 (1976).

67. Y. Margoninski, *Contemp. Phys.*, **27**, 203 (1986).

68. J.J. Barton, *Phys. Rev. Lett.*, **61**, 1356 (1988).

69. D.A. Wesner, F.P. Coenen and H.P. Bonzel, *Surf. Sci.*, **199**, L419 (1988).

70. J.J. Barton, C.C, Bahr, Z. Hussain, S.W. Robey, L.E. Klebanoff and D.A. Shirley, *J. Vac. Sci. Technol. A*, **2**, 847 (1984).

71. S. Thevuthasan, R.X. Ynzunza, E.D. Tober, C.S. Fadley, A.P. Kaduwela and M.A. Van Hove, *Phys. Rev. Lett.*, **70**, 595 (1993).

72. S.W. Robey, J.J. Barton, C.C. Bahr, G. Liu and D.A. Shirley, *Phys. Rev. B*, **35**, 1108 (1987).

73. L.J. Terminello, K.T. Leung, Z. Hussain, T. Hayahsi, X.S. Zhang and D.A. Shirley, *Phys. Rev. B*, **41**, 12787 (1990).

74. L.-Q. Wang, Z. Hussain, Z.Q. Huang, A.E. Schach von Wittenau, D.W. Lindle and D.A. Shirley, *Phys. Rev. B*, **44**, 13711 (1991).

75. M.E. Davila, M.C. Asensio, D.P. Woodruff, K.-M. Schindler, Ph. Hofmann, K.-U. Weiss, R. Dippel, P. Gardner, V. Fritzsche, A.M. Bradshaw, J.C. Conesa and A.R. Gonzalez-Elipe, *Surf. Sci.*, **311**, 337 (1994).

76. M. Kittel, M. Polcik, R. Terberg, J.-T. Hoeft, P. Baumgartel, A.M. Bradshaw, R.L. Toomes, J.-H. Kang, D.P. Woodruff, M. Pascal, C.L.A. Lamont and E. Rotenburg, *Surf. Sci.*, **470**, 311 (2001).

77. T. Shimida, H. Kondoh, I. Nakai, M. Nagasaka, R. Yokota, K. Amemiya and T. Ohta, *Chem. Phys. Lett.*, **406**, 232 (2005).

78. M. Kittel, D.I. Sayago, J.-T. Hoeft, M. Polcik, M. Pascal, C.L.A. Lamont, R.L. Toomes and D.P. Woodruff, *Surf. Sci.*, **516**, 237 (2002).

79. J. Robinson and D.P. Woodruff, *Surf. Sci.*, **556**, 193 (2004).

80. K.-U. Weiss, R. Dippel, K.-M. Schindler, P. Gardner, V. Fritzsche, A.M. Bradshaw, D.P. Woodruff, M.C. Asensio and A.R. Gonzalez-Elipe, *Phys. Rev. Lett.*, **71**, 581 (1993).

9 Scanning Probe Microscopy

GRAHAM J. LEGGETT

Department of Chemistry, University of Sheffield, Sheffield, UK

9.1 Introduction

The development of the scanning tunnelling microscope (STM) and its close relative the atomic force microscope (AFM) or scanning force microscope (SFM) has proved to be one of the most exciting developments in the past two decades, not only in surface science, the field from which they emerged, but also much more widely in science, particularly through the explosion of interest in AFM in a wide range of fields of research. Both techniques offer, in principle, the possibility for obtaining very highly resolved images with, in many cases, a minimum of prior preparation of samples. The ease of construction and operation of scanning probe microscopes (SPMs) (a generic term covering both techniques, and a range of other related techniques), has led to a proliferation of companies capable of providing effective instruments, some at relatively inexpensive prices. The 'observation' of individual atoms has become a commonplace event in many modern laboratories (for example, see Figure 9.1), and it is fair to say that the general impact of SPM on our understanding has been revolutionary. The consequence of this has been the application of SPM techniques across a broad spectrum of scientific research, from biological processes to solid state physics. This wide-ranging potential was undoubtedly responsible for the very rapid award of the Nobel Prize to the inventors of the STM, Binnig and Rohrer, in 1986, just five years after its invention.

The objectives of this chapter are three-fold: firstly, to provide a basic understanding of the physical principles underlying SPM techniques;

Surface Analysis – The Principal Techniques 2nd Edition Edited by John Vickerman and Ian Gilmore
© 2009 John Wiley & Sons, Ltd

Figure 9.1 Atomic resolution STM image of the 7×7 reconstruction of a silicon(III) surface. Reproduced by permission of Dr P.H. Beton, Department of Physics, The University of Nottingham

secondly, to identify relevant practical and experimental considerations; and thirdly, to give a flavour of the kinds of data which SPM techniques are capable of providing. Because of the vast literature which has arisen in the field, this overview is far from comprehensive in scope. Instead, we will provide some illustrative examples of the sort of capability that probe microscopy offers, hoping to provide a starting point for more detailed reading on specific subjects. We will spend most time looking at the most widely used instruments, the scanning tunnelling microscope and the atomic force microscope, and mention scanning near field optical microscopy more briefly. While SPM is properly regarded as a branch of microscopy, and thus finds important applications in the characterization of surface morphology (i.e. surface contours and surface structure), a particular focus in this chapter will be to highlight methods for the use of SPM-based methods for surface analysis: some methods are now emerging that enable the quantitative measurement of molecular composition and chemistry with nanometre scale resolution. These are significantly extending the reach of SPM techniques beyond the traditional domain of microscopy.

9.2 Scanning Tunnelling Microscopy

The STM and the AFM are both stylus-type instruments, in which a sharp probe, scanned raster-fashion across the sample, is employed to detect changes in surface structure on the atomic scale. One of the most commonly employed analogies is with the operation of a record-player, in which a stylus moves up and down with the topography of the rotating record to generate a signal. In the case of the AFM, this simple analogy is not far from the truth: the interaction force between the probe and surface structural features is measured to reveal the surface topography in a fairly direct fashion. In the case of the STM, the analogy is less accurate: it is, in fact, the surface electron density that is measured. However, for surfaces which have relatively uniform electronic properties (and in the limits of dimensions >10 Å), the STM image effectively represents the surface topography. For electronically inhomogeneous surfaces, for example at high resolution where individual bonding states may be imaged, the interpretation of the image requires more sophistication. A lateral resolution of 0.1 Å and a vertical resolution of 0.01 Å are attainable with the STM. In principle, comparable lateral resolution is also possible for the AFM, although in practice, one generally finds that atomic resolution is more readily accessible with the STM – given a conducting sample, that is. It is rare to obtain images with genuine atomic resolution in common commercial AFM systems, but using non-contact techniques, impressive advances have been made in this direction in certain more specialized cases.

9.2.1 BASIC PRINCIPLES OF THE STM

Quantum Tunnelling.   The STM is an example of the practical exploitation of a strictly quantum mechanical phenomenon: quantum tunnelling. Tunnelling processes play an important role in a number of phenomena which occur at surfaces and interfaces. For example, in SIMS, incident ions may be neutralized prior to penetration of the solid surface by electron tunnelling, and, similarly, ejected secondary ions may be neutralized by tunnelling processes. Quantum mechanical tunnelling involves the penetration of a potential barrier by an electron wave function. The potential barrier may be a layer of insulating material (for example, an oxide layer in the case of a metallic electrode) or a vacuum gap (in the case of the STM). The basic physical process is best understood by consideration of the simple case of one-dimensional tunnelling (widely discussed in textbooks on quantum mechanics or solid state physics; for example see Atkins [1]). Consider the simple system (Figure 9.2(a)) in which an electron is incident upon an infinitely thick potential barrier of height V. The Schrödinger equation has two components. Where $x < 0$:

$$H = -(\hbar^2/2m)(\mathrm{d}^2/\mathrm{d}x^2) \tag{9.1}$$

Inside the barrier ($x > 0$):

$$H = -(\hbar^2/2m)(\mathrm{d}^2/\mathrm{d}x^2) + V \tag{9.2}$$

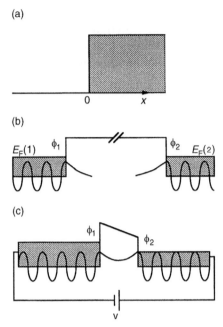

Figure 9.2 (a) An infinitely thick potential barrier. Potential $= V$ for $x>0$ and 0 for $x < 0$. (b) Two potential wells which are spatially separated. (c) Two potential wells separated by a small distance, and with an applied potential difference

The solutions of these equations are:

$$\psi = Ae^{ikx} + Be^{-ikx}, \quad k = (2mE/\hbar^2) \text{ inside the well, and} \qquad (9.3)$$
$$\psi = Ce^{ik'x} + De^{-ik'x}, \quad k = (2m(E - V)/\hbar^2)^{1/2} \text{ inside the barrier} \quad (9.4)$$

The wave function inside the barrier has an imaginary part (which rises to infinity and is thus discounted) and a real part which decays exponentially with distance inside the barrier. This is a very important result, because where penetration is classically forbidden (for $E < V$), the quantum mechanical wavefunction is non-zero: the electron may tunnel into the potential barrier. There is, therefore, a finite probability that the electron will be found inside the barrier. A simple extension of this case involves the consideration of two potential wells close together (in other words, separated by a potential barrier of finite thickness). If the potential barrier is relatively narrow, there is a probability that an electron may penetrate it [1] and pass from one well to the other. Consider two nearby metallic electrodes with a work function ϕ, separated by a large distance (Figure 9.2(b)). The effective overlap of the Fermi level wavefunctions is negligible, because of the exponential decay of the two separate wavefunctions. If these electrodes are brought close together, with some separation d (Figure 9.2(c)), then the overlap of the wavefunctions may be sufficiently great to facilitate quantum mechanical tunnelling and, under the influence of an applied potential difference, the passage of a measurable current. The magnitude of the tunnelling current I is a measure of the overlap of the two wavefunctions, and is given by:

$$I \propto e^{(-2\kappa d)} \tag{9.5}$$

where κ is related to the local work function by:

$$\kappa = (2m\phi/\hbar^2)^{1/2} \tag{9.6}$$

Unfortunately, the simple one-dimensional model is not really adequate for a full description of the STM. The electronic structures of the tip and the surface are involved in a complex fashion which really demands a three-dimensional treatment. Tersoff and Hamann have provided a more general treatment [2]. Here, the full general expression for the tunnelling current is:

$$I = (2\pi e/\hbar)\, e^2 V \sum_{\mu,\nu} |M_{\mu,\nu}|2\delta(E_\nu - E_F)\delta(E_\mu - E_F) \tag{9.7}$$

where E_F is the Fermi energy, E_μ is the energy of the state ψ_μ in the absence of tunnelling, and $M_{\mu\nu}$ is the tunnelling matrix element between states ψ_μ of the probe tip and ψ_ν of the sample, given by:

$$M_{\mu,\nu} = (\hbar^2/2m)\int dS(\psi_\mu^* \nabla \psi_\nu - \psi_\nu \nabla \psi_\mu^*) \tag{9.8}$$

Working on the assumption that the STM tip is spherical, one arrives at the following expression for the tunnelling current [2]:

$$I = 32\pi^3 \hbar^{-1} e^2 V \varphi_0^2 D_t(E_F) R_t^2 \kappa^{-4} e^{2\kappa R_t} \sum |\psi_\nu(r_0)|\, 2\delta(E_\nu - E_F) \tag{9.9}$$

in which ϕ_o is the work function and $D_t(E_F)$ is the density of states at the Fermi level per volume of the tip. Thus the tunnelling current is proportional to the local density of states at the Fermi level and at the centre of the STM tip. This means that the STM can provide a direct image of quantum mechanical electronic states at the surface and, therefore, Equation (9.9) provides a basis for the application of STM to atomic-scale surface spectroscopy (discussed below).

Equation (9.9) preserves the important result that the tunnel current depends exponentially upon the separation between the STM tip and the sample surface. This exponential dependence on the tip–sample separation, or the tunnelling gap, provides the basis for the astonishing resolution of the technique. To a rough approximation, the tunnelling current decreases by an order of magnitude for every increase of 1 Å in the tunnelling gap. For a sufficiently sharp tip, the bulk of the tunnelling current will flow through the very end of the tip (illustrated in Figure 9.3) and thus the effective diameter of the tip becomes very small (of the order of atomic dimensions [3]).

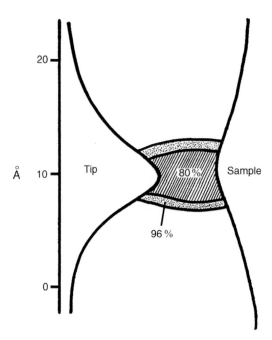

Figure 9.3 Schematic illustration of the current density distribution for tunnelling from a tip to a corrugated surface. After Binnig and Rohrer [3]

The Role of Tip Geometry.   Ideally, we require a tip which has a single atom situated at its apex. This sounds like a very stringent constraint; however, tips sharp enough to generate high resolution images may be obtained fairly straightforwardly. Perhaps the simplest (and most surprising) approach to tip preparation is the use of mechanical methods: cutting a piece of fine wire (for example platinum–iridium or tungsten) at an oblique angle with a pair of sharp wire-cutters yields a very sharp probe with a high frequency. Alternatively, a chemical agent (for example, NaOH) may be used to etch the wire; the lower portion of the wire drops off leaving a sharp tip. Electrochemical sharpening methods may be used. In addition to these methods, *in situ* methods of tip sharpening can be useful. Again, a number of methods exist. One technique involves applying a voltage pulse between the tip and the substrate [4], with the consequence that, according to one possible explanation [5], the electrical field in the gap draws atoms (for example, tungsten atoms on a tungsten tip) towards the apex of the tip. Often, this may result in a sharper tip profile, so that a damaged tip may be regenerated; however, the technique is also useful as a means of removing contaminants which may have become adsorbed to the tip during scanning of a contaminated or molecular sample. An alternative approach involves gently colliding the tip with a surface. Usually, this would result in terminal damage to the tip (known as 'crashing' the tip). However, it has been reported that gentle collisions between an STM tip and a silicon surface can result in the removal of a silicon cluster by the tip, leaving a small hole in the silicon surface. In a similar fashion, the wetting of tips by gold atoms has been reported [6] although these are of little value for imaging purposes.

Despite the ease with which tips capable of providing a very high level of resolution may be prepared, it nevertheless remains the case that tip structure plays an important role in determining the nature of the STM image. Variations in tip atomic structure may cause variations in the STM image. Equation (9.7), which gives an expression for the tunnelling current, involves the wavefunctions for both the tip and the substrate, and in reality the STM image represents a complex convolution of the structures of the tip and the substrate surface. As the required resolution is increased, the tip structure comes, increasingly, to determine the STM image. At the most fundamental level, the detailed atomic structure of the tip may determine the nature of the image at high resolution. Different atomic arrangements (crystal structures) at the apex will result in different electronic states being able to interact with the sample and this will result in changes to the resulting image. Figure 9.4 shows an illustration of this [7]. There are two non-equivalent carbon atoms in the top-most layer of graphite. A second carbon atom lies directly underneath carbon atom A, in the next lowest plane of the solid, but the next carbon atom to lie directly beneath carbon atom B is in the second plane below the surface. In this respect, graphite provides an important example of the way in which the corrugation in the STM image reflects the local density of states (LDOS) at the Fermi level, and not the positions of atoms. The LDOS is lower at site A than it is at site B. Consequently, the honeycomb lattice is not seen and a hexagonal lattice is observed instead. For a typical W_{10} [111] tip such as that illustrated in Figure 9.4(b), the maximum of the tunnelling current contour map therefore lies over carbon B, with the minimum at the centre of the hexagonal ring of carbon atoms. However, if the atom at the apex of the tip is removed (Figure 9.4(c)), the image changes. Now the weakest current region is located at site B, and the maximum in the current contour map lies at carbon atom A. Tsukada *et al.* suggested the following explanation: in the second case, the tunnelling current reaches a maximum when the combined current contribution from the three equivalent topmost tungsten atoms is the largest, which will occur when the centre of the tip is positioned over site A. When the tip is centred over an A site, the three tungsten atoms nearest the surface are positioned over three B carbon atoms. When the tip is centred over site B, the three topmost tungsten atoms are located over the centres of three graphite hexagons, and the current is minimal.

Steps and defects in graphite, and grain boundaries, can perturb the electron density giving rise to the observation of superstructures. For example, large scale hexagonal superstructures have been observed and attributed to Moiré effects due to rotational misorientation of two basal

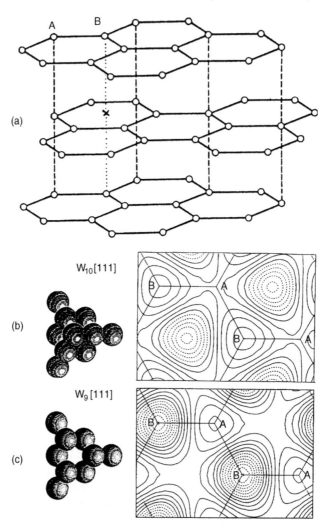

Figure 9.4 (a) Structure of graphite, showing the characteristic layered structure. Note the two non-equivalent lattice sites, marked A and B. (b) Simulation of the image of graphite surface obtained with a W_{10} [111] tip. (c) Simulation of the expected image after removal of the atom from the apex of the tip. Reproduced with permission of Elsevier Science from Tsukada *et al.* [7]

planes near the surface [8]. Such phenomena make the interpretation of 'atomic resolution' images much less straightforward than one might like. The positions of the points may in fact match the expected positions of the points in an authentic atomic resolution image, but, for example, with a much larger corrugation amplitude. Initially this was not fully appreciated and many images that were ascribed to atomic resolution imaging in the early days of the technique resulted from such effects.

On larger scales, the geometry of the tip may be convoluted with the surface structure, as illustrated rather simplistically in Figure 9.5. An atomically sharp tip scanning a region of the surface which contains an atomic protrusion would provide an accurate representation of the surface morphology (Figure 9.5(a)). However, if the tip is blunt, so that the surface feature has dimensions smaller than those the apex of the tip (Figure 9.5(b)), then the surface feature may effectively image the tip. Under such circumstances, it is not clear what the surface topography really is. Double-tip effects may also occur, where two asperities are formed in close proximity at the apex of the probe.

In summary, it is clear that an STM image, even at atomic resolution, is far from being a simple visual representation of the spatial locations of atoms. What we see instead is a complex image, determined by a convolution of the electronic structures of the tip and the sample surface. However, when it is interpreted correctly, the STM image is capable of providing us with data on the structure and bonding of atoms and molecules at surfaces which are quite unique in nature.

9.2.2 INSTRUMENTATION AND BASIC OPERATION PARAMETERS

Figure 9.6 illustrates the operation of an STM in a very schematic fashion. A wide variety of STM designs are in use, but certain features are common

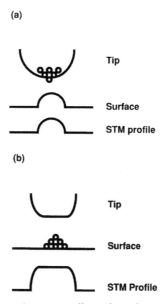

Figure 9.5 (a) Sharp tip imaging a small surface feature, yielding an STM image which accurately represents the surface topography. (b) Blunt tip imaging a sharp surface feature, with the consequence that an image of the tip (rather than the surface topography) is formed

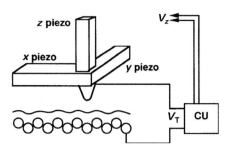

Figure 9.6 Operation of the STM, where CU is the control unit, V_T is the sample bias voltage and V_z is the voltage applied to the z piezo to maintain constant tunnelling current

to all of them. The key components are the tip, some means of achieving very delicate movements in the x-, y- and z-directions and a computer system to control the whole operation. Fine control of the tip position (both laterally and vertically) is achieved by the use of a piezoelectric crystal, on which either the STM tip or the sample is mounted. The piezoelectric crystal moves in a well defined fashion in each of the three spatial directions as the applied potential difference in each direction is varied.

There are three basic modes of operation: constant current, constant voltage and spectroscopic modes. We will discuss spectroscopic operation in a separate section; for the meantime we concern ourselves with the two imaging modes. In both modes, the tip is brought close to the surface (a few Ångströms) and a potential difference is applied, with the consequence that a current begins to flow. This current is the tunnel current, and it is monitored as the tip begins to scan across the surface. In constant voltage mode, the potential between the tip and the surface (the bias voltage) is maintained at a constant value, and the image represents the variation in the measured tunnelling current with position. The more commonly employed mode is constant current mode, however. In this mode, the instantaneous tunnelling current is measured at each position and the bias voltage is adjusted via a computer-controlled feedback loop such that the tunnelling current re-assumes some pre-set value. The STM image thus represents the variation in the z-voltage with coordinate (x_i, y_i). Adjustments to the bias voltage cause the piezoelectric crystal to move up and down, and if the displacement of the crystal is known for a given change δV in the bias voltage, then the image may be plotted as tip displacement (height), z_i against surface coordinate (x_i, y_i). Although the contours of this image are really formed by variations in electron density, in the limit of dimensions >10 Å, the image will provide a very good approximation to the surface topography.

There are two key experimental parameters: the tunnelling current and the tunnelling voltage. It is also sometimes helpful to think about the resistance of the tunnelling gap. The gap resistance clearly gives a measure of the distance between the tip and the surface. If the tip is close to the surface, the gap resistance will be relatively small; as the tip moves away from the surface, the gap resistance will increase. The distance of separation between the tip and the surface is in turn determined by the set-point for the tunnelling current: for large tunnelling current set-point values, the gap will be small; for low set-point values, the gap will be large. Typically, the tunnelling current lies in the range 10 pA–1 nA. At values much greater than 1 nA, there is an increasing likelihood that tip–surface interactions will become strong enough to change the surface morphology, although this is dependent upon the nature of the sample. For weakly bound adsorbates, the tip–surface interactions may become significant at much lower values, posing a number of problems which we will discuss in a little more detail below. Tip-induced damage to the surface can be quite severe, and thus sets upper limits on the magnitude of the tunnelling current which may be used. On the other hand, resolution is often better at higher tunnelling currents, and clearly a balance needs to be struck between the resolution of the recorded image on the one hand, and the likelihood of damage to the surface on the other. While tunnelling currents of the order of tens of pA may be employed for molecular and biological systems, currents of the order of several nA may be acceptable (and even necessary – see Section 9.2.3) when imaging metallic surfaces.

Variations in the bias voltage also have a strong effect on the nature of the image recorded. At very high bias voltage values, alterations to the surface structure may be induced. It becomes possible to etch the sample surface, and if the tip is moved in a controlled fashion across the surface whilst a high bias voltage is maintained, it is possible to create nanometre scale surface structures. Even at low bias voltages, the electric field gradient in the tunnel gap is substantial; for a tip–surface separation of about 10^{-10} m, and a sample bias potential of 0.1 V, $E = 10^9$ V m^{-1}.

9.2.3 ATOMIC RESOLUTION AND SPECTROSCOPY: SURFACE CRYSTAL AND ELECTRONIC STRUCTURE

It was STM images of individual atoms at surfaces and, still more astonishing, images in which surface electronic structure was resolved, that first fired the imagination of the scientific community. These early studies revealed the enormous potential of scanning probe techniques, and provided the impetus for the exploration of other materials. Consequently, we begin our survey of the applications of the STM with a brief overview of some of the important milestones in high resolution imaging of crystal surfaces. Studies of surface crystal and electronic structure still provide a fertile field of application for the STM, so our interest here is not merely historical.

Generally speaking, a crystalline solid organizes itself in such a way as to minimize its total energy. The resulting structure is usually accessible through X-ray diffraction analysis. When the crystal is cleaved, the atoms in the newly formed surface rarely remain in their original positions. For a metal, smoothing of the surface electronic charge density leaves surface atoms out of electrostatic equilibrium with the newly asymmetrical screening distribution. Relaxation of the surface atoms occurs in order to re-establish equilibrium. Whereas metallic bonding is non-directional, the bonding in a semiconductor is highly directional. When a semiconductor crystal is cleaved, for example to give the Si(111) surface shown in Figure 9.7, the atoms in the surface layer are left with dangling bonds directed away from the surface. Considerable reconstruction can occur in order to facilitate the formation of fresh bonds. The reconstructed surface may be very complex, and the determination of its structure can be an extremely difficult task.

Several techniques do exist by which the surface crystal structure may be studied, chief amongst them being low energy electron diffraction, LEED. However, the possibility of being able to visualize directly the arrangement of atoms at a surface has considerable appeal, and it was in the determination of surface crystal structure that the first successes of the STM were achieved. The ability of the STM to provide atomic-scale topographic information and atomic-scale electronic structure information (via spectroscopic operation) has enabled it to make a real and profound impact in this area.

Studies of Gold Surfaces. STM images which exhibited atomic rows, with atomic resolution perpendicular to the rows, were obtained first

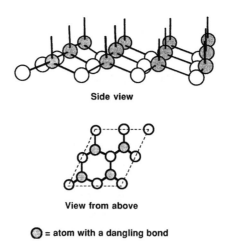

Side view

View from above

◯ = atom with a dangling bond

Figure 9.7 Ideal surface of Si(111) seen from the side and above

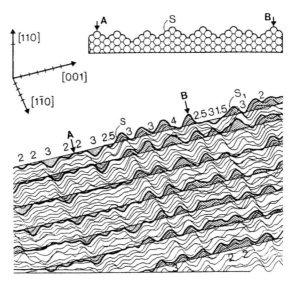

Figure 9.8 STM image of an Au(110) surface, exhibiting disorder. The straight lines help to visualize the terraced structure with monolayer steps (e.g. at S); below each line, the missing rows, and above each line, the remaining rows, are enhanced. The numbers in the top scan give distances between maxima in units of the bulk lattice spacing. The inset shows the proposed structural model for the observed corrugation between A and B. Reproduced with permission of Elsevier Science from Binnig *et al.* [9]

by Binnig *et al.* for the Au(110) surface [9]. The STM images showed parallel hills, running for several hundred Å in the [110] direction, and formed from a number of different facets, each of which exhibited a characteristic reconstruction. Figure 9.8 shows an STM image of a region of an Au(110) surface together with the authors' interpretation of the data (in the inset). In a subsequent paper, Binnig *et al.* reported observations of the reconstruction of the Au(100) surface [10]. A large-scale STM image of a clean 1×5 reconstructed Au(100) surface revealed a flat topography with monatomic steps (see Figure 9.9). However, higher resolution images revealed inhomogeneities in the detailed structure of the five periodicities, and the authors were able to postulate a detailed structure based upon their STM data.

Although these studies reported images of atomic rows, the individual atoms within the rows were not resolved. For a long while, this was thought to be impossible because of the intrinsic smoothness of electron density functions at the surfaces of metals. Au(111) was another close-packed surface to which this consideration was thought to apply. Although images had been obtained which exhibited extensive terraces separated by monolayer steps, no corrugation had been observed along the terraces. However, important progress was made in 1987, when Hallmark *et al.* demonstrated that individual gold atoms could be resolved on an Au(111) surface with

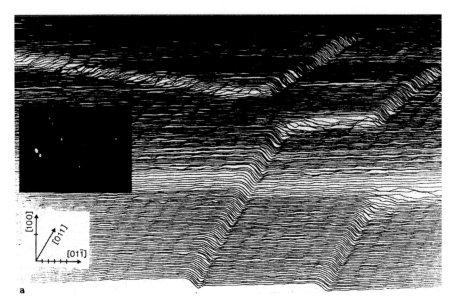

Figure 9.9 STM image of a clean 1×5 reconstructed Au(100) surface, showing monatomic steps. Reproduced with permission of Elsevier Science from Binnig *et al.* [10]

the STM [11]. Not only was this possible under UHV conditions, but under ambient conditions, too. This was principally because of the resistance of gold surfaces to oxidation in air. Most metals are rapidly oxidized on exposure to atmospheric oxygen, with the result that a passive oxide layer is formed. Besides the change in surface crystal structure which would be expected to accompany oxidation, the oxide layer formed on the surfaces of most metals under ambient conditions is sufficiently thick to preclude electron tunnelling.

Figure 9.10 shows a 25×25 Å region of an Au(111) thin film in air, prepared by epitaxial evaporation of gold onto a heated mica substrate at $300°C$. The atomic spacing in the image is 2.8 ± 0.3 Å, which compares well with the known gold interatomic spacing of 2.88 Å. Note that the image in Figure 9.10 is a tunnelling current image, rather than the usual z-voltage image. It was recorded with a DC level of 2 nA and a bias potential of $+50$ mV, with the gap resistance estimated to be ca. $10^7 \, \Omega$. Under these conditions, the tunnelling gap is relatively small. The authors attributed a good part of their success to the use of such a low gap resistance, compared to the larger values of around $10^9 \Omega$ employed by other workers, and noted that increases in the gap resistance to around $2 \times 10^8 \Omega$ led to the non-observation of atomic corrugation [11].

In fact, Binnig and Rohrer experienced considerable difficulties in their first studies of gold surfaces, and with hindsight, gold seems to have been

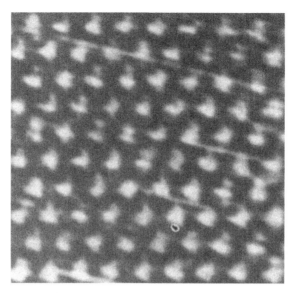

Figure 9.10 Atomic resolution image of an Au(111) surface. Reproduced with kind permission from Hallmark *et al.* [11], Copyright 1987, American Institute of Physics

a difficult choice to begin work on. The difficulties were not simply due to the smoothness of the surface electron density, either, but to the mobility of gold, which is so great that rough surfaces tend to smooth themselves out. Binnig *et al.* found that gold atoms were transferred to the STM tip, and the self-smoothing of these gold atoms led to a blunting of the tip with a concomitant loss of resolution. On occasions, they observed the resolution to jump unpredictably from high to low values, and attributed this to migrating adatoms locating themselves temporarily at the apex of the tip [12].

The effects of this gold mobility have been observed directly with the STM, using time-lapse imaging of the surface topography [13]. Indentations were created in an Au(111) surface by gently colliding an STM tip with it, creating a distinctive 'footprint'. This was achieved by applying a short-duration voltage pulse to the tip. At $30°C$, steps were observed to move, and recessed regions were filled in. Figure 9.11 shows an initial image of the gold surface showing three steps which run vertically along the {112} direction, followed by fourteen images taken at successive intervals of eight minutes after colliding the tip with the surface.

The stability of gold in air clearly makes it an ideal material on which to examine molecular samples, provided it is possible to prepare surfaces which are atomically flat over large enough areas. A number of methods have been investigated [14]. For example, the preparation of gold spheres by flame-annealing gold wire. Schneir *et al.* have described the formation of Au(111) facets on the surfaces of gold spheres formed by the melting of

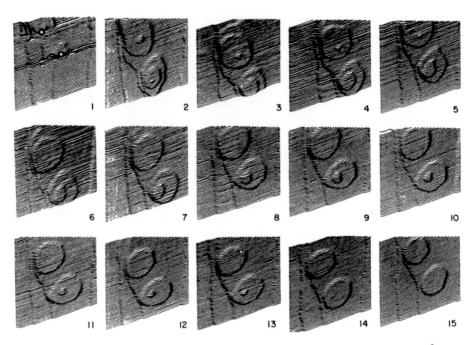

Figure 9.11 A series of time-lapse STM topographic images showing a 400×400 Å area of Au(111) after cleaning and annealing (1), creation of a 'footprint' with the STM tip (2) and thereafter at intervals of 8 min. During the 2-h period, the smaller craters in the footprint are filled in by diffusing gold atoms. Reproduced with kind permission from Jaklevic and Elie [13], Copyright 1988, American Institute of Physics

gold wire into an oxy-acetylene flame [15]. These facets are large enough to be observed quite straightforwardly using an optical microscope, and STM images confirmed them to be Au(111) surfaces which are atomically flat (or atomically flat with steps) over considerable distances (often extending as far as hundreds of nanometres). Although the facets do not cover the surface of the ball, they may be prepared routinely without any elaborate preparation procedures.

Graphite Surfaces.   The ability of the STM to obtain images in air (noted in Section 9.3.1 above) was demonstrated by Baro *et al.* [16] and subsequently exploited to good effect by Park and Quate [17] who obtained images of graphite surfaces with atomic resolution under ambient conditions. They studied the surface of highly oriented pyrolytic graphite (HOPG), supplied as a block which may easily be cleaved to yield a clean surface which is atomically smooth. The most popular method of preparation of HOPG surfaces is by removal of the top few layers of an HOPG block using adhesive tape. The clean surface which is exposed is stable in air for several days.

HOPG was initially a very popular substrate for STM studies, because of its relatively low cost and the ease of preparation of surfaces suitable for STM studies. However, the observation of a variety of graphite features which bear a close resemblance to helical molecules and other artefactual features has created interest in other substrates in certain applications – especially biological applications (see Section 9.2.3) – although HOPG is still widely used.

Early on, Sonnenfeld and Hansma showed that atomic resolution images could be obtained with the sample mounted in water [18]. Furthermore, it proved possible to operate the microscope in a saline solution. This latter capability indicated the potential role for the STM in studies of biological systems, where the possibility of performing high resolution microscopy in physiological-type buffer solutions became a reality for the first time. In order to facilitate studies under water, it was necessary to coat the bulk of the length of the STM tip with an insulating material, in order to minimize the area of the tip which could conduct current through the water and into the sample surface. This is a less demanding operation than might at first be expected, and many groups operate the STM and AFM under water, with the consequence that, for example, *in situ* studies of interfacial electrochemical processes are possible. A variety of tip-coating procedures have been reported, some of which are astonishingly simple (for example, coating the tip with wax or nail varnish).

The Silicon (111) 7 × 7 Reconstruction.   Shortly after their studies of the Au(110) reconstruction, Binnig *et al.* tackled a truly demanding problem which had been the subject of intense interest for some time: the 7×7 reconstruction of the Si(111) surface [19]. A number of models had been postulated for the structure of this surface, but none of these had been confirmed. In fact, the STM images did not fit any of the models exactly, but they matched one particular model, the adatom model, more closely than the others. It is a modified version of the adatom model which has ultimately come to be accepted, supported by data from other techniques, including transmission electron microscopy [20].

The STM images exhibited the expected rhombohedral 7×7 unit cell, bounded by lines of minima. These minima corresponded to empty adatom positions. Twelve maxima were observed inside each cell. These correspond to the positions of twelve adatoms, each bonded to three silicon atoms in the second layer of the reconstructed surface. This second layer is in fact formed from the topmost layer of the crystal during reconstruction. In the process of reconstruction, seven atoms are lost from the top layer (the reconstructed surface has only 42 atoms in its second layer). The unit cell of the reconstructed surface is customarily divided into two. In one half, the atoms in the second layer exhibit a stacking fault; the other half is unfaulted. It is along the row where these two halves of the unit cell are matched that the seven atoms are lost. Some of these atoms may occupy the adatom sites; alternatively, the adatoms may come from elsewhere in the crystal. In any case, the energetic gain is readily seen: 36 silicon atoms previously in the topmost layer of the solid are now incorporated into the second layer. Thus the number of dangling bonds in the surface is substantially reduced, even though the adatoms still possess dangling bonds, and this means an accompanying gain in energetic terms. The structure of the reconstructed surface is shown in Figure 9.12. It was the adatoms which were imaged by the STM; these are atoms which still possess dangling bonds in the reconstructed surface. The constant-current topograph was predominantly thought to be formed by tunnelling from the dangling bonds. An STM image of the Si(111)-(7 × 7) reconstruction is shown in Figure 9.13.

Scanning Tunnelling Spectroscopy.   Conventional surface-sensitive spectroscopic techniques (such as X-ray photoelectron spectroscopy (Chapter 3), Auger spectroscopy (Chapter 4) and vibrational spectroscopies (Chapter 7)) generate data which are averaged over an entire surface. However, many surface phenomena are strictly local in nature (for example, those associated with steps and defects). The images generated by the STM are determined by the electronic states at the surface and in the tip and, therefore, the STM can, in principle, map the electronic structure of a surface with atomic resolution. It is possible, in other words, to perform spectroscopic measurements on single atoms, facilitating detailed studies of local surface phenomena.

The STM image is dependent upon the bias potential in a complex fashion. It is this dependence which forms the basis for the spectroscopic operation of the STM. To see this, consider the diagrams in Figure 9.14 for a hypothetical one-dimensional system. When the tip potential is negative

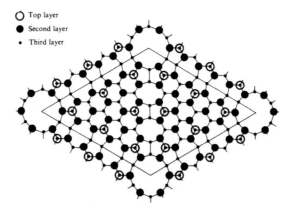

Figure 9.12   The structure of the Si (111)–(7 × 7) reconstruction. Reproduced with permission of the Cambridge University Press from A. Zangwill, *Physics at Surfaces*, Cambridge University Press, Cambridge, UK, 1988

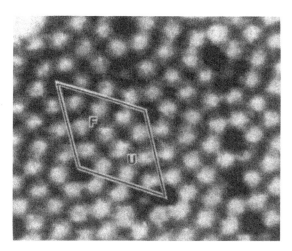

Figure 9.13 Topographic image of the Si (111)–(7 × 7) reconstruction, showing the unit cell. The faulted and unfaulted halves are marked *F* and *U*, respectively. Reprinted with permission of the American Institute of Physics from Hamers *et al.* [21]

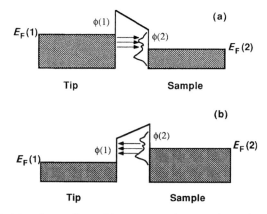

Figure 9.14 (a) Tip biased negative with respect to the sample, so that tunnelling is from the tip to the sample. (b) Polarity and direction of tunnelling reversed

with respect to the sample, electrons tunnel from occupied states of the tip to unoccupied states of the sample (Figure 9.14(a)). When the tip potential is positive with respect to the sample (negative sample bias), electrons tunnel from occupied states of the sample to unoccupied states of the tip (Figure 9.14(b)). Since states with the highest energy have the longest decay lengths into the vacuum, most of the tunnelling current arises from electrons lying near the Fermi level of the negatively biased electrode.

There are several spectroscopic modes of operation of the STM. The simplest of these is known as voltage-dependent STM imaging, which involves acquiring conventional STM images at different bias voltages, and which provides essentially qualitative information. For example, consider the case of the Si(111)-(7 × 7) reconstruction. Images recorded at positive sample biases reveal 12 adatoms of equal height in each unit cell [19]. However, as we noted in Section 9.3.1, the unit cell of this reconstruction may be divided into two non-equivalent halves, one of which exhibits a stacking fault with the underlying atomic planes. Imaging of the surface at negative sample bias voltages results in the adatoms in the faulted half of the unit cell appearing higher than those in the unfaulted half [22]. Furthermore, in each half of the unit cell, the adatoms nearest the deep corner holes appear higher than the central adatoms.

In scanning tunnelling spectroscopy (STS), quantitative information is created through the use of a constant DC bias voltage with a superimposed high-frequency sinusoidal modulation voltage between the sample and the tip. The component of the tunnelling current that is in phase with the applied voltage modulation is measured, whilst maintaining a constant average tunnelling current. This enables the simultaneous measurement of dI/dV and the sample topography.

A third spectroscopic mode of operation involves the local measurement of tunnelling $I–V$ curves. These $I–V$ curves must be measured with atomic resolution at well defined locations, with a fixed tip–sample separation in order to correlate the surface topography with the local electronic structure. The first such measurements were made by Hamers *et al.* [21] who termed their technique current-imaging-tunnelling spectroscopy (CITS). Other variants have been described (for further discussion and a review of spectroscopic operation of the STM, see Hamers [23]). Using CITS, Hamers *et al.* were able to map out the electronic structure of the Si(111)-(7 × 7) unit cell with a lateral resolution of 3 Å [22, 23]. At voltages between −0.15 and −0.65 V, most of the current arises from dangling bond states of the twelve adatoms (see Figure 9.15(a)). More current comes from the adatoms in the faulted half of the unit cell (cf. the example of voltage-dependent imaging above) and, in each half-cell, more current comes from the adatoms adjacent to a corner hole than comes from the other adatoms. A differential current image recorded with bias voltages between −1.0 and −0.6 V (Figure 9.15(b)) revealed the positions of the six dangling bonds on atoms in the second layer (the Si(111)-(7 × 7) unit cell has a total of 18 dangling bonds).

These dangling bonds exhibited a reflection symmetry which was attributed to the presence of the stacking fault.

A third occupied state was imaged as the differential current between −2.0 and −1.6 V (Figure 9.15(c)), revealing regions of higher current density, corresponding to Si—Si backbonds, and bright spots at the corner holes, where additional back bonds are exposed. These quite astonishing images exemplify the remarkable capabilities of the STM for the determination of surface electronic structure, and illustrate the power which the technique

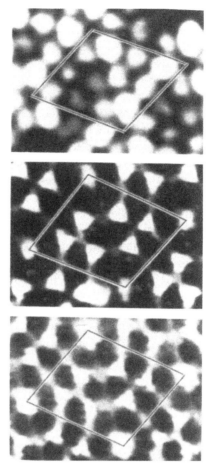

Figure 9.15 CITS images of occupied Si (111)–(7 × 7) surface states: (a) adatom state at −0.35 V; (b) dangling-bond state at −0.8 V; (c) backbond state at −1.7 V. Reproduced with permission of the American Institute of Physics from Hamers *et al.* [21]

makes available to the surface scientist for the determination of surface structure.

Metal oxides are important in both electronic devices and catalytic systems. In metal–oxide field effect transistors (MOSFETs), the basis for modern electronics, the decreasing thickness of the gate oxide layer is of critical importance in ensuring the continued performance of devices as their dimensions shrink. Dielectric layers may be characterized by STM and STS if they are thin enough. For example, for layers of NaCl on Al(111) and of CoO and NiO on Ag(001), it has been estimated that the maximum thickness is three layers [24]. The growth of MgO layers on silver was studied by Schintke *et al.* [25]. They evaporated Mg under a partial oxygen pressure of 10^{-6} mbar. After decomposition of 0.3 monolayers of MgO, two-dimensional islands approximately 10–15 nm in size were observed to have nucleated. After deposition of two monolayers, the silver surface was completely covered with MgO, forming terraces typically 50 nm wide and pyramidal islands. Scanning tunnelling spectra at one monolayer exhibited two structures in the LDOS at 1.7 and 2.5 eV at positive bias (unoccupied states). The feature at 1.7 eV was attributed to MgO–Ag interface states, while the peak at 2.5 eV was attributed to the onset of the MgO(001) empty surface state band. It was concluded that the electronic structure of an MgO single crystal developed within the first three monolayers.

Carbon nanotubes have attracted enormous interest because of their potential utility in nanostructured electronic devices (for example, a number of workers have reported the fabrication of carbon nanotube transistors). As a result there has been a great deal of interest in measuring the electronic properties of carbon nanotubes. STS measurements on chiral single walled nanotubes have enabled the LDOS to be measured in two different ways, either simply as the differential dI/dV, or as the normalized differential, $(V/I)(dI/dV)$. Singularities, thought to be the so called van Hove singularities associated with the one-dimensional electronic structure of the nanotubes [26, 27], have been observed near the Fermi level where the tunnel current falls rapidly to 0. There are a variety of problems associated with the interpretation of such data [28]. For example, under the conditions used, the tip should be very close to the surface and may apply a significant load, which would be expected to lead to an alteration of the electronic structure of the nanotubes. Paradoxically, however, measured bandgaps are similar to those of undistorted nanotubes.

Emission of tunnelling electrons by an STM tip may lead to inelastic processes that result in the emission of photons. Berndt et $al.$ studied C_{60} molecules adsorbed onto a reconstructed Au(110) surface [29]. STM images were able to resolve individual molecules in a hexagonal array. Photon emission was measured simultaneously, and it was found that each C_{60} molecule yielded a bright spot due to photon emission. Emission appeared to be maximized when the tip was situated directly above a molecule.

Despite the impressive nature of these successes, they revolve around the analysis of surface electronic structure rather than chemical structure: they provide information on electronic states (via measurements of the local density of states) but not on specific bond types and chemistries. One approach is the method known as inelastic electron tunnelling spectroscopy (STM-IETS) [30, 31]. In this technique, I/V measurements are made, but the underlying principle is that a small, sharp increase in the tunnelling conductance may be observed when the energy of the tunnelling electrons reaches that of a vibrational mode for molecules at the junction. The increase is a result of electrons losing their energies to the vibrational mode, giving rise to an inelastic tunnelling channel, which is forbidden when tunnelling electrons have energies below the quantized vibrational energy. Stipe et $al.$

were able to acquire IET spectra from single acetylene molecules adsorbed on single crystal surfaces at 8 K [30, 31]. The requirement to operate at such low temperatures constitutes a significant limitation to the technique.

STS is a powerful tool. For the right systems, it can provide unique insights into the electronic structure of a surface at very high resolution. However, it does present challenges. Not only can the interpretation of the data be complex, but it does not provide any direct access to chemical structure information, and it is generally regarded as a UHV method. While it can be extremely valuable, it is not likely to become a tool that sees widespread application.

Molecules at Surfaces. There are many situations in which one might wish to visualize the interactions of molecules with surfaces, and many of these lie outside of the conventional boundaries of surface science. The STM and the AFM have made dramatic impacts on a number of fields in which it is molecular material which is of interest. In a large number of these applications, the STM and the AFM are able to provide structural data under circumstances in which other techniques would be rendered useless. Here we examine two examples in a little more detail.

Liquid crystals – Smith *et al.* investigated the interaction of liquid crystals with graphite surfaces [32]. They examined alkylcyanobiphenyl molecules of the type 4′-*n*-alkyl-4-cyanobiphenyl, known by the abbreviated names *m*CB, where *m* is the number of carbons in the alkyl tail attached to the biphenyl ring. The liquid crystal molecule 8CB has the structure:

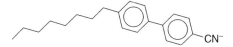

At first sight it might seem unlikely that it would be possible to obtain STM images of something so poorly conducting as a liquid crystalline film of organic molecules. In fact, however, a low intrinsic conductivity does not constitute an obstacle provided the sample is in the liquid state; the STM tip simply travels through the bulk of the liquid until it is sufficiently close to the surface for electron tunnelling to reach a measurable level. Practically, this effectively restricts the probe to the interfacial region, and it therefore means that the molecules which are imaged are those molecules closest to the substrate surface. The key constraint in the case of liquid crystals is that the samples are close to their liquid to solid-crystalline transition temperature. At higher temperatures, the films become completely liquid-like and no order is observable in the STM images. At lower temperatures, the films no longer wet the graphite.

With a tunnelling current of 100 pA and a bias voltage of 0.8 V, images of the liquid crystal/graphite interface revealed a complex interlocking

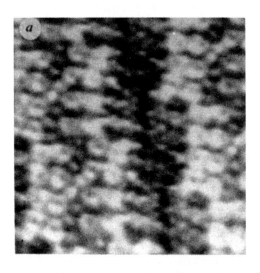

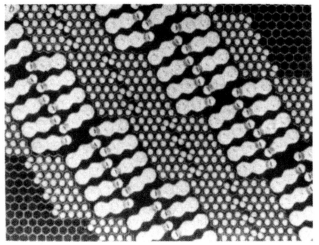

Figure 9.16   (a) 56 × 56 Å STM image of 8CB on graphite and (b) a model of the 8CB lattice showing the hydrogens of the alkyl chains registered with the hexagonal centres of the graphite lattice. Reproduced with permission from the Nature Publishing Group from Smith *et al.* [32]

arrangement of molecules (see Figure 9.16). Bright spots were observed which corresponded to the locations of the cyanobiphenyl headgroups. The alkyl chains were also observed, although they showed up less brightly, as patterns of points which corresponded to methylene groups. The cyano groups were observed as a small bright point at the opposite end of each molecule. Examination of high resolution images indicated that this bright spot was separated from the phenyl group to which it was joined by some distance, suggesting that it corresponded to the nitrogen atom of the cyano group in particular, and that the triply bonded carbon atom of the cyano group therefore has a relatively low tunnelling conductance. The cyano groups pointed inwards towards each other, interdigitating slightly and increasing the packing density.

It was suggested that the two-dimensional lattice formed by the molecules was primarily the result of registry of the alkyl tails with the graphite lattice. By decreasing the gap resistance, and so driving the tip closer to the surface, it was possible to penetrate the film and to image the underlying substrate. The directions of the graphite lattice vectors were therefore determined, and it was found that the alkyl chain of each molecule aligned along a particular graphite lattice vector, whilst the cyano group aligned along a different lattice vector. At grain boundaries, the tails were often observed to rotate by $60°$, matching the graphite symmetry.

Hara *et al.* have performed STM measurements on 8CB, but instead of using graphite as a substrate, they used molybdenum disulfide [33, 34]. MoS_2 is electrically conducting, and clean surfaces may be prepared by cleaving an MoS_2 single crystal. In contrast to the bilayer structure which 8CB forms on graphite, it forms a periodic single row structure on MoS_2 in which cyanobiphenyl head groups and alkyl tails alternate in each row (see Figure 9.17). Each row has a width of around 21 Å, whereas the width of the bi-layer formed on graphite is around 38 Å. The driving force for adsorption in this arrangement is thought to be the strong interaction between the 8CB molecules and the substrate, and, in fact, the periodicities observed in the liquid crystal film correlate closely with the lattice spacing of the MoS_2

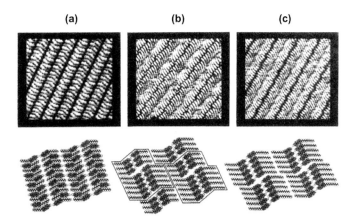

(a)   (b)   (c)

Figure 9.17   STM images and models showing the anchoring structures of *m*CB on MoS_2: (a) homogeneous 8CB single-row (15×15 nm^2); (b) inhomogeneous (mixed) double-row (20×20 nm^2); (c) homogeneous 12CB double-row (20×20 nm^2). Reproduced with permission of RIKEN (The Institute of Physical and Chemical Research) from Hara [34]

substrate. The single row structure is thus associated with the formation of the anchoring phase of the nematic phase of these liquid crystals.

In contrast, 12CB, which exhibits a smectic phase, forms a double row structure on MoS$_2$, in which cyano groups face each other. This structure is also associated with the formation of an anchoring phase at the interface when the isotropic-to-smectic transition occurs in the liquid crystal bulk. When mixtures of 8CB and 12CB were formed, it was found that the molecules organized into an inhomogeneous double row structure, effectively a phase-segregated structure in which nanoscopic domains of 8CB and 12CB were formed side-by-side.

Self-assembled monolayers – Alkanethiols, $HS(CH_2)_nX$, where X could be a variety of organic functional groups, including methyl, hydroxyl, carboxylic acid and amine, adsorb spontaneously onto gold surfaces to form ordered assemblies in which the alkyl chains are oriented at an angle of ca. $30°$ from the surface normal [35]. Early work on these systems using low energy electron diffraction indicated that they exhibited quite long-range order, and in many ways exhibited common structural characteristics with Langmuir–Blodgett (LB) films – ordered, effectively two-dimensionally crystalline assemblies of amphiphilic molecules adsorbed onto a solid surface [36]. However, LB films are formed by using a Langmuir trough to compress a film of amphiphiles at the air–water interface until surface pressure compels them to stand perpendicular to the water surface and adopt a close-packed arrangement; in a second subsequent step the LB film is transferred to a solid surface. In contrast, monolayers of alkanethiols assemble themselves spontaneously on gold surfaces, organizing themselves to form close-packed, ordered structures. As a result, this class of materials has become known as 'self-assembled' monolayers, or SAMs and their ease of preparation, compared to LB films, combined with their versatility (for example, the polarity of the surface may readily be changed by varying the nature of the tail group, X), have made them attractive for both fundamental investigations (of diverse phenomena, including wetting, adhesion and biological interfacial interactions) and for applications in sensors and electronic devices and structures [37].

The rapidly burgeoning interest in SAMs during the early 1990s led to a great deal of interest in understanding their structures, and STM provided some particularly powerful insights into what turned out to be a rich variety of film structures. Initial adsorption of alkanethiols onto gold surfaces was demonstrated early on to be a rapid process, in which hydrogen was lost from the head group and a gold thiolate adsorption complex resulted [38]:

$$Au + HS - R \rightarrow AuSR + \tfrac{1}{2}H_2$$

Equivalent products resulted from the adsorption of dialkyldisulfides, RSSR, via cleavage of the disulfide linkage. However, what was a great deal more uncertain was the time that was taken by the system to come to equilibrium and, indeed, the nature of the equilibrium structure. For example, when mixed monolayers were formed by exposing a gold-coated substrate to a solution containing two contrasting thiols, the composition of the monolayer that formed did not exactly match that of the solution, indicating that the SAM composition was not simply determined by the rapid adsorption kinetics.

As noted above, early work using LEED suggested that the adsorbates adopted an ordered, hexagonal arrangement on the gold surface, thought to be a $(\sqrt{3} \times \sqrt{3})R30°$ structure (i.e. the surface mesh was $\sqrt{3}$ times the size of the Au surface mesh and rotated by 30°). However, LEED measurements on delicate, electron-sensitive organic monolayers are difficult. Direct evidence was needed to support this assignment. In a series of elegant studies, Poirier and co-workers established much of the fundamental understanding we now have through careful application of STM to monolayers formed by the adsorption of thiols in vacuum.

Delamarche *et al.* [39] and Poirier and Tarlov [40] published the first work pointing to a more complex phase behaviour in 1994. Both groups published images in which individual adsorbate molecules were resolved. The basic arrangement of molecules was certainly hexagonal, in agreement with the predicted $(\sqrt{3} \times \sqrt{3})R30°$ structure. However, small height differences were observed that indicated additional complexity. A $p(3 \times 2\sqrt{3})$ unit mesh was identified by Poirier and Tarlov, and both groups identified a $c(4 \times 2)$ superlattice. Delamarche *et al.* proposed a model in which rotations clockwise and anticlockwise about the molecular axis led to differences in the angle of orientation of the terminal C—H bonds with respect to the surface normal. These subtle changes in adsorbate geometry led to a height difference of 0.6–0.7 Å: a small but detectable difference that yielded the two different sites observed in the STM images.

Figure 9.18 shows a large area STM image of an SAM of dodecanethiol $(HS(CH_2)_{11}CH_3$, or C12). It is clear that while the surface certainly exhibits extensive crystalline regions, it is not ordered over macrosopic distances, at odds with the rather idealized representations for SAMs often found in the literature. The monolayer is instead a patchwork of domains, each some 5–20 nm in diameter. Within each domain the adsorbates are highly ordered, and may be described as adopting two-dimensional crystalline structures. However, there are breakdowns in order along domain boundaries, as domains in which rows of adsorbates exhibit different directionality meet. The dark patches in Figure 9.18 are not, in fact, holes in the monolayer, but are depressions, where removal of Au atoms from the surface has made the surface a little lower than in surrounding areas. The depressed regions are filled with adsorbates. Figure 9.19 shows high resolution STM images showing two regions that exhibit slightly different packing arrangements. The locations of individual adsorbate molecules are clearly resolved. Two

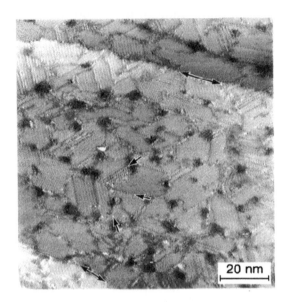

Figure 9.18 STM image of a self-assembled monolayer of dodecanethiol on Au(111) acquired using a Pt/Ir tip at 1.2 V with a tunnel current of 3 pA. The greyscale range is 5 Å from black to white. Reproduced with permission of the American Chemical Society from Delamarche *et al.* [39]

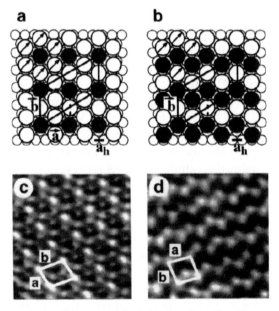

Figure 9.19 STM images (c,d) and models (a,b) showing c(4 × 2) superlattice structures of dodecanethiol on Au(111). Reproduced with permission of the American Chemical Society from Delamarche *et al.* [39]

different superstructures are shown, both with the same sized unit mesh. In each case the superlattice cell is marked on the STM image.

Studies of monolayers of butanethiol on gold [41, 42] hinted at the existence of a larger and more complex phases space. The order in SAMs was found in early studies to depend upon the length of the adsorbate molecule: while long-chain (more than 10 carbon atoms) adsorbates were found to be present in all-*trans* conformations, and to exhibit high degrees of order, increasing alkyl chain mobility occurred, leading to increasing numbers of *gauche* defects, as the adsorbate chain length decreased below 10. When Poirier and co-workers studied SAMs of butanethiol using STM, they acquired comparatively featureless images initially. However, they noticed that after extended periods, structure started to become visible. So-called striped phases were observed, within which adsorbate molecules were clearly resolved. These consisted of rows of molecules (pronounced stripes in STM images) that, on imaging at high resolution, were found to be composed of adsorbate molecules lying parallel to the substrate surface. The initial featureless morphology was presumably the consequence of chain mobility in the disordered butanethiol layers; it was concluded that desorption of adsorbates occurred and led subsequently to the formation of the striped phases. The implication of these observations was that chain–chain interactions provided a substantial element of the stabilization of an SAM; in adsorbates as short as butanethiol, chain–chain interactions were weak and the upright orientation of the adsorbate regarded previously as normal, was not so clearly favoured compared to a flat orientation in which the alkyl chain interacted with the gold surface – which in fact exhibits a rather strong Lifshitz dispersion force.

A subsequent study of the slightly longer adsorbate, hexanethiol, led to dramatic data revealing the opposite process – the transformation of flat-lying adsorbates into close-packed monolayers of upright alkyl chains [43]. Hexanethiol experiences slightly more interchain stabilization than butanethiol, and moreover, rather than beginning with a fully formed monolayer, Poirier and Pylant began with a clean gold surface, on which the characteristic herringbone reconstruction could be resolved, and adsorbed thiol molecules. At low coverages, adsorption led to the nucleation of domains of adsorbates that clearly exhibited the striped morphology. As coverage increased, the sizes of these striped domains grew. At an exposure of 600 L, a little more than half of the surface was covered with striped phase domains, and the herringbone reconstruction remained visible on the rest of the surface. Gold vacancy islands, presumably formed by the thiol-induced reconstruction of the gold surface, had begun to form. As the exposure was increased, surface coverage increased but at exposures of ca. 1000 L, a new phase was observed, consisting of small islands of upright molecules. As the coverage increased, the area occupied by the upright phase domains increased, with large domains being observed at 2500 L. This process was accompanied by growth in the sizes of the gold vacancy islands. Eventually, the upright phase came to dominate the surface structure.

A great deal of work has followed that has been directed towards the explication of the process of monolayer assembly, and on the basis of which Poirier was able to propose a detailed description of six discrete structural phases [44]. While scattering techniques and other methods have been important in exploring mechanisms of SAM formation and restructuring, STM has undoubtedly played a pivotal role in building the basis for our understanding of the structures of these systems at a fundamental level.

Applications of STM in Biology.   In the biological sphere, scanning probe techniques hold tremendous promise, and there is likely to be considerable further development in this area in the next few years. However, some very difficult practical problems are also encountered and much effort is currently being directed towards resolving them.

We have already noted, in Section 9.3.1, that the STM is capable of operating in a liquid environment. Not only is it possible to operate in water, but it is also possible to operate in salt solution. This is very important in a biological context. Although electron microscopy has been developed into a very powerful technique for the study of biomolecular structure, it is not capable of imaging molecules without some form of sample pre-treatment. Biological molecules are highly sensitive to environmental changes (for example, proteins are highly sensitive to changes in pH, and are easily denatured resulting in severe structural modifications), and many of the procedures employed to prepare samples for electron microscopy (for example, drying and freezing) are potentially the causes of structural modification in biomolecules.

The first STM images of DNA were obtained by Binnig and Rohrer [45]. Considerable interest was provoked, and early hopes were that the STM might provide a probe for structure more than capable of competing with existing biophysical techniques. There were, indeed, even hopes that DNA sequencing might become possible using the STM. In 1989, Lindsay *et al.* [46] obtained the first images of DNA under water. The way had opened up for the STM to provide the capability for visualizing biomolecules and their interactions in physiological buffer solutions. The realization of this possibility would constitute a major step forward in our ability to understand biological molecules and their interactions.

However, a re-evaluation of the situation was called for when it was discovered that features of graphite surfaces could mimic DNA (and other helical molecules, for that matter) [47]. Clemmer and Beebe presented images of freshly cleaved HOPG which bore an astonishing resemblance to helical molecules (Figure 9.20). Heckl and Binnig [48] have recorded images at graphite grain boundaries which also resemble DNA quite closely. Comparison with a computer-generated model (Figure 9.21) reveals just how

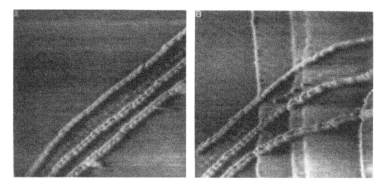

Figure 9.20 Graphite features which resemble helical molecules. Reproduced by permission of the American Association for the Advancement of Science from Clemmer and Beebe [47]

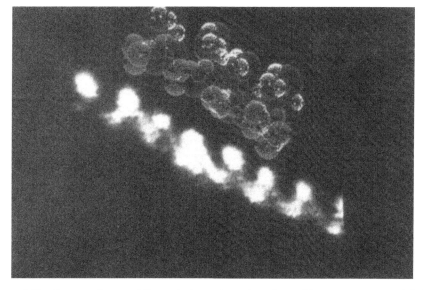

Figure 9.21 Image of a graphite grain boundary, together with a computer model of a DNA molecule. Reproduced by permission of Elsevier Science from Heckl and Binnig [48]

close the resemblance is. Besides these problems with surface features which mimic biomolecules, it was also being recognized that the tip–sample interaction force during imaging was sufficiently large to move biomolecules. The consequence of this tip-induced movement is that the sample molecules are pushed outside of the scanned area of the surface, a phenomenon illustrated by Roberts *et al.* in the case of mucin proteins [49]. Moving the tip to a new spot simply reproduces the problem. Ultimately, it is often only possible to image molecules which are prevented from movement by some physical obstacle (for example, a step edge), but then it can become difficult to deconvolute the two structures.

A number of approaches have been developed to overcome these problems. The first has been a shift of interest away from graphite substrates to other substrates that present fewer interpretational problems (for example, gold and mica). More importantly, a number of groups have begun to attempt to develop novel methodologies for sample preparation in STM. The most successful methodologies can be roughly divided into two: immobilization by the application of a conducting coating [50], and immobilization by coupling the sample molecule to a chemically functionalized surface [51–55]. In either case, the objective is to secure the biomolecule so that it is not moved by tip–sample interactions during imaging, and to generate a uniform surface coverage.

The very fact that STM images may be obtained of uncoated biological molecules poses some very interesting problems. A protein may be of the order of a few nanometres thick, and yet one may still be able to record a tunnelling current. Tunnelling directly through the biomolecule in the conventional sense is ruled out, because the tunnelling gap would simply be too great (remember that the tunnelling current is reduced by an order of magnitude for every 1 Å increase in the tip–surface separation). A number of competing models have been proposed, and the issue remains the subject of a great deal of contention. One explanation, due to Lindsay [56], is that deformation of the sample molecule by the STM tip alters its electronic structure, with the consequence that states are created at the Fermi level of the substrate. Under these conditions, resonant tunnelling can occur and an enhancement of the tunnel current will be observed. Although it has clear applications in explaining data obtained under UHV conditions, such a process would result in a substantial disruption of the molecular structure, with a concomitant reduction in the usefulness of the data obtained.

Alternatively, it has been suggested that water, present on the surface of the biomolecule, provides the means for conduction. This latter hypothesis is supported by evidence provided by Guckenberger et al. [57], in studies of two-dimensional protein crystals. They found that image acquisition was difficult if the relative humidity (RH) inside the STM chamber was less than about 33 %, implying that hydration of a biological sample was necessary for the operation of the STM in air. Significantly, they also reported that very low tunnel currents (typically no larger than a few pA, giving an extremely large gap resistance) were necessary. A rationalization for these observations was provided by Yuan et al. [58] who proposed an essentially electrochemical image formation mechanism in which ions were transported through the film of water covering the protein molecule. A water bridge linked the STM tip to the sample. This model was tested further by Leggett et al. who examined the effects of dehydration of the

STM chamber on image contrast. Protein molecules were immobilized by covalent attachment to thiol monolayers on gold substrates, and imaged under ambient conditions (RH about 33 %). A desiccant was then inserted into the STM chamber, and a series of images was recorded as the relative humidity fell, over several hours to 5 % [59]. The image contrast gradually changed until at 5 % RH, a reversal of contrast had occurred. The protein structures, which under ambient conditions had appeared to be raised some 6 nm above the substrate, now appeared as troughs. The STM measured a sharp drop in conductance as the tip traversed the dehydrated protein molecules, as compared to a rise under ambient conditions. This observation lends strong support to the hypothesis that water provides the conducting path in the STM. However, it also provides a salutory reminder that seeing is not necessarily believing in the STM: image contrast does not necessarily bear a simple relationship to surface topography, and without an adequate theoretical basis for image interpretation, the STM data for biological systems must be treated with extreme caution.

9.3 Atomic Force Microscopy

9.3.1 BASIC PRINCIPLES OF THE AFM

Forces at Surfaces. We have already noted that at high tunnelling currents, the STM tip may interact physically with the surface in such a way that disruption of the surface structure occurs. In fact, there is usually a finite interaction force between the tip and the sample surface. Even at relatively low tunnelling currents, the interaction force may be substantial when measured against the strengths of molecular interactions. Knowledge of the existence of these forces led Binnig *et al.* [60] to develop the AFM, in which the probe becomes a cantilever, placed parallel to the surface rather than normal to it. The cantilever of the AFM has a sharp, force-sensing tip at its end, and it is this that interacts with the surface. As the interaction force between the cantilever tip and the surface varies, deflections are produced in the cantilever. These deflections may be measured, and used to compile a topographic image of the surface. The process is illustrated schematically in Figure 9.22. Microscopes have been designed which can monitor interactions due to a range of forces, including electrostatic and magnetic forces. For example, the magnetic force microscope has a tip which possesses a magnetic moment and which therefore responds to the magnetic field of a magnetized sample, while the electrostatic force microscope senses surface charge; it is the electrostatic interaction between the charged tip and the sample which is measured. Scanning force microscopes (SFMs) generally measure forces in the range 10^{-9}–10^{-6} N, although the measurement of forces as low as 3×10^{-13} N has been reported [61].

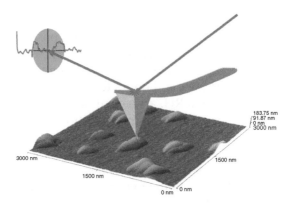

Figure 9.22 Schematic illustration of the operation of the AFM

Clearly, because the AFM is based upon force measurement, there is no longer any need for the sample to be an electrical conductor. This is perhaps the most important distinction from the STM and the reason that AFM has attracted such widespread interest. Effectively, it offers high resolution microscopy (with a resolution comparable to that of an electron microscope in many cases) for insulating samples under ambient and liquid conditions. This represents an enormous spread of new capability.

The measurement of forces between atoms and molecules can tell us much about their structures and the nature of their interactions. The forces between atoms may be described by the Lennard–Jones potential (see Figure 9.23):

$$V(r) = 4E\left[\left(\frac{\sigma}{r}\right)^{12} - \left(\frac{\sigma}{r}\right)^{6}\right] \qquad (9.10)$$

The energy of interaction has a minimum value E at an equilibrium separation r_0, and the separation is σ at $V(r) = 0$. At separations greater than r_0, the potential is dominated by long-range attractive interactions that decay as a function of $1/r^6$, while at shorter distances, the interaction becomes increasingly dominated by short-range repulsive interactions that vary with $1/r^{12}$. These are quantum-mechanical in nature and arise from the interpenetration of the electron shells of the interacting atoms at small separations.

For macroscopic objects, single atomic interactions are replaced by interactions between larger ensembles of molecules or atoms, and the mathematical description is slightly different [62]. For a molecule interacting with a planar surface at a distance D, the interaction energy is given by:

$$W = -\pi C\rho/6D^3 \qquad (9.11)$$

where C is a collection of constants and ρ is the density of the material. Already it may be seen that the interaction energy varies with the inverse

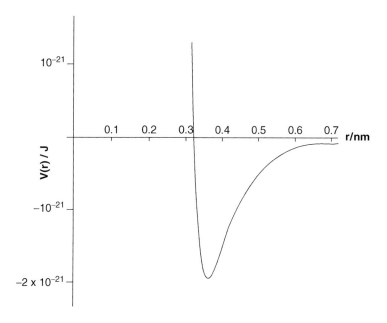

Figure 9.23 The Lennard–Jones pair potential for two argon atoms

cube root, rather than the inverse sixth power, of the separation. If the molecule is replaced by a hemisphere – a reasonable approximation to the tip of a conventional AFM probe – then the relationship becomes:

$$W = -AR/6D \qquad (9.12)$$

where R is the radius of the sphere and A is the Hamaker constant. Now, the interaction energy simply varies with the reciprocal of the separation. The fact that the interaction energy varies so much less sharply with the separation, when compared to the Lennard–Jones potential, is of critical importance in facilitating measurement of the interaction force as a function if distance – if it varied with $1/r^6$ then it would be a much more challenging problem to measure the force accurately as a function of separation.

For two crossed cylinders the force is:

$$W = -A\sqrt{R_1 R_2}/6D \qquad (9.13)$$

Equation (9.13) has particular importance because for over three decades, accurate measurement of forces at surfaces has been possible using the surface forces apparatus (SFA), in which crossed mica cylinders are allowed to interact while the force of interaction is measured. Space does not permit a detailed examination of the SFA here, and the reader is referred to Israelachvilli's excellent textbook on surface forces for further details [62]. However, it is useful to note that while there are conceptual similarities between the SFA and the AFM (for example, both utilize a spring to measure the interaction force), there are some important differences, perhaps most significantly that the SFA is not an imaging tool, while the AFM clearly is. It is probably fair to say that force measurement in the SFA is more precise than in the AFM. However, the SFA is subject to the limitation that the interacting surfaces must be formed onto atomically smooth cylinders. This places substantial constraints on the range of experimental systems that may be studied. Hence while the SFA is a very powerful, quantitative tool, the AFM, with its potentially very broad range of applicability, offers unique capabilities that have made it an important tool for the fundamental study of molecular interactions (in addition to its utility for imaging).

Force–Distance Measurement.   Forces are measured quantitatively in a variety of ways. The deflection of the cantilever both perpendicular to the surface and parallel to the plane of the surface may be measured. The force–distance measurement (also known as a 'force curve' or 'force spectroscopy') is the most basic type of quantitative measurement. Figure 9.24 shows what is involved qualitatively.

The probe is positioned above the sample surface, with the tip not touching the sample. The cantilever is then lowered towards the surface (A). As the probe comes very close to the surface, a mechanical instability causes the tip to 'snap into contact' with the surface (B). The tip is now touching the surface. As the piezoelectric scanner is advanced further, the tip is pushed into the surface (C), leading to the measurement of a repulsive force. At some pre-determined point, the cycle is stopped and the tip is retracted (D). If the tip adheres to the surface, hysteresis will occur: the retraction path will not be the same as the approach path. The

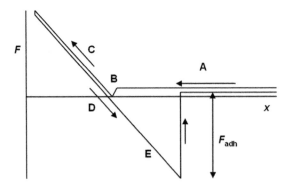

Figure 9.24 Variation in the interaction force as a function of distance during a force–distance measurement

consequence of this is that the tip must be lifted further than the point at which it originally contacted the surface to separate it (E). During this phase of the force–distance measurement, the force has a negative sign – i.e. it is attractive: the tip is effectively pulling on the sample. Eventually, the tip 'snaps off' the surface and returns to a position of non-contact. The separation between the minimum in the force–distance plot and the position of non-contact is the adhesion force, F_{adh}, which is also called the pull-off force.

There has been a great deal of effort directed towards the development of functionalized probes, to enable the capacity of the AFM for force–distance measurement to be exploited in such a way that specific molecular interactions may be studied. Two important areas of activity are provided by chemical force microscopy (Section 9.3.3) and the measurement of biological recognition and unfolding phenomena (Section 9.3.5).

Contact Mechanics.   It is useful to be able to model the interaction between the tip and the surface, in order to calculate the contact area, or to explore relationships between the interaction force and other parameters. Given that with a standard commercial AFM probe, which typically has a radius of curvature of tens of nm, the contact area is large compared to molecular dimensions, quantum-mechanical effects may safely be ignored, and the tip–sample interaction treated using continuum mechanics (classical) models. This is a very useful approximation to make. Note that some workers have used specialized sharp tips and under such circumstances, continuum mechanics may no longer apply. However, the assumption is safe with 'conventional' probes.

We noted above that the AFM tip may be thought to resemble a hemisphere. The simplest contact mechanics model for such a situation is that due to Hertz in which the relationship between the load F_N and the area of contact A is:

$$A = \pi \left(\frac{R}{K} F_N \right)^{2/3} \tag{9.14}$$

in an elastic contact where K is the elastic modulus and R is the radius of the hemisphere.

While the Hertz model has proved useful, particularly for inorganic materials, many materials exhibit significant adhesion to the probe. The Hertz model may be modified to take account of this. Two models, in particular, have attracted a great deal of interest. In the Johnson–Kendall–Roberts (JKR) model, adhesion is introduced in the form of the interfacial free energy γ at the tip–sample contact. According to this model, adhesive forces may cause deformation of the hemispherical tip even in the absence of an applied external force. The radius, a, of the contact area between the tip and the surface is given by:

$$a = \left(\frac{R}{K}\right)^{1/3} \left(F_N + 6\pi\gamma R + \sqrt{12\pi\gamma RF_N + (6\pi\gamma R)^2}\right)^{1/3} \tag{9.15}$$

and at zero load it is:

$$a_o = (12\pi R^2 \gamma_{SV}/K)^{1/3} \tag{9.16}$$

As the tip is lifted away from the surface during a force measurement, hysteresis (non-reversibility) means that the tip will remain attached to the surface after the point at which it originally snapped into contact. A neck forms between the tip and gradually narrows as the tip is lifted from the surface. The neck remains stable until the radius of the neck reaches a critical value as, related to the radius of the contact area at zero load by:

$$a_s = a_o/4^{1/3} = 0.63a_o \tag{9.17}$$

At this radius, the neck becomes unstable and the tip separates from the surface. The load at which this occurs is:

$$F_S = -3\pi R\gamma_{SV} \tag{9.18}$$

This is equal to the adhesion force measured in the force–distance experiment. Hence, from Equation (9.18) it can be seen that the pull-off force is proportional to (i) the area of contact between the tip and the sample and (ii) the interfacial free energy at the tip–sample contact.

The JKR model is thought to apply in circumstances where adhesive interactions are comparatively strong and act at short-range. For weaker, long-range interactions, the Deraguin–Muller–Toporov (DMT) model is used. Here the relationship between the radius of the contact area and the load is given by:

$$a = \left(\frac{R}{K}\right)^{1/3} (F_N + 4\pi\gamma R)^{1/3} \tag{9.19}$$

The selection of an appropriate model for a given set of experimental circumstances will be based upon careful consideration of a variety of factors and it is not possible to be prescriptive. Suffice to say that for quantitative work, it is important that careful consideration be given to the most appropriate approach to use in modelling the data.

Quantification.   In its simplest mode of operation, known as contact mode, the AFM is operated in such a way that the tip of the probe always remains in mechanical contact with the surface. The cantilever is treated as a Hookean spring, and hence a simple relationship may be assumed between the deflection of the lever, x, and the force F acting on the tip:

$$F = -kx \tag{9.20}$$

The constant of proportionality is the spring constant, or force constant, k.

The mechanical properties of the cantilever are important in controlling the performance of a force microscope. The cantilever deforms in response to the interaction forces between the probe tip and the surface. It is this deformation which determines the performance of the microscope. Ideally, the cantilever must have as small a force constant as possible, in order to ensure the largest possible deflection for a given force. However, this requirement must be balanced against the need to minimize the sensitivity of the cantilever to thermal noise from its environment and to keep the response time small. These latter considerations suggest that a stiff material should be used (silicon nitride, silicon or silicon oxide). The solution to these apparently contradictory demands is to microfabricate the cantilever from a stiff material such as silicon nitride. Typical commercially produced cantilevers have lengths of the order of 100–200 μm with thicknesses of the order of 1 μm. Typical spring constants lie in the range 0.1–1 Nm^{-1}, while resonant frequencies are in the range 10–100 kHz (see Figure 9.25).

The tip itself has an important impact on the nature of the data acquired. Clearly, the resolution may be expected to be enhanced when sharper tips are used. The cheapest contact-mode tips have radii of curvature of ca. 50 nm, but there are now a number of manufacturers who can supply a range of probes with significantly smaller radii of curvature for both contact mode and other modes of operation. A variety of sharpened probes boast radii of approaching 1 nm. Carbon nanotube-functionalized tips have been prepared, and offer small tip radii combined with significant rigidity. Diamond probes are available for indentation work, and stiff probes for tapping experiments. It is important to bear in mind that the pressure exerted at a given load increases with the square of the radius of curvature; hence

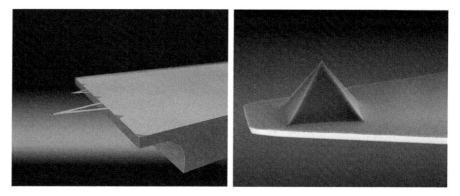

Figure 9.25 (a) 3-d sketch of a triangular silicon nitride contact-mode AFM probe. (b) SEM image of a tip at the apex of a probe like the one shown in (a). Images reproduced with kind permission from NanoWorld AG (http://nanoworld.com).

while sharpened tips offer the attractive prospect of enhanced resolution, they greatly increase the likelihood of tip-induced damage during imaging.

In order to acquire quantitative data, it is necessary to calibrate the spring constant of the cantilever. The photodetector response must first be calibrated. This is accomplished by measuring the response as a function of z-piezo displacement while the tip is approached against a hard sample (for example, mica). If the cantilever is significantly less stiff than the sample, then sample deformation will be negligible (all of the deformation will be in the cantilever) and the cantilever deflection Δz may be assumed to equal the distance moved by the z-piezo. The photodetector response for a given movement of the cantilever may thus be determined. To convert these cantilever deflections into forces, the cantilever spring constant normal to the sample surface, k_N, must be determined. The force normal to the surface is given by:

$$F_N = -k_N \Delta z \qquad (9.21)$$

It is in the determination of k_N that the complexity arises. There are a variety of approaches, none of which is perfect. One approach is to calculate the value of k_N. For rectangular cantilevers, the normal and lateral spring constants are given by $k_N = Ewd^3/4l^3$ and $k_L = Gwd^3/3h^2l$, respectively, where w, d and l are the cantilever width, thickness and length, h is the tip height and E and G are the Young's modulus and the shear modulus of the material from which the cantilever is fabricated [63]. However, many commercial microfabricated cantilevers are 'V'-shaped, and here the mechanical analysis is much more complex. For example, some groups have utilized finite element analysis [64, 65]. Moreover, whatever the cantilever geometry, it is necessary to determine accurate dimensions and to know the relevant moduli of elasticity in order to calculate the value of k_N. The measurement of cantilever length, width and thickness (typically by electron microscopy) is subject to experimental error and while the mechanical constants for a particular material may well be known accurately for bulk samples, their values may not necessarily be the same for the microfabricated material. Inhomogeneity in the cantilever, as a result of the microfabrication process, or the deposition of a gold coating onto the back face (commonly done to enhance its reflectivity), may lead to a significant deviation in mechanical behaviour from that of a cantilever fabricated from, for example, pure silicon nitride. It is not difficult to see that the combined effect of these uncertainties may be substantial.

Triangular levers were initially introduced because it was thought that they would be less prone to certain types of buckling behaviour that were associated with rectangular levers. The sacrifice was clearly much more complex mechanics. Recently, Sader has demonstrated that there is no need to use a triangular lever to address these problems: proper design of rectangular levers can ensure that they have the appropriate mechanical properties and at the same time, render theoretical analysis of data more tractable [66, 67].

Given the complexities of theoretical treatment of the bending of triangular levers, there are good reasons for seeking an experimental approach. However, the uncertainties can still be large. One of the most accurate approaches, due to Cleveland and co-workers, involves the attachment of a known mass to the cantilever and the measurement of the change in resonance frequency [68]. Other approaches include the determination of k from the observation of cantilever oscillations due to thermal noise [69, 70], the measurement of the resonance frequency of a mechanically driven cantilever [71] and the measurement of cantilever deflection in contact with a second, reference cantilever [72]. Clearly, this latter approach is only as accurate as the uncertainty in the spring constant of the reference. However, it does provide a method for ensuring the accurate standardization of a set of measurements. A final, and particularly promising approach, is due to Cumpson *et al.* [73]. They used a microfabricated array of reference springs (MARSs) supporting a mirrored polycrystalline silicon disc. Measurement of the interaction between the cantilever and the measuring system yields the value of the force constant very rapidly. The only drawback at present is the complexity of fabrication of the device, a process that requires specialized equipment and appropriate expertise. Commercialization may enable this ingenious approach to be adopted by a larger number of researchers.

Contaminants at the surface can exert a considerable influence on the nature of the tip–surface interactions. For example, thin films of water at a surface can be the cause of attractive capillary forces (as large as 4×10^{-7} N on mica in air). Consequently, the instrument is often used with the probe and sample immersed in liquid. Not only does this eliminate the contribution of capillary forces, but it also provides scope to control the tip–sample interaction in other significant ways. For example, operation in an aqueous environment can reduce the strength of dispersion interactions, and the surface charge which is resident on the probe tip is screened both by the attraction of counter-ions from solution, and by dielectric lowering of the interaction energy [56]. Conversely, the use of a medium with a small dielectric constant leads to an enhancement in the strength of dispersion interactions. Against this, however, the presence of the liquid can create its own problems, through the adsorption of ions onto the sample surface, dissolution of material from the surface, or through the creation of polarization forces where the dielectric constant of the tip differs from that of the solvent [56].

Modes of Operation.   The first requirement for the construction of a force microscope is some suitably sensitive means by which the deflections of the cantilever may be measured. The first AFM utilized, effectively, an STM tip to monitor deflections using electron tunnelling. The most widely utilized approach in commercial instruments is to measure the deflection of a laser beam which is reflected onto a photodetector from the back of the cantilever. However, other methods are also used, including ones based on interferometry and electrical detection methods based on piezoelectric devices. The measured signal is then used to control movements of a piezoelectric crystal, on which either the cantilever or the sample is mounted, via a feedback system – in just the same way that the movements of the STM tip are controlled – both to regulate the tip–sample distance and to scan the microscope in the x, y plane.

The microscope can operate in constant force mode or in constant height mode, in the same way that the STM may operate in constant height or constant current mode. In constant height mode, variations in the cantilever deflection are measured as the tip scans the surface, while in constant force mode, the cantilever height is effectively adjusted continuously so that a constant tip–sample interaction force (and hence a constant cantilever deflection) is maintained. It is the constant force mode that is the more widely used and which provides the most accurate data on sample topography.

In contact mode, the tip always exerts a mechanical load on the sample. This can lead to damage. While, in principle, non-contact operation offers a solution to this, it is, in practice, rather complex to apply. An alternative approach was provided by the development of tapping mode imaging. Tapping mode seeks to reduce the rate of energy dissipation at the sample surface by utilizing an oscillating tip, driven at its resonant frequency with a high amplitude. Typically a stiff silicon probe ($k = $ ca.50 N m^{-1}) is used. The probe makes intermittent contact with the surface, at the bottom of the oscillation cycle: it 'taps' the surface. Tapping mode data provide topographical information with much superior resolution on soft materials such as polymers. However, one should be wary of assuming that the method is always non-destructive: mechanical contact still occurs, leading to the dissipation of energy, and damage may still occur. One may distinguish different regimes of tapping by comparing the amplitude of oscillation during tapping (the set-point oscillation, which is maintained at a constant value by adjustment of the z-piezo) with the amplitude of free oscillation, when the tip does not contact the sample. If the ratio of these quantities is close to 1, then the amount of energy dissipated in the contact is small and the conditions may be referred to as 'light tapping'. At a ratio of ca. 0.5, medium tapping conditions, the energy dissipation rate is increased – however, the resolution will also be increased, provided the sample is not damaged. Decreasing the amplitude of the set-point oscillation further may yield further improvements in resolution, or increase the extent of sample degradation, depending on the properties of the material under study. As always, the trade-off between resolution and damage needs to be explored carefully.

An important by-product of tapping mode imaging is that phase images may also be acquired. In phase imaging, the phase lag between the driving oscillation and the cantilever response is measured. The magnitude of the lag provides an indication of the amount of energy dissipated in the tip–sample interaction: if the contact is elastic, with small amounts of energy being dissipated, then the phase lag will be small, but as the sample becomes more viscous and the amount of energy dissipation increases, the phase-lag also increases. For materials that are heterogeneous in their mechanical properties, this provides a useful tool for the characterization of surface structure. For example, Figure 9.26 shows a tapping mode image of a sample of polyester film and the corresponding phase image. The sample concerned is a biaxially oriented polymer film the surface of which has been modified by the incorporation of silicate additives. The tapping mode image provides a picture of the topography with high resolution, but the phase image provides a wealth of further detail that is not evident in the topography. Small features may be observed that are thought to be nanocrystallites [74]. These crystalline domains are, like the silicate additives, mechanically stiffer than the intervening polymeric material and consequently exhibit brighter contrast, indicating a smaller phase lag. Amorphous polymeric material, in contrast, behaves in a more viscous fashion and exhibits darker contrast.

Non-Contact Mode and Atomic Resolution.   It was initially hoped that the AFM would generate data with the same kind of high spatial resolution provided by the STM. There were apparently some initial early successes, with contact mode images of HOPG being reported which appeared very similar to those obtained using the STM. However, the origin of these images soon became shrouded in doubt. The popular image of a single atom at the apex of the tip in contact with the sample was rightly questioned: under these circumstances, with a load of a few nN, the pressure exerted by the tip would be sufficient to significantly disrupt the structure of the sample surface (even supposing the tip material was strong enough to sustain the load). In reality, contact areas are typically larger, and may be a few nm in diameter at loads of tens of nN (and larger at higher loads).

If multi-atom contact was occurring, what might the mechanism of contrast formation be in these 'atomic resolution' images? Significantly, atomic defects were conspicuous by their absence from these early high resolution images. There was much speculation concerning their origin. For example, in the case of HOPG, it has been suggested that a graphite flake was dislodged and, sliding across the sample surface under the tip, gave rise to an interference effect which reflected the periodicity of the graphite surface structure. There was, for a while, doubt that true atomic resolution was possible using the AFM; the doubt was dispersed in 1993 when Ohnesorge and Binnig reported atomically resolved images of inorganic

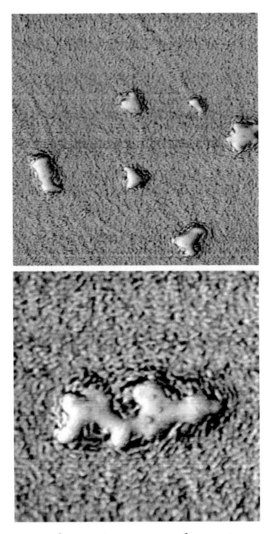

Figure 9.26 $1500 \times 1500 \, \mathrm{nm}^2$ (left) and $750 \times 750 \, \mathrm{nm}^2$ (right) phase images of regions of Mylar D, a free-standing, biaxially oriented poly(ethylene terephthalate) film material

crystals, including atomic defects, using the non-contact mode [75]. They employed a maximum attractive force of only -4.5×10^{-11} N.

   These were impressive findings. Significant effort was subsequently put into exploring such non-contact imaging more fully. The mechanism of actuation of the microscope is critical. It is now clear that imaging under static conditions (as in the common imaging modes of a conventional AFM) leads to significant experimental complexities due to the instability of the cantilever and the potential for the formation of strong bonds between the probe and the sample, and the use of some kind of probe modulation is thus key to realizing the potential of the technique. One approach is to use amplitude modulation, which bears some similarity to tapping mode and utilizes a fixed drive frequency. Changes in the amplitude of oscillation are measured as the tip approaches the surface and these are used as the feedback signal. The main difference between this and tapping mode imaging is that the tip is very much closer to the surface in tapping mode and experiences a much more substantive mechanical interaction. However, a more widely used approach is frequency modulation [76], which yields very high resolution. The technique is most usefully applied to well-defined surfaces (crystals and highly ordered thin films on well-defined substrates) which enable full advantage to be taken of its high spatial resolution. The frequency modulation AFM yielded the first atomic resolution images of the 7×7 reconstruction of Si(111) [77], and has grown steadily in use over the intervening decade or so. A recent text has provided a useful overview of applications of the technique [78].

Non-contact mode imaging enables the acquisition of force spectroscopy data that have exquisite sensitivity to electronic structure and bonding at the surface. By controlling the oscillation of the tip, it may be caused to come close enough to participate in the initial stages of bond formation or, alternatively, to participate transiently in bond formation. Bond formation leads to a frequency shift, which may be mapped as a function of position to yield information on the distribution of particular surface states. A tip may be also prepared under controlled conditions in UHV with a specific electronic structure, enabling its use as a probe for specific interactions with a sample surface. For example, suppose an Si tip may be prepared with the atom at its apex having a dangling bond. On interaction with an Si adatom with a dangling bond on an Si crystal surface, covalent bond formation takes place. Bonding and antibonding orbitals are formed, the energy of the bonding state being lower than the dangling bond energy by an amount ΔE, while the energy of the antibonding state is raised by an equal amount. Because two electrons enter the bonding state, the total energy of the dangling bond electrons will be reduced and there will be a corresponding interaction force. By mapping the interaction force between the tip and the surface it may thus be possible to map the distribution of specific bonding states.

While the spatial resolution of the non-contact AFM, and its sensitivity to chemical state, are very impressive, the physics of the tip–sample interaction is nevertheless extremely complex. Thus despite its enormous potential, it remains generally much less widely used than the other modes described above.

9.3.2 CHEMICAL FORCE MICROSCOPY

The functionalization of the tip of an AFM provides a means by which a probe with both chemical (or molecular) specificity and nanometre scale spatial resolution may be created. A range of different approaches have been employed, including the attachment of beads to a tip [79–81] and the attachment of a variety of molecules in order to probe particular recognition interactions [82, 83]. For studies of nanoscale friction and adhesion between molecular materials, the most appealing approach is the adsorption of a monolayer of functionalized organic molecules. Nakagawa *et al.* first demonstrated the feasibility of this [84, 85]. They adsorbed monolayers of alkylsilanes onto silicon nitride cantilevers and measured adhesion forces between the modified tips and monolayers of methyl terminated and perfluorinated alkylsilanes on silicon with different alkyl chain lengths. They found that while the adhesion force was almost negligible for a bare silicon nitride tip, it was larger for the functionalized tips and increased with the alkyl chain length of the molecules adsorbed to the silicon substrate. This was attributed to non-covalent interactions between the alkyl chains of interacting molecules attached to tip and substrate. Ito *et al.* demonstrated that a wider range of tip functionalities could be achieved by adsorption of alkenyltrichlorosilanes, which could subsequently be converted to polar functionalities [86].

An alternative approach was adopted by Lieber and co-workers and termed chemical force microscopy (CFM). They evaporated layers of gold onto silicon nitride cantilevers and immersed them in dilute ethanolic solutions of alkanethiols, resulting in the formation of self-assembled monolayers on the tip surfaces [87, 88]. While this surrendered the advantage enjoyed by Nakagawa *et al.* of minimal modification to the cantilever, because silanes could be attached directly to the silicon nitride tip whereas adsorption of alkanethiols required the prior deposition of a gold layer, it did bring the benefits associated with the better-defined nature of alkanethiol SAMs. Adhesion forces, determined from force–distance measurements carried out in ethanol were in the order $COOH$—$COOH$ > CH_3—CH_3 > CH_3—$COOH$, in accordance with expectations based on simple consideration of intermolecular forces. In other words, interactions between similar functionalities are stronger than interactions between dissimilar ones, and polar interactions are stronger than dispersion interactions. Frisbie *et al.* presented friction force microscopy (see Section 9.3.4 below) images of patterned SAMs acquired using tips with different functionalities [87]. For patterns composed of carboxylic acid and methyl terminated regions, an inversion of contrast was observed on switching from carboxylic acid to methyl terminated tips. With carboxylic acid terminated tips, acid terminated regions exhibited brighter contrast (higher frictional forces) than methyl terminated regions, while the reverse was true when methyl terminated tips were used, indicating stronger adhesion when like pairs of terminal groups interacted than when dissimilar functionalities were involved. Subsequently, Noy *et al.* attempted to quantify these data [88].

The capacity of the AFM to yield data for samples immersed in fluid has enabled liquid–solid interfacial interactions to be studied with exquisite spatial resolution. For example, Van der Vegte and Hadziioannou measured adhesion forces for a range of tip–sample functional group pairs [89]. At low pH, acid–acid interaction forces were large because the functional groups existed predominantly in the undissociated state. As the pH increased, and dissociation into carboxylate anions occurred, so the adhesion force declined sharply due to repulsion between like species. The reverse was true for amine groups, which were predominantly positively charged at low pH (leading to repulsive interactions) and uncharged at high pH. Interactions between hydroxyl groups were unaffected by the pH. Vezenov *et al.* also measured adhesion forces as a function of pH [90]. Like Van der Vegte and Hadziioannou, they found that interaction forces between acid groups were large at low pH while those between amine groups were large at high pH. Both groups also used 'force titrations' to determine the pK_a of carboxylic acid terminated SAMs. Van der Vegte and Hadziioannou determined that the pK_a was 4.8, while Vezenov *et al.* obtained a value of 5.5. Both of these are quite close to the values typically reported for aqueous organic acids. Vezenov *et al.* recorded the friction force as a function of load for different pH conditions. They found that values for the friction coefficient measured at pH values less than 5.5 were significantly larger than those determined at higher pH values, consistent with their interpretations of the adhesion force data in terms of specific intermolecular interactions.

It is not only acid–base and charge–charge interactions that may be influenced by the liquid medium. Sinniah *et al.* measured adhesion forces for a range of alkanethiol systems in water, ethanol and hexadecane [91], and observed significant differences in the values obtained in the different liquids, although their data for experiments carried out under ethanol differ from those reported by other workers. Feldman *et al.* carried out force measurements for a range of polymer surfaces, and showed that careful consideration needed to be given to the liquid medium [92]. In particular, account needed to be made of the dielectric constant of the liquid medium, in addition to any tendency for electrical double-layer formation (of particular importance in water). Clear and Nealey also studied the effect of the liquid medium [93]. For interactions between methyl functionalized tips and a methyl terminated monolayers, they recorded the largest adhesion force in water, with the adhesion force declining in the order water > 1,3-propanediol > 1,2-propanediol > ethanol, hexadecane. In contrast, for interactions between a carboxylic acid terminated tip and an oxidized (i.e. polar) monolayer of an alkene-terminated tricholorosilane, they observed the weakest adhesion force in water and the largest in hexadecane. Moreover, in hexadecane, the methyl–methyl interaction was weaker than the acid–oxidized silane interaction, while in water the relationship was very much the opposite. However, the differences in the friction data were less pronounced. Friction–load plots appeared to be linear for all sample/liquid combinations, but while the friction coefficients were larger when measured in hexadecane than in water, and the difference between the methyl/methyl and acid/oxidized monolayers contacts was smaller in hexadecane, the polar/polar contact yielded the largest friction coefficient in all liquid media.

There is now a growing body of literature on chemical force microscopy and space does not permit us to provide a comprehensive overview here. The interested reader is referred to two excellent recent reviews for a more detailed analysis [94, 95].

9.3.3 FRICTION FORCE MICROSCOPY

Measurement of lateral forces, in the plane of the sample surface, is straightforwardly possible using typical commercial SFM instruments which employ four-quadrant photodetectors. The light intensity reflected from the back of the cantilever is measured on each quadrant. In conventional constant force mode, the force normal to the sample surface is monitored by measuring the difference between the signals falling on the top and bottom halves of the photodetector. However, measurement of the difference between the signals falling on the left and right halves of the detector is usually possible simultaneously, yielding the force acting on the tip parallel to the sample surface. This lateral force provides a measure of the frictional interaction between the tip and the surface. Care must be exercised, however, because the lateral force typically contains a component due to the sample topography as well as the frictional force [96]. This topographical component may be identified by comparing images recorded in the forwards and backwards directions [97]. By subtracting the forwards and backwards images, the frictional force may be calculated and the resulting image shows the spatial distribution of surface friction (see Figure 9.27). This approach is referred to as lateral force microscopy (LFM) or, more commonly, friction force microscopy (FFM).

The measurement of surface friction by FFM is, perhaps unexpectedly, a very powerful aide to nanometre scale surface analysis. Indeed, in many ways, FFM is to date the only widely accessible tool to offer sensitivity to variations in chemical composition on sub-100 nm length scales. While localized Raman measurements using near-field optical methods offer the capacity for nanoscale surface spectroscopy, they have proved so difficult in practice that their exploitation has been quite limited. In contrast, FFM is accessible on any commercial AFM instrument and is easily implemented.

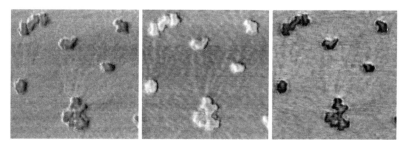

Figure 9.27   Forward (left) and backward (right) FFM images of Mylar D together with (right) the subtracted (forwards – reverse) image which gives the friction contrast free of topographical contributions

Figure 9.28 illustrates the power of FFM for mapping the distribution of different surface functional groups. A patterned self-assembled monolayer, consisting of narrow bands of carboxylic acid terminated adsorbates separated by broader bands of methyl terminated adsorbates, has been imaged. Brighter contrast is observed on the polar regions than on the non-polar ones. The explanation for this is the difference in the intermolecular forces acting at the interface between the tip and the sample. As a general rule, in intermolecular forces like attracts like (as opposed to electrostatics, where opposites attract). The outer surface of the probe (in this case a commercial silicon nitride probe) is composed of a thin layer of silicon dioxide, which is polar. Consequently, there is a relatively strong attractive interaction between the tip and the carboxylic acid functionalized regions of the surface, and a much weaker interaction with the methyl terminated regions.

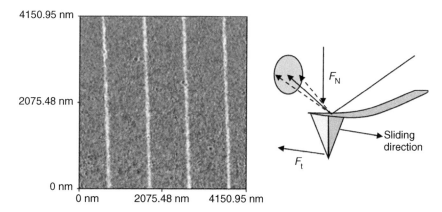

Figure 9.28   Friction force microscopy image of a patterned self-assembled monolayer showing 50 nm-wide regions of carboxylic acid terminated adsorbates separated by broader regions functionalized by methyl terminated adsorbates

Consequently, the tip adheres more strongly to the carboxylic acid terminated regions, leading to a higher rate of energy dissipation, and hence a larger friction force, as the tip slides across those regions of the sample.

FFM is not only capable of providing qualitative data, however. It is possible to quantify the strength of the frictional interaction between the tip and the sample by analysing the relationship between the friction force and the load. In order for this to be done, it is necessary to select an appropriate contact mechanics model. In situations where there is strong tip–sample adhesion, it may seem likely that the JKR model is more appropriate (see Section 9.3.1 above). According to Tabor, the frictional force F_F between two sliding surfaces should be proportional to the area of contact, or, from the JKR analysis:

$$F_F \propto \pi \left(\frac{R}{K}\right)^{2/3} (F_N + 3\pi \gamma R + \sqrt{6\pi \gamma R F_N + (3\pi \gamma R)^2})^{2/3} \qquad (9.22)$$

The JKR model has been widely used in FFM measurements on inorganic systems, but for molecular materials it has been less widely used. For organic monolayers and polymers, Amontons' law has been used much more widely. According to Amontons' law, the friction force is proportional to the load applied perpendicular to the surface F_N and the constant of proportionality is the coefficient of friction, μ:

$$F_F = \mu F_N \qquad (9.23)$$

Thus a plot of the friction force as a function of the load should yield a straight line fit, with gradient μ and an intercept with the y-axis at the origin. In fact, in many cases where a linear friction–load relationship is observed, the line that fits the data does not pass through the origin, and this is generally attributed to the influence of adhesion between the tip and the sample. A modification, first proposed by Deraguin, is thus made to Equation (9.23):

$$F_F = F_0 + \mu F_N \qquad (9.24)$$

Here, F_0 is the friction force at zero load.

The contact mechanics associated with FFM have been puzzling in many ways. Amontons' law is generally presented as a macroscopic law. The argument behind it is that at the microscopic scale, materials are generally rough. They consist of an array of many peaks, or asperities. As two macroscopic surfaces slide against each other, these microscopic asperities rub against each other and deform, giving rise to friction. As the load increases, the asperities deform and the contact area, and hence the friction force, increase. Clearly, an AFM tip is in many ways an idealized single asperity. According to the generally received wisdom, it thus seems superficially surprising that FFM data should obey Amontons' law. Surely it would be more likely that they would be modelled by a single asperity approach, such as the Hertz or JKR theories.

Recently, there has been some new insight into these issues. A very important contribution was made by Gao *et al.* [98], who conducted a systematic investigation of frictional forces on length scales ranging from the nanoscopic to the macroscopic. They examined the relationship between the friction force and the contact area. While it is often assumed that the friction force is proportional to the area of contact, they argued that the area of contact is not, in fact, a fundamental physical quantity; while the contact area often provides a useful guide to the density of intermolecular interactions at an interface, it is actually the sum of all of the molecular interactions at the interface that determines the strength of the frictional interaction. They provided evidence that the JKR model may apply to situations where sliding is adhesion-controlled, while Amontons' law applies to situations in which non-adhesive sliding occurs. They also reported a transition from JKR-type behaviour to a linear friction–load relationship following damage in a system that initially exhibits adhesive sliding. Recent data from the author's laboratory supports this hypothesis [99]. Measurements made for poly(ethylene terephthalate), PET, in perfluorodecalin, a medium with a very small dielectric constant which yields strong tip–sample adhesion, fit the behaviour predicted by the JKR model. In contrast, measurements in ethanol, a medium with a high dielectric constant, yield a linear friction–load relationship, consistent with Amontons' law (or Deraguin's modification of it). Given that ethanol is a widely used medium for FFM experiments on molecular materials, this perhaps explains why linear friction–load relationships have been so widely reported.

The coefficient of friction is a quantity with intuitively obvious meaning, and the means of its determination, by linear regression analysis of a friction–load plot, is straightforward. It thus presents a simple and attractive means by which to try to quantify surface compositions – it is authentically a nanometre scale surface analysis tool with potentially broad applicability. Figure 9.29 shows friction–load data for two monolayer systems, hydroxyl and methyl terminated self-assembled monolayers on gold, acquired using a tip that has been coated with a thin film of gold and functionalized with a carboxylic acid terminated thiol. The vertical axis plots the photodetector signal, which is proportional to the friction force. It may be seen that the gradient of the friction–load plot is in each case linear, and that the gradient is significantly steeper for the adsorbate with the polar terminal group (i.e. the coefficient of friction is larger for that adsorbate).

Such analyses have been employed by a number of authors in studies of a wide range of systems. Monolayers (often self-assembled monolayers of alkanethiols on gold) have attracted significant interest. Space does not permit a comprehensive review and more detailed treatments may be found elsewhere; here we mention briefly a few illustrations.

Importantly, as already indicated, friction coefficients vary with the nature of the terminal group (the one at the uppermost end of the adsorbate)

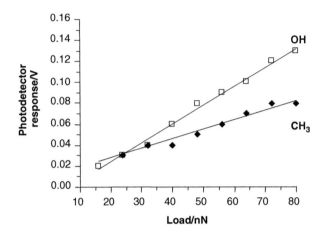

Figure 9.29 Friction–load plots for hydroxyl and methyl terminated self-assembled monolayers acquired using an AFM probe functionalized with a carboxylic acid terminated adsorbate

in self-assembled monolayers [100–110], suggesting a straightforward role for FFM in comparison of surface composition. In mixed SAMs formed by the adsorption of hydroxyl and methyl terminated adsorbates, for example, the coefficient of friction varies in a linear fashion with the composition. There have been a number of studies in which FFM has been used to probe molecular organization [111–115] and mechanical properties [116–118]. The susceptibility of SAMs to deformation during sliding interactions varies with the length of the adsorbate molecule: short chain adsorbates are comparatively mobile (liquid-like) but the density of *gauche* defects decreases with chain length as the multiple dispersion interactions between adjacent methylene groups start to become substantial; as a consequence, the coefficient of friction decreases as the length of the adsorbate alkyl chain increases. Friction measurements may also be made under fluids, including in aqueous media, and some workers have utilized FFM to study the acid–base characteristics of SAMs [90, 119].

To conclude this section we provide one detailed illustration of the use of FFM to carry out nanoscale surface analysis. When exposed to UV light in the presence of oxygen, monolayers of alkanethiols are oxidized to yield alkylsulfonates:

$$Au - S(CH_2)_nX + \tfrac{3}{2}O_2 + e^- \rightarrow Au + X(CH_2)_nSO_3^- \qquad (9.25)$$

Although alkylthiolates are strongly adsorbed onto gold surfaces, the alkylsulfonate oxidation products are only weakly adsorbed. If a carboxylic acid terminated SAM is exposed to UV light, and then immersed in a solution of a methyl terminated thiol, any oxidized adsorbates will be displaced by the solution-phase thiols leading to the formation of an SAM with reduced

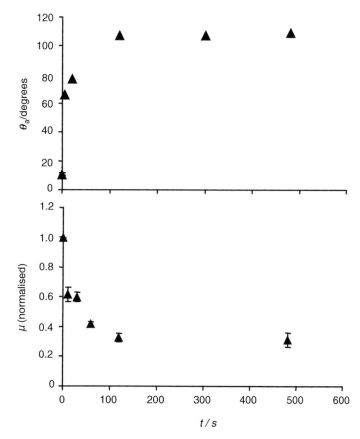

Figure 9.30 (a) Variation in the contact angle and (b) variation in the normalized coefficient of friction for carboxylic acid terminated self-assembled monolayers following exposure to UV light for varying periods of time and insertion in a solution of a methyl terminated thiol

surface free energy. Figure 9.30(a) shows the variation in the advancing contact angle of water as a function of the exposure time. It may be seen that the contact angle increases from an initial low value to a limiting value after *ca* 2 min. As the surface energy is reduced, there should be a corresponding change in the coefficient of friction of the SAM. Figure 9.30(b) shows that the change in the coefficient of friction (with values normalized to the coefficient measured for the as-received carboxylic acid terminated monolayer) correlates very closely with the variation in the contact angle. Clearly, however, the friction measurement is capable of being carried out with a spatial resolution of a few nm, in contrast to the macroscopic contact angle measurement or, indeed, conventional surface spectroscopic techniques.

To illustrate this, Figure 9.31(a) shows a patterned sample formed by exposing the monolayer to UV light through a mask. In regions exposed to

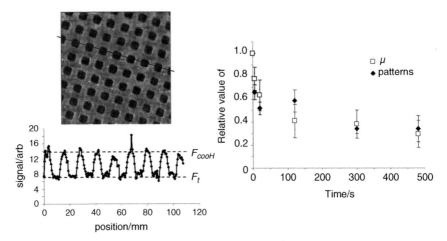

Figure 9.31 (a) Line section through a patterned self-assembled monolayer formed by exposing a carboxylic acid terminated monolayer to UV light and immersing it in a solution of a methyl terminated thiol. (b) Variation in the normalized coefficient of friction of the exposed areas with time as determined from analysis of line sections through patterned samples and also by measurement of the coefficient of friction of unpatterned samples

the UV light from the fibre, the adsorbates were oxidized to alkylsulfonates, which were displaced by immersion of the sample in a solution of a methyl terminated adsorbate. A line section may be drawn through the image, and the ratio of the friction signals from the exposed and masked areas related to the respective coefficients of friction by the following simple expression:

$$\frac{\mu_t}{\mu_{COOH}} = \frac{F_t}{F_{COOH}} \qquad (9.26)$$

The resulting coefficients of friction for the exposed areas relative to the masked areas (equal to the coefficients of friction of the exposed areas normalized to the value obtained for the unexposed monolayer) are shown in Figure 9.31(b). It is clear that the agreement with the data shown in Figure 9.30 is very good. To emphasize this, when an analysis of the kinetics of the oxidation reaction was carried out using the two sets of friction data, rate constants of 0.78 s^{-1} and 0.69 s^{-1}, respectively, were obtained [120]. Not only were these values arrived at by different methods, but also the two sets of data were acquired on different microscopes, and in view of this, it may be concluded that FFM data are a very reliable source of quantitative surface compositional data with high spatial resolution.

9.3.4 BIOLOGICAL APPLICATIONS OF THE AFM

There have been impressive successes in the application of the AFM to biological problems. Under the most favourable circumstances, exceptional spatial resolution has been achieved using the AFM. However, in contact mode, the load exerted by the tip upon the sample is substantial on a molecular scale. Moreover, there is a significant frictional interaction as the tip slides across the sample surface. The forces involved are usually more than adequate to displace biomolecules. A variety of approaches have been explored to solving these problems, including the use of covalent coupling schemes to tether biomolecules in place. Another approach is to crystal-lize the sample into a periodic array, and rely upon the cohesive forces within the close-packed molecular assembly to counterbalance the disrup-tive influence of the tip. Although not all biomolecules may be crystallized, this approach has led to some spectacular successes. Engel and co-workers have been studying membrane proteins for a number of years, and their data have provided breathtaking insights into molecular structure [121]. In studies of bacterial surface layers, or S-layers (the proteins that constitute the outermost layer of the cell wall), they were able to examine the effects of enzymatic digestion with a spatial resolution better than 1 nm [122]. The S-layers were deposited onto mica substrates and found to form bilayers or multilayers. The topmost layer exhibited a triangular structure when images at low forces (100 pN); however, imaging at elevated loads (600 pN) led to removal of the top layer and exposure of a hexameric flower-shaped morphology. S-layers that had been enzymatically digested were found to be present as single layers which exhibited each type of surface with equal probability. In another study [123], AFM data on several membrane proteins were presented with a resolution of less than 0.7 nm. Importantly, in these studies raw AFM data were presented that clearly exhibited substructural details of individual protein molecules. In contrast, electron micrographs with the best resolution typically represent averaged data from a large num-ber of molecules. The AFM data enable the observation of crystal defects, or molecule-to-molecule variability in structure. Nevertheless, computational analysis of AFM images of large assemblies is still possible, leading to image averaging or more sophisticated analyses. Fotadis *et al.* studied one of the components of the photosynthetic apparatus of the bacterium rhodospiril-lum rubrum [124], which consisted of a ring structure (the light-harvesting complex LH1) containing within it the reaction centre (RC). RC receives energy from LH1. Two dimensional crystals of the complex were formed and deposited onto mica. Contact mode images revealed patterns of alter-nating bright and dark rows (Figure 9.32). These resulted from the existence of two distinct orientations for the RC–LH1 complex. On the cytoplasmic side of the complex, the reaction centre protrudes, and is observed as a feature with bright contrast, while on the periplasmic side, dark contrast is observed over the centre of the complex. A small number of crystal defects were observed, in which the LH1 complex adopted a different morphology; these would have been lost in electron microscopy investigations due to averaging. In some cases, the reaction centre was observed to be missing,

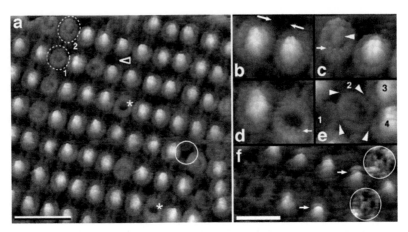

Figure 9.32 (a) High resolution AFM image of a two dimensional crystal of RC—LH1 complexes. The broken circle (1) and the ellipse (2) mark complexes that lack the RC—H subunit. The asterisks denote 'empty' LH1 complexes that completely lack the RC seen from the cytoplasmic side. The arrow denotes a missing RC seen from the periplasmic side. (b–e) A variety of complexes at higher magnification. (f) The periplasmic side of the RC—LH1 complex imaged at higher load. The scale bars are 40 nm in (a) and 15 nm in (f). Reproduced from Fotiadis *et al.* [124], Copyright 2004, the American Society for Biochemistry and Molecular Biology

even at low loads, possibly attributable to its removal by the tip as it traversed the crystal. Imaging at loads of 200–300 pN was found to yield the best resolution. On the periplasmic side of the complex, an X-shaped structure was observed, which was attributed to the periplasmic face of the RC.

Tapping mode has recently been facilitating the imaging of proteins adsorbed onto solid surfaces. In contrast to the beautiful images acquired by Engel and co-workers for protein crystals, these data are typically less well resolved but do provide data for isolated molecules. Marchant and co-workers have imaged von Willebrand factor (VWF), a large multimeric protein that adheres rapidly to biomaterial surfaces upon exposure to blood [125]. Protein–surface interactions play a key role in regulating thrombus formation, a phenomenon of great importance when biomaterials are placed in contact with the blood because it can lead to failure of the prosthetic device. In their study, Marchant and co-workers compared VWF adsorbed to hydrophobic monolayers of octadecyltrichlorosilane (OTS) adsorbed on glass, and hydrophilic mica. On the OTS monolayers, VWF was found to exhibit a coiled conformation, while on mica, the polypeptide chains were observed to adopt extended conformations which exhibited much larger end-to-end dimensions.

Fibronectin (FN) is another protein of considerable importance in the development of prosthetic biomaterials. FN plays an important role in cellular attachment, and is recognized by integrin receptors in cell membranes, which regulate the mechanism of attachment. FN is a dimeric protein, consisting of two polypeptide chains joined by disulfide linkages. A specific region of the molecule, containing the tripeptide sequence arginine–glycine–aspartic acid (RGD), is recognized by the integrin receptors. FN undergoes surface-specific conformational changes, and these changes in conformation lead to differences in the orientation of the cell binding domain of the molecule with respect to the solid surface on which the molecule adsorbs. The characterization of the conformations of adsorbed proteins is very challenging, and many techniques, such as infrared spectroscopy, provide only limited information. Lin *et al.* used the tapping mode AFM to image FN adsorbed onto the surface of mica [126]. Single FN molecules were observed. FN was exposed to heparin-functionalized gold nanoparticles. Bound nanoparticles could be resolved as bright features situated part-way along the FN polypeptide chain, enabling the binding site to be estimated. It was concluded that there were two binding sites, based on the AFM data, attributed to the Hep I and the Hep II sites previously identified using biochemical means. A difference in binding affinity for the two sites was postulated, based on the observation that twice as many functionalized nanoparticles were observed to bind to Hep I than to Hep II.

While these papers report impressive findings it is important to note that for studies of biomolecular structure, electron microscopy remains a very powerful technique. Electron microscopists have developed a suite of powerful sample preparation and analysis methods and it remains doubtful, for many systems, that use of the AFM may even match, let alone improve upon, the resolution that electron microscopy is capable of providing. It is important therefore that the appropriate approach is selected for the problem in hand. Where the AFM is particularly powerful is in the analysis of securely immobilized materials, such as the two dimensional crystals that Engel and co-workers have addressed so successfully, where the ability to image single molecules negates the requirement for averaging often required in electron microscopy, and also in the analysis of biomolecules on difficult substrates (for example, polymers) where electron microscopy would be very difficult.

Another unique feature of the AFM when compared with electron microscopy is its capacity to acquire images in biological environments and, importantly, in real time. An impressive early example was provided by Drake *et al.* who used the AFM to image the thrombin-induced polymerization of fibrin in a physiological buffer in real time [127]. Although the resolution was poor, individual molecules could be observed forming chains as the reaction proceeded. Haberle *et al.* examined the behaviour of monkey kidney cells infected with pox viruses, under normal cell growth conditions, using the AFM [128]. They observed an initial and short-lived softening of the cell membrane immediately after infection. Subsequently, protrusions were observed to grow out of the cell membrane and then disappear again. Ultimately, after about 20 h, a different kind of event was observed, which was interpreted as the exocytosis (birth) of progeny viruses. This latter process was observed over a period of 7 min, to yield the astonishing images seen in Figure 9.33. An unusually long finger-like structure is visible, at the end of which a protrusion appears and then disappears again after about 3 min. The size of the protrusion, the time-scale of the event, and comparison with data from electron microscopy suggested that this process was indeed the exocytosis of a virus.

The AFM also yields important and unique quantitative data on the forces involved in biological processes and interactions. In an early land-mark study, Lee *et al.* adsorbed biotinylated albumin to glass microspheres and measured the variation in force as these microspheres were approached towards, and withdrawn from, mica surfaces functionalized with strep-tavidin [129]. Biotin and streptavidin exhibit a very strong molecular recognition interaction. In a similar study, Florin *et al.* used a biotiny-lated agarose bead and a streptavidin-coated tip [130]. They measured

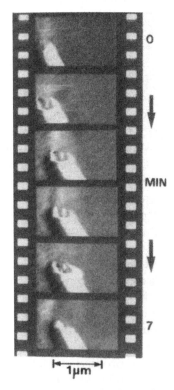

Figure 9.33 Exocytic process seen about 19 h after infection of a monkey kidney cell with pox virus. Reproduced with permission of Elsevier Science from Haberle *et al.* [128]

pull-off forces and subsequently investigated the relationship between the interaction force and enthalpy [131]. Lee *et al.* studied the recognition forces between complementary strands of DNA, thiolated at their 3′ and 5′ ends for attachment to alkylsilane monolayers attached to a silica probe and a planar surface [132]. By attaching bases to tip and surface, Boland and Ratner set out instead to measure single base-pair interactions [133], an approach adapted by Lioubashevski *et al.* as the basis for a a sensor [134]. They adsorbed cysteine-modified peptide nucleic acids (PNAs) onto gold-coated AFM tips, and then probed interaction forces with alkanethiol monolayers before and after hybridization with PNAs or RNA. Hybridization reduced the pull-off force because the hybridized nucleic acids could not bind to the probe. The recognition force can be increased by increasing the interaction area. Mazzola *et al.* modified a latex microparticle to functionalize it with DNA; this could then be employed as a probe for arrays of immobilized oligonucleotides [135]. These illustrations suggest the potential usefulness of techniques based upon highly sensitive recognition measurements by use of the AFM.

Many of the phenomena that are involved in the measurement of biomolecular interactions by force spectroscopy are very complex. Protein unfolding, for example, presents significant theoretical problems and the application of force spectroscopy, while very useful, is technically challenging. A benefit of this has been that problems have been addressed that may have been overlooked in simpler systems. One important example is provided by rate-dependent phenomena [136, 137]. Evans and co-workers demonstrated that the application of an external mechanical force to a protein effectively tilts the 'energy landscape' for the unfolding process, reducing the activation energy for the unfolding process. It has been recognized that the use of a variety of unloading rates is an important component in the investigation of the unfolding process. These insights have facilitated quantitative investigation of a range of phenomena [138–140].

The AFM certainly has a great deal to contribute to the understanding of biological interactions, and it has now become an accepted biophysical tool.

9.4 Scanning Near-Field Optical Microscopy

9.4.1 OPTICAL FIBRE NEAR-FIELD MICROSCOPY

Surface analysis revolves around the acquisition of information about surface composition and chemical structure. The 'ideal' nanoscale surface analysis tool would provide spectroscopic information with a resolution of a few tens of nm. Scanning near-field optical microscopy, SNOM, also known as near-field scanning optical microscopy, NSOM, is one member of the SPM family that provides spectroscopic information in the sense most familiar to chemists. In particular, it is capable of providing Raman spectroscopy data with a spatial resolution that may be as good as 20 nm.

SNOM developed as a tool for the optical characterization of materials. The idea behind SNOM was to surpass the so-called diffraction limit by utilizing near-field excitation. The resolution r of a conventional optical method is defined by Rayleigh's equation:

$$r = \frac{0.61\lambda}{n_r \sin \alpha} \tag{9.27}$$

where λ is the wavelength and $n_r \sin \alpha$ is the numerical aperture. When light propagates through a small aperture (or lens), it undergoes diffraction leading to the formation of an Airy disc. Rayleigh's equation defined the minimum separation at which two Airy discs may be resolved. It is often approximated to $\lambda/2$, so that with visible light, one may generally assume that the best resolution achievable is no better than 200 nm. However, there is associated with the aperture an additional optical excitation, a non-propagating *evanescent* field (from the Latin word for decay). This evanescent, or near, field is not subject to diffraction, but it decays very rapidly away from the aperture – hence under normal conditions we are not able to interact with the optical near field. The absence of diffraction effects means that the near field does offer the potential for optical excitation (and hence measurement) with a resolution that is, in principle, unlimited. The possibility of exploiting such phenomena was first grasped by Synge, in 1928 [141]. He proposed a very simple and elegant means to achieve near-field imaging, in which a screen, containing a nanoscopic aperture, was scanned across a sample. The screen and sample were held in very close proximity, so that the sample could interact with the optical near-field. Under such circumstances, he proposed, optical characterization could be performed with a resolution much better than Rayleigh's limit (Figure 9.34).

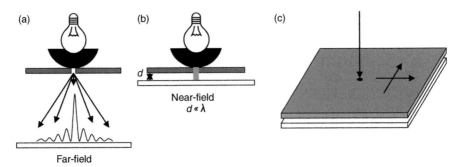

Figure 9.34 Schematic illustrations of (a) the formation of an Airy disc, (b) imaging the evanescent field to generate a near-field image and (c) Synge's concept for a near-field microscope

The practical obstacle that prevented the realization of this elegant concept was that the presence of even a small number of dust particles between the screen and the sample – practically impossible to prevent, even in a clean room – would place the sample outside the optical near-field. Consequently, Synge's proposal remained on the shelf for nearly sixty years, when the SNOM first began to be developed based upon SPM technology. In the most common realization, a SNOM utilizes an optical fibre probe as the light source (although cantilever probes, with hollow tips that contain apertures, and bent optical fibres are also used in commercial instruments). This solves the problem encountered by Synge's device: the aperture is formed at the apex of a sharp tip formed at the end of the probe, meaning that the effective area of the 'screen' that is translated across the surface is very small indeed and the problem of maintaining the sample at all times within the near-field ceases to be a significant one. In the most common realization of the technique (Figure 9.35), the probe is attached to a tuning fork (similar to the ones familiar to musicians, but very much smaller) which is caused electrically to oscillate [142]. As the probe interacts with the sample, the oscillation is damped, through the action of shear forces (hence the method is typically referred to as shear-force modulation) and this damping may be detected and used as the signal to control a feedback system, in which the sample is raised or lowered to maintain the tuning fork oscillation at some pre-set frequency, or set-point. Shear-force imaging exerts a significant pressure on the sample, which may be damaging under certain circumstances, but if the probe has a sharp apex (as is often the case) then

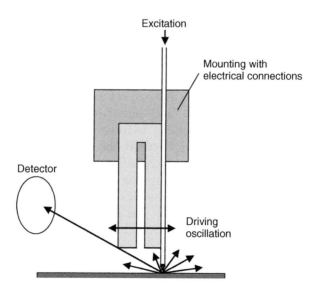

Figure 9.35 Essential features of a shear-force optical fibre-based SNOM

shear-force topographical images of surprisingly high resolution may often be obtained.

The probe may be prepared by sharpening an optical fibre into a point (most commonly by the use of chemical etching [143, 144] or by heating the fibre with a laser whilst pulling it, so that the eventual fracture of the fibre yields a sharp tip [145, 146]). There are a variety of factors that influence the quality of the resulting probe. For example, pulled probes are generally sharper, but etched probes have larger cone angles resulting in higher transmission, with less energy being dissipated as heat in the metal coating at the end of the probe. The tip is then coated with an opaque material (commonly aluminium although other metals, such as gold, are used). The optical properties of the material are important: the coating must be thick enough to prevent lateral penetration of the electric field through the probe, but at the same time, not so thick that it becomes difficult to form an aperture at the apex. The aperture may be formed in a variety of ways, most simply by colliding the probe with a surface and in a more controlled fashion by using focused ion beam milling (FIB) to drill a hole through the metal coating.

There are a variety of ways of arranging the collection of a signal. The sample may be irradiated, and a signal collected through the fibre. Alternatively, and more commonly, the excitation (from a suitable laser light source) is delivered through the fibre, and an optical signal collected either in transmission (through an objective placed directly beneath the probe) or in reflection mode. In reflection mode, fluorescence, for example, may be collected using a sensitive photodetector. Given that fluorescent emission will occur in all directions, and the detector will be located at a specific location, it must be extremely sensitive. However, single photon counting systems, based on avalanche photodiodes, are comparatively inexpensive and provide a realistic means of achieving detection.

Fluorescence images may be acquired by SNOM with sub-diffraction-limit resolution. Not surprisingly, biological systems have been the focus of a great deal of attention in this respect. For example, Dunn et al. measured the fluorescence lifetimes of single light harvesting complexes in photosynthetic membranes [147]. The ability to function under water is also important for biological systems [148]. Recently the feasibility of studying liquid–liquid interfaces was demonstrated [149], opening the possibility of carrying out in vitro measurements. There has also been a great deal of interest in the characterization of dye molecules often embedded in polymeric matrices, such as latex particles [150]. The photobleaching of a single latex particle may be studied by selectively exposing it to light through the SNOM probe, and comparing images before and after exposure. Single molecule fluorescence has also been measured using SNOM [151].

Vibrational spectroscopy is a powerful tool for the characterization of bonding at surfaces. While fibre materials that are compatible with mid-IR excitation are not readily available, there have been some significant successes using Raman methods. For example, Batchelder and co-workers coupled a SNOM to a commercial Raman spectrometer, and despite the low scattering cross-section, managed to carry out submicron measurements of stress in silicon [152, 153]. Raman scattering cross-sections may be increased if surface enhancement (SERS) is employed [154] and Ziesel *et al.* have deposited dye molecules onto silver substrates to facilitate the acquisition of near-field SERS spectra from as few as ca. 300 molecules [155]. They were able to achieve a resolution of 100 nm [156] during imaging.

However, despite such enormous potential, and widespread excitement, it is probably fair to say that SNOM has unfortunately largely failed to realize its potential as a nanoscale characterization tool. Impressive data have been acquired by SNOM (for a useful review, see Dunn [157]) and a strong community of users continues to exploit the technique. They have developed ingenious techniques that have provided unique insights into the optical properties of nanostructures. However, SNOM has failed to have the widespread impact that was at one time anticipated. There are a variety of reasons for this. Perhaps the most significant obstacle has been a general perception that SNOM is a difficult technique. There is, perhaps, some justification for this. Given advances in other techniques, including the spectacular development of new optical methods such as stimulated emission depletion microscopy (STED) [158], and the fact that a good confocal microscope can achieve a resolution of 200 nm – hard to better by SNOM in studies of difficult systems – SNOM has failed to become widely used.

9.4.2 APERTURELESS SNOM

In addition to methods based upon optical aperture probes, it is also possible to excite near-field phenomena in the absence of an aperture by using a metal tip as the probe. There has recently been a growth of interest in these so-called apertureless SNOM techniques, in the hope of delivering superior resolution to that provided by aperture probes. They rely upon the fact that when a nanoscale metallic asperity is held in close proximity to a surface and illuminated with polarized light, the electric field associated with the excitation may be significantly enhanced in a small region directly beneath the tip [159]. The magnitude of the enhancement in the field strength may be very large – several orders of magnitude under optimal conditions. The phenomenon has become known rather evocatively as 'the lightning-rod effect'. The field enhancement is associated with scattering from the tip, and consists mainly of non-propagating (evanescent) components. The advantages of apertureless approaches are that they deliver significantly enhanced resolution compared to techniques based on aperture probes and, moreover, tips are significantly easier to fabricate than optical fibre probes.

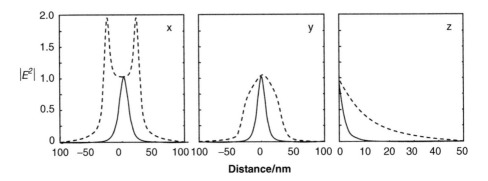

Figure 9.36   Field enhancements in ASNOM after Novotny *et al.*, ref. [159]

Figure 9.36 shows the extent of the resolution improvement that is possible in principle by comparing the spatial variation in the electric field strength associated with a conventional aperture probe with that associated with a plasmon excited apertureless from a silver probe.

Apertureless approaches offer the same attractive capability for carrying out spectroscopic characterization of materials. Apertureless Raman microscopy has been used to carry out spectroscopic characterization of single carbon nanotubes by Novotny and co-workers, for example [160]. In addition, a variety of more sophisticated optical measurements is possible. It has been demonstrated that the magnitude of the field enhancement under a silver asperity is sufficient to enable the excitation of non-linear optical processes. For example, two-photon absorption experiments have been reported [161] and second harmonics have been excited from metal tips [162]. Fluorescence measurements are also possible. By attaching a donor chromophore to the tip and an acceptor on the surface, for example, it is possible to carry out a kind of localized, scanning fluorescence resonant energy transfer (FRET) experiment. These capabilities are all very exciting. The best resolution achieved to date using apertureless techniques does appear to exceed significantly the achievements of aperture-based SNOM. For example, Novotny and co-workers have reported a resolution of 20 nm in their apertureless Raman studies of carbon nanotubes [160] (see Figure 9.37). However, against this it must be noted that apertureless SNOM is, if anything, even more technically challenging than aperture-based measurements. It remains to be seen how widely adopted it will become, therefore.

9.5 Other Scanning Probe Microscopy Techniques

Because of the limited space available to us here we have been able only to examine the three most widely used scanning probe microscopy techniques.

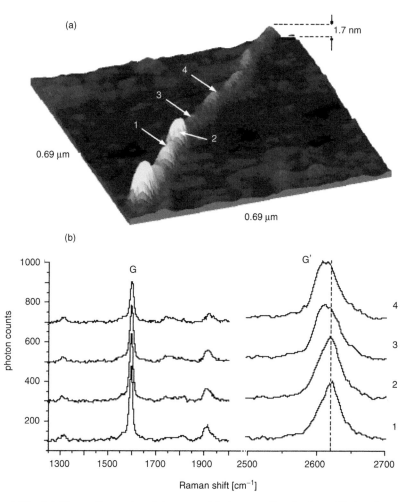

Figure 9.37 (a) Three dimensional topographic image of a single-walled carbon nanotube on a glass surface. (b) Near-field Raman spectra acquired at the marked locations 1 to 4 in (a). The distance between positions 2 and 3 is 35 nm.

However, there are a great many others. For example, there are techniques designed to measure electrostatic interactions at surfaces, including scanning Kelvin probe microscopy (which measures the surface potential in a localized fashion) and scanning electrochemical microscopy; there are thermal methods (including scanning thermal microscopy, which yields a kind of scanning probe analogue of differential scanning calorimetry, and scanning photothermal microscopy, which offers access to vibrational spectroscopic data). The fact that we have been unable to treat these methods here is merely a reflection on the lack of space available: there appears to be almost no limit to the malleability of the basic concept of scanning probe microscopy and doubtless many more techniques have yet to be discovered.

9.6 Lithography Using Probe Microscopy Methods

While this is strictly a book about surface *analysis*, it is nevertheless appropriate to mention briefly the capabilities that scanning probe instruments offer for the patterning of materials.

9.6.1 STM LITHOGRAPHY

We noted in Section 9.2.1 that the interactions between the STM tip and the sample surface can be quite substantial, and can cause alterations in the sample surface structure. These interactions may be utilized in order to modify the surface in a controlled fashion, giving rise to the possibility for nm-scale manipulation of surface structure. However, besides these mechanical interactions, there are a number of more subtle ways in which the STM may be used to modify a surface structure on the nanometre scale.

Lyo and Avouris have also used field-induced effects (generated by applying a voltage pulse of +3 V to the sample) to manipulate atoms on an Si(111)-(7 × 7) reconstructed surface [163]. They studied the formation of structures at varying field strengths and tip–sample separations. A lower threshold field strength was determined at which small mounds could be generated. At higher field strengths, the top atomic layer was removed from a small area of the sample. At small tip–sample separations, small mounds were created which were surrounded by moats. It proved possible to move these mounds. Application of an initial pulse of +3 V was required to form the mound (Figure 9.38(a)); a second subsequent pulse of +3 V resulted in the mound being picked up by the STM tip. The STM tip was then moved a short distance and the mound deposited by applying a pulse of −3 V (Figure 9.38(b)). Considerable fine control of the process was possible, to the extent that single atoms could be manipulated. In Figure 9.39(a), a region of the Si(111)-(7 × 7) surface is shown. A +1 V pulse was applied to the STM tip, removing one of the silicon atoms and creating the vacancy visible in Figure 9.39(b). Finally, the atom was redeposited on the surface at its original location (Figure 9.39(c)).

Perhaps the most extreme example of the manipulation of matter was provided by Eigler and Schweizer, who utilized a different technique to manipulate xenon atoms on a nickel (110) surface under UHV conditions, with the entire STM cooled to 4 K [164]. They reported that, at this temperature, the contamination rate of the surface was decreased, and the stability of the microscope increased, to such a degree that experiments could be performed on a single atom for days at a time.

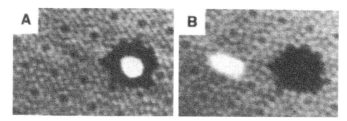

Figure 9.38 Creation and movement of a small mound on a silicon surface by pulsing the STM tip. Reproduced with permission of the American Association for the Advancement of Science from Lyo and Avouris [163]

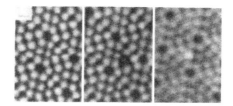

Figure 9.39 Removal (b) and replacement (c) of a single silicon atom on an Si(111) surface. Reproduced with permission of the American Association for the Advancement of Science from Lyo and Avouris [163]

The manipulative procedure involved sliding atoms across the surface of the nickel substrate using the STM tip. The tip was scanned across the surface until it was positioned directly above an atom. The tip–sample separation was then decreased by increasing the tunnel current, lowering the tip onto the atom. The atom was then dragged to its desired location and tip withdrawn, allowing the atom to be imaged in its new position. The process is illustrated in Figure 9.40. In Figure 9.40(a), we see a nickel (110) surface after dosing with xenon. In Figure 9.40(b–f), the atoms are rearranged to form the letters 'IBM'.

Zeppenfeld *et al.* have shown that atomic manipulation using this sliding process is not restricted to the very weakly bound xenon atoms, but may be applied to CO, and even Pt atoms, on a Pt(111) surface [165]. Several structures were created by positioning individual CO molecules on the Pt surface, including the letters 'CO', a small hexagonal island, and a Molecular Man, built from 28 CO molecules and measuring 45 Å from head to foot (see Figure 9.41).

9.6.2 AFM LITHOGRAPHY

The results described above are extremely impressive, but with process times of several days at temperatures of a few K, they are unlikely ever

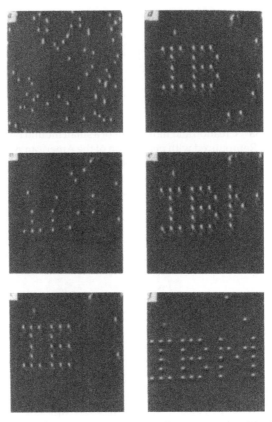

Figure 9.40   Clean nickel (110) surface dosed with xenon atoms (a) which are arranged using the STM tip to form the letters 'IBM'. Reproduced with permission of the Nature Publishing Group from Eigler and Schweizer [164]

to find widespread application. However, methods based upon atomic force microscopy have recently shown enormous promise, and are proving to be capable of straightforward implementation using commonly available equipment. An important motivation for the development of such techniques is the widespread interest in the organization of molecules on nanometre length scales. Outside of the electronic device industry, there are a plethora of problems in nanoscale science that may be solved using nanostructured assemblies of molecules. These include both fundamental investigations of molecular behaviour, and much more application-driven research, for example into the development of novel types of ultra-sensitive systems for biological analysis. This interest has led to the development of a variety of scanning probe lithography systems. Three of the most widely used are illustrated in Figure 9.42. For a detailed treatment of these methods, the reader is directed to the excellent review by Kramer *et al.* [166]. Here we simply sketch the basic principles.

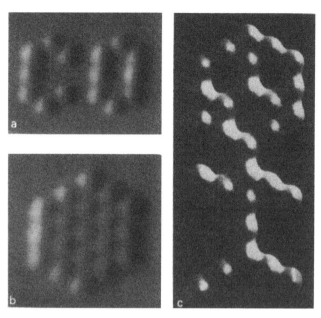

Figure 9.41 Features formed by sliding CO molecules across a Pt(111) surface. Reproduced with permission of Elsevier Science from Zeppenfeld *et al.* [165]

In dip-pen nanolithography, or DPN (Figure 9.42(a)), an AFM probe is 'inked' in a solution of a molecular adsorbate and traced across a suitable substrate [167, 168]. Under ambient conditions, a liquid bridge forms between the tip and the surface, facilitating transfer of molecular material. DPN has been used to create molecular patterns in a variety of ways and has realized very high spatial resolution (ca. 15 nm) under optimal conditions (which typically include an atomically flat substrate). A variety of factors influence the resolution of DPN, including the wetting properties of the ink on the substrate of the surface, the ambient humidity (which influences the diameter of the liquid bridge across which the ink molecules are transferred – decreasing the humidity improves the resolution but if the humidity becomes too low, then lines become incomplete), the roughness of the substrate and the writing rate (increasing the rate decreases the line width, but if the rate is too fast the lines are incomplete). A wide variety of 'inks' have been reported. Initially, work focused on the deposition of alkanethiols onto gold surfaces. For example, DPN offers several approaches to the deposition of biomolecules to form miniaturized arrays. One approach is to pattern the surface to introduce chemistries with specific organizing influences on biomolecules. For example, Lee *et al.* fabricated arrays of spots of mercaptohexadecanoic acid (MHA) by DPN, and then passivated the regions in between them using a highly protein-resistant oligo(ethylene glycol) (OEG) terminated thiol, before exposing the samples to a solution

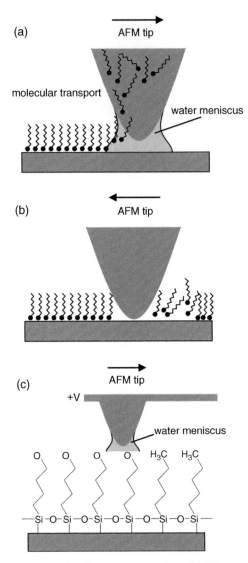

Figure 9.42   Techniques for molecular nanopatterning. (a) Dip-pen nanolithography, (b) nanoshaving/nanografting and (c) local oxidation lithography

of protein [169]. Protein adsorbed to the regions functionalized by MHA, but not to the OEG functionalized regions resulting in structures as small as 100 nm. Subsequently, a wide range of other inks have been reported, including proteins [170], oligonucleotides [171], precursors to semiconducting polymers [172] and metal salts, for electrochemical conversion to metallic nanowires [173].

In contrast, nanoshaving/nanografting (Figure 9.42(b)) relies upon the selective *removal* of adsorbates, usually thiols. An AFM probe is scanned across the sample at elevated load. If the load is high enough (ca. 100 nN for a monolayer of alkanethiols on gold) the adsorbates will be displaced from the surface, leaving a bare region. The bare region of the surface may be filled with a second adsorbate [174, 175]. Nanografting is a variation of the same method in which the operation is carried out in a solution of an alkanethiol. As the surface is cleared of adsorbates, thiols adsorb from the solution phase. Impressive spatial resolution has been achieved by nanoshaving/nanografting. Using this approach, a patch of octade-canethiol only 2×4 nm^2 in size was grafted into a decanethiol monolayer. However, its major limitation is its reliance upon mechanical forces for surface modification – adsorbates must be used that may be readily displaced, in a selective fashion, at loads accessible in an AFM system while maintaining a small contact area.

In neither DPN nor nanografting is the direct, selective initiation of a chemical reaction a possibility. The technique that provides the best capability in this direction is local oxidation lithography (Figure 9.42(c)), where a potential difference is applied between an AFM tip and a sample under ambient conditions. A liquid bridge again forms between tip and sample, providing a kind of one-dimensional electrochemical cell. Remarkable spatial resolution has been achieved in this way, particularly for semiconducting surfaces, on which oxide nanostructures with very small widths (better than 10 nm) can be fabricated in a controlled fashion [176], and the apparatus required is straightforward. Despite the fact that one would expect the specificity of the chemistry accessible using such an approach to be limited, because the driving force for bond-breaking is a presumably a locally high electric field gradient, some sophisticated surface functionalization processes have nevertheless been reported [177–180]. Sagiv and co-workers have used local oxidation methods to selectively oxidize the tail groups of methyl terminated monolayers of alkylsiloxanes adsorbed on silicon dioxide [177]. The resulting oxygen containing functionalities are significantly more reactive and may be derivatized with a further siloxane molecule. If this is terminated by a double bond, then addition across the double bond provides a means of creating new functionalities at the surface.

9.6.3 NEAR-FIELD PHOTOLITHOGRAPHY

It was realized soon after its initial development that an aperture-based SNOM system provides, effectively, a very small light source and, therefore, that it should be possible to carry out lithography with it. Early efforts to write structures into photoresists proved disappointing, however. The reason was probably the finite thickness of the films (a few tens of nm): the electric field associated with the near-field probe is known to diverge reasonably fast in the dielectric medium beneath it and for such films the area of excitation probably extended far outside the original region of excitation.

The problem was solved by reducing the thickness of the resist layer. In particular, by utilizing a self-assembled monolayer of alkanethiols adsorbed on a gold surface, it was possible to confine the photochemical modification to a monatomic layer – the adsorbate sulfur atoms – which could be selectively photo-oxidized to yield weakly bound sulfonates which could be displaced from the surface. Using this approach, termed scanning near-field photolithography (SNP), a resolution comparable to that achievable by electron beam lithography has been achieved [181, 182]. The best result to date, a resolution of 9 nm in an alkanethiol monolayer, corresponds to a resolution of nearly $\lambda/30$. Importantly, while SNOM is regarded as a difficult tool for carrying out surface characterization, lithography using a SNOM system is very much easier because most of the problems are in fact associated with signal detection.

The key criterion for the exceptional resolution of SNP was the excitation of a specific photochemical reaction in a group distributed with monolayer coverage on a solid surface. However, there are a wide variety of systems that fit this basic description. Some of these are illustrated in Figure 9.43. They include monolayers of photoactive siloxanes [183] adsorbed on the native oxide of silicon, monolayers of phosphonic acids formed on the native oxide of aluminium, structures etched into metallic substrates through photopatterned monolayers [184], protein and DNA nanostructures and nanowires formed by the oxidative modification of nanoparticles [185]. Such methodologies reflect the broad range of chemical transformations that may be effected by photochemical means and suggest a lithographic tool that combines high spatial resolution with chemical selectivity and versatility.

9.6.4 THE 'MILLIPEDE'

The argument has always been made in the past that the drawback of scanning probe lithographic techniques is their intrinsically serial nature (the probe writes only one feature at a time). In contrast, conventional photolithographic processes, such as are used to manufacture semiconductor chips, are parallel processes – large numbers of lithographic operations are executed simultaneously. However, this criticism has been addressed powerfully and elegantly by Binnig and co-workers at IBM [186]. They suggested that a massively parallel device, in which a large array of probes all functions simultaneously, will be functionally equivalent to a parallel fabrication process. Their device, named the "Millipede", consists of over a thousand AFM-type cantilevers, all of which may be operated simultaneously in parallel. As a result, they have been able to demonstrate the capacity to write data at enormous density and high speed – rendering portable Tb

Figure 9.43 Examples of structures fabricated by scanning near-field photolithography. (a) 20 nm-wide lines of methyl-terminated thiols formed in a carboxylic acid-terminated self-assembled monolayer on gold. (b) Concentric rings of carboxylic acid groups formed at the surface of a chloromethylphenyltrichlorosilane (CMPTS) monolayer on the native oxide of silicon. (c) Array of DNA nanospots formed on a CMPTS monolayer. (d) 60 nm-wide gold nanowires formed by photo-oxidation of thiol-stabilized gold nanoparticles

information storage systems a practical possibility. This represents a major triumph, and with their development of massively parallel approaches to AFM lithography, Binnig and co-workers have provided a new paradigm for probe lithography. We may thus hope to see a growth of applications for such methodology in the future.

9.7 Conclusions

Scanning probe microscopy has made a significant impact in a broad range of scientific disciplines where the structures of surfaces, the adsorption of molecules and interfacial interactions are of interest. Early naivety regarding the potential difficulties surrounding image interpretation and sample preparation is being displaced by a developing understanding of both the fundamental theoretical principles and the nature of the experimental operation of scanning probe microscopes. Quantitative methods for surface analysis with nanometre scale are developing that are capable of application to a wide range of materials. While the SPM has already made enormous impact, the ripples are still travelling rapidly outwards and as the SPM family matures, we may expect these techniques to become even more firmly embedded as key tools for the analysis of surfaces.

📖 References

1. P.W. Atkins, *Molecular Quantum Mechanics*, Oxford University Press, Oxford, p. 41 (1983).

2. J. Tersoff and D.R. Hamann, *Phys. Rev. B* **31**, 805 (1985).

3. G. Binnig and H. Rohrer, *IBM J. Res. Develop.* **30**, 355 (1986).

4. J. Winterlin, J. Wiechers, H. Burne, T. Gritsch, H. Hofer and R.J. Behm, *Phys. Rev. Lett.* **62**, 59 (1989).

5. C.J. Chen, *J. Vac. Sci. Technol. A* **9**, 44 (1991).

6. U. Landman and W.D. Luedtke, *Appl. Surf. Sci.* **60/61**, 1 (1992).

7. M. Tsukada, K. Kobayashi, I. Nobuyuki and H. Kageshima, *Surf. Sci. Rep.* **8**, 265 (1991).

8. X. Yang, Ch. Bromm, U. Geyer and G. von Minnigerode, *Ann. Physik.* **1**, 3 (1992).

9. G. Binnig, H. Rohrer, Ch. Gerber and E. Weibel, *Surf. Sci.* **131**, L379 (1983).

10. G. Binnig, H. Rohrer, Ch. Gerber and E. Stoll, *Surf. Sci.* **144**, 321 (1984).

11. V.M. Hallmark, S. Chiang, J.F. Rabolt, J.D. Swalen and R.J. Wilson, *Phys. Rev. Lett.* **59**, 2879 (1987).

12. G. Binnig and H. Rohrer, *Rev. Mod. Phys.* **59**, 615 (1987).

13. R.C. Jaklevic and L. Elie, *Phys. Rev. Lett.* **60**, 120 (1988).

14. C.R. Clemmer and T.P. Beebe, Jr, *Scanning Microsc.* **6**, 319 (1992).

15. J. Schneir, R. Sonnenfeld, O. Marti, P.K. Hansma, J.E. Demuth and R.J. Hamers, *J. Appl. Phys.* **63**, 717 (1988).

16. A.M. Baro, R. Miranda, J. Alaman, N. Garcia, G. Binnig, H. Rohrer, Ch. Gerber and J.L. Carrascosa, *Nature* **315**, 253 (1985).

17. S. Park and C.F. Quate, *Appl. Phys. Lett.* **48**, 112 (1986).

18. R. Sonnenfeld and P.K. Hansma, *Science* **232**, 211 (1986).

19. G. Binnig, H. Rohrer, Ch. Gerber and E. Weibel, *Phys. Rev. Lett.* **50**, 120 (1983).

20. K. Takayanagi, Y. Tanishiro, M. Takahashi and S. Takahashi, *J. Vac. Sci. Technol. A* **3**, 1502 (1985).

21. R.J. Hamers, R.M. Tromp and J.E. Demuth, *Phys. Rev. Lett.* **56**, 1972 (1986).

22. R.M. Tromp, R.J. Hamers and J.E. Demuth, *Phys. Rev. B* **34**, 1388 (1986).

23. R.J. Hamers, *Ann. Rev. Phys. Chem.* **40**, 531 (1989).

24. W. Hebenstreit, J. Redinger, Z. Horozova, M. Schmid, R. Podloucky, and P. Varga, *Surf. Sci.* **424**, L321 (1999).

25. S. Schintke, S. Messerli, M. Pivetta, F. Patthey, L. Libioulle, M. Stengel, A.De. Vita, and W.D. Schneider, *Phys. Rev. Lett.* **87**, 276801 (2001).

26. J.W.G. Wildoer, L.C. Venema, A.G. Rinzler, R.E. Smalley and C. Dekker, *Nature* **391**, 59 (1998).

27. T.W. Odom, J. Huang, P. Kim and C. Lieber, *Nature* **391** 62 (1998).

28. Ph. Redlich, D.L. Carroll and P.M.N. Ajayan, *Curr. Opinion Solid State Mater. Sci.* **4**, 325 (1999).

29. R. Berndt, R. Gaisch, J.K. Gimzewski, B. Reihl, R.R. Schlittler, W.D. Schneider, and M. Tschudy, *Science* **262**, 1425 (1993).

30. B.C. Stipe, M.A. Rezaei and W. Ho, *Science* **280**, 1732 (1998).

31. B.C. Stipe, M.A. Rezaei and W. Ho, *Phys. Rev. Lett.* **82**, 1724 (1999).

32. D.P.E. Smith, J.K.H. Horber, G. Binnig and H. Nejoh, *Nature* **344**, 641 (1990).

33. M. Hara, Y. Iwakabe, K. Tochigi, H. Sasabe, A.F. Garito and A. Yamada, *Nature* **344**, 228 (1990).

34. M. Hara, *RIKEN Rev.* No. 37, **48** (2001).

35. R.G. Nuzzo, L.H. Dubois and D.L. Allara, *J. Am. Chem. Soc.* **112**, 558 (1990).

36. L. Strong and G.M. Whitesides, *Langmuir* **4**, 546 (1988).

37. J.C. Love, L.A. Estroff, J.K. Kreibel, R.G. Nuzzo and G.M. Whitesides, *Chem. Rev.* **105**, 1103 (2005).

38. C.D. Bain, E.B. Troughton, Y.-T. Tao, J. Evall, G.M. Whitesides and R.G. Nuzzo, *J. Am. Chem. Soc.* **111**, 321 (1989).

39. E. Delamarche, B. Michel, Ch. Gerber, D. Anselmetti, H.-J. Guntherodt, H. Wolf and H. Ringsdorf, *Langmuir* **10**, 2869 (1994).

40. G.E. Poirier and M.J. Tarlov, *Langmuir* **10**, 2853 (1994).

41. G.E. Poirer, M.J. Tarlov and H.E. Rushmeier, *Langmuir* **10**, 3383 (1994).

42. G.E. Poirier and M.J. Tarlov, *J. Phys. Chem.* **99**, 10966 (1995).

43. G.E. Poirier and E.D. Pylant, *Science* **272**, 1145 (1996).

44. G.E. Poirier, *Langmuir* **15**, 1167 (1999).

45. G. Binnig and H. Rohrer, in J. Janka and J. Pantoflicek (Eds), *Trends in Physics*, European Physical Society, The Hague, p. 38 (1984).

46. S.M. Lindsay, T. Thundat, L. Nagahara, U. Knipping and R.L. Rill, *Science* **244**, 1063 (1989).

47. C.R. Clemmer and T.P. Beebe, Jr, *Science* **251**, 640 (1991).

48. W.M. Heckl and G. Binnig, *Ultramicroscopy* **42–44**, 1073 (1992).

49. C.J. Roberts, M. Sekowski, M.C. Davies, D.E. Jackson, M.R. Price and S.J.B. Tendler, *Biochem. J.* **283**, 181 (1992).

50. M. Amrein, A. Stasiak, H. Gross, E. Stoll and G. Travaglini, *Science* **240**, 514 (1988).

51. W.M. Heckl, K.M.R. Kallury, M. Thompson, Ch Gerber, H.J. Horber and G. Binnig *Langmuir*, **5**, 1433 (1989).

52. L.A. Bottomley, J.N. Haseltine, D.P. Allison, R.J. Warmack, T. Thundat, R.A. Sachlebenm, G.M. Brown, R.P. Woychik, K.B. Jacobson and T.L. Ferrell, *J. Vac. Sci. Technol. A* **10**, 591 (1992).

53. Y.L. Lyubchenko, S.M. Lindsay, J.A. DeRose and T. Thundat, *J. Vac. Sci. Technol. B* **9**, 1288 (1991).

54. D.K. Luttrull, J. Graham, J.A. DeRose, D. Gust, T.A. Moore and S.M. Lindsay, *Langmuir*, **8**, 765 (1992).

55. G.J. Leggett, C.J. Roberts, P.M. Williams, M.C. Davies, D.E. Jackson and S.J.B. Tendler, *Langmuir*, **9**, 2356 (1993).

56. S.M. Lindsay in D.A. Bonnell (Ed.), *Scanning Tunnelling Microscopy: Theory, Techniques and Applications*, VCH, New York, NY, pp. 335–408 (1993).

57. R. Guckenberger, W. Wiegrabe, A. Hillebrand, T. Hartmenn, Z. Wang and W. Baumeister, *Ulramicroscopy* **31**, 327 (1989).

58. J.-Y. Juan, Z. Shao and C. Gao, *Phys. Rev. Lett.* **67**, 863 (1991).

59. G.J. Leggett, M.C. Davies, D.E. Jackson, C.J. Roberts, S.J.B. Tendler and P.M. Williams, *J. Phys. Chem.* **97**, 8852 (1993).

60. G. Binnig, C.F. Quate and C. Gerber, *Phys. Rev. Lett.* **56**, 930 (1986).

61. Y. Martin, C.C. Williams and K. Wickramsinghe, *J. Appl. Phys.* **61**, 4723 (1987).

62. J.N. Israelachvili, *Intermolecular and Surface Forces*, 2nd Edition, Academic Press, London (1992).

63. E. Gnecco, R. Bennewitz, T. Gyalog and E. Meyer, *J. Phys., Condens. Matter* **13**, R619 (2001).

64. J.M. Neumeister and W.A. Ducker, *Rev. Sci. Instrum.* **65**, 2527 (1994).

65. J.E. Sader, *Rev. Sci. Instrum.* **66**, 4583 (1993).

66. J.E. Sader, *Rev. Sci. Instrum.* **74**, 2438 (2003).

67. J.E. Sader and R.C. Sader, *Appl. Phys. Lett.* **83**, 3195 (2003).

68. J.P. Cleveland, S. Manne, D. Bocel and P.K. Hansma, *Rev. Sci. Instrum.* **64**, 403 (1993).

69. J.L. Hutter and J. Bechhoeffer, *Rev. Sci. Instrum.* **64**, 1868 and 3342 (1993).

70. H.-J. Butt and M. Jaschke, *Nanotechnology* **6**, 1 (1995).

71. J.L. Hazel and V.V. Tsukruk, *Trans ASME* **120**, 814 (1998).

72. M. Tortonese and M. Kirk, *Proc. SPIE* **53**, 3009 (1997).

73. P.J. Cumpson, P. Zhdan and J. Hedley, *Ultramicroscopy* **100**, 241 (2004).

74. B.D. Beake, G.J. Leggett and P.H. Shipway, *Surf. Interface Anal.* **27**, 1084 (1999).

75. F. Ohnesorge and G. Binnig, *Science* **260**, 1451 (1993).

76. T.R. Albrecht, P. Grutter, D. Horn, and D. Rugar, *J. Appl. Phys.* **69**, 668 (1991).

77. F.J. Giessigl, *Science* **267**, 68 (1995).

78. S. Morita, R. Wiesendanger and E. Meyer (Eds), *Noncontact Atomic Force Microscopy*, Springer-Verlag, Berlin (2002).

79. G.U. Lee, L.A. Chrisey and R.J. Colton, *Science* **266**, 771 (1994).

80. G.U. Lee, D.A. Kidwell and R.J. Colton, *Langmuir* **10**, 354 (1994).

81. K. Hu and A.J. Bard, *Langmuir* **13**, 5114 (1997).

82. E.-L. Florin, V.T. Moy and H.E. Gaub, *Science* **264**, 415 (1994).

83. H. Schonherr, M.W.J. Beulen, J. Bugler, J. Huskens, F.C.J.M. van Veggel, D.N. Reinhoudt and G.J. Vancso, *Langmuir* **122**, 4963 (2000).

84. T. Nakagawa, K. Ogawa, T. Kurumizawa and S. Ozaki, *Jpn J. Appl. Phys.* **32**, L294 (1993).

85. T. Nakagawa, K. Ogawa and T. Kurumizawa, *J. Vac. Sci. Technol. B* **12**, 2215 (1994).

86. T. Ito, M. Namba, P. Buhlmann and Y. Umezawa, *Langmuir* **13**, 4323 (1997).

87. C.D. Frisbie, L.F. Rozsnyai, A. Noy, M.S. Wrighton and C.M. Lieber, *Science* **265**, 2071 (1994).

88. A. Noy, C.D. Frisbie, L.F. Rozsnyai, M.S. Wrighton and C.M. Lieber, *J. Am. Chem. Soc.* **117**, 7943 (1995).

89. E.W. van der Wegte and G. Hadziioannou, *Langmuir* **13**, 4357 (1997).

90. D.V. Vezenov, A. Noy, L.F. Rozsnyai and C.M. Lieber, *J. Am. Chem. Soc.* **119**, 2006 (1997).

91. S.K. Sinniah, A.B. Steel, C.J. Miller and J.E. Reutt-Robey, *J. Am. Chem. Soc.* **118**, 8925 (1996).

92. K. Feldman, T. Tervoort, P. Smith and N.D. Spencer, *Langmuir* **14**, 372 (1998).

93. S.C. Clear and P.F. Nealey, *J. Coll. Interface Sci.* **213**, 238 (1999).

94. D.V. Vezenov, A. Noy and P.J. Ashby, *Adhes. Sci. Technol.* **19**, 313 (2005).

95. G.J. Vancso, H. Hillborg and H. Schonherr, *Adv. Polym. Sci.* **182**, 55 (2005).

96. S. Grafstrom, M. Neizert, T. Hagen, J. Ackerman, R. Neumann, O. Probst and M. Wortge, *Nanotechnology* **4**, 143 (1993).

97. G. Haugstad, W.L. Gladfelter, E.B. Weberg, R.T. Weberg and R.R. Jones, *Langmuir* **11** 3473 (1995).

98. J. Gao, W.D. Luedtke, D. Gourdon, M. Ruths, J.N. Israelachvili and U. Landman, *J. Phys. Chem. B* **108**, 3410 (2004).

99. C.R. Hurley and G.J. Leggett, *Langmuir* **22**, 4179 (2006).

100. C.D. Frisbie, L.F. Rozsnyai, A. Noy, M.S. Wrighton and C.M. Lieber, *Science* **265**, 2071 (1994).

101. S. Akari, D. Horn, H. Keller and W. Schrepp, *Adv. Mater.* **7**, 549 (1995).

102. J.-B.D. Green, M.T. McDermott, M.D. Porter and L.M. Siperko, *J. Phys. Chem.* **99**, 10960 (1995).

103. A. Noy, C.D. Frisbie, L.F. Rozsnyai, M.S. Wrighton and C.M. Lieber, *J. Am. Chem. Soc.* **117**, 7943 (1995).

104. B.D. Beake and G.J. Leggett, *Phys. Chem. Chem. Phys.* **1**, 3345 (1999).

105. S.C. Clear and P.F. Nealey, *J. Coll. Interface Sci.* **213**, 238 (1999).

106. L. Zhang, L. Li, S. Chen and S. Jiang, *Langmuir* **18**, 5448 (2002).

107. L. Li, S. Chen and S. Jiang, *Langmuir* **19**, 666 (2004).

108. N.J. Brewer and G.J. Leggett, *Langmuir* **20**, 4109 (2004).

109. H.I. Kim and J.E. Houston, *J. Am. Chem. Soc.* **122**, 12045 (2000).

110. K.S.L. Chong, S. Sun and G.J. Leggett, *Langmuir* **21**, 2903 (2005).

111. B.D. Beake and G.J. Leggett, *Langmuir* **16**, 735 (2000).

112. S.C. Clear and P.F. Nealey, *J. Chem. Phys.* **114**, 2802 (2001).

113. N.J. Brewer, T.T. Foster, G.J. Leggett, M.R. Alexander and E. McAlpine, *J. Phys. Chem. B* **108**, 4723 (2004).

114. X. Yang, and S.S. Perry, *Langmuir* **19**, 6135 (2003).

115. S. Li, P. Cao, R. Colorado, X. Yan, I. Wenzl, O.E. Schmakova, M. Graupe, T.R. Lee and S.S. Perry, *Langmuir* **21**, 933 (2005).

116. J.A. Hammerschmidt, B. Moasser, W. Gladfelter, G. Haugstad and R.R. Jones, *Macromolecules* **29**, 8996 (1996).

117. J.A. Hammerschmidt, W.A. Gladfelter and G. Haugstad, *Macromolecules* **32**, 3360 (1999).

118. F. Dinelli, C. Buenviaje and R.M. Overney, *J. Chem. Phys.* **113**, 2043 (2000).

119. A. Marti, G. Hahner and N.D. Spencer, *Langmuir* **11**, 4632 (1995).

120. K.S.L. Chong and G.J. Leggett, *Langmuir* **21**, 3903 (2006)

121. D. Fotadis, S. Scheuring, S.A. Muller, A. Engel and D.J. Muller, *Micron* **33**, 385 (2002).

122. S. Scheuring, H. Stahlberg, M. Chami, C. Houssin, J.-L. Rigaud and A. Engel, *Mol. Microbiol.* **44**, 675 (2002).

123. S. Scheuring, D.J. Muller, H. Stahlberg, H.-A. Engel and A. Engel, *Eur. Biophys. J.* **31**, 172 (2002).

124. D. Fotiadis, P. Qian, A. Philippsen, P.A. Bullough, A. Engel and C.N. Hunter, *J. Biol. Chem.* **279**, 2063 (2004).

125. M. Raghavachari, H.-M. Tsai, K. Kottke-Marchant and R.E. Marchant, *Coll. Surf. B: Biointerfaces* **19**, 315 (2000).

126. H. Lin, R. Lal and D.O. Clegg, *Biochemistry* **39**, 3192 (2000).

127. B. Drake, C.B. Prater, A.L. Weishorn, S.A.C. Gould and T.R. Albrecht, *Science* **243**, 1586 (1989).

128. W. Habarle, J.K.H. Horber, F. Ohnesorge, D.P.E. Smith and G. Binnig, *Ultramicroscopy* **42–44**, 1161 (1992).

129. G.U. Lee, D.A. Kidwell and R.J. Colton, *Langmuir* **10**, 354 (1994).

130. E.-L. Florin, V.T. Moy and H.E. Gaub, *Science* **264**, 415 (1994).

131. V.T. Moy, E.-L. Florin and H.E. Gaub, *Science* **266**, 257 (1994).

132. G.U. Lee, L.A. Chrisey and R.J. Colton, *Science* **266**, 771 (1994).

133. T. Boland and B.D. Ratner, *Proc. Natl Acad. Sci. USA* **92**, 5297 (1995).

134. O. Lioubashevski, F. Patolsky and I. Willner, *Langmuir* **17**, 5134 (2001).

135. L.T. Mazzola, C.W. Frank, S.P.A. Fodor, C. Mosher, R. Lartius and E. Henderson, *Biophys. J.* **76**, 2922 (1999).

136. R. Merkel, P. Nassoy, A. Leung, K. Ritchie and E. Evans, *Nature* **397**, 50 (1999).

137. E. Evans, *Biophys. Chem.* **82**, 83 (1999).

138. E. Evans and K. Ritchie, *Biophys. J.* **76**, 2439 (1999).

139. E. Evans, A. Leung, D. Hammer and S. Simon, *Proc. Natl Acad. Sci. USA* **98**, 3784 (2001).

140. P.M. Williams, S.B. Fowler, R.B. Best, J.L. Toca-Herrera, K.A. Scott, A. Steward and J. Clarke, *Nature* **422**, 446 (2003).

141. E.H. Synge, *Phil. Mag.* **6**, 356 (1928).

142. E. Betzig and J.K. Trautman, *Science* **257**, 189 (1992).

143. P. Hoffman, B. Dutoit and R.-P. Salthé, *Ultramicroscopy* **61**, 195 (1995).

144. R. Stockle, C. Fokas, V. Deckert, R. Zenobi, B. Sick, B. Hecht and U.P. Wild, *Appl. Phys. Lett.* **75**, 160 (1999).

145. A. Harootunian, E. Betzig, M.S. Isaacson and A. Lewis, *Appl. Phys. Lett.* **49**, 674 (1986).

146. B. Hecht, B. Sick, U.P. Wild, V. Deckert, R. Zenobi, O.F. Martin and D.W. Pohl, *J. Chem. Phys.* **112**, 7761 (2000).

147. R.C. Dunn, G.R. Holton, L. Mets and X.S. Xie, *J. Phys. Chem.* **98**, 3094 (1994).

148. P.J. Moyer and S.B. Kämmer, *Appl. Phys. Lett.* **68**, 3380 (1996).

149. M. De Serio, A.N. Bader, M. Heule, R. Zenobi and V. Deckert, *Chem. Phys. Lett.* **380**, 47 (2003).

150. M. Rücker, P. Vanoppen, F.C. De Schryver, J.J. Ter Horst, J. Hotta and H. Masuhura, *Macromolecules* **28**, 7530 (1995).

151. E. Betzig and R.J. Chichester, *Science* **262**, 1422 (1993).

152. S. Webster, D.N. Batchelder and D.A. Smith, *Appl. Phys. Lett.* **72**, 1478 (1998).

153. S. Webster, D.A. Smith and D.N. Batchelder, *Vib. Spectrosc.* **18**, 51 (1998).

154. D.A. Smith, S. Webster, M. Ayad, S.D. Evans, D. Fogherty and D. Batchelder, *Ultramicroscopy* **61**, 247 (1995).

155. D. Ziesel, V. Deckert, R. Zenobi and T. Vo-Dinh, *Chem. Phys. Lett.* **283**, 381 (1998).

156. V. Deckert, D. Ziesel, R. Zenobi and T. Vo-Dinh, *Anal. Chem.* **70**, 2646 (1998).

157. R.C. Dunn, *Chem. Rev.* **99**, 2891 (1999).

158. T.A. Klar, S. Jakobs, M. Dyba, A. Egner and S.W. Hell, *Proc. Natl Acad. Sci. USA* **97**, 8206 (2000).

159. L. Novotny, D.W. Pohl and B. Hecht, *Ultramicroscopy* **61**, 1 (1995).

160. A. Hartschuh, E.J. Sanchez, X.S. Xie and L. Novotny, *Phys. Rev. Lett.* **90**, 095503 (2003).

161. E.J. Sánchez, L. Novotny and X.S. Xie, *Phys. Rev. Lett.* **82**, 4014 (1999).

162. A. Bouhelier, M. Beversluis, A. Hartschuh and L. Novotny, *Phys. Rev. Lett.* **90**, 013903 (2003).

163. I.-W. Lyo and P. Avouris, *Science* **253**, 173 (1991).

164. D.M. Eigler and E.K. Schweizer, *Nature* **344**, 524 (1990).

165. P. Zeppenfeld, C.P. Lutz and D.M. Eigler, *Ultramicroscopy* **42–44**, 128 (1992).

166. S. Kramer, R.R. Fuierer and C.B. Gorman, *Chem. Rev.* **103**, 4367 (2003).

167. R.D. Piner, J. Zhu, S. Hong and C.A. Mirkin, *Science* **283**, 661 (1999).

168. S. Hong, J. Zhu and C.A. Mirkin, *Science* **286**, 523 (1999).

169. K.-B. Lee, S.-J. Park, C.A. Mirkin, J.C. Smith and M. Mrksich, *Science* **295**, 1702 (2002).

170. K.-B. Lee, J.-H. Lim and C.A. Mirkin, *J. Am. Chem. Soc.* **125**, 5588 (2003).

171. L.M. Demers, D.S. Ginger, S.-J. Park, Z. Li, S.-W. Chung and C.A. Mirkin, *Science* **296**, 1836 (2002).

172. B.W. Maynor, S.F. Filocamo, M.W. Grinstaff and J. Liu, *J. Am. Chem. Soc.* **124**, 522 (2002).

173. (a) Y. Li, B.W. Maynor and J. Liu, *J. Am. Chem. Soc.* **123**, 2105 (2001). (b) B.W. Maynor, Y. Li and J. Liu, *Langmuir* **17**, 2575 (2001).

174. S. Xu and G.-Y. Liu, *Langmuir* **13**, 127 (1997).

175. G.-Y. Liu and N.A. Amro, *Proc. Natl Acad. Sci. USA* **99**, 5165 (2002).

176. M. Calleja and R. Garcia, *Appl. Phys. Lett.* **76**, 3427 (2000).

177. R. Maoz, E. Frydman, S.R. Cohen and J. Sagiv, *Adv. Mater.* **12**, 725 (2000).

178. S. Hoeppener, R. Maoz, S.R. Cohen, L. Chi, H. Fuchs, and J. Sagiv, *Adv. Mater.* **14**, 1036 (2002).

179. S. Liu, R. Maoz, G. Schmid and J. Sagiv, *Nano Lett.* **2**, 1055 (2002).

180. Z.M. Fresco and J.M.J. Frechet, *J. Am. Chem. Soc.* **127**, 8302 (2005).

181. S. Sun and G.J. Leggett, *Nano Lett.* **4**, 1381 (2004).

182. G.J. Leggett, *Chem. Soc. Rev.* **35**, 1150 (2006)

183. S. Sun, M. Montague, K. Critchley, M.-S. Chen, W.J. Dressick, S.D. Evans and G.J. Leggett, *Nano Lett.* **6**, 29 (2006).

184. R.E. Ducker and G.J. Leggett, *J. Am. Chem. Soc.* **128**, 392 (2006).

185. S. Sun, P. Mendes, K. Critchley, S. Diegoli, M. Hanwell, S.D. Evans, G.J. Leggett, J.A Preece and T.H. Richardson, *Nano Lett.* **6**, 345 (2006).

186. P. Vettiger, M. Despont, U. Drechsler, U. Durig, W. Haberle, M.I. Lutwyche, H. Rothuizen, R. Stutz, R. Widmer and G.K. Binnig, *IBM J. Res. Devel.* **44**, 323 (2000).

Problems

1. The magnitude of the tunnel current in the STM is exponentially dependent upon the tip–sample separation d, and is given approximately by the equation:

$$I \propto e^{(-2\kappa d)}$$

(see Section 9.2.1). The high spatial resolution of the STM is a consequence of this exponential dependence. The total measured tunnelling current is the sum of all the tunnelling interactions between tip and sample; however, the tunnelling interactions between the tip apex and the sample have much higher probabilities than those between other regions of the tip and the surface, and so contribute a greater proportion of the total measured current.

Explore this dependence in the following way. Plot the following tip profiles:

(a) $y = 2 + x^2$

(b) $y = 7 - (25 - x^2)^{1/2}$

(c) $y = 2 + x^2/4$.

Now assume that the tunnel current is the sum of a number of elemental tunnelling currents δI between regions dx of the tip surface and the sample. Assume that the current density is uniform across the surface, and arbitrarily assigned the value 1. Assume that the elemental tunnelling current between any point x and the tip is proportional to e^y where y is the separation between tip and sample. Now plot, for each tip, the elemental tunnelling currents. Which tip will provide the highest resolution?

2. The STM tips above represent very simplistic models; real tips rarely exhibit such idealized geometries and atomic asperities are quite possible. AFM tips are typically much broader, however, and may have curved profiles which approximate much better to a spherical shape. The radius of curvature of the AFM tip may be substantial, and the consequence of this is that small features may have an apparent size which is somewhat larger than their real size. A number of authors have derived expressions which it is hoped can quantify the extent of broadening of small features in the AFM image. Many of these treatments are based upon a simple geometrical analysis of the tip–sample interaction. One example is the following expression [91] which, for features with a circular cross-section smaller than the tip radius of curvature R_t, relates the real radius r of a feature to its radius in the AFM image R (see Figure 9.37):

$$r = R^2/(4R_t)$$

(a) Calculate the measured radius of a DNA molecule (real radius 2 nm) when imaged by tips of radius of curvature 45 nm and 100 nm. Evaluate the percentage broadening for each tip.

(b) For a tip of radius of curvature 45 nm, calculate the measured radius of features with diameters 20 and 5 nm. Evaluate the percentage broadening for each feature.

3. From Probem 2 it will be clear that far from being a vanishing quantity, the area of contact between the FM and a flat sample is substantial on the atomic scale. For some flat materials, 'atomic' resolution images have been recorded, showing arrays of points with atomic dimensions which appear very similar to atomic resolution STM images. However, even if a sufficiently sharp tip could be produced, there would be theoretical difficulties in attempting to image individual atoms. Suppose that a minimal load of 1 nN is applied to the FM tip, with a contact area of radius 10 pm. Estimate the mean pressure applied. Compare this figure with the yield strength for silicon nitride, 550 MPa. What is the likelihood of single atom contacts giving rise to the AFM image?

4. A cantilever with a force constant of $0.1\,\mathrm{N\,m^{-1}}$ is scanned across a surface in constant height mode. What will be the change in force as it passes over a bump of height 100 nm?

5. For a silicon nitride AFM probe of radius 50 nm, calculate the contact area at zero load, the pull-off force and the contact area at separation of the tip from the surface, assuming that JKR mechanics are obeyed and that the interfacial free energy (γ) is $40\,\mathrm{mN\,m^{-1}}$. The elastic modulus of silicon nitride is 64 GPa.

6. The strength of the dispersion interaction between two methyl-terminated self-assembled monolayers is to be studied by chemical force microscopy. A pull-off force of 1.2 nN is measured. The interfacial free energy is 2.5 mN $\mathrm{m^{-1}}$. Use the JKR model to determine the tip radius and the contact area at separation, and hence calculate the force per molecular pair. Assume that the modulus of the tip is 64 Gpa and that the area occupied by one molecule is $0.25\,\mathrm{nm^2}$.

7. Liquid-supported films of two substances A and B were transferred to solid substrates. The transfers were assumed to occur without significant alteration to the structures of the films. In friction force microscopy experiments carried out on these films, the data shown in the following table were acquired.

	Load (nm)			
	10	20	30	40
Frictional signal for film A (V m^{-1})	0.95	1.51	2.02	2.49
Frictional signal for film B (V m^{-1})	0.60	0.81	0.99	1.18

Calculate the coefficients of friction of A and B. Given the surface pressure–area plots for A and B shown in Figure 9.44, suggest a reason why their coefficients of friction are different.

8. A friction force microscopy experiment was carried out on a sample of polyester film, and the data shown in the table below were acquired. Determine whether Amonton's law or the Hertz model best fits the data.

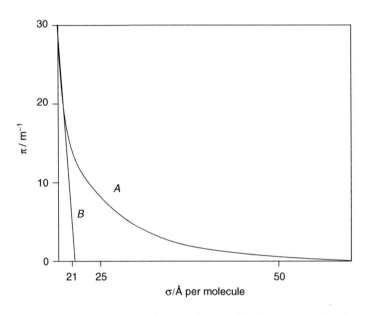

Figure 9.44   Surface–pressure area plots for the two film-forming molecules A and B

Load (nN)	0	10	20	30	40	100
Lateral force (nN)	0.00	0.33	0.52	0.68	0.83	1.52

9. An optical microscope uses a lens with a numerical aperture of 1.5. What is the resolving power when green light (wavelength 488 nm) is used to illuminate the sample?

10 The Application of Multivariate Data Analysis Techniques in Surface Analysis

JOANNA L.S. LEE
National Physical Laboratory, Teddington, UK

IAN S. GILMORE
National Physical Laboratory, Teddington, UK

10.1 Introduction

Multivariate analysis was developed in the 1950s with its roots founded in the study of behavioural science in the 1930s. It is now widely used in analytical chemistry to provide identification and quantification for a range of spectroscopic techniques [1]. Multivariate analysis is often applied in the field of chemometrics, which is the science of relating measurements made on a chemical system to the state of the system via the application of mathematical or statistical methods [2]. Multivariate analysis involves the use of simultaneous statistical procedures for two or more variables in a data set. An essential aspect of multivariate analysis is the statistical study of the dependence (covariance) between different variables. By summarizing the data using a small number of statistical variables, the interpretation of complex data sets involving a large number of dependent variables can quickly be simplified.

Multivariate analysis has been used for a number of years in surface analysis, most notably in the techniques of secondary ion mass spectrometry

Surface Analysis – The Principal Techniques 2nd Edition Edited by John Vickerman and Ian Gilmore
© 2009 John Wiley & Sons, Ltd

(SIMS), X-ray photoelectron spectroscopy (XPS) and Raman spectroscopy, as the raw data obtained from these techniques are multivariate in nature. For example, in X-ray photoelectron spectroscopy (XPS), the intensities on more than one variable (i.e. binding energy) are recorded during each measurement, whereas in time-of-flight secondary ion mass spectrometry (ToF–SIMS) a complete and detailed mass spectrum, containing detected ion intensities over a million mass channels, can be obtained. The number of publications on surface chemical analysis using different multivariate methods since 1990 is shown in Figure 10.1. The dramatic growth in recent years reflects the increased power and throughput of modern instruments, and the increasing requirements for fast, robust methods of data analysis. Multivariate analysis has numerous advantages over traditional (manual) analysis. It provides an objective and statistically valid approach using all available information in a data set. The need for manual identification and selection of key peaks and features for analysis is eliminated or significantly reduced, thereby reducing the need for *a priori* knowledge about the system under study and minimizing the potential for bias. By correlating data across a number of variables, an improved signal to noise ratio can be obtained. Multivariate analysis can also be fast and automated. A typical analysis takes only a few minutes on a modern desktop computer, and therefore it has potential for on-line analysis of real time processes. However, for many scientists, there have been significant ambiguities, confusion in terminology and jargon, low confidence in the results, and a need for an improved understanding of basic and practical aspects. In particular, there is widespread confusion over the choice of the most appropriate multivariate

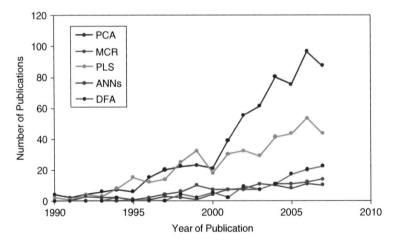

Figure 10.1 The number of published papers since 1990 on the use of multivariate methods in surface chemical analysis, including XPS, SIMS, AFM, Raman spectroscopy or SERS, from the ISI Web of Knowledge publications database [3]

technique and data preprocessing for each application. This situation needs resolution, since the procedures are well established mathematically and can be very helpful indeed for many analytical situations.

We begin this chapter by outlining the basic concepts in Section 10.2, covering the matrix representation of data and its relation to multivariate analysis. We will then review the principle and theory behind many popular multivariate analysis methods, focusing on the objectives and characteristics of each method rather than detailed mathematical derivation. This includes identification using the factor analysis methods of principal component analysis (PCA) and multivariate curve resolution (MCR) in Section 10.3, calibration and quantification using the regression methods of principal component regression (PCR) and partial least squares regression (PLS) in Section 10.4, and classification and clustering using discriminant function analysis (DFA), hierarchical cluster analysis (HCA) and artificial neural networks (ANNs) in Section 10.5. Throughout the chapter, recent examples from the literature will be illustrated, with a focus on applications in surface analysis.

10.2 Basic Concepts

When data are stored in a computer for processing, it is often stored as a matrix. A matrix is a rectangular table of numbers, usually denoted by a bold capital letter. Matrix algebra is manipulation of data stored in these matrices. Therefore, knowledge of matrix algebra is required in understanding the theory behind most multivariate techniques. Here we shall assume a basic knowledge of vector and matrix operations. This includes addition, multiplication, transpose and matrix inverse; the ideas of collinearity and orthogonality between vectors; vector projections; and the eigenvectors decomposition of matrices. Readers unfamiliar with these concepts should consult references [4, 5].

10.2.1 MATRIX AND VECTOR REPRESENTATION OF DATA

Figure 10.2 shows an example data matrix obtained from a generic experiment, consisting of three samples measured over five variables. In statistics, 'sample' denotes any individual measurements made on a system and 'variable' denotes the channels over which the measurements are made. For example, in secondary ion mass spectrometry (SIMS) the variables refer to the mass or time of flight of secondary ions, and in X-ray photoelectron spectroscopy (XPS) the variables refer to the binding energies of photoelectrons detected.

The matrix X has three rows and five columns, and is called a 3×5 data matrix. Each row of the data matrix represents experimental results

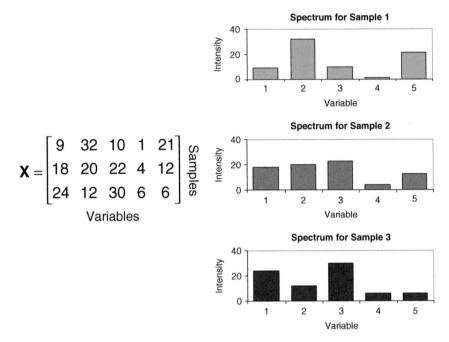

Figure 10.2 Example data matrix obtained from a spectroscopic experiment. Each row of the matrix represents an individual measurement, which is shown on the right

from an individual measurement, shown by the spectrum on the right hand side, respectively. We can see that each individual measurement can be represented by a *vector*, which is a matrix with a single row or column. Vectors, normally denoted with a bold lower case letter, can also be represented by directions in space. A simple vector in two-dimensional space is shown in Figure 10.3. The vector **a** can be expressed by either of the following equations:

$$a = 2\hat{x} + 3\hat{y}$$
$$a = 3.2\hat{x}^* + 1.6\hat{y}^* \tag{10.1}$$

where $\hat{x}$ is a vector of unit length in the direction of axis x, etc. In matrix notation, Equation (10.1) becomes:

$$a = [2 \quad 3] \begin{bmatrix} \hat{x} \\ \hat{y} \end{bmatrix}$$

$$\tag{10.2}$$

$$a = [3.2 \quad 1.6] \begin{bmatrix} \hat{x}^* \\ \hat{y}^* \end{bmatrix}$$

Here the normal rule of matrix multiplication applies, and **a** is represented by a row vector representing the magnitude of the projections, multiplied by a column vector representing the direction of the axes. As we can see from

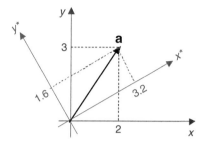

Figure 10.3 An example vector **a** in two-dimensional space

Equation (10.2), the vector **a** can be written in terms of any set of axes by finding the relevant projections. The rotated axes x^* and y^* can themselves be written in terms of x and y in matrix notation, as follows:

$$\begin{bmatrix} \hat{x}^* \\ \hat{y}^* \end{bmatrix} = \begin{bmatrix} 0.87 & 0.5 \\ -0.5 & 0.87 \end{bmatrix} \begin{bmatrix} \hat{x} \\ \hat{y} \end{bmatrix} \tag{10.3}$$

The same ideas can be applied to individual measurements represented by the vectors which exist in the rows of the matrix. Instead of the directions x and y in physical space, the vectors representing the measurement exist in a K-dimension variable space (also called 'data space'), which has the axes of variable 1, variable 2, ..., variable K, where K is the number of variables recorded for each measurement. When more than one measurement is carried out, the data can be recorded in an $I \times K$ data matrix, where I is the number of sample measurements.

10.2.2 DIMENSIONALITY AND RANK

The dimensionality of the data is of vital importance in multivariate analysis. This is represented by the *rank* of the data matrix. The rank of a matrix is the maximum number of rows or columns that are linearly independent. A matrix where all the rows or columns are linearly independent is said to have 'full rank'. The rank of a data set represents the number of independent parameters that are needed to fully describe the data. This is analogous to the degrees of freedom in a physical system or the number of independent components in a chemical system. For an experimental raw data matrix with random, uncorrelated noise, the rank of the data, R, is equal to the number of samples I or the number of variables K, whichever is smaller. However, the true rank of the data in the absence of noise, which is the subject of interest, is often much smaller. For example, a data matrix containing 100 different spectra may have a true rank of only three, if all the spectra consist of linear mixtures of three independently varying chemical components only. The concept of rank and dimensionality is therefore very important in multivariate analysis as we seek to describe a large, complex set of data in the simplest and most meaningful way.

10.2.3 RELATION TO MULTIVARIATE ANALYSIS

In traditional, univariate analysis, each variable is treated as independent from each other and analysed separately. In terms of the data space, univariate analysis is equivalent to dealing with the projection of the data on each of the K dimensions independently. Multivariate analysis concerns the simultaneous analysis on all the dimensions of one or more sets of data, which can easily be accomplished by vector and matrix algebra. One important feature of multivariate analysis is the description of data on a different set of axes (or 'basis'), which can be advantageous in many analyses. For example, variables obtained from an experiment are often highly correlated, and it is possible to describe the data more efficiently using a smaller number of rotated axes, which describes independent contributions from each group of correlating variables. Axes may also be chosen to reflect interesting features about the system under study. For example, in multivariate curve resolution (MCR), which is covered in Section 10.3.4, the axes are chosen to align with contributions from pure chemical components, such that each measurement can be described as a mixture of these components. Finally, by using a different set of axes, entire dimensions which describe only noise variations may be removed from the data altogether in order to achieve noise rejection. This is a major element of principal component analysis (PCA) and will be discussed in Section 10.3.3.

10.2.4 CHOOSING THE APPROPRIATE MULTIVARIATE METHOD

Before undertaking any data analysis, it is crucial to have a clear aim and hypothesis. Figure 10.4 shows the typical questions an analyst may ask

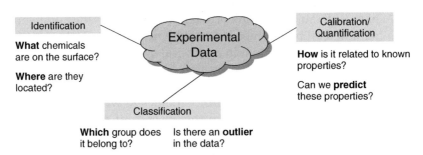

Figure 10.4   Typical questions involved in the analysis of data obtained from surface analysis. These can be broadly split into three categories

when confronted with a data set. Broadly, multivariate analysis methods fall into the following three categories:

1. *Exploratory analysis and identification (Section 10.3).* This involves the examination of the data and aids the identification of important features without prior knowledge of the sample, and includes factor analysis techniques such as principal component analysis (PCA), multivariate curve resolution (MCR) and maximum autocorrelation factors (MAF).

2. *Calibration and quantification (Section 10.4).* This involves analysing the relation between two or more sets of independent measurements made on the same samples, and can be used for quantitative predictions, and includes multivariate regression methods such as principal component regression (PCR) and partial least squares regression (PLS).

3. *Classification and clustering (Section 10.5).* This involves the classification of data either using predetermined groups or via unsupervised clustering, and includes discriminant function analysis (DFA), hierarchical cluster analysis (HCA) and artificial neural networks (ANN).

A diagram summarizing the objectives of multivariate analysis methods reviewed in this chapter, arranged by their typical order of application, is shown later in Figure 10.25 on page 607. Most multivariate methods (with the exception of artificial neural networks) provide analysts with statistical results that summarize the relationships within the data. The role of the analyst is to utilize these results in order to draw valid conclusions about the data. It is crucial that multivariate analysis is not treated as a 'black box' approach to data analysis. An understanding of the assumptions and validity of each multivariate method, and expert knowledge of the physical and chemical properties of the system, are vital in obtaining valid, physical interpretation of the results. With this in mind, the theories and applications of the multivariate methods outlined above will be presented in detail in the following sections.

10.3 Factor Analysis for Identification

Factor analysis [1] is a broad field that has been in continual development for over 70 years, and today has an extensive range of applications in fields such as spectroscopy [6], remote sensing [7], social sciences [8] and economics [9]. It is a technique for reducing matrices of data to their lowest meaningful dimensionality by describing them using a small number of factors, which are directions in the data space that reflect useful properties of the data set. This is equivalent to a transformation so that the new axes (factors) used to describe the data are a linear combination of the original variables.

Table 10.1 The factor analysis terminology adopted in this chapter in relation to those commonly used in literature for different multivariate factor analysis techniques

Term used here	Symbol	Definition	Term commonly used in PCA	Term commonly used in MCR
Factor	–	An axis in the data space representing an underlying dimension that contributes to summarizing or accounting for the original data set	Principal component	Pure component
Loadings	p	Projection of a factor onto the variables, reflecting the covariance relationship between variables	Loadings, eigenvector	Pure component spectrum
Scores	t	Projection of the samples onto a factor, reflecting the relationship between samples	Scores, projections	Pure component concentration

10.3.1 TERMINOLOGY

As a result of its history and scope, factor analysis is laden with confusing terminology, with different names given to similar or equivalent terms depending on the technique and the field of application. Here, we seek to clarify the terminologies in order to ensure clarity and consistency and emphasize the relationship between different multivariate analysis techniques. We refer only to factors, loadings and scores in this chapter. Table 10.1 shows the definitions of these terms, and conversion between this and the various terminologies commonly used in the literature. These definitions, developed in close consultation with international experts, are being incorporated into ISO 18115 Surface Chemical Analysis–Vocabulary [see J.L.S. Lee, B.J. Tyler, M.S. Wagner, I.S. Gilmore and M.P. Seah, *Surf. Interface Anal.* 41, 76 (2009)].

10.3.2 MATHEMATICAL BACKGROUND

A typical factor analytical model containing N factors can be written in matrix notation as:

$$\mathbf{X} = \mathbf{TP}' + \mathbf{E} \Leftrightarrow \mathbf{X} = \sum_{n=1}^{N} \mathbf{t}_n \mathbf{p}'_n + \mathbf{E} \Leftrightarrow x_{ik} = \sum_{n=1}^{N} t_{in} p_{nk} + e_{ik} \qquad (10.4)$$

where letters in upper case, bold font, denote matrices and letters in the lower case, bold font, denote vectors. Letters in unbold, italic font, denote scalars. The matrix transpose is denoted by an apostrophe. All indices are taken to run from one to their capital versions, e.g. $i = 1, 2, \ldots I$. $\mathbf{X}$ is the 'data matrix' and is an $I \times K$ matrix containing experimental data obtained for I samples over K variables, after suitable data preprocessing. $\mathbf{P}$ is the 'loadings matrix', with dimensions $K \times N$, whose rows are the correlation between the variables and the factors. $\mathbf{T}$ is called the 'scores matrix' and is an $I \times N$ matrix whose rows are the projections of the samples onto the factors. $\mathbf{E}$ is the error between the factor analysis model and the experimental data and is called the 'residuals matrix'. It has the dimensions of $I \times K$, and is usually assumed to contain noise only.

Different factor analysis techniques differ in the way in which the factors are extracted. Rotational and scaling ambiguities mean that there are no unique solutions to Equation (10.4). The minimum number of factors needed to reproduce the original data matrix satisfactorily within experimental error is the true rank of the data. Often a factor analysis model would not be optimal, and the number of factors in the model, N, would be larger than the true rank. By excluding factors larger than N, and hence subtracting the residuals matrix from the data, we can construct a 'reproduced data matrix' $\overline{\mathbf{X}}$, such that:

$$\overline{\mathbf{X}} = \mathbf{T}\mathbf{P}' \approx \mathbf{X} \tag{10.5}$$

Using this, we can gauge the success of different factor analytical models based on its ability to reproduce interesting features in the data using the fewest number of factors. A detailed explanation of factor analysis may be found in reference [1].

10.3.3 PRINCIPAL COMPONENT ANALYSIS

Principal component analysis (PCA) [10] is a multivariate technique for reducing matrices of data to their lowest dimensionality by describing them using a small number of orthogonal factors. The goal of PCA is to extract factors, or 'principal components', that capture the largest amount of variance within the multi-dimensional data set. PCA is perhaps the most popular and widely used multivariate analysis method, with applications ranging from face recognition [11] to behavioural sciences [12]. Often, PCA is used as a first step for data reduction prior to other methods of statistical analysis.

Basic Principles.   A two dimensional graphical representation of PCA, applied after mean centering of the data, is shown in Figure 10.5. This shows data for 28 samples measured over two variables, x_1 and x_2. The first

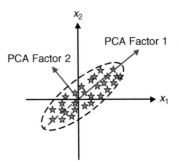

Figure 10.5 A two dimensional graphical representation of principal component analysis (PCA)

PCA factor describes the direction of the largest variance, or spread, of data points in the data set. The second PCA factor is the direction orthogonal (i.e. at a right angle) to the first that captures the largest remaining variation. It is obvious that PCA factors can be interpreted as rotated axes in the data space that optimally describe the variance within the data. Using PCA, we have transformed the correlated variables x_1 and x_2 into a new basis which are uncorrelated. At this stage, two factors describe all features in the data set. However, it may be useful to assume that x_1 varies linearly with x_2, and the scatter in the data set arises only from experimental noise. It is then beneficial to discard information in PCA Factor 2, so that all the relevant chemical information would be provided by the projection of the data onto PCA Factor 1. In doing so, we have achieved the dimensionality reduction desired in factor analysis, and the data set, originally containing two variables, can now be described solely using one factor. This ability for PCA to transform variables into an optimal basis and achieve dimensionality reduction is extremely important in a large data set with many variables which are highly correlated, as is the case for many practical analyses.

Formulation. PCA follows the factor analysis equation (Equation (10.4)). The main steps of PCA are shown in a schematic diagram in Figure 10.6. PCA factors are computed using the eigenvector decomposition of matrix **Z**, where:

$$\mathbf{Z} = \mathbf{X}'\mathbf{X} \tag{10.6}$$

Here **X** is the data matrix containing experimental data, after suitable data preprocessing. If **X** is mean centered, then **Z** is called the covariance matrix and often denoted as $\mathbf{Z}_{cov}$. If **X** is auto scaled, then **Z** is referred to as the correlation matrix and often denoted as $\mathbf{Z}_{corr}$. An eigenanalysis of **Z** gives:

$$\mathbf{Z}\mathbf{q}_r = \lambda_r \mathbf{q}_r \tag{10.7}$$

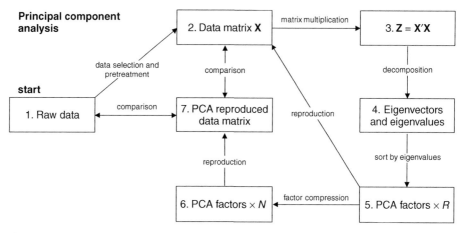

Figure 10.6 A schematic diagram illustrating typical steps in principal component analysis (PCA), after Malinowski [1]

where $\mathbf{q}_r$ is the rth eigenvector of $\mathbf{Z}$ and λ_r is its associated eigenvalue. Due to the properties of eigenvalue decomposition of a symmetric matrix, the eigenvectors are orthonormal (i.e. orthogonal and normalized), and the eigenvalues can only have positive or zero values. At this stage, the total number of non-zero eigenvectors and eigenvalues obtained is equal to R, the rank of the data matrix. Because $\mathbf{Z}$ contains information about the variances of the data within the data set, the eigenvectors are special directions in the data space that are optimal in describing the variance of the data. The amount of variance accounted for by each eigenvector is given by the eigenvalues.

PCA factors consist of the eigenvectors of $\mathbf{Z}$, sorted in descending order by their associated eigenvalues, such that PCA Factor 1 describes the direction of the largest variance and has the largest associated eigenvalue. The eigenvectors matrix $\mathbf{Q}$ is given by:

$$\mathbf{Q} = (\mathbf{q}_1 \quad \mathbf{q}_2 \cdots \mathbf{q}_R) \tag{10.8}$$

where $\mathbf{q}_r$ are column vectors of the sorted eigenvectors such that $\mathbf{q}_1$ is associated with the largest eigenvalue λ_1. To rewrite the data using the PCA factors as new axes, we calculate the scores matrix $\mathbf{T}_{\text{full}}$ which contains the projections of the data onto all the factors. This can be written in matrix notation:

$$\mathbf{T}_{\text{full}} = \mathbf{XQ} \tag{10.9}$$

We can now manipulate Equation (10.9) to obtain an expression for data matrix $\mathbf{X}$, by post multiplying the equation by $\mathbf{Q}^{-1}$, which is the matrix inverse of $\mathbf{Q}$. This gives:

$$\mathbf{X} = \mathbf{T}_{\text{full}} \mathbf{Q}^{-1} \tag{10.10}$$

By comparison of Equation (10.10) with Equation (10.4), the factor analysis equation is satisfied if we set the transpose of the loadings matrix, $\mathbf{P}'_{full}$, to be equal to $\mathbf{Q}^{-1}$. However, since $\mathbf{Z}$ is a symmetrical matrix in PCA, the columns of $\mathbf{Q}$ are orthogonal and $\mathbf{Q}^{-1}$ is simply equivalent to $\mathbf{Q}'$. We therefore obtain the PCA solution to the factor analysis equation:

$$\mathbf{P}_{full} = \mathbf{Q} \tag{10.11}$$

$$\mathbf{X} = \mathbf{T}_{full}\mathbf{P}'_{full} = \sum_{n=1}^{R} \mathbf{t}_n\mathbf{p}'_n \tag{10.12}$$

At this stage, all PCA factors have been included in the factor analysis model, and the scores $\mathbf{T}_{full}$ and loadings $\mathbf{P}_{full}$ reproduce the original data matrix $\mathbf{X}$ fully. Often, it is preferable to discard higher PCA factors in order to reduce the dimensionality of the data. This is referred to as factor compression. If we assume that variances in the data arising from N chemical features are greater than the variances arising from random noise, then all chemical information can be accounted for using the first N PCA factors. Therefore, by carrying out the summation in Equation (10.12) for the first N factors only, we obtain:

$$\overline{\mathbf{X}} = \sum_{n=1}^{N} \mathbf{t}_n\mathbf{p}'_n = \mathbf{T}\mathbf{P}' \tag{10.13}$$

where the scores and loadings matrices $\mathbf{T}$ and $\mathbf{P}$ now contain only N columns for the first N factors. Finally, $\overline{\mathbf{X}}$ is the PCA reproduced data matrix and contains a noise-filtered version of the data matrix $\mathbf{X}$, reconstructed using variances described in the first N PCA factors only. $\overline{\mathbf{X}}$ differs from the original data matrix $\mathbf{X}$ by an amount accounted for in the residuals matrix $\mathbf{E}$, namely:

$$\mathbf{X} = \overline{\mathbf{X}} + \mathbf{E} = \mathbf{T}\mathbf{P}' + \mathbf{E} \tag{10.14}$$

This gives the full PCA solution to the factor analysis equation.

Number of Factors. If we assume that variances in the data arising from N chemical features are greater than the variances arising from random noise, then all chemical information can be accounted for using the first N factors. There are many ways of determining the number of factors that one should retain in a PCA model. Often a combination of methods, together with the experience of the analyst, produces the best results.

Figure 10.7 shows a simulated data set consisting of eight SIMS spectra created by mixing the library spectra of three reference materials. The eigenvalues associated with each factor are plotted. Without noise, only three factors with non-zero eigenvalues exist. This is equal to the rank of the data set and the number of independent components. Therefore only

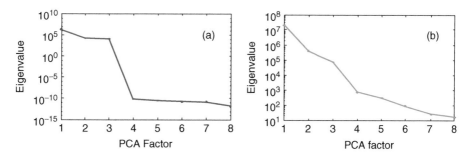

Figure 10.7 Eigenvalue diagram obtained from eight synthetic SIMS spectra produced using the library spectra of three reference materials, (a) before addition of noise (b) after addition of Poisson noise

three factors are needed to explain all the features of the data set. With random, Poisson noise, added to simulate the ion counting statistics of the SIMS detector, the number of factors one needs to retain is not so clear. One of the most popular ways to determine the number of factors required to describe the data satisfactorily is by inspection of the eigenvalue plot in what is known as the scree test [13]. This is so-called as the plot visually resembles the scree, or debris, that accumulate at the base of a cliff. The scree test assumes that eigenvalues decrease in a steady manner for factors that describe variations arising from noise. Often, a turn off point would be visible on the eigenvalue plot, where the factors describing large variances due to chemical features ('the cliff') stops and the factors describing smaller variances due to noise ('the debris') appear. Using the scree test on Figure 10.7(b), we recover that three factors are needed to describe the data. Often, the scree test is used in conjunction with the percentage of total variance captured by the N eigenvectors, which is given by:

$$\% \text{ variance captured} = \frac{\text{sum of eigenvalues up to factor } N}{\text{sum of all eigenvalues}} \times 100\% \quad (10.15)$$

The total variance captured provides a good guidance on the number of factors one should retain in order to describe and reproduce the data satisfactorily. This will depend on the level of noise in the data set and the number of minor features, such as contamination, non-linear and other effects that one wants to include in the PCA model. Finally, inspection of the residuals matrix **E**, and associated lack-of-fit statistics such as Q residuals, can be helpful in determining if any meaningful structure has been excluded from the PCA model.

Data Preprocessing.   Since all multivariate analysis techniques seek to describe the underlying structure of the data, they are sensitive to data preprocessing and transformations [14]. Data preprocessing can enhance the application of PCA by bringing out important variance in the dataset, but because it makes assumptions about the nature of the variance in the data, it can distort interpretation and quantification and therefore needs to be applied with care [15]. Prior to multivariate analysis, data selection and binning are often performed to reduce the size of the data set. The following is a description of data preprocessing methods common in the field of surface analysis.

In **mean centering**, each variable is centred by the subtraction of its mean value across all samples. In the case of spectral data, such as XPS or SIMS, this is equivalent to subtracting the mean spectrum of the data set from each sample. Mean centering is a common practice for PCA, so that the first PCA factor would go through the centre of the data rather than the origin. This allows for the effective description of the differences between samples rather than their variations from zero intensity, as can be seen in Figure 10.8.

In **normalization**, the data are scaled by a constant for each sample, which could be the value of a specific variable, the sum of selected variables or the sum of all variables for the sample. Normalization preserves the shape of the spectra data and is commonly used in SIMS, assuming chemical variances can solely be described by the relative changes in ion intensities. Normalization therefore removes the gross variations in total ion intensities caused by topography, sample charging, changes in primary ion dose and other effects.

In **variance scaling**, each variable is divided by its standard deviation in the data set. Variance scaling is referred to 'auto scaling' when it is followed by mean centering. Variance scaling equalizes the importance of each variable, and is important for multivariate data sets where data from different techniques are concatenated and variables therefore have different units and magnitudes. Variance scaling is often used in SIMS to emphasize high mass, low fragmentation ions which often have lower intensities. However,

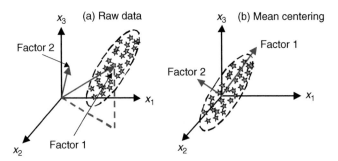

Figure 10.8 The effect of mean centering on PCA. (a) Without mean centering, PCA Factor 1 goes from the origin to the centre of gravity of the data. (b) With mean centering, PCA Factor 1 goes from origin and accounts for the highest variance within the data set

it can be problematic for weak peaks with variations arising mostly from background signal, noise or minor contaminants, and therefore variance scaling is commonly used on a selection of strong characteristic peaks only. Finally, in **Poisson scaling** [16], it is assumed that the statistical uncertainty of each variable is dominated by the counting statistics from the detector, which are Poisson in nature. This is a good approximation for SIMS and XPS raw data where the detector is operating within linearity, and Poisson scaling cannot be applied in conjunction of other data scaling methods. Note that for multidetector XPS systems, the counting statistics are close to Poissonian but considerably lower [17]. Poisson scaling weights the data by their estimated uncertainty, using the fact that the noise variance arising from Poisson statistics is equal to the average counted intensity. Since multivariate methods generally assume uniform uncertainty in the data, Poisson scaling has been shown to provide greater noise rejection in the multivariate analysis of SIMS [15, 16, 18] and XPS data [19]. Poisson scaling thus emphasizes the weak peaks which vary above the expected counting noise, over intense peaks with large variance solely accounted for using Poisson statistics. Poisson scaling is especially valuable for image data sets, which can have low counts per pixel and can therefore be dominated by Poisson noise. A detailed explanation of Poisson scaling is given in Keenan and Kotula [16].

Example – protein Characterization by SIMS.   PCA has been applied successfully to the SIMS characterization and quantification of a variety of materials including inorganic materials [20, 21], polymers [22, 23], polymer additives [24], organic thin films [25–27] and proteins [28–32]. In an example obtained from Graham *et al.* [28], SIMS spectra were obtained for 42 samples, consisting of three different protein compositions (100 % fibrinogen, 50 % fibrinogen/50 % albumin and 100 % albumin) absorbed onto poly(DTB suberate). Since all proteins consist of identical amino acids arranged in different sequences, the SIMS spectra of different proteins are similar, with main differences being the relative intensities of amino acid related fragment peaks. PCA is therefore ideal at summarizing the main differences in the data and discriminating between different proteins. Prior to PCA analysis, the data selection is applied to include amino acid related fragment peaks only. This ensures that PCA is effective at capturing small variances in the data due to variations in amino acid composition, rather than other gross variations such as changes in surface coverage between samples. The data are then normalized to the sum of selected peaks and mean centred before analysis. Figure 10.9 shows the PCA results obtained. Factor 1, describing 62 % of total variance in the data set, successfully distinguishes the two proteins. The loadings plot (left) shows positive peaks for amino acids that are more abundant in fibrinogen and negative peaks for those more abundant in albumin. The scores plot (right) shows positive

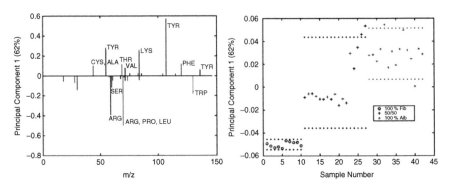

Figure 10.9 Loadings (left) and scores (right) on PCA Factor 1 from SIMS data of protein adsorption onto a poly(DTB suberate) [28]. Reproduced by permission of Elsevier Science from Graham *et al.* [28]

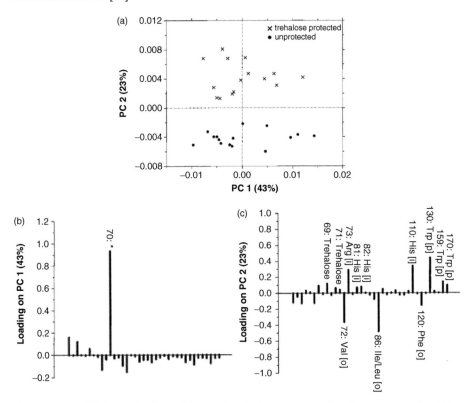

Figure 10.10 PCA results from 0.01 wt % trehalose-protected and unprotected antiferritin adsorbed and characterized by static SIMS [29]. (a) Scores on PCA Factor 1 against scores on PCA Factor 2. Factor 2 is able to discriminate between protected and unprotected samples. (b) Loadings on PCA Factor 1, dominated by a large peak at $m = 70$ u attributed to sample heterogeneity. (c) Loadings on PCA Factor 2, showing peaks consisting with trehalose preserving protein conformation. Reproduced with permission from the American Chemical Society from Xia *et al.* [29]

scores for pure fibrinogen samples and negative scores for pure albumin samples. Samples with 50/50 protein compositions have scores close to zero, indicating a successful mixture of the two proteins.

PCA have also been applied to study conformation changes in adsorbed protein films by Xia *et al.* [29]. SIMS spectra were acquired for antiferritin with and without a trehalose coating, which preserves the conformation of proteins during sample preparation. As before, amino acid peaks were selected, normalized and mean centred before analysis. Figure 10.10(a) shows scores on PCA Factor 1 against scores on PCA Factor 2. Separation between protected and unprotected samples is obtained on Factor 2. The largest variance in the data, as described in PCA Factor 1, arises from sample heterogeneity which exists regardless of trehalose protection. Inspection of the loadings show that this heterogeneity is typified by the large variation at $m = 70$ u which, unlike other peaks, cannot be attributed uniquely to a single amino acid. The second PCA factor, accounting for the direction of the biggest variance orthogonal to the first, is able to discriminate between protected and unprotected samples. The loadings on Factor 2 shows that protected samples contain higher intensities of polar and hydrophilic amino acid fragments, labelled with [p] and [i], respectively, and lower intensities of hydrophobic amino acid fragments, labelled with [o]. This is consistent with trehalose preserving protein conformation, since it is known that hydrophilic amino acids locate preferentially in the exterior of the protein molecule as they interact with the aqueous surroundings. This, combined with the extreme surface sensitivity of SIMS, enables PCA to discern small but important changes in protein orientation during sample preparation for analysis of proteins in ultrahigh vacuum.

10.3.4 MULTIVARIATE CURVE RESOLUTION

Multivariate curve resolution (MCR) [33–35] belongs to a family of methods generally referred to as 'self modelling curve resolution' [36], which are designed for the recovery of pure components from a multi-component mixture, when little or no prior information is available. MCR uses an iterative least-squares algorithm to extract solutions to the factor analysis equation (Equation (10.4)), while applying suitable constraints.

Basic Principles.   A two dimensional graphical representation of MCR is shown in Figure 10.11. MCR assumes that each spectrum can be described as a linear sum of contributions (MCR scores) from individual chemical components, each associated with a particular spectral profile (MCR loadings). This is true for many systems such as absorption spectroscopy, where Beer's law dictates that absorbance is proportional to concentration. However, this is only a first approximation in SIMS and XPS, where many factors other

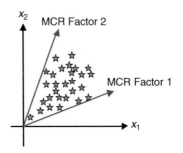

Figure 10.11 A two dimensional graphical representation of multivariate curve resolution (MCR)

than chemical composition can affect the position and the intensity of peaks, including matrix effects, topography, detector saturation and sample degradation during analysis. Unlike PCA, MCR factors are not required to be mutually orthogonal. An advantage of MCR is the application of constraints during the solution process. By applying non-negativity constraints to the loadings and scores matrices using optimization, MCR solutions obtained resemble SIMS or XPS spectra and chemical contributions more directly, as these must have positive values. For example, in Figure 10.11, all data points can be expressed as a positive mixture of MCR Factors 1 and 2, and the factors themselves are positive combinations of the original variables x_1 and x_2. Other constraints can also be applied, including equality constraint, where using *a priori* knowledge of the system, one or more columns of the loadings or scores matrices can be fixed to known spectral or contribution profiles prior to the resolution of unknown components. However, MCR is also more computationally intensive than PCA and requires more input prior to analysis. Most importantly, unlike PCA which produces a unique solution for each data set, MCR results are not unique and are strongly dependent on initial estimates, constraints and convergence criterion. The accuracy of the resolved spectra depends on existence of samples with chemical contribution from one component only, and features from intense components can often appear in the spectral profiles resolved for weak components [33]. Therefore, careful application of MCR and interpretation of the outputs is required to obtain optimal results.

Formulation.   MCR follows the factor analysis equation (Equation (10.4)) The main steps of MCR are shown in a schematic diagram in Figure 10.12. In the first stage, the number of factors, N, to be resolved is determined independently, either by prior knowledge of the system or via the application of PCA and the inspection of the eigenvalue diagram (scree test). An initial estimate of either the scores matrix $\mathbf{T}$, representing the contribution profiles of each component, or the loadings matrix $\mathbf{P}$, representing the spectra of

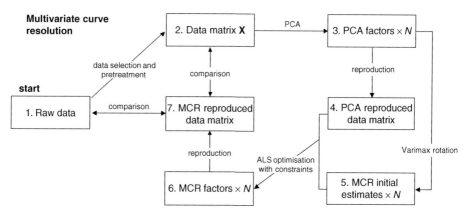

Figure 10.12 A schematic diagram illustrating typical steps in multivariate curve resolution (MCR)

each component, is then required as an input to the alternating least squares (ALS) algorithm. The initial estimates can be obtained in many ways, for example by the use of 'pure variable' detection algorithms, which find variables with contributions from single components only, or by the Varimax rotation [37] of PCA factors, which simplifies PCA factors by an orthogonal rotation such that each factor only has a small number of variables with large loadings. MCR then uses an iterative algorithm to extract solutions to the factor analysis equation, Equation (10.4), by the ALS minimization of error matrix **E**. To increase the stability of the algorithm, PCA is applied to the data as a first step and the ALS fitting is done on the noise filtered PCA reproduced data matrix, $\overline{\mathbf{X}}$, rather than the original data matrix. Assuming an initial estimate of the loadings matrix **P**, a least squares estimate of the scores matrix **T** can be obtained by:

$$\mathbf{T} = \overline{\mathbf{X}}(\mathbf{P}')^{+}$$

(10.16)

where $(\mathbf{P}')^{+}$ is the pseudoinverse of matrix $\mathbf{P}'$. A new estimate of the loadings matrix **P** can then be obtained:

$$\mathbf{P}' = \mathbf{T}^{+}\overline{\mathbf{X}}$$

(10.17)

In each stage of the fitting, suitable constraints are applied to the loadings and scores matrices **P** and **T**, such as non-negativity. Finally, Equations (10.16) and (10.17) are repeated until **T** and **P** are able to reproduce $\overline{\mathbf{X}}$, within an error specified by the user, i.e. convergence is achieved.

Example – resolution of Buried Interfaces in ToF–SIMS Depth Profiling. MCR has been successfully applied to a large range of data sets obtained from SIMS [15, 18, 35, 38–40], XPS [41–43] and Raman spectroscopy [35].

Here, we will illustrate one example for the ToF–SIMS depth profiling of multilayer interfaces by Lloyd [40]. In this example, a thin copper film was grown on a TaN-coated silicon wafer surface and a depth profile was obtained in dual beam mode. The data set consists of a series of complete time-of-flight spectra recorded at different sputter times. Manual analysis of the data by an expert involves the inspection of individual ion species, as shown in Figure 10.13(a). The difficulty lies in the interpretation of the data, for example, the Si^- signal can arise from SiO_x^-, SiN^- or the silicon substrate, and therefore the careful inspection and correlation of a large number of ion peaks is needed to obtain reliable compositional information for each layer. MCR with a non-negativity constraint is applied to the data, after unit mass binning of the spectra and removal of peaks with mass < 20 u. Poisson scaling is then applied to the data to equalize the noise variance of each data point and improve multivariate analysis. The scores from the eight-factor MCR model of the data are shown in Figure 10.13(b). Examples of MCR scores and loadings on three of the factors are shown in Figure 10.13(c). MCR aids the analysis of depth profiles by resolving the data into factors, where loadings resemble the spectra of individual chemical phases and the scores resemble the contribution of these phases to the measured spectrum. A straightforward interpretation of the factors can be made using the loadings, which gives information on peaks in the depth profile that are correlated and therefore belong to the same phase. This is a considerable improvement from the interpretation of single peak profiles in manual analysis and represents a major benefit of MCR. Since each MCR factor contains information from a number of correlated peaks, improved signal to noise is obtained on the contribution profiles. The MCR results shown here agree with the results obtained from manual analysis by an expert analyst, but provide additional information that would not be obvious, such as the existence of an SiO_x-rich layer near the surface. MCR therefore enables the rapid and unbiased identification of features within a complex depth profile without any prior knowledge of the system.

10.3.5 ANALYSIS OF MULTIVARIATE IMAGES

With the advances in instrumentation, many spectroscopy or spectrometry instruments are now capable of generating images where a whole spectrum is recorded at each pixel [44]. This paves the way for the study of spatially localized features, but provides an acute problem for data analysis and interpretation due to the vast amount of information recorded. For example, in a typical ToF–SIMS image consisting of 256×256 pixels with the spectra nominally binned to unit mass up to 400 u, the data set would contain 400 individual images or 26 million data points, occupying 200 megabytes of computer memory. In addition, spectroscopic images obtained from these techniques often suffer from low signal-to-noise ratio due to the need to

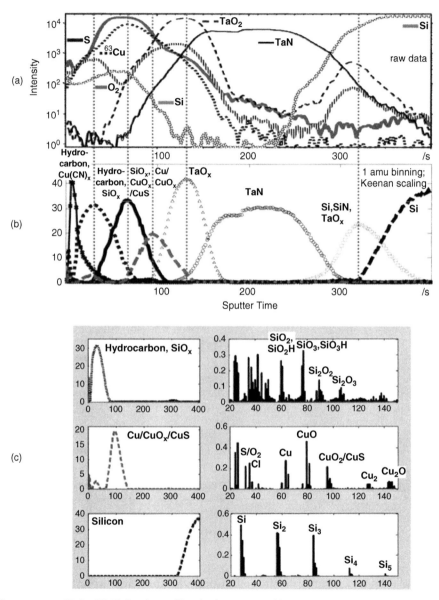

Figure 10.13   ToF–SIMS depth profile of a thin copper film on a TaN-coated silicon wafer [40]. (a) Results from manual analysis, showing the variation of intensity of selected secondary ions with sputter time. (b) Results from MCR analysis, showing the scores on eight distinct factors resolved from the data, after Poisson scaling. (c) Examples of MCR scores and loadings on three of the factors. Reproduced from Lloyd [40] with permission of AVS, The Science and Technology Society

minimize acquisition time as well as surface damage caused by the primary ion in SIMS or X-rays in XPS [45, 46]. As a result, the available signal deteriorates rapidly with increased spatial resolution, and noise arising from the counting statistics of the detector becomes significant. Traditional data analysis involves the manual selection and comparison of key images, which is slow and requires *a priori* knowledge of the compounds on the surface. Consequently, small but chemically significant features can easily be overlooked, for example, any localized contamination that covers only a small area on the surface of the sample. Factor analysis methods, which take into account the whole data set, are therefore ideal for exploring complex image data sets obtained from XPS and SIMS.

PCA and MCR Image Analysis. So far, we have demonstrated the application of PCA and MCR on the analysis of sets of spectra. This can be easily extended to the analysis of multivariate images, as follows. A multivariate image data set contains a spatial raster of $I \times J$ pixels, where each pixel contains a complete spectrum consisting of K variables. Prior to PCA or MCR, the spatial information is discarded, and each pixel is treated as a separate sample. The image data cube, with dimensions of $I \times J \times K$, is 'unfolded' into a two-dimensional data matrix with dimensions $IJ \times K$, and factor analysis is then carried out to obtain scores and loadings as described previously. On completion of the analysis, the scores matrix $\mathbf{T}$ can be 'folded' so that it has the dimensions of $I \times J \times N$ and can therefore be displayed as a series of N images, one for each of the N factors in the model.

Example – ToF–SIMS image of an immiscible PC/PVC polymer blend. PCA has been used to study ToF–SIMS images obtained for immiscible PC/PVC polymer blends by Lee *et al.* [15]. Figure 10.14 shows one example, which has on average only 42 counts per pixel and therefore suffers from an extremely low signal to noise ratio. Poisson scaling and mean centering of the data is applied prior to analysis, and two factors are clearly identified using the scree test. As expected, the biggest variance in the image arises from the phase separation of the polymers, and the first PCA factor successfully distinguishes the two phases, showing positive PVC peaks ($^{35}Cl^-$ and $^{37}Cl^-$) and negative PC peaks ($O^- + OH^-$). The second PCA factor, describing the remaining variance not explained by the first, reveals a relative loss of $^{35}Cl^-$ intensity compared to all other ions at the brightest PVC areas of the image. This is due to the saturation of the detector caused by the intense $^{35}Cl^-$ peak. Since PCA models the data using linear combinations of factors, an extra factor is required to model any non-linear variations in ion intensities. PCA therefore enables the detection of the detector saturation, which may be easily missed if ion images were analysed individually.

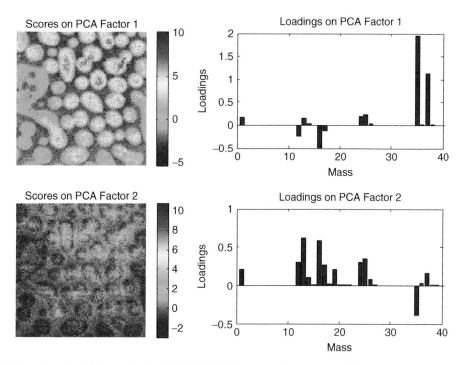

Figure 10.14 PCA results for the ToF–SIMS image of an immiscible PC/PVC polymer blend

Example-ToF–SIMS image of hair fibres with multi-component pretreatment. The identification of unknown chemical components in a complex real-life image can be demonstrated using a ToF–SIMS image of hair fibres pre-treated with a multi-component formulation. Prior to PCA and MCR analysis, the data are binned to unit mass resolution and the intense sodium peak is removed. Poisson scaling is employed as it gives the clearest eigenvalue diagram, and four factors are clearly identified using the scree test. Using this, PCA successfully highlights major trends and variations in the scores images, as well as important peaks in the loadings, as shown in Figure 10.15. The first PCA factor describes the overall intensity variation throughout the image arising from the differences between the hair sample and the background area, and PCA Factors 2–4 describe the chemical variations on different areas of the hair surface. However, it is very difficult to obtain direct information about the identity and distribution of individual chemical components using the PCA results. Characteristic peaks arising from different chemical components (labelled A, B and C) appear in more than one set of PCA loadings, and it is not possible to observe directly their distribution on the hair surface using the PCA scores. Furthermore,

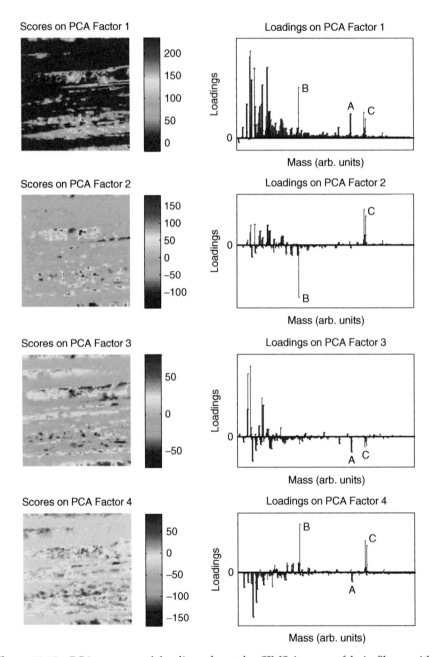

Figure 10.15 PCA scores and loadings from the SIMS images of hair fibres with a multi-component pretreatment, showing the first four factors after Poisson scaling. Raw data courtesy of Dr Ian Fletcher of Intertek MSG

peaks that are anti-correlated in one factor are correlated in another (e.g. peaks B and C in Factors 2 and 4), making interpretation of the scores images very difficult. These problems arise because PCA factors are constrained to be orthogonal and optimally capture the largest variance in the data. The loadings on PCA are therefore abstract combinations of chemical spectra that reflect the largest correlation and anti-correlation of various components on different areas of the image. As a result, although PCA is successful in identifying major trends and variations in the data, information relating to the identity and distribution of individual components are often spread over several factors, and direct chemical interpretation can be difficult.

MCR with non-negativity constraints is also applied to the hair fibres image, resolving four factors as identified by the scree test (Figure 10.16). Three chemical components on the hair surface (MCR Factors 1–3) are identified, in addition to one component showing the hair fibre itself (MCR Factor 4). Since MCR seeks factors that describe individual chemical components, it is more intuitive and easier to interpret than PCA, whose factors describe the correlations and anti-correlations between different components. For each MCR factor, the loadings resemble the complete SIMS spectrum of the component, showing its characteristic peaks (labelled A, B or C) and fragmentation pattern, while the scores reveal directly the distribution of the species on the surface. MCR therefore enables the straightforward identification of the chemical components using their complete resolved spectra without any prior knowledge of the system. This represents a significant improvement from manual analysis, where identification is based on the manual correlation of several key ions only. MCR results take under 10 min to compute, representing significant timesaving to the analysts.

Example – Quantitative analysis of XPS images.   Recently, factor analysis has also been applied successfully to a range of XPS image data sets [19, 41-43, 47]. Similar to the ToF–SIMS examples discussed so far, factor analysis methods can be used in XPS for the resolution of chemically relevant spectral components in wide-scan energy spectra using MCR [42] and in the detailed study of overlapping peak shapes in narrow-scan energy spectra using PCA [47]. Different to ToF–SIMS examples, however, PCA has also been employed recently as a noise reduction method in the quantitative analysis of XPS images by Walton *et al.* [19, 48]. Since most XPS instruments are not capable of parallel energy spectra and image acquisition, multivariate XPS images are usually obtained by acquiring a series of XPS images, each at energy increments over a range of energy channels. However, XPS images obtained this way often suffer from extremely low signal-to-noise ratio, limited by the acquisition time required at each energy channel. This causes a problem for obtaining quantitative chemical-state images, which requires

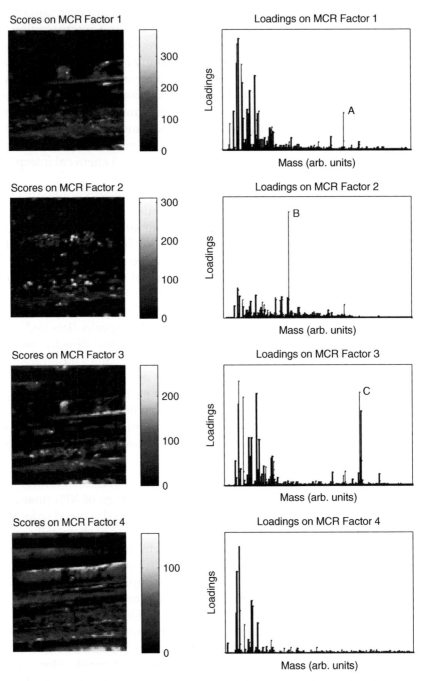

Figure 10.16 MCR scores and loadings from the SIMS images of hair fibres with a multi-component pretreatment, showing the first four factors after Poisson scaling. Raw data courtesy of Dr Ian Fletcher of Intertek MSG

curve fitting to the spectra at each pixel and are therefore subject to noise. Walton *et al.* [48] have shown that by applying PCA with suitable data scaling to the raw data and keeping only the first few significant factors, a noise reduced PCA reproduced data matrix can be obtained. This is shown in Figure 10.17(a) and 10.17(b) for a sample consisting of silicon dioxide islands on silicon, showing the Si 2p and O 1s photoelectron regions for a single pixel. The noise is greatly reduced after PCA reproduction and curve fitting becomes possible. Using the PCA reproduced data, quantitative XPS images showing atomic concentrations of each chemical state can be obtained from applying curve fitting to each pixel, giving the quantified maps shown in Figure 10.17(c).

Maximum Autocorrelation Factors (MAFs).   In the factor analysis methods (PCA and MCR) described above, each pixel from an image is treated as an independent sample and spatial information is discarded prior to analysis. However, the spatial information of an image can be extremely valuable in multivariate analysis. For example, the identification of small, localized contamination can be of huge importance in surface analysis. However, if the area coverage of the contaminant is small, this contributes to only a small amount of total variance in the data. The localized feature may therefore be overshadowed by noise variations, which exist throughout the image and therefore can have a larger total variance compared to the localized feature.

   A factor analysis method that is specifically designed for the extraction of spatially interesting variance is the maximum autocorrelation factor (MAF). MAF is similar to PCA in that it involves the eigenvector decomposition of a matrix. In MAF, however, this matrix consists of information on the spatial correlation of neighbouring pixels, so that MAF extracts factors which maximize variation across the entire image while minimizing the variation between neighbouring pixels. Detailed description and examples of MAF can be found in Larsen [7] and Tyler [49, 50]. The formulation of MAF broadly follows that of PCA in Section 10.3.3. MAF factors are computed using the eigenvector decomposition of matrix **B**, where:

$$\mathbf{B} = \mathbf{A}^{-1}\mathbf{X}'\mathbf{X} \tag{10.18}$$

Here **X** is the data matrix and **A** is a matrix consisting of information on the spatial correlation of the original image with a copy of itself that has been shifted by one pixel horizontally or vertically. **A** is calculated by taking the covariance matrix of the difference between the original and shift image. The eigenvectors of **B** are extracted and sorted by their associated eigenvalues to form the eigenvectors matrix **Q** ((Equation(10.8)). The MAF scores matrix $\mathbf{T}_{\text{full}}$ is obtained by projecting the data matrix onto the MAF eigenvectors (Equation(10.9)), and the expression is rewritten in the form of the factor analysis equation (Equation(10.10)). However, since **B** is not symmetrical,

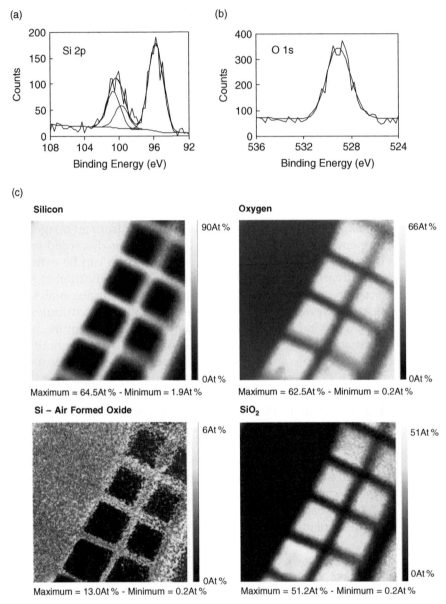

Figure 10.17 Multivariate XPS image consisting of silicon dioxide islands on silicon, acquired with 0.4 eV and 0.25 eV steps for the Si 2p and O 1s peaks, respectively, from Walton and Fairley [48]. (a) Spectra for the Si 2p photoelectron region, showing raw data and following PCA smoothing. The curve-fit peaks, envelope and background are also shown. (b) Spectra for the O 1s photoelectron region, showing raw data and following PCA reproduction. (c) Quantified XPS images for silicon, oxygen and two oxidized states of silicon, showing percentage atomic concentrations for each pixel. Reproduced with permission from John Wiley & Sons, Ltd from Walton and Fairley [48]

the eigenvectors $\mathbf{Q}$ obtained from MAF are not orthogonal and $\mathbf{Q}^{-1}$ is no longer equivalent to $\mathbf{Q}'$. The MAF loadings matrix is therefore given by:

$$\mathbf{P}_{full} = (\mathbf{Q}^{-1})' \qquad (10.19)$$

Finally, factor compression can be carried out by retaining only the first N MAF factors (Equations (10.12) and (10.13)). After this, the interpretation of MAF scores and loadings follows directly from PCA, except that the first MAF factors will successively account for the variance that displays the largest spatial correlation between neighbouring pixels, allowing for improved detection of localized features from noise in a multivariate image.

10.4 Regression Methods for Quantification

Multivariate regression [51, 52] involves the analysis of the relationship between two or more measurements on the same entities. While factor analysis methods provide descriptive models of a single data set, multivariate regression produces models that relate multiple sets of variables for quantitative predictions. There are two primary objectives for carrying out a multivariate regression. The first goal is to gain an understanding of the relation between the two sets of variables on the same set of samples via a study of their covariance. This is useful in, for example, relating results obtained from different surface analysis techniques, or in studying the effect of different sample preparation parameters on the surface chemistry of the samples. The second goal of multivariate regression is to compute a predictive model which can be used for the quantification of the samples. By using a calibration data set where both sets of variables are measured and known, the regression model can be applied to future samples in order to predict the response based only on one set of measurements, for example, quantifying surface composition or coverage on a set of samples using only their surface spectra. In the following, we shall describe two methods of multivariate regression, principal component regression (PCR) and partial least squares regression (PLS).

10.4.1 TERMINOLOGY

Like factor analysis, multivariate regression methods can suffer from confusing terminology related to different techniques and the field of application. To ensure clarity and consistency, Table 10.2 shows the definitions of terms used in this chapter, and conversion between this and the various terminologies commonly used in the literature.

Table 10.2   The multivariate regression analysis terminology adopted in this chapter in relation to those commonly used in the literature

Term used here	Symbol	Definition	Term commonly used in literature
Predictor variables	**X**	Variables containing measurements or parameters which can be used to predict the response variables	X block, Independent variables, Observed variables
Response variables	**Y**	Variables containing measurements or parameters which can be predicted by the predictor variables	Y block, Dependent variables,
Factor	–	An axis in the data space representing an underlying dimension that contributes to summarising or accounting the predictor variables while simultaneously predicting the response variables.	Latent variables, Latent vectors, LV, Component
Loadings	**p**	Projection of a factor onto the variables, reflecting the covariance relationship between variables	Loadings
Scores	**t**	Projection of the samples onto a factor, reflecting the relationship between samples	Scores

10.4.2 MATHEMATICAL BACKGROUND

Multivariate regression methods are based on an extension of linear regression, which seek to relate a 'response' variable y using a linear combination of 'predictor' variables x_k measured on the same sample, such that:

$$y = b_1 x_1 + b_2 x_2 + \cdots + b_K x_K + e = \mathbf{x}\mathbf{b} + e \qquad (10.20)$$

Here, $\mathbf{x} = (x_1, x_2, \cdots, x_K)$ is a row vector containing data on K predictor variables obtained on the sample, $\mathbf{b} = (b_1, b_2, \cdots, b_K)$ is a column vector consisting of regression coefficients, and e is the residual representing the amount of y that is not explained by the regression model. Often, in surface analysis, $\mathbf{x}$ contains a spectrum obtained from SIMS or XPS on an individual sample, and y contain the value of an independently measured property to which we wish to relate the spectrum. We can generalize Equation(10.20) to a model covering I samples and M different response variables, which gives:

$$Y = XB + E \tag{10.21}$$

Equation (10.21) is referred to as the regression equation. X is an $I \times K$ matrix containing experimental data obtained for I samples over K predictor variables. Y is an $I \times M$ matrix containing experimental data obtained for each of the I samples over a different set of M response variables. The data in X and Y are related by B, the matrix of regression coefficients with dimensions of $K \times M$. E is the 'residuals matrix' containing the error between the regression model and the experimental data, and has the dimensions of $I \times M$.

Multivariate regression seeks a correlation between the response data Y and the predictor data X via a linear combination of the X (predictor) variables. This is equivalent to relating Y to the projection of the X data onto the regression matrix B. Once the regression matrix B is calibrated on a set of samples with known X and Y data, it can be used for future prediction of new samples provided only their predictor variables, X_{new}, are measured, such that:

$$Y_{new} = X_{new} B \tag{10.22}$$

Different regression techniques differ in the way in which matrix B is calculated for a given set of X and Y. In the simplest case, a least squares solution to Equation (10.21), which minimizes the residuals E, can be found using multiple linear regression (MLR). This gives:

$$B = X^+ Y = (X'X)^{-1} X'Y \tag{10.23}$$

where X^+ is the matrix pseudoinverse of X. The MLR solution produces well-defined regression coefficients provided that a matrix inverse of $X'X$ can be found, i.e. X does not have any linearly dependent rows or columns. However, this is not true in many multivariate data in surface analysis, including SIMS and XPS spectra where the number of variables is large, and the intensities at many variables (ion mass or photoelectron energy) are highly correlated. MLR therefore produces a solution that is overfitted to the many variables in the data, reducing the robustness of the model and making it unsuitable for use in calibration and prediction. Fortunately, the study of the covariance between variables is of major importance in multivariate analysis, and there are several methods that can circumvent the problem with collinearity in MLR. Two methods, principal component regression (PCR) and partial least squares regression (PLS), will be described in the sections below.

10.4.3 PRINCIPAL COMPONENT REGRESSION

Principal component regression (PCR) is based on the use of principal component analysis (PCA) to reduce the dimensionality of the data $\mathbf{X}$ and remove problems with collinearity in the solution of the regression equation (Equation (10.21)). PCA is described in detail in Section 10.3.3. Instead of describing the data $\mathbf{X}$ using a large number of correlated variables, PCA uses a small number of factors, expressed as linear combinations of the original variables. PCA factors are mutually orthogonal and therefore independent from each other. As a result, the PCA scores $\mathbf{T}$ (with dimensions $I \times N$), which are the projection of the data onto the factors, are also orthogonal and do not suffer from the problem of collinearity in $\mathbf{X}$. We can therefore replace matrix $\mathbf{X}$ (containing K correlated variables) in MLR with the PCA scores matrix $\mathbf{T}$ (containing N uncorrelated scores) in the linear regression. This gives the following regression equation for PCR:

$$Y = TB^* + E \tag{10.24}$$

$$B^* = T^+Y = (T'T)^{-1}T'Y \tag{10.25}$$

Since the matrix $\mathbf{T'T}$ is guaranteed to be invertible, a well-defined solution for regression coefficients $\mathbf{B^*}$ (with dimensions $N \times M$) can now be obtained. The regression can be written in terms of the original predictor variables $\mathbf{X}$ in the form of Equation (10.21), by setting:

$$B = PB^* \tag{10.26}$$

where $\mathbf{P}$ is the PCA loadings matrix with dimensions $K \times N$.

PCR estimates the values of response variables on each sample using linear combinations of their PCA scores. As a result, the regression results are strongly dependent on the number of factors retained in the PCA model. Indeed, if all factors are included, then PCR is simply equivalent to MLR. The use of validation and cross-validation methods to determine the optimal number of factors will be discussed in Section 10.4.5. An advantage of PCR is the reduction in noise achieved by discarding higher PCA factors that describe small variances in the data $\mathbf{X}$. PCR results are therefore much more robust than MLR and the regression can be modelled on aspects of the data $\mathbf{X}$ that describes meaningful variance rather than experimental noise. However, it is possible that the PCA factors describing the largest variances in $\mathbf{X}$ may not contain any information that relates it to $\mathbf{Y}$. This is the case if the first few PCA factors describe matrix effects, or if features in $\mathbf{X}$ that correlates with $\mathbf{Y}$ accounts for a small amount of total variance and therefore appear only in higher PCA factors excluded in the regression. To avoid this problem and to obtain a regression model based on the covariance between $\mathbf{X}$ and $\mathbf{Y}$ rather than the variance of $\mathbf{X}$ only, partial least squares regression (PLS) can be used.

10.4.4 PARTIAL LEAST SQUARES REGRESSION

Partial least squares regression [1, 51, 52] was developed in the 1970s by Herman Wold as a generalization of multiple linear regression (MLR) which can analyse data containing large number of variables that are strongly correlated. An excellent tutorial to PLS regression is given by Wold *et al.* [52]. PLS finds factors in the predictor variables **X** that also capture a large amount of variance in the response variables **Y**, using the simultaneous decomposition of the two matrices. A schematic diagram of PLS, for a one-factor ($N = 1$) example, is shown in Figure 10.18. There are numerous different formulations of PLS, differing in notations and normalization conditions. The formulation below follows those by Wold *et al.* [52]. PLS involves finding a set of PLS scores **T**, common to the data **X** and **Y**, that is a good descriptor of both sets of data so that associated factor models (see Section 10.3.2) can be found where residuals **E** and **F** are small. For a PLS model with N factors, we can thus write:

$$\mathbf{X} = \mathbf{TP}' + \mathbf{E} \tag{10.27}$$
$$\mathbf{Y} = \mathbf{TC}' + \mathbf{F} \tag{10.28}$$

Here, **T** is an $I \times N$ matrix of PLS scores common to **X** and **Y**, **P** is an $K \times N$ matrix of x-loadings and **C** is an $M \times N$ matrix of y-loadings used to construct the factor models. **E** and **F** are x- and y-residuals which give the part of the data not explained by the models. The scores matrix **T** is calculated as the projection of the data **X** onto a weights matrix **W***, with dimensions $K \times N$, which is directly related to eigenvectors of **X**'**YY**'**X** such that successive PLS factors account for the largest covariance between **X** and **Y**:

$$\mathbf{T} = \mathbf{XW}^* \tag{10.29}$$

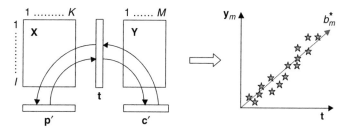

Figure 10.18   A schematic diagram of one-factor PLS analysis, adapted from Malinoski [1]. Factor decomposition (left) is used to obtain scores **t** common to **X** and **Y** that reflects their covariance. Regression (right) is then applied to relate **t** to each response variables $\mathbf{y}_m$ via regression coefficient b_m^*, which can be transformed to reflect the original **X** variables and give the regression matrix **B** for Equation (10.21)

PLS solutions are constrained so that the scores on each factor, given by the columns of $\mathbf{T}$, are orthogonal, and the columns of $\mathbf{W}^*$ are orthonormal. Unlike PCA, the x- and y-loadings in PLS, $\mathbf{P}$ and $\mathbf{C}$, are no longer orthogonal due to the constraints in finding common scores $\mathbf{T}$ in the decomposition. Using this, we can finally rearrange Equations (10.27)–(10.29) to resemble the regression equation in Equation(10.21). This gives the following PLS solution to the regression:

$$\mathbf{Y} = \mathbf{XW}^*\mathbf{C}' + \mathbf{F} = \mathbf{XB} + \mathbf{F} \tag{10.30}$$
$$\mathbf{B} = \mathbf{W}^*\mathbf{C}' \tag{10.31}$$

Essentially, PLS estimates the values of response variables $\mathbf{Y}$ using linear combinations of the PLS scores $\mathbf{T}$, obtained through simultaneous decomposition of $\mathbf{X}$ and $\mathbf{Y}$ and therefore reflects the largest covariance between them. PLS removes redundant information from the regression, such as factors describing large amounts of variance in $\mathbf{X}$ that does not correlate with $\mathbf{Y}$. PLS therefore tends to produces more viable, robust solutions using fewer factors compared to PCR. Since PLS scores $\mathbf{T}$ are orthogonal, the problem of collinearity in MLR is also removed. As with PCR, the result of the regression is strongly dependent on the number of factors retained in the PLS model, and the proper use of validation and cross validation methods is extremely important. This is described in the section below.

10.4.5 CALIBRATION, VALIDATION AND PREDICTION

The regression matrix, $\mathbf{B}$, obtained from PCR or PLS above, allows us to evaluate the relationship between the predictor and response variables measured on the same set of samples. Since the model is computed to fit the data provided, it is said to be calibrated on the data. As with all statistical models used for calibration and quantification, careful validation must be applied to PCR and PLS models to ensure that it is robust [53, 54]. In particular, the appropriate number of factors to be retained in the model needs to be determined carefully. Similar to PCA, the number of factors in PCR or PLS can be considered as the number of independent parameters that are responsible for the linear relationship between the predictor and response data sets. Since successive PCR and PLS factors describe the most important variance (or covariance) in the data, increasing the number of factors retained in the model increases information in the regression that are less relevant and more susceptible to experimental noise, thus increasing the risk of overfitting the model to the data. Indeed, if all factors are included, then PCR and PLS are simply equivalent to MLR. Figure 10.19 shows the effect of the number of factors on the accuracy of a PLS calibration, for an example data set. The root mean square error of calibration (RMSEC) is calculated for each model, using the residuals matrix $\mathbf{E}$ in Equation (10.21).

It is clear that as the number of factors goes up, the RMSEC value goes down and the model fits the calibration data increasingly accurately. However, with too many factors the model will not be robust.

To guard against overfitting and determine the appropriate number of factors in the model, cross validation should be used. For a small data set typically obtained in surface analysis, leave-one-out cross validation is the most popular approach. It works as follows:

1. Exclude sample i from calibration data, and calculate regression model for n factors.

2. Apply the model to sample i and predict its response, using Equation (10.22).

3. Record the difference between predicted and measured response value for sample i.

4. Replace sample i. Repeat steps 1–3 for a different sample until all samples have been used.

5. Calculate the root mean square error of cross validation (RMSECV).

6. Repeat steps 1–5 for different number of factors n.

The root means square error of cross validation (RMSECV) for the example data set is shown in Figure 10.19. With a small number of factors, the value of RMSECV is slightly higher than RMSEC, since the excluded samples do not contribute to the calibration. As the number of factors increases, the models are increasingly overfitted to the calibration data and become unsuitable for predicting the excluded samples, giving a large RMSECV value. Figure 10.19 therefore indicates that the model with two factors, giving the minimum RMSECV value, should be used. After determining the number of factors using cross validation, a final regression model can be calculated on the full data set or in order to study the relationship between the predictor and response variables.

While leave-one-out cross validation is suitable for a small data set, it tends to over estimate the number of factors required when the data set is large. This is because, for a large data set with many samples, the likelihood is great that the excluded sample is very similar to a sample retained in the calibration set, giving a lower RMSECV value than expected if the excluded test sample is truly independent from the calibration. A more appropriate method for large data sets is the leave-many-out approach, using a number of randomly drawn subsets of the data for cross validation [54]. Alternatively, to save computation time, the data can be split into separate calibration and validation sets, and the number of factors determined using the root mean square error of validation (RMSEV). In all cases, it is very important to ensure that the validation or cross validation data are statistically independent from

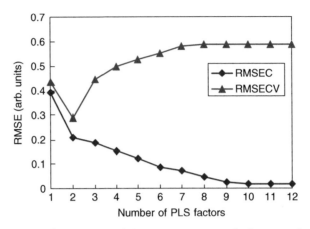

Figure 10.19   Results of PLS cross validation on an example data set, showing the root mean square error of calibration (RMSEC) and the root mean square error of cross validation (RMSECV)

the calibration data, for example, repeat spectra should be treated as one and either included or excluded together.

Finally, while cross validation ensures the regression model is self-consistent with the data and not overfitted to noisy features, an independent validation with a separately acquired experimental data set is essential before the model can be used for future predictions [55]. This is important in surface analysis when relatively small differences in contamination, sample preparation or instrumental settings may give rise to relatively large changes in the data. A self-consistent regression model can therefore be obtained for data from a single experiment, which may not be predictive for future samples and experiments. Independent validation is therefore absolutely essential to assess the prediction accuracy and estimate RMSEP, the root mean square error of prediction, before the regression models are used for prediction of unknown samples.

10.4.6 EXAMPLE – CORRELATING TOF–SIMS SPECTRA WITH POLYMER WETTABILITY USING PLS

PLS has been successfully applied to a number of ToF-SIMS data [23, 26, 27, 30]. One example is a recent study, by Urquhart *et al.* [23], on the relationship between ToF–SIMS spectra of copolymers and their experimental water contact angles (WCA). 576 copolymers are synthesized in a micropatterned array using a 70:30 ratio of 24 different acrylate-based monomers mixed pairwise and printed onto glass slides. Both positive and

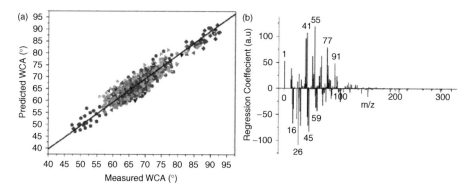

Figure 10.20 A PLS model correlating ToF–SIMS spectra with wettability of 576 polymers [23]. (a) The predicted water contact angles (WCA) against measured WCA. (b) The regression coefficients, consistent with high WCA correlating with polar, hydrophilic fragments and vice versa. Reproduced with permission from the American Chemical Society from Urquhart *et al.* [23]

negative ion ToF–SIMS spectra are obtained on the polymers, and WCA values are measured with a contact angle goniometer on separately printed microarrays not analysed by SIMS. Data selection and binning is carried out separately on positive and negative ion data, using comprehensive peak lists generated from the spectra of pure polymers, each composed of a single monomer. The positive and negative ion data for each sample are then normalized separately, and concatenated (combined) so both sets of data are contained in a single data matrix. Concatenation allows different fragmented species, both positively and negatively ionized, to be analysed simultaneously, therefore increasing the relevant information available to the regression. Using the concatenated SIMS data as the predictor X, and the WCA values as the response Y, a PLS model is calculated after mean centering of both sets of data. Six factors are retained, as determined using leave-one-out cross validation. Figure 10.20(a) shows the predicted WCA values against their measured values, displaying a good fit of the model to the experimental data and a low calibration error. The resulting regression coefficients are shown in Figure 10.20(b). By analysing the regression coefficients, Urquhart *et al.* showed that nonoxygenated hydrocarbon fragments $C_nH_m^+$, which have positive regression coefficients, are correlated with higher WCA values, while linear oxygenated fragments ($C_nH_mO^+$) and amine/amide fragments (CN^-, CNO^- and CH_4N^+), with negative regression coefficients, are correlated with low WCA values. This agrees with the conventional understanding that nonpolar species are hydrophobic and polar species are hydrophilic, and demonstrates the potential of using ToF–SIMS for building models between surface chemistry and resulting surface properties for a large number of materials in a high-throughput fashion.

10.5 Methods for Classification

In many analytical situations, there is a requirement to classify a sample into one of a number of possible groups of materials that exhibit similar characteristics: for example, does a particular surface have good or bad cell growth properties? A further requirement may be to investigate particular aspects of the sample that drives the relationship, i.e. are the good cell growth properties related to the surface concentration of hydrophilic groups? Multivariate methods provide a powerful route to this type of classification and interpretation of spectral information from surface analytical techniques. In this section we consider three popular methods: discriminant function analysis (DFA), hierarchal cluster analysis (HCA) and artificial neural networks (ANNs).

10.5.1 DISCRIMINANT FUNCTION ANALYSIS

Earlier in Figure 10.10, we have seen an excellent example of how PCA may be used to categorize between trehalose-protected and unprotected antiferritin in SIMS analysis [29]. In this case, it is actually the second PCA factor that distinguishes between the two groups of samples, accounting for approximately half the variance of the first factor which describes large variations due to sample heterogeneity. Clearly, for classification purposes, it is desirable if the first few factors capture the largest variations between the groups rather than other variations within the data set, which may arise from contamination. This is the purpose of discriminant function analysis (DFA). The objective of DFA is to separate measurement groups based on the measurements of many variables [56]. Usually, DFA is conducted following PCA as this reduces the dimensionality of the data and removes any collinearity in the data so that the data matrix is of full rank. The factors extracted from DFA are called 'discriminant functions' and they are linear combinations of the original variables for which the Fisher ratio, F, [57] is maximized [56]. For two groups, the Fisher ratio is given by:

$$F = \frac{(m_1 - m_2)^2}{\sigma_1^2 + \sigma_2^2} \tag{10.32}$$

where m_1 and σ_1^2 is the mean and variance of group 1, respectively, and similarly for group 2. Maximizing the Fisher ratio therefore minimizes the variations within each group while maximizing the variations between different groups. DFA factors, or discriminant functions, allow us to study features in the data that accounts for the differences between groups of samples. Analogous to PCA, the discriminant functions successively capture the most significant variations between groups. They are effectively a rotation of the principal components so that the rotated factors describe the largest differences between groups rather than the largest variance in the data. By constructing a DFA model using a calibration data set where *a priori* information of group membership is known, the model can be used for future classification of unknown samples. DFA is therefore a supervised classification method, and careful validation with independent data is needed before it can be used for future prediction, as previously discussed in Section 10.4.5. For interested readers a detailed introduction to DFA is given in Manly [56].

Example – Discrimination of Bacteria Using SERS and DFA. DFA has been used effectively for the classification of bacteria using a variety of analytical techniques including pyrolysis mass spectrometry [57], surface enhanced Raman spectroscopy (SERS) [58, 59], attenuated total reflectance Fourier transform infrared spectroscopy (ATR-FTIR) [60] and SIMS [61]. Jarvis and Goodacre [58] demonstrated the use of SERS to discriminate bacteria to strain level with the use of DFA. Their study provides a good example of careful cross-validation, which is essential for effective and robust classification. In their study, 21 bacterial isolates were obtained from patients with urinary tract infections. This included species and strains of *Klebsiella*, *Citrobacter*, *Enterococcus* and *Proteus*. To improve the repeatability of SERS spectra, 36 median spectra were selected from 50 repeat spectra and summed to provide an average spectrum. This significantly reduced the high variability commonly seen in SERS spectra. The spectra were binned to 1.015 cm^{-1} width and a linear base line was subtracted from the spectra so that the minimum spectral intensity was zero. The spectra were subsequently smoothed and normalized to a maximum intensity of unity. For each bacterial isolate, five replicates were analysed. Four were used as the calibration set to develop the DFA model. Importantly, the validation of the model was done using a fifth replicate acquired on a different day. It is worth highlighting this here. It is critical for the validation data set to be independent to the calibration data, for example data acquired on a different day on a separate batch of samples, rather than repeat spectra of the same batch, which improve repeatability but do not take into account instrumental parameters such as constancy of the spectrometer and natural variations between samples. Owing to the large differences between spectra for the different bacteria species the DFA analysis was conducted separately on sub-groups of bacteria strains. Figure 10.21 shows the classification of the *Enterococcus* isolate data on the first three discriminant functions together with the projections of data from the validation data set. It is clear that SERS, combined with analysis by DFA, is capable of giving strain level classification of bacteria using a robust model, as shown by the proximity of the calibration and validation data for each group.

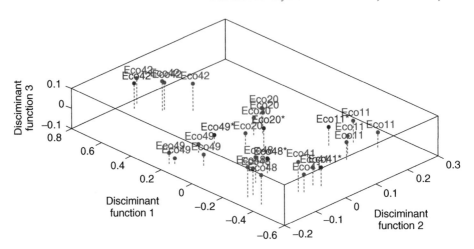

Figure 10.21 DFA of SERS spectra from six *Enterococcus* isolates from patients with urinary tract infections, labelled Eco11, Eco20, Eco41, Eco42, Eco48 and Eco49, replotted from 'Figure 8b' in Jarvis and Goodacre [58]. The calibration set data from four replicate measurements are shown in red and the projection of the validation set is shown in blue. Note the strain level separation and the consistent grouping of the validation set. Reproduced with permission from the American Chemical Society from Jarvis and Goodacre [58]

10.5.2 HIERARCHAL CLUSTER ANALYSIS

Cluster analysis refers to a family of unsupervised methods that partitions the data into distinct, un-predefined groups based on the similarities or differences between the samples. Unlike DFA, no *a priori* information on group membership is needed to perform cluster analysis, and the resulting model can be used for classification and prediction. However, cluster analysis does not provide direct information on the ways in which samples from different groups are related to each other, unlike DFA where discriminant functions can be studied directly. There are many methods for cluster analysis and these are reviewed in detail in Manly [56]. One of the most popular methods is hierarchal cluster analysis (HCA), which uses a metric to measure the distance between samples. All samples below a threshold distance away from each other are classified as being in the same group. The threshold value is then increased so that groups are agglomerated until all the data belong to a single group. This produces a dendrogram (tree diagram) structure, as shown in Figure 10.22. The threshold distance can be specified in order to classify the data into the desired number of groups. Generally, the simple Euclidean distance between data values is used for the agglomeration algorithm, but other distance measures can also be used. It is usual for HCA to be performed after data analysis by PCA or DFA, as the dimensionality of the data is significantly reduced.

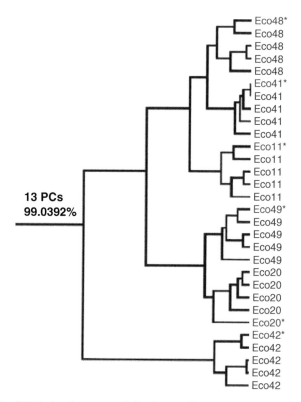

13 PCs
99.0392%

Figure 10.22 An HCA dendrogram of the five replicate SERS measurements from six *Enterococcus* isolates from patients with urinary tract infections from the discriminant function space shown in Figure 10.21. Reproduced with permission of the American Chemical Society from part of 'Figure 8' in Jarvis and Goodacre [58]

Example – Classification of Bacteria Using SERS and HCA. To continue the example of bacterial strain classification above, Jarvis and Goodacre [58] applied HCA to the projection of the bacteria samples in DFA space, shown in Figure 10.21, using Euclidean distances for the agglomerative algorithm [56]. The resulting dendrogram is shown in Figure 10.22. This clearly shows the strain level classification of bacteria and the robustness of the model to the validation data, which falls tightly within the clusters formed by the calibration data.

10.5.3 ARTIFICIAL NEURAL NETWORKS

Artificial neural networks (ANNs) are popularly used in data analysis and modelling, including pattern recognition, classification and regression of data. ANNs are completely distinct from the factor-based analysis methods discussed so far in this chapter. They are a mathematical formalism of the biological organ for cognitive processing, the brain. The brain is enormously powerful for processing complex information for the purposes of pattern recognition. This power is due to massive connectivity between some 10^{11} simple processing units known as neurons. Each neuron connects to other neurons via electrical pulses along long filaments known as axons. The connection between the end of the axon and the input of the other neuron is called the synapse, and each synapse has a threshold level that must be exceeded before the electrical signal can pass.

ANNs are developed in a similar way to benefit from this parallel processing architecture. Each network comprises a pattern of simple processing units called neurons or nodes. Figure 10.23 shows a schematic of a neuron with three inputs and one output. A multiplication weight is applied to each input, equivalent to the strength of the electrical signal in the organic version, and the neuron sums the values and adds a bias value. Conversion of the input to an output is made via a transfer function. For the purpose of classification this is usually a sigmoidal function that gives a stable output of between 0 and 1 for inputs ranging from ± infinity. The addition of a bias value means that neurons act like the synapse, and its threshold can be altered. Neurons are arranged in a number of different architectures. For pattern recognition and classification, the most popular is the multilayer feed forward network also known as the multilayer perceptron [62]. An input layer of neurons maps to a number of so called hidden layers which finally map onto an output layer. An example of such a network is shown in Figure 10.23(b).

There are many examples of the use of ANNs, especially in the biological fields such as detection of the adulteration of virgin olive oil [63] and the adulteration of orange juice [64], identification of amino acids in glycogen

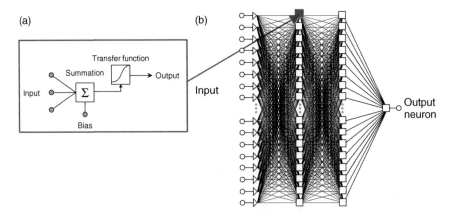

Figure 10.23 A schematic of an ANN. (a) A single neuron with a sigmoidal transfer function and (b) a network of neurons arranged as a multilayer feed forward network consisting of an input layer, a middle or 'hidden layer' and a single output neuron

[65], discrimination of bacteria sub-species [66], quantification of binary mixtures [67], bacteria identification [68] and the classification of adsorbed proteins by static SIMS [31]. It is interesting to see from Figure 10.1 that the published use of ANNs in surface chemical analysis was growing at around the same rate as PCA in the early nineties, but soon reached a steady plateau, whilst the use of PCA has accelerated such that it is now almost ten times more popular. This is largely owing to the difficulty in training or calibrating a network that is robust and based on significant chemical features rather than background signal or contamination. However, a great advantage of ANNs is their ability to work with non-linearity in the data. The multivariate methods discussed thus far all assume that the data can be expressed as a linear sum of factors and that those factors relate to chemical contributions in the data. For many techniques, such as optical methods and gas-phase mass spectrometries, this is a reasonable assumption. Unfortunately, for surface analysis many complications can arise that can lead to strong non-linearities, such as contamination overlayers, matrix effects and topographical effects. ANNs may therefore have utility in these applications.

Example – An Artificial Neural Network to Classify SIMS Spectra. Gilmore *et al.* [69] developed a multilayer feed forward network for pattern recognition of static SIMS spectra and classification using library data [70]. The network used, as shown in Figure 10.23(b), consists of three layers of neurons to give the required flexibility. Static SIMS spectra are binned to unit mass intervals and limited to the 'fingerprint' region (1 to 300 u), which reduced the number of inputs to the network to 300. Each spectrum is normalized to give a maximum intensity of unity. A single output is required from the network, which is trained to give an output of one for a correct match and an output of zero otherwise. It is necessary to define how many neurons should be in each layer. If a network is made too powerful it may be over-fitted to the calibration data in a similar manner to PLS, as shown in Figure 10.19. There are no precise rules to determine the number of neurons in each layer but a simple rule [71] states that the number of neurons in the first layer should be approximately half of the number of inputs. Therefore, the first layer is set to have 150 neurons and the second layer is chosen to have 50 which all map onto one final output neuron. Input weights and biases are determined as the network learns from examples. This is known as supervised learning. Examples are presented to the network, which then tries to match the answer at the output. The method used to calculate the values of input weights and biases is called back propagation [62] which is a gradient descent algorithm. A number of different minimization algorithms are available with different computational requirements for both memory and speed. For this network, the conjugate gradient algorithm is found to give a reliable and fast convergence.

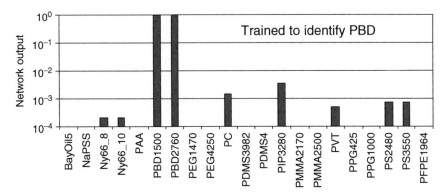

Figure 10.24 The output of an ANN trained to identify the static SIMS spectrum of PBD for library spectra of a wide variety of polymers and molecular weight distributions [69]. Note the logarithmic ordinate scale

The network was taught to recognize a wide variety of polymers including poly(methyl methacrylate) (PMMA), polybutadiene (PDB), poly(ethylene glycol) (PEG), polystyrene (PS), polycarbonate (PC) and polydimethylsiloxane (PDMS) from a static SIMS library [70] consisting of fifteen different polymers with a further eight different molecular weight distributions. Figure 10.24 shows the outputs of the trained to identify PBD when exposed to spectra from the library. The network demonstrated an excellent ability to classify the spectra with the lowest correct value of 0.9899 and the highest incorrect value of 0.0109.

As we have discussed earlier, the key test to any methods of classification is the ability of the network to work with independent data. To test this, Gilmore *et al.* acquired spectra of polycarbonate using a different instrument to that used for the library data [70], using Ar^+, Cs^+ and SF_5^+ primary ions with energies between 4 and 10 keV. It is well known that different primary ions produce significant differences in the fragmentation patterns and spectral intensities in static SIMS [72]. The network correctly identified the material as PC with an average output of 0.9962. The largest misidentification was for PS with an output of 0.113 for 10 keV SF_5^+ primary ions but values were typically < 0.002. Notwithstanding the variety of primary ion species and energies the network has provided a clear and correct identification with high confidence.

10.6 Summary and Conclusion

In this chapter, we have reviewed the most popular multivariate analysis techniques used in surface analysis. Figure 10.25 shows the objectives and typical application of different multivariate methods, arranged by their

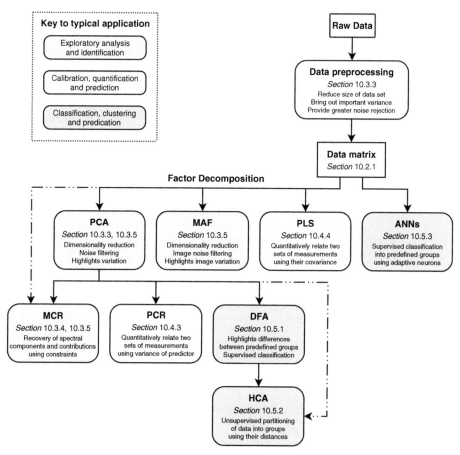

Figure 10.25   A diagram summarizing the objectives of multivariate analysis methods reviewed in this chapter, arranged by their typical order of application

typical order of application. We have shown that multivariate methods are powerful for identification, quantification and classification of a wide range of data obtained from surface analysis, including spectra, images and depth profiles. The speed, automation and accuracy of multivariate analysis, and its ability to obtain information that would be difficult to extract without prior knowledge of the system, make it advantageous to traditional analysis methods for many types of data. Current research is actively addressing many issues and challenges in multivariate analysis, from the fundamentals such as data scaling for optimal discrimination of chemical features from noise, to practical applications such as the analysis of fibre images with large topography. Terms and definitions, developed in consultation with leading industry and academic experts, are now being incorporated into the next version of ISO 18 115 – surface chemical analysis – vocabulary [73].

In addition, an ISO guide to multivariate analysis, recently identified as a high priority by industry analysts, is currently in active development. As a result, clear recommendations and guidelines on multivariate analysis are becoming available, making the methods easier to use as well as giving confidence to the analysis results. With the increased power and throughput of modern analytical instruments, and the growing analytical requirements in novel research areas, for example the surface analysis of biomaterials and innovative devices, multivariate analysis is becoming an increasingly indispensable tool in extracting the maximum information from data with a fast, robust and unbiased approach.

Acknowledgements

The authors would like to thank D.G Castner, M.S. Wagner, B.J. Tyler and S.J. Pachuta for valuable discussions on terminology for multivariate analysis. This work forms part of the Chemical and Biological Programme of the National Measurement System of the UK Department of Innovation, Universities and Skills (DIUS).

References

1. E.R. Malinowski, *Factor Analysis in Chemistry*, 3rd Edition, John Wiley & Sons, Inc., New York, NY (2002).

2. S. Wold, *Chemomet. Intell. Lab Syst.* **30**, 109 (1995).

3. Thomson Scientific, Philadelphia, PA, http://www.isiknowledge.com/ (2008).

4. M.L. Boas, *Mathematical Methods in the Physical Sciences*, 3rd Edition, John Wiley & Sons, Inc., New York, NY (2006).

5. B.M. Wise, N.B. Gallagher, R. Bro, J.M. Shaver, W. Windig and R.S. Koch, *PLS_Toolbox Version 3.5 Manual*, Eigenvector Research Inc., Wenatchee, WA (2005).

6. P. de B. Harrington, P.J. Rauch and C. Cai, *Anal. Chem.* **73**, 3247 (2001).

7. R. Larsen, *J. Chemomet.* **16**, 427 (2002).

8. J. Stevens, *Applied Multivariate Statistics for the Social Sciences*, Lawrence Erlbaum Associates, Inc., Mahwah, NJ (2001).

9. R.S. Tsay, *Analysis of Financial Time Series*, John Wiley & Sons, Inc., New York, NY, p. 335 (2001).

10. I.T. Jollife, *Principal Component Analysis*, 2nd Edition, Springer-Verlag, New York, NY (2002).

11. M. Turk and A. Pentland, *J. Cognitive Neurosci.* **3**, 71 (1991).

12. A. Beuzen and C. Belzung, *Physiol. Behav.* **58**, 111 (1995).

13. R.B. Cattell, *Multivariate Behav. Res.* **1**, 245 (1966).

14. S.N. Deming and J.A. Palasota, *J. Chemomet.* **7**, 393 (1993).

15. J.L.S. Lee, I.S. Gilmore and M.P. Seah, *Surf. Interface Anal.* **40**, 1 (2008).

16. M.R. Keenan and P.G. Kotula, *Surf. Interface Anal.* **36**, 203 (2004).

17. M.P. Seah and P.J. Cumpson, *J. Electron Spectrosc.* **61**, 291 (1993).

18. M.S. Wagner, D.J. Graham and D.G. Castner, *Appl. Surf. Sci.* **252**, 6575 (2006).

19. J. Walton and N. Fairley, *J. Electron Spectrosc.* **148**, 29 (2005).

20. S.J. Pachuta, *Appl. Surf. Sci.* **231–232**, 217 (2004).

21. M.L. Pacholski, *Appl. Surf. Sci.* **231–232**, 235 (2004).

22. X. Vanden Eynde and P. Bertrand, *Surf. Interface Anal.* **25**, 878 (1997).

23. A.J. Urquhart, M. Taylor, D.G. Anderson, R. Langer, M.C. Davies and M.R. Alexander, *Anal. Chem.* **80**, 135 (2008).

24. N. Médard, C. Poleunis, X. Vanden Eynde and P. Bertrand, *Surf. Interface Anal.* **34**, 565 (2002).

25. M.C. Biesinger, P.-Y. Paepegaey, N. Stewart McIntyre, R.R. Harbottle and N.O. Petersen, *Anal. Chem.* **74**, 5711 (2004).

26. L. Yang, Y.-Y. Lua, G. Jiang, B.J. Tyler and M.R. Lindord, *Anal. Chem.* **77**, 4654 (2005).

27. M.S. Wagner, D.J. Graham, B.D. Ratner and D.G. Castner, *Surf. Sci.* **570**, 78 (2004).

28. D.J. Graham, M.S. Wagner and D.G. Castner, *Appl. Surf. Sci.* **252**, 6860 (2006).

29. N. Xia, C.J. May, S.L. McArthur and D.G. Castner, *Langmuir* **18**, 4090 (2002).

30. S. Ferrari and B.D. Ratner, *Surf. Interface Anal.* **29**, 837 (2000).

31. O.D. Sanni, M.S. Wagner, D. Briggs, D.G. Castner and J.C. Vickerman, *Surf. Interface Anal.* **33**, 715 (2002).

32. M.S. Wagner and D.G. Castner, *Langmuir* **17**, 4649 (2001).

33. A. de Juan and R. Tauler, *Anal. Chim. Acta* **500**, 195 (2003).

34. A. de Juan and R. Tauler, *Crit. Rev. Anal. Chem.* **36**, 163 (2006).

35. N.B. Gallagher, J.M. Shaver, E.B. Martin, J. Morris, B.M. Wise and W. Windig, *Chemomet. Intell. Lab. Syst.* **73**, 105 (2004).

36. J.-H. Jiang, Y. Liang and Y. Ozaki, *Chemomet. Intell. Lab. Syst.* **71**, 1 (2004).

37. H.F. Kaiser, *Psychometrika* **23**, 187 (1958).

38. V.S. Smentkowski, M.R. Keenan, J.A. Ohlhausen and P.G. Kotula, *Anal. Chem.* **77**, 1530 (2005).

39. V.S. Smentkowski, S.G. Ostrowski, E. Braunstein, M.R. Keenan, J.A. Ohlhausen and P.G. Kotula, *Anal. Chem.* **79**, 7719 (2007).

40. K.G. Lloyd, *J. Vac. Sci. Technol. A* **25**, 878 (2007).

41. D.E. Peebles, J.A. Ohlhausen, P.G. Kotula, S. Hutton and C. Blomfield, *J. Vac. Sci. Technol. A* **22**, 1579 (2004).

42. M.R. Keenan and P.G. Kotula, *Microsc. Microanal.* **11**, 36 (2005).

43. K. Artyushkova and J.E. Fulghum, *J. Electron Spectrosc.* **121**, 33 (2001).

44. P. Geladi and H. Grahn, *Multivariate Image Analysis*, John Wiley & Sons, Ltd, Chichester, UK (1996).

45. I.S. Gilmore and M.P. Seah, *Surf. Interface Anal.* **24**, 746 (1996).

46. M.P. Seah and S.J. Spencer, *Surf. Interface Anal.* **35**, 906 (2003).

47. K. Artyushkova and J.E. Fulghum, *Surf. Interface Anal.* **33**, 185 (2002).

48. J. Walton and N. Fairley, *Surf. Interface Anal.* **36**, 89 (2004).

49. B.J. Tyler, *Appl. Surf. Sci.* **252**, 6875 (2006).

50. B.J. Tyler, *Biomaterials* **28**, 2412 (2007).

51. P. Geladi and B.R. Kowalski, *Anal. Chim. Acta* **185**, 1 (1986).

52. S. Wold, M. Sjöström and L. Eriksson, *Chemomet. Intell. Lab Syst.* **58**, 109 (2001).

53. H.A. Martens and P. Dardenne, *Chemomet. Intell. Lab Syst.* **44**, 99 (1998).

54. J. Shao, *J. Am. Stat. Assoc.* **88**, 486 (1993).

55. A.M.C. Davies, *Spectrosc. Eur.* **16/4**, 27 (2004).

56. B.F.J. Manly, *Multivariate Statistical Methods*: *A Primer*, Chapman & Hall, London (1995).

57. W. Windig, J. Haverkamp and P.G Kistemaker, *Anal. Chem.* **55**, 81 (1983).

58. R.M. Jarvis and R. Goodacre, *Anal. Chem.* **76**, 40 (2004).

59. R.M. Jarvis, A. Brooker and R. Goodacre, *Faraday Discuss.* **132**, 281 (2006).

60. C.L. Winder and R. Goodacre, *Analyst* **129**, 1118 (2004).

61. J.S. Fletcher, A. Henderson, R.M. Jarvis, N.P. Lockyer, J.C. Vickerman and R. Goodacre, *Appl. Surf. Sci.* **252**, 6869 (2006).

62. C.M. Bishop, *Neural Networks for Pattern Recognition*, Oxford University Press, Oxford (1995).

63. R. Goodacre and D.B. Kell, *J. Sci. Food Agric.* **63**, 297 (1993).

64. R. Goodacre, D. Hammond and D.B. Kell, *J. Anal. Appl. Pyrol.* **40–41**, 135 (1997).

65. R. Goodacre, A.N. Edmonds and D.B. Kell, *J. Anal. Appl. Pyrol.* **26**, 93 (1993).

66. R. Goodacre, A. Howell, W.C. Noble and M.J. Neal, *Zbl. Bakt.* **284**, 501 (1996).

67. M.L. Ganadu, G. Lubinu, A. Tilocca and S.R. Amendolia, *Talanta* **44**, 1901 (1997).

68. J. Schmitt, T. Udelhoven, D. Naumann and H.-C. Flemming, *SPIE*, **3257**, 236 (1998).

69. I.S. Gilmore and M.P. Seah, *Static SIMS: Methods of Identification and Quantification using Library Data – An Outline Study*, NPL Report, CMMT(D)268 (2000).

70. B.C. Schwede, T. Heller, D. Rading, E. Niehius, L. Wiedmann and A. Benninghoven, *The Münster High Mass Resolution Static SIMS Library*, ION-TOF, Münster, Germany (1999).

71. D.A. Cirovic, *Trends Anal. Chem.* **16**, 148 (1997).

72. I.S. Gilmore and M.P. Seah, *Appl. Surf. Sci.* **161**, 465 (2000).

73. M.P. Seah, *Surf. Interface Anal.* **31**, 1048 (2001).

Problems

1. Outline the advantages and disadvantages of using multivariate analysis compared to traditional analysis for:

   (a) Research.

   (b) Routine quantitative analysis and quality control.

   What are the limitations of multivariate analysis?

2. The factor analysis model can be written simply as:

$$\mathbf{X} = \sum_{n=1}^{N} \mathbf{t}_n \mathbf{p}'_n + \mathbf{E} \tag{10.33}$$

   Explain the meaning and significance of each term in the equation. Summarize the differences between principal component analysis (PCA) and multivariate curve resolution (MCR), which are both based on this model.

3. Deduce the rank of the data matrix $\mathbf{X}$ shown in Figure 10.2. Calculate the matrix $\mathbf{Z} = \mathbf{X}' \mathbf{X}$ and compute the eigenvectors and associated eigenvalues of $\mathbf{Z}$. Convert the eigenvalues into percentage variance captured in PCA.

4. Apply mean centering to the data matrix $\mathbf{X}$ shown in Figure 10.2. Calculate the covariance matrix $\mathbf{Z}_{cov}$, and compute the eigenvectors and associated eigenvalues of $\mathbf{Z}_{cov}$. Convert the eigenvalues into percentage variance captured in PCA. Explain why these results differ from those in Problem 3 above.

5. Figure 10.14 shows the results of PCA applied to the ToF–SIMS image of an immiscible polycarbonate/poly(vinyl chloride) polymer blend, after mean

centering. Explain the results you would expect to see if MCR is applied to the raw data, resolving two factors using non-negativity constraints to the scores and loadings. Discuss the relative merits of using PCA or MCR for this data set.

6. Explains the differences between validation and cross-validation in multivariate regression. Devise an experimental plan and outline the steps you would take to produce a partial least squares (PLS) regression model relating the surface chemistry of seven samples with their bacterial cell growth properties, including details of any validation or cross-validation procedures. If the model is used to predict the cell growth properties of future samples, how would you deduce the error of prediction and what could be done to improve this?

7. Figure 10.21 shows the scores of thirty bacterial samples on three discriminant functions, showing good separation between different bacterial strains using discriminant function analysis (DFA). The scores on each discriminant function are plotted on the same scale – why is this useful? What other multivariate method does this also apply to?

8. What are the advantages and disadvantages of using artificial neural networks compared to factor decomposition based multivariate methods?

Appendix 1
Vacuum Technology for Applied Surface Science

ROD WILSON

A1.1 Introduction: Gases and Vapours

A prerequisite of the vast majority of surface analysis techniques is a
'vacuum' environment in which the particular technique can be applied.
The reasons for needing a high vacuum environment are manifold, the most
fundamental being the requirement for long mean free paths for particles
used in studying surfaces. High vacuum conditions will mean that the
trajectories of particles such as ions and electrons used in surface analysis
will remain undisturbed when in a surface analysis system. It is often
necessary also to keep a surface free from absorbed gases during the course
of a surface analysis experiment; this requires tighter vacuum constraints
to keep the so-called monolayer time long enough to gather data from a
surface. The need to be able to sustain high voltages in a surface analysis
system without breakdown or glow discharges being created also imposes
vacuum constraints. An important part therefore of understanding surface
analysis and its applications is understanding vacuum technology and what
'vacuum' physically means. Since vacuum technology deals with gases and
vapours it is important to develop a physical picture of a gas or vapour.

Gases are a low density collection of atoms or molecules which can often
be pictured as simple hard spheres and are generally treated as having no
forces acting between them, except at the instant of collision. A picture of the
instantaneous structure of a small volume of gas is shown in Figure A1.1. The
molecules are usually a large distance apart compared with their diameter
and there is no regularity in their arrangement in space. The molecules
are distributed at random throughout the whole volume they occupy and
are moving randomly and, at room temperature, will have a mean velocity

Surface Analysis – The Principal Techniques 2nd Edition Edited by John Vickerman and Ian Gilmore
© 2009 John Wiley & Sons, Ltd

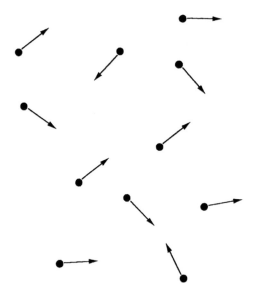

Figure A1.1   An instantaneous picture of a small volume of gas, where the arrows show the random nature of the motion of the particles. (NB If the relative sizes of the particles and their separation was to scale then this picture would correspond to a cube of $\approx$30 Å at a pressure of $\approx$20 atm)

typically of the order of $10^2\,\mathrm{ms}^{-1}$. It is worth noting that the noble gases present a close physical realisation of this ideal behaviour.

   This postulate that a gas consists of a number of discrete particles between which no forces are acting led to a series of theoretical considerations which are referred to as 'the kinetic theory of gases'. This theory tries to explain the macroscopic properties of gases, such as pressure and temperature by considering the microscopic behaviour of the molecules of which they consist. One of the first and most important results from this type of treatment was to relate the pressure, p of a gas to the gas density, ρ and the mean square velocity, $\langle c^2 \rangle$ of the gas molecules each of mass, m. This relation is given by:

$$p = \tfrac{1}{3}\rho\langle c^2 \rangle \tag{A1.1}$$

or:

$$p = \tfrac{1}{3}nm\langle c^2 \rangle \tag{A1.1a}$$

where:

$$\langle c^2 \rangle = 3\frac{kT}{m} \tag{A1.2}$$

and n is the number density of particles (m^{-3}), k is the Boltzmann constant (JK^{-1}) and T is the absolute temperature (K).

The gas molecules will travel with a distribution of velocities in straight lines and collide with the walls of the container they are in and also collide elastically with each other. The average numbers of collisions per second between particles is called the collision rate, z and the path which each particle makes on average between collisions is called the *mean free path, λ*. Both these parameters are functions of the mean particle velocity, $\langle c \rangle$, the particle diameter, $2r$ and the number density of particles. To a good approximation they are given by:

$$z = \frac{\langle c \rangle}{\lambda} \tag{A1.3}$$

where:

$$\langle c \rangle = \sqrt{\frac{8kT}{pm}} \tag{A1.4}$$

and using simple geometry to show that if particles cross such that their diameters are less than $2r$ apart then a collision will occur it can be shown that:

$$\lambda = \frac{1}{\sqrt{2}n\pi(2r)^2} \tag{A1.5}$$

The quantity $\pi(2r)^2$ is called the collision cross-section of the molecule and is often denoted by the symbol σ. It follows from this that the mean free path is inversely proportional to the number density of molecules and therefore the gas pressure p. At constant temperature, for every gas, the product $p\lambda$ is a constant.

Understanding the concept of mean free path is an important stage in understanding what is happening on a molecular level inside your vacuum system and often the need for a vacuum is governed by a need to increase the mean free path of the molecules in the system. The main aim of vacuum technology in itself is to simply reduce the number density of atoms or molecules in a defined volume. Before the methodology for achieving this reduction in number density can be explained it is important to define some of the terminology and fundamental quantities used in vacuum technology and the principles involved.

The most common term used in vacuum technology is pressure, and is most often denoted by the symbol p. This is defined as 'the quotient of the perpendicular force on a surface and the area of this surface'. More simply it can be considered as the force per unit area applied to a surface by a fluid and is given the units of force per unit area. In the SI system of units this is given as newtons per metre^2 (Nm^{-2}) or pascals (P). In the field of vacuum technology, however, it is often more convenient to refer to pressure in the units of millibars (mbar). Previously in vacuum technology it was common to refer to pressure in the units 'torr'. This is no longer a recommended unit

Table A1.1 Table showing conversion of pressure units

mbar	Pa	torr[a]	bar	atm[b]	at[c]	% vac[d]	mmH$_2$O[e]
1013	101 300	760.00	1.01	1.0	1.03	0	1.03×10^4
1000	100 000	750.00	1.00	0.987	1.02	1.3	1.02×10^4
981	98 100	735.75	0.981	0.968	1	3.2	10^4
900	90 000	675.00	0.90	0.888	0.918	11.1	9 177
600	80 000	600.00	0.80	0.789	0.816	21.0	8 157
700	70 000	525.00	0.70	0.691	0.714	30.9	7 137
600	60 000	450.00	0.60	0.592	0.612	40.8	6 118
500	50 000	375.00	0.50	0.494	0.510	50.6	5 098
400	40 000	300.00	0.40	0.395	0.408	60.5	4 078
300	30 000	225.00	0.30	0.296	0.306	70.4	3 059
200	20 000	150.00	0.20	0.197	0.204	80.2	2 039
100	10 000	75.00	0.10	0.099	0.102	90.1	1 019
90	9 000	67.50	0.09	0.089	0.092	91.1	918
80	8 000	60.00	0.08	0.079	0.082	92.1	816
70	7 000	52.50	0.07	0.069	0.071	93.1	714
60	6 000	45.00	0.06	0.059	0.061	94.1	612
50	5 000	37.50	0.05	0.049	0.051	95.1	510
40	4 000	30.00	0.04	0.040	0.041	96.1	408
30	3 000	22.50	0.03	0.030	0.031	97.0	306
20	2 000	15.00	0.02	0.020	0.020	98.0	204
10	1 000	7.50	0.01	0.010	0.010	99.0	102
5	500	3.75	0.005	0.005	0.005	99.5	51
1	100	0.75	0.001	0.001	0.001	99.9	10
0.5	50	0.375	5×10^{-4}	5×10^{-4}	5×10^{-4}	99.9	5
0.1	10	0.075	1×10^{-4}	1×10^{-4}	1×10^{-4}	99.99	1
1×10^{-2}	1	7.5×10^{-3}	1×10^{-5}	1×10^{-5}	1×10^{-5}	99.99	0.1
1×10^{-n}	$1 \times 10^{-(n+2)}$	$7.5 \times 10^{-(n+1)}$	$1 \times 10^{-(n+3)}$	$1 \times 10^{-(n+3)}$	$1 \times 10^{-(n+3)}$		$1 \times 10^{-(n-1)}$
0	0	0	0	0	0	100	0

[a] 1 torr = 1 mmHg.
[b] 1 atm (= standard atmosphere) = 101 325 Pa (standard reference pressure).
[c] 1 at = 1 technical atmosphere.
[d] α_0 vacuum: a pressure increase (or decrease) of ca. 1 mbar corresponds to a change of vacuum of 0.1%.
[e] 1 mm H$_2$O (column) = 10^{-4} at = 9.8 mbar.

but should be mentioned as it is descriptive of the physical picture. 1 torr is understood as the fluid pressure which is able to balance, at $0°C$, a column of mercury 1 mm in height (for conversion factors see Table A1.1).

Although pressure is by far the most common term used in vacuum technology it is in itself of little relevance when applied to the field of surface science. The chief objective of vacuum technology here is to reduce the number density of particles, n in a given volume, V. The number density of particles is however related to the gas pressure p and the thermodynamic temperature, T by the laws of the kinetic theory of gases. This relationship is most often expressed as:

$$p = nkT \tag{A1.6}$$

where k is the Boltzmann constant.

It can be seen that, at constant temperature, a reduction in the number density of particles is always equivalent to a reduction in pressure. It must be emphasized, however, that a pressure decrease (keeping the volume constant) may also be attained by a reduction in the temperature of the gas. This must always be taken into account if the same temperature does not prevail throughout the volume of interest.

It can also be seen from Equation (A1.6) that, at a given temperature, the pressure depends only on the number density of molecules and not on the nature of the gas (i.e. the pressure of a gas is independent of chemical species for a given temperature and number density of particles).

When refering to pressure in vacuum technology it is invariably the absolute pressure, not referenced to any other pressure (analogous to absolute temperature); however, in certain cases it is necessary to be more specific and add an index to the symbol p. Several examples of this follow:

(i) The total pressure, p_{tot} in a container is equal to the sum of the partial pressures of all the gases and vapours within it.

(ii) The partial pressure, p_{part} of a given gas or vapour in a container is that pressure that the gas or vapour would have if it were present alone in the container.

(iii) The vapour pressure, p_d is the pressure due to vapours in a system as opposed to gases.

(iv) The pressure of a saturated vapour is called the saturation vapour pressure, p_s and for a given material it is only a function of temperature.

(v) The ultimate pressure, p_{ult} in a vacuum container is the lowest attainable pressure in that container for a given pumping speed.

At this stage it is also necessary to explain what distinguishes a gas from a vapour: 'gas' generally refers to material in the gaseous state which is not condensable at the operating temperature; 'vapour' likewise refers to material in the gaseous state which is, however, condensable at the ambient temperature. In the following text, the distinction between gases and vapours will only be made when it is required for understanding.

The volume, V, expressed in litres or metre^3 (l, m^3), is another frequently referred to term in vacuum technology. Physically, this is the purely geometric volume of the vacuum container or whole vacuum system. Volume, however, can also be used to indicate the pressure dependent volume of a gas which is, for example, transferred by a vacuum pump or cleaned up by sorption materials. The volume of gas which flows through a conducting element per unit time, at a specified temperature and pressure, defines the volume flow rate, q_v of that conducting element. It must be made clear that the number of particles transferred at a given volume flow rate is different for different temperatures and pressures.

The idea of volume flow rate leads to a parameter which is very important in defining the performance of a pump or pumping system, the pumping speed, S. The pumping speed of a pump (usually expressed in $l\,s^{-1}$ or $m^3 s^{-1}$ or similar) is the volume flow rate though the inlet aperture of the pump, expressed as:

$$S = dV/dt \tag{A1.7}$$

or, if S is constant during the pumping process, as is the case over the operating ranges of most high vacuum pumps, the differential quotient:

$$S = \partial V/\partial t \tag{A1.7a}$$

As stated previously, the volume flow rate does not indicate directly the number of particles flowing per unit time. It will also be a function of both the temperature and the pressure. Therefore it can often be more informative to define the flow rate of a quantity of gas through an element. A quantity of gas can be specified in terms of its mass G in the unit of grams or kilograms; in vacuum technology however the product pV is often of far more relevance. It can be seen from the following equation:

$$G = pV\frac{M}{RT} \tag{A1.8}$$

(which is a simple re-arrangement of the ideal gas equation, where G is the mass (kg), M is the molar mass (kg mol^{-1}) and R is the molar gas constant ($R = 8.314\,J\,mol^{-1}K^{-1}$)) that the mass can be readily calculated from a knowledge of the nature of the gas and the temperature.

The quantity of gas flowing through an element can therefore be expressed as the mass flow rate, q_m where:

$$q_m = G/\Delta t \tag{A1.9}$$

or as the throughput, q_{pV} where:

$$q_{pV} = p(\Delta V/\Delta t) \tag{A1.10}$$

(units $= Pa\ m^3 s^{-1}$ or more commonly, mbar $l\,s^{-1}$).

The throughput of a pump or pumping system is often used to define its performance and is the throughput through the intake aperture of the pump where p is the pressure at the intake side of the pump. If p and ΔV are constant, which they approximately are after pumpdown, then the throughput of the pump is given by the simple relation:

$$q_{pV} = pS \tag{A1.11}$$

where S is the pumping speed of the pump at the intake pressure p.

The concept of pump throughput is of great importance in understanding how pumping systems work and how to design one. It is important that the concepts of pump throughput and pumping speed be fully understood and not confused with each other.

The ability of a pump to remove gas from a system will not only be determined by the throughput of the pump, but also and often more importantly, by the ability of the elements in the pumping system to transmit the gas. The throughput of gas through any conducting element, for example, a hose or a valve or an aperture etc., is given by:

$$q_{pV} = C(p_1 - p_2) \tag{A1.12}$$

Here $(p_1 - p_2)$ is the difference between the pressures at the intake and exit of the conducting element. The constant of proportionality C is called the conductance of the element and is determined by the geometric nature of the element.

This conductance is analogous to electrical conductance and can be treated in a similar way using an analogy of Ohm's law in vacuum technology. For example, if several elements, A, B, C, etc. are connected in series then their total conductance is given by:

$$\frac{1}{C_{tot}} = \frac{1}{C_A} + \frac{1}{C_B} + \frac{1}{C_C} \tag{A1.13}$$

and if connected in parallel, it is given by:

$$C_{tot} = C_A + C_B + C_C \tag{A1.14}$$

A1.2 The Pressure Regions of Vacuum Technology and their Characteristics

As the pressure of a gas changes, then some of the characteristics of its behaviour also change. These changes can often determine the pressure conditions required for a certain experiment. It has therefore become customary in vacuum technology to subdivide the pressure region we call 'vacuum' into smaller regions in which the behaviour of the gas would have similar characteristics. In general these subdivisions are:

Rough vacuum	1000–1 mbar
Medium vacuum	$1-10^{-3}$ mbar
High vacuum	$10^{-3}-10^{-7}$ mbar
Ultra-high vacuum	10^{-7} mbar and below

These divisions are somewhat arbitrary, and boundaries between the regions cannot be considered sharp. The characteristic which changes most strikingly between the regions is the nature of gas flow.

In the rough vacuum region, viscous flow prevails almost exclusively. Here the mutual interaction of the particles with one another determines the character of the flow, i.e. the viscosity or inner friction of the streaming material plays the dominant role. If vortex motion appears in the streaming fluid, then it is referred to as turbulent flow. If however, the various layers of the streaming medium slide over each other it is referred to as laminar flow. The criterion for viscous flow is that the mean free path of the particles is smaller than the diameter of the conducting tube, i.e. $\lambda < d$.

In the high and ultra-high vacuum regions, the particles can move freely, virtually without any mutual hindrance and the type of flow dominant here is called molecular flow. The criterion here is that the mean free path of the particles is greater than the diameter of the conducting tube, i.e. $\lambda > d$.

In the medium vacuum region there is a transition from the viscous type flow to the molecular type flow – this type of flow is called Knudsen flow. For this type of flow the mean free path of the particles must be of the same order as the diameter of the conducting tube, i.e. $\lambda \approx d$.

It is often important to have a mental picture of what the behaviour of the gas on a molecular level is like in the different pressure regions with the different types of flow. In the viscous flow region the preferred direction of travel for all the gas molecules is the same as the macroscopic direction of flow of the streaming gas. The particles forming the gas are densely packed and they will collide much more frequently with each other than with the boundary walls of the containing vessel. In the region of molecular flow, however, the collisions of the particles with the vessel walls predominate. As a result of elastic reflections from the walls and desorption of gas particles off the vessel walls, gas particles in the high vacuum region can have any random direction – it is incorrect to think of streaming of the gas in the macroscopic sense. This is why in the molecular flow region the conductance of a pumping system is controlled totally by the geometry of the system, since gas particles only arrive at the apertures or openings by chance. It is important to understand this characteristic of gases in a high vacuum environment as it is frequently misunderstood.

These different characteristics of flow allow one to distinguish easily between the rough, medium and high vacuum regions. To distinguish between the high and ultra-high vacuum regions, however, requires the introduction of another parameter called the *monolayer time*, τ. In the high and ultra-high vacuum regions it is the nature of the container walls which is of most significance, since at pressures below 10^{-3} mbar there are more gas molecules on the surfaces of the vacuum vessel or chamber than there are in the gas space itself. It is therefore an important parameter in characterizing this pressure region, to consider the time it takes to form a single molecular or atomic layer, a so-called monolayer, on a gas-free surface in the vacuum. The value of this monolayer time is calculated with the assumption that every particle that impinges on the surface remains bonded to it. This monolayer time is obviously related to the number of particles which are incident upon unit surface area per unit time, the so-called impingement rate, Z_A. In a gas in the static state the impingement rate is related to the number density of particles and the mean velocity of particles as given by:

$$Z_A = \frac{n \langle c \rangle}{4} \tag{A1.15}$$

If a surface has a number of free places per unit surface area, a then the monolayer lifetime is given by:

$$\tau = \frac{a}{Z_A} = \frac{4a}{n \langle c \rangle} \tag{A1.16}$$

Assuming therefore that a monolayer is absorbed on the inner wall of an evacuated sphere of 1 litre volume, then the ratio of the number of absorbed particles to the number of free particles in the space is:

at 1 mbar	10^{-2}
at 10^{-6} mbar	10^4
at 10^{-11} mbar	10^9

Using this information and the concept of monolayer time, we can understand the need for the boundary between the high vacuum and ultra-high vacuum regions. In the high vacuum region the monolayer time amounts to only a fraction of a second; in the ultra-high vacuum region however, it is of the order of minutes or hours.

To achieve and maintain a surface which is free of absorbed gas for any practical length of time it is obviously necessary to keep the surface in an ultra-high vacuum environment. In many areas of surface analysis, for example when studying adsorption processes, it is obvious that such conditions are required.

Flow and monolayer time are not the only properties which change significantly as the pressure changes: other physical properties, for example thermal conductivity and viscosity of gases, are also strongly dependent on pressure. It is understandable therefore that the pumps needed for the production of vacuum in the different regions employ various physical methods, as do the vacuum gauges which are applicable to the measurement of these pressures.

A1.3 Production of a Vacuum

A1.3.1 TYPES OF PUMP

In order to reduce the number density of gas particles, and thereby the pressure, in a vessel, gas particles must be removed from it. This is the purpose of the vacuum pump. Vacuum pumps come in many shapes and sizes and with many different mechanisms of operation; however, fundamentally a distinction can be made between two classes of pump:

(i) Those which remove gas particles from the pumped volume and convey them to the atmosphere in one or more stages of compression. These are called compression or gas transfer pumps.

(ii) Those which condense or chemically bind the particles to be removed to a solid wall, which is often part of the vessel being pumped. These are called entrapment pumps.

Within these two classes there are subsections which further distinguish the method of operation of a pump. In the class of compression pumps there are:

(a) Pumps which operate by creating periodically increasing and decreasing chamber volumes.

(b) Pumps which transfer gas from a low pressure to a high pressure side in which the pump chamber volume is constant.

(c) Pumps in which the pumping action is due to diffusion of gases in a stream of high velocity particles.

In the class of entrapment pumps there are:

(a) Pumps which pump vapours by condensation or remove gases by condensation at very low temperatures.

(b) Pumps which bind or embed gases at extensive gas-free surfaces by adsorption or absorption.

To give detailed descriptions of all types of pump is beyond the scope of this book (for more details see Foundations of Vacuum Science and Technology, James M. Lafferty (Editor), John Wiley & Sons Ltd, 1998. [1]); however, an understanding of the operation of those most commonly found in surface science is essential to anyone wishing to work in this field and therefore an outline of the mechanisms of operation of these will follow.

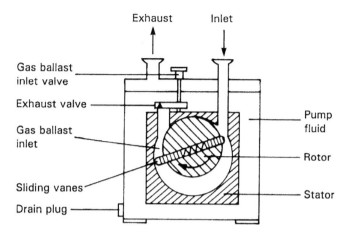

Figure A1.2   Cross-section of a single-stage sliding vane rotary pump

Examples of specific applications in the field of surface analysis are given in earlier chapters.

One of the most common types of pump found in vacuum technology is the rotary pump. These are mechanical pumps which belong to the group of compression pumps operating by creating a periodically increasing and decreasing chamber volume. These again come in a few different designs with the most common of these found in surface science being the rotary vane pump (see Figure A1.2). This consist of a cylindrical housing or stator in which rotates an eccentrically mounted, slotted rotor. The rotor contains vanes which are forced apart either by centrifugal force or, in some models, by springs. These vanes slide along the walls of the stator and thereby push forward the low pressure gas drawn in at the inlet and eject it finally at increased pressure through the outlet or discharge valve. The oil charge of the rotary vane pump not only serves for lubrication and cooling but also as the sealing medium, filling up dead space and any gaps in the pump. Rotary vane models come in single (as in Figure A1.2) and two-stage models (as is represented in Figure A1.3). Lower ultimate pressures can be produced by the two stage model, at the expense of cost and to a certain extent reliability.

There are several other types of rotary pump used in vacuum technology, for example, rotary piston pumps, trochoid pumps and the high pumping speed and lower ultimate pressure roots pumps. However, these are less common in surface analysis and for descriptions of these the reader should see reference [1].

An important part of the modern rotary pump is the **gas ballast**. If vapours are pumped by a rotary pump they can only be compressed to their saturation vapour pressure at the temperature of the pump. For example, if water vapour is pumped at a pump temperature of 70°C it can only be compressed to 312 mbar. On further compression the water

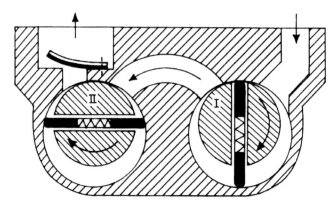

Figure A1.3 Cross-section of a two-stage sliding vane rotary pump: I is the high vacuum
stage and II is the rough vacuum stage

vapour condenses without increase in pressure. This is insufficient pressure
to open the discharge valve of the pump and the water (in liquid form)
remains in the pump and emulsifies with the pump oil. As a result of this
the lubricating properties of the pump decrease rapidly and the overall
performance of the pump is affected. To overcome this problem the gas
ballast device, as developed by Gaede in 1935, can be used, preventing
possible condensation of water vapours in the pump. This device works as
is shown in Figure A1.4. Before the compression action can begin, an exactly
regulated quantity of air, namely the 'gas ballast', is let into the pump's
compression stage. Now the vapours can be compressed with the gas ballast
before their condensation point is reached and they can be ejected out of the
pump.

   The main application of such rotary pumps is to achieve pressures
in the rough and medium vacuum regions or to act as backing pumps,
removing the gas compressed by high vacuum pumps which will then
achieve pressures in the high and ultra-high vacuum regions.

   Probably the dominant pump used for achieving pressures in the high
and ultra-high pressure regions in surface analysis instrumentation is the
turbomolecular pump. This kind of pump falls into the classification of
compression pumps which transport gas from a low pressure to a high
pressure region where the chamber volume is constant. A sectional drawing
of a typical turbomolecular pump is shown in Figure A1.5 The principle
of operation of molecular pumps has been well known since 1913 and
depends on the fact that the gas particles to be pumped will receive,
through impact with the rapidly moving surface of a rotor, an impulse in
the required flow direction. Early molecular pumps which used rotor blades
simply in the form of discs, suffered from constructional difficulties and a
high susceptibility to mechanical failure. More recently the blading of the

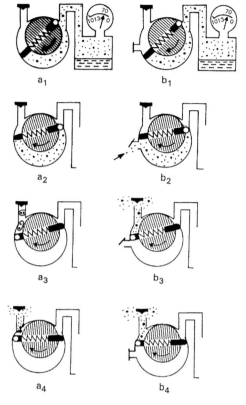

Without gas ballast

a₁) The pump is connected to the vessel which is already almost empty of air (approx. 70 mbar). It must therefore transport mostly vapour particles. It works without gas ballast.

a₂) The pump chamber is separated from the vessel. Compression begins.

a₃) The content of the pump chamber is already so far compressed that the vapour condenses to form droplets. Overpressure is not yet reached.

a₄) The residual air only now produces the required overpressure and opens the discharge outlet valve. But the vapour has already condensed and the droplets are precipitated in the pump.

With gas ballast

b₁) The pump is connected to the vessel which is already almost empty of air (approx. 70 mbar). It must therefore transport mostly vapour particles.

b₂) The pump chamber is separated from the vessel. Now the gas-ballast valve opens, through which the pump chamber is filled with additional air from outside. This additional air is called "gas ballast".

b₃) The discharge outlet valve is pressed open; particles of vapour and gas are pushed out. The overpressure required for this to occur is reached very early because of the supplementary gas-ballast air, as at the beginning of the whole pumping process. Condensation cannot occur.

b₄) The pump discharges further air and vapour.

Figure A1.4 Illustration of the pumping process in a rotary vane pump, without (left) and with (right) a gas ballast device when pumping condensable substances

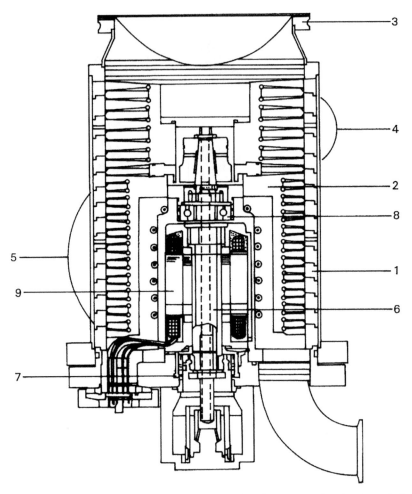

Figure A1.5 Cross-sectional picture of a turbomolecular pump of a single-ended axial flow design: 1, strator blades; 2, rotor body; 3, intake flang; 4, blades of the suction stage; 5, blades of the compression stage; 6, drive shaft; 7 and 8, ball bearings; 9, high-frequency motor

rotors in molecular pumps was constructed in the form of a turbine, which allowed easier construction and greater reliability of operation, and in this form has developed into the turbomolecular pump of today (as shown in Figure A1.5).

The gas to be pumped arrives directly through the aperture of the pump inlet flange, giving maximum possible conductance. The top blades on the rotor, the so-called 'vacuum stage', are of large radial span, allowing a large annular inlet area. The gas captured in the high vacuum stages is transferred to the lower 'compression stages' which have blades of shorter radial span; here the gas is compressed to the backing pressure. From here

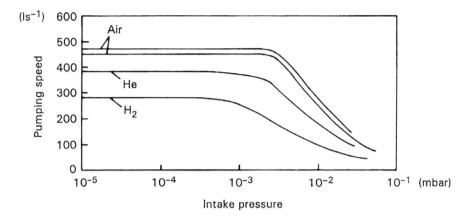

Figure A1.6 The pumping speed characteristic curves for a nominally 450 l s^{-1} turbo-molecular pump for different gases

it is removed by a backing pump, which is most commonly a rotary vane pump.

The pumping speed characteristics of turbomolecular pumps are shown in Figure A1.6. The pumping speed is constant over the whole working pressure range. It decreases, however, at intake pressures greater than 10^{-2} mbar, where the transition between molecular flow and laminar viscous flow takes place. Although a turbomolecular pump backed by a rotary pump can pump a chamber directly from atmosphere, at pressures above 10^{-2} mbar, due to the viscous nature of the gas, the pump will be operating under strain and should therefore be only exposed to these pressures for a short period of time.

Figure A1.6 also shows that the pumping speed very much depends on the type of gas. Due to its high pumping speed for high mass hydrocarbon molecules, a turbomolecular pump can be fitted directly to a vacuum chamber without any need for cooled baffles or traps. When these pumps are switched off, however, they must be vented to atmosphere, or oil from the pump and backing pump will be sucked back into the vacuum system. If venting fails during shut-down or does not operate correctly, then oil vapours can get through into the vacuum chamber. To impede this isolation, valves are often fitted between these pumps and the vacuum chamber so that the pump can be vented independently of the chamber.

In recent times, some sub-species of the turbomolecular pump have been developed for specific applications. Of these the most used is a turbo-molecular pump fitted with magnetically suspended blades – the so-called 'Maglev' pump – these have the advantages of very low vibration levels for imaging applications and lower oil vapours than conventional turbomolecular pumps.

Another common pump used often in surface science for achieving high and ultra-high vacuum is the **diffusion pump** although these have been superseded in recent years by the turbomolecular pump. These come under the classification fluid entrainment pumps where the pumping action is due to the diffusion of gases in a stream of high velocity particles. In these pumps the pumped gas molecules are removed from the vessel into a fast moving stream of pump fluid, most often in vapour form (most typically oil or mercury). This conveys the gas in the pumping direction by impact; thereby the pumped gas is transported into a space at higher pressure. The pump fluid itself after leaving the nozzle in the form of a vapour, condenses on the cooled outer walls of the pump.

The are several types of fluid entrainment pumps used depending on the pressure conditions required; the most common in surface science, however, are low vapour density oil diffusion pumps, with a working pressure region below 10^{-3} mbar. A schematic diagram showing the mode of operation of such a pump is shown in Figure A1.7 These pumps consist basically of a pump body with a cooled wall and a three- or four-stage nozzle system. The low vapour density oil is in the boiler and is vaporized here by electrical heating. The oil vapour streams through the chimneys and emerges with supersonic speed through the different nozzles. The so-formed jet of oil vapour widens like an umbrella until it reaches the walls of the pump where it condenses and flows back in liquid form to the boiler. Diffusion pumps have high pumping speed over a wide pressure range; it is also practically constant over the whole region lower than 10^{-3} mbar (data for this are presented in Figure A1.8).

The cooling of the walls of diffusion pumps is critical to their operation and almost all large diffusion pumps are water cooled, but some smaller ones are air cooled. Thermally operated protection switches or water flow switches are often fitted to diffusion pumps which will switch off the pump heater if there is a cooling failure.

Since a certain pressure is needed in the pump before the vapour will form, the pump providing the backing (most typically a rotary pump) must have attained a certain pressure before the heater can be switched on, a backing pressure of typically 10^{-1} or 10^{-2} mbar. Because of this it is important that these pumps are protected against running with a pressure in the backing stage which is too high, and they should be fitted with a vacuum trap from the backing pressure gauge which operates such that if the backing pressure exceeds 10^{-1} mbar the pump heater will switch off. The failure to fit such a trip can have serious consequences to the pump, because if the pump runs hot with too high a backing pressure, the pump will stall and backstreaming of oil vapours into the vacuum system will occur. If this condition continues for a period of a few tens of minutes, the oil in the pump can pyrolize and is then useless; this can only be

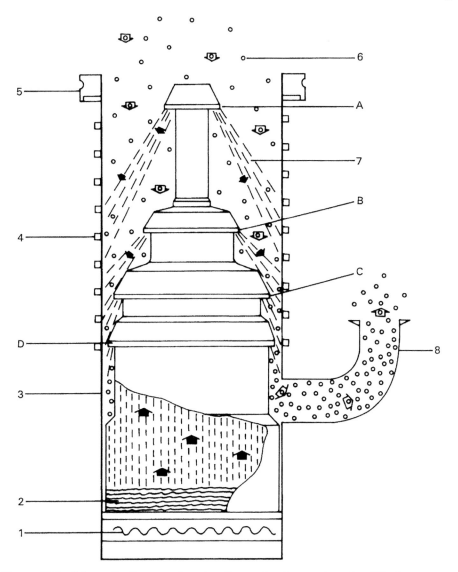

Figure A1.7 Schematic diagram showing the mode of operation of a diffusion pump: 1, heater; 2, boiler; 3, pumping body; 4, cooling coil; 5, high vacuum flange connection; 6, gas particles; 7, vapour jet; 8, backing vacuum connection port: A–D, nozzles

remedied by completely stripping down the pump and putting in a new fill of oil.

For the lowest ultimate pressure from such pumps, backstreaming of the pump fluid into the vacuum chamber should be reduced as far as possible. For this purpose, therefore, cold traps are used which are cooled with liquid nitrogen so that the temperature of a baffle between the pump and the

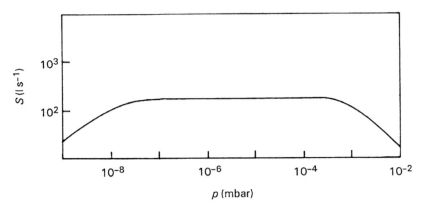

Figure A1.8 Characteristic pumping speed curve for a nominally $150 \, l \, s^{-1}$ diffusion pump (without cryo-trap)

vacuum chamber is maintained at $-196°C$. The vapours will condense on the baffle at these temperatures and therefore will be effectively removed from the vacuum system until the temperature of the baffle rises.

A large classification of pumps which are less generally applicable than those mentioned previously but none the less have areas of application in surface science are sorption pumps. Within the concept of sorption pumps we include all arrangements where the pumped gas particles are removed from the space by being bound to a surface or embedded in the surface. This process can either be on the basis of temperature dependent adsorption by the so-called 'van de Waals forces', chemisorption, absorption or by the particles becoming embedded in the course of the continuous formation of a new surface. By comparing their operation principles we can therefore further distinguish between adsorption pumps in which the sorption of gas takes place by simple temperature dependent adsorption processes and getter pumps where the sorption and retention of gases is due to the formation of chemical compounds.

Adsorption pumps work by the adsorption of gases on the surface of molecular sieves or other materials. Materials are chosen which possess an extraordinarily high surface area for a given mass of the material. This can be of order of $10^3 \, m^2 \, g^{-1}$ of the solid substance. A typical such material is Zeolite 13X. In practice the pump is connected to the vacuum chamber through a valve and it is only on immersing the pump in liquid nitrogen that the sorption effect becomes useful and the valve can be opened. These pumps are most often used for the clean roughing of a system prior to the use of a high vacuum pump such as a cryo or ion pump.

The pumping speeds of such pumps are very dependent on the gases being pumped; typically the best values being achieved for nitrogen, carbon dioxide, water vapour and hydrocarbon vapours, where light noble gases are hardly pumped at all. For pumping down a vessel which contains atmospheric air, where light noble gases are only present at a few ppm, a pressure of $<10^{-2}$ mbar can be obtained without difficulty. After the pumping process, the pump has to be warmed to room temperature for the adsorbed gases to be given off and the zeolite be regenerated for use.

One of the more common types of sorption pumps used in surface science are sublimation pumps. These are a form of getter pump in which the getter material is evaporated and deposited on the inner wall of the vacuum chamber as a getter film. Titanium is almost exclusively the getter material in sublimation pumps and is evaporated from a wire of high titanium content alloy by resistance heating. Particles from the gas which impinge on the getter film become bound to it by chemisorption and form stable compounds with the titanium, which have immeasurably low vapour pressures.

These pumps are not used continuously but are switched on in short bursts of a few minutes and are often controlled by a timer such that they will come on at regular intervals, say typically every few hours. As such, these pumps are used to supplement the pumping from other pumps on the system; their incredibly high pumping speeds for active gases means that they can be used periodically to keep a system clean from such gases or where a sudden evolution of such gases needs to be pumped away quickly.

A similar pump to the sublimation pump but one which can be used continuously is the sputter ion pump. The pumping action of these pumps is due to two processes:

(i) Ions impinge on the cathode surface of a cold cathode discharge system and sputter material off it (the material of the cathode is again titanium). The titanium deposits on surfaces in the pump and acts as a getter film as before.

(ii) The energy of the ions is high enough on sputtering incidence for them to become deeply embedded in the cathode where they are in essence adsorbed by ion implantation. The process can pump all types of ions, including rare gases.

The ions are created in a Penning gas discharge between the cathode and the anode in the pump. In the pump there are two parallel cathodes made of titanium between which are arranged a system of cylindrical anodes made of stainless steel (see Figure A1.9). The cathode is maintained at high negative electrical potential, of the order of a few kV, with respect to the anode. The whole electrode assembly is maintained in a strong homogeneous magnetic field (flux density ≈ 0.1 T). Electrons near the anode are trapped in the high magnetic field and set up a region of high electron density in the anode cylinders ($n_e \approx 10^{13}$ cm^{-3}). Here the electrons will collide with gas particles in the system and ionize them. Due to their greater mass, these ions will remain relatively unaffected by the magnetic field and will be accelerated

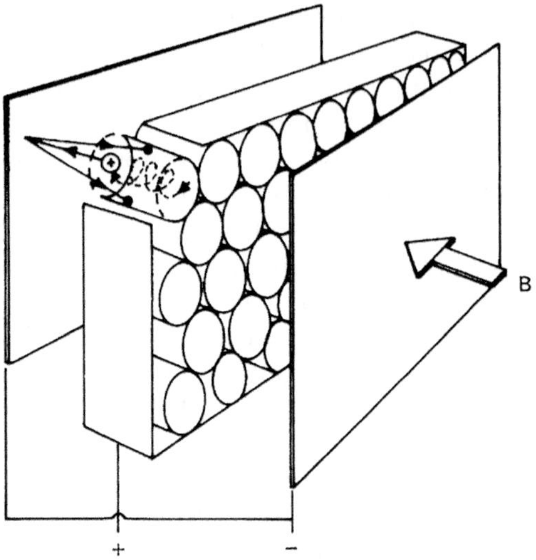
B
+
–
O Direction of motion of the ionized gas particles
→● Direction of motion of the sputtered titanium
--→ Direction of motion of the electrons

Figure A1.9 Schematic showing the mode of operation of a sputter ion pump (diode type)

toward and impinge on the cathode. At pressures of 10^{-3} mbar or below such a discharge is self-sustaining and does not require a hot cathode. The discharge current will obviously be proportional to the number density of gas particles in the system and therefore such a pump can also be used to measure the gas pressure in a system.

The pumping speed of sputter ion pumps will depend on the pressure and on the type of gas being pumped. For air, N_2, CO_2 and H_2O the pumping speeds are practically the same; compared with these gases, the pumping speeds for other gases are about:

Hydrogen	150–200 %
Methane	100 %
Other light hydrocarbons	80–120 %
Oxygen	80 %
Argon	30 %
Helium	28 %

These pumps will only operate at pressures below 10^{-2}–10^{-3} mbar, but for them to work all that is required is that a high electric potential is supplied to the pump. One problem with these pumps can be encountered when pumping large volumes of inert gases because as the pump operates, the buried layers of these gases can be re-exposed by the sputtering of the cathode, and therefore the pump can re-emit these gases back into the system. More complicated designs of ion pump reduce this problem.

Care must be taken, when using these pumps on your vacuum system, that stray magnetic fields from the pump do not interfere with the rest of the operation of your system. This may be a particular problem in electron spectroscopy-type apparatus. Also stray ions can escape from the pump and care must be taken to put shielding in your system to prevent these ions from interfering with any experiments being undertaken.

Cryopumps are a different type of sorption pump used frequently in MBE systems or ion implanter systems and to some extent in surface science. They consist inherently of a surface cooled down to a temperature of <120 K so that gases and vapours condense or bond to this surface; the cooled surface is situated in the vacuum itself.

The liquid nitrogen cooled vapour traps used frequently with oil diffusion pumps (mentioned previously) are in themselves a type of cryopump. If cooling media of even lower boiling points are used, for example liquid helium with a boiling point of 4.2 K, then gases like O_2, N_2 and H_2 can also be pumped. In order to use such a pump to attain high and ultra-high vacuum conditions the cryosurfaces have to be cooled down to below 20 K.

Various mechanisms are effective when bonding gases to cold surfaces. As well as condensation, cryotrapping and cryosorption also play a part. In order for a cryopump to pump by condensation, the vapour pressure of the solid condensate has to be significantly lower than the working pressure one wishes to achieve. If a working pressure of 10^{-9} mbar is required, then for gases like air, O_2 and N_2 this requires temperatures below 20 K; for gases like Ne and He, however, this requires the lower temperature only attainable with liquid helium. Hydrogen is of particular interest since it constitutes the major part of the residual gas composition in high and ultra-high vacuum environments. This is very difficult to condense out of a system requiring temperatures of 3.5 K. Therefore to remove hydrogen from a system with such a pump, another mechanism of pumping must be utilized.

One such mechanism which is used for the 'condensation' of difficult gases is 'cryotrapping'. Here a gas which is easily condensable is let into the system such that a mixture of gases is produced. Typical gases which are used are Ar, CO_2, CH_4, NH_3 and other heavier hydrocarbons. The condensate mixtures produced have vapour pressures several orders of magnitude lower than that of pure hydrogen. This mechanism is automatically present when cryopumping any gas mixture and is of course not confined to the pumping of hydrogen.

Another such mechanism used in the pumping of difficult gases is cryosorption. Here a pre-introduced sorption material, for example, activated charcoal, is cooled and the gases adsorb on this. This has the advantage that no continuous admittance of trapping gas into your system is required.

In principle, a cryopump could be switched on at atmospheric pressure; however, in practice this would lead to the creation of a very thick layer of condensate on the pump at the beginning of the pump-down process, considerably reducing the pumping capacity of the pump. General practice when high and ultra-high vacuums are required, is to rough pump out the chamber to a pressure of $\approx 10^{-3}$ mbar before switching on the cryopump.

In practice the mechanism for achieving the cold surface can be different. For example, there are liquid pool cryopumps, continuous flow cryopumps and refrigerator cryopumps (see reference [1] for more information) but for all, the principle of pumping is the same.

Cryopumps suffer from problems with vibration and are therefore generally not used on systems where imaging applications are being used.

A1.3.2 EVACUATION OF A CHAMBER

One of the first considerations when designing a pumping system is what size of pump is required. If you pick too large a pump you will waste money; too small a pump and you will not achieve the required conditions for your experiment. Basically, two questions arise when choosing the size of pumping system:

(i) How large must the effective pumping speed of the system be so that the pressure will be reduced to the desired value in a given time?

(ii) How large must the pumping speed be so that gases released into the vacuum system during operation can be pumped away quickly enough so that the required pressure is not exceeded?

This brings us to the idea of the effective pumping speed, S_{eff}, of the system. This is understood to be the pumping speed of the whole pump arrangement that actually prevails at the vessel, taking into account the conductances of the components between the pump and vessel; for example, valves, apertures baffles, coldtraps, etc. If the conductances of these components are known, and the actual pumping speed of the pump itself is known (this is referred to as the nominal pumping speed, S, of the pump) then the effective pumping speed can be determined. The relationship between the effective pumping speed and the nominal pumping speed of a system is given by the formula:

$$\frac{1}{S_{\text{eff}}} = \frac{1}{S} + \frac{1}{C}$$

where C is the the total flow conductance of the tubulation between the pump and the vacuum chamber (baffles, pipes, etc.).

This can be broken down into the conductances of the individual components as:

$$\frac{1}{C} = \frac{1}{C_1} + \frac{1}{C_2} + \cdots \frac{1}{C_N}$$

The conductances of simple tubes can be calculated in the different pressure regimes (see reference [1]) but the conductances of geometrically more complicated components such as cold traps, valves and baffles must usually be determined experimentally.

A1.3.3 CHOICE OF PUMPING SYSTEM

The choice of pumping system will be dependent on the processes which are to be undertaken in the vacuum system and also on the available budget. In general, in vacuum technology the choice of pumping is initially governed by whether these processes fall into the categories of wet or dry processes. Dry processes are those in which there are no significant amounts of vapour to be pumped, whereas wet processes will evolve a significant amount of water vapour which must be pumped away.

In surface science we are almost solely concerned with dry processes and therefore, in this book, limit ourselves to discussing pumping systems relating only to such processes. In most surface science applications, the required vacuum is produced prior to the experimental measurement. Furthermore, in such systems the degassing of the vacuum system itself and the components in it is a critical stage.

When working pressures in the rough and medium vacuum regimes are needed, rotary vane pumps are often ideal. They are especially suitable for pumping down vessels from atmospheric pressure to pressures below 0.1 mbar, in order to work continuously in this lower pressure region. A very common need for medium vacuum is when evacuating a vessel to a pressure such that other high and ultra-high vacuum pumps can be used and subsequently as a backing pump for such pumps. Here two-stage (ultimate pressure $= 10^{-3}$ mbar) rotary vane pumps are ideal.

Pressures in the high vacuum and ultra-high vacuum (UHV) regions can be achieved using diffusion pumps, sputter ion pumps, turbomolecular pumps and cryopumps, all fitted with a suitable roughing or backing pump and often used in conjunction with sublimation pumps. Pumping alone, however, will not allow true ultra-high vacuum conditions to be fulfilled, since in this pressure region the major contribution to the pressure comes from gas evolved off the container walls. To achieve such pressures, therefore, the whole vacuum chamber must be baked, whilst pumping it, to temperatures of about 250–350°C in order to desorb the gases off the walls of the chamber and allow them to be pumped away. UHV chambers, therefore, are almost invariably made of stainless steel and are fitted with all-metal seals (described in Section A1.3.5). When a system has been made leak-tight and leaked test with a helium leak detector, then baking is undertaken. This can extend over several hours or even days. Before the system is allowed to cool fully all components in the system which may desorb gases during their operation must be degassed. This will include hot cathodes or filament assemblies in the system, for example sublimation pump filaments, or filaments in ion or electron guns.

A1.3.4 DETERMINATION OF THE SIZE OF BACKING PUMPS

The size of a backing pump is determined by the fact that the quantity of gas or vapour transported by the high vacuum pump must also be handled by the backing pump such that the maximum permissible backing pressure is not exceeded. If Q is the quantity of gas pumped by the high vacuum pump, with an effective pumping speed S_{eff} at an inlet pressure p_A, this quantity of gas must also be transported by the backing pump of speed S_V and backing pressure p_V. Therefore:

$$Q = p_A S_{eff} = p_V S_V$$

and thus the minimum pumping speed of the backing pump can be calculated to be:

$$S_V = \frac{p_A}{p_V} S_{eff}$$

A1.3.5 FLANGES AND THEIR SEALS

Demountable joints in metallic vacuum components are invariably provided with flanges which are sealed by means of a gasket which is compressed or deformed in some way as the flange is tightened and the seal made. Commonly used flange sizes up to 200 mm outer diameter are built to internationally standardized dimensions and come in several different sizes.

Flanges which are suitable for use in the rough, medium and high pressure regions are usually made from a black rubber type material called Viton. Such seals can be used for pressures down to 10^{-8} mbar and are

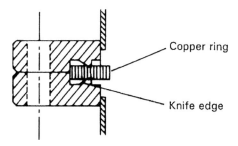

Figure A1.10 A cross-section of a 'Conflat$^{\circledR}$' (or knife-edge seal) flange

bakeable to $\approx 200\,°\mathrm{C}$. For true UHV pressures, the seals are made from metal and all components should be bakeable to $350\,°\mathrm{C}$. The most common type of flange is the so-called 'knife edge' or 'ConFlat$^{\circledR}$' (Varian) flanges which are sealed with copper gaskets. A cross-section through such a flange and gasket is shown in Figure A1.10. On larger or non-circular flanges, other metal seals which use a loop of soft metal, typically gold, indium or aluminium, are sometimes used. The seal is made by simply compressing the soft metal loop between the two flat metal surfaces of the flange. Great care must be taken with all sealing surfaces in a UHV system to prevent scratching or damage which will impair the leak-tightness of the seal.

A1.4 Measurement of Low Pressures

In modern vacuum technology, pressure measurements have to be made over a range of 16 orders of magnitude from 10^3 mbar to 10^{-13} mbar. It is impossible on fundamental physical grounds to build a gauge which will measure quantitatively over this whole pressure region. Therefore a series of gauges have been developed which have characteristic pressure ranges, typically extending over a few orders of magnitude.

The types of gauge fall into two categories: those which measure the pressure directly, that is those which measure the pressure in accordance with its definition as the force which acts on a unit area, and those which measure it indirectly where the pressure is measured as a function of a pressure-dependent property of the gas (for example, thermal conductivity, ionization probability, electrical conductivity). Only in the case of the direct or absolute pressure measurement is the reading independent of the nature of gas (in accordance with Equation (A1.6)) it will, however, be dependent on the temperature. In the case of indirect pressure measurement the properties measured are almost invariably dependent on the molar mass of the gas studied as well as the pressure. Consequently such a pressure reading will be dependent on the nature of the gas. On such gauges the scale is always calibrated with air or nitrogen as the test gas. For other gases, correction factors must be used.

The measurement of pressure in the rough vacuum region can be undertaken by gauges with direct pressure measurement, and the pressure can be determined to quite a high level of accuracy. Measurement in the lower pressure regimes, however, is invariably undertaken by indirect methods. This usually means that the accuracy of the measurement is limited by certain fundamental errors. These inaccuracies affect the measurement to such a degree that to make a pressure measurements in the medium and high vacuum regions with an error of less than 50 % would take special care from the experimenter. The inaccuracies are even more severe in the ultra-high vacuum region, such that to achieve measurements in these low pressure regimes which are accurate to within a few % requires special measuring equipment and great care. For this reason the reliability of pressure readings in an experiment has to be treated with careful consideration.

Furthermore, if one wants to make a statement about the pressure in a vacuum chamber recorded by a gauge, firstly, the location of the gauge has to be taken into account. For example, in the pressure regions where laminar flow prevails in the system, a pressure gradient due to the pumping process will be present in the system such that a gauge which is situated near the inlet to a pump will record a lower pressure than that which is actually present in the chamber. Also, in this pressure region the conductances of the tube in the chamber can introduce pressure gradients and lead to false readings. In the high and ultra-high vacuum regions the situation is even more complicated, where outgassing of the walls of the chamber and of the gauge itself can have a significant effect on the accuracy of the reading.

A1.4.1 GAUGES FOR DIRECT PRESSURE MEASUREMENT

Gauges for direct pressure measurement work on mechanical principles, measuring the force which the gas particles exert on a surface by virtue of the thermal velocities. This is usually achieved by measuring the displacement of an interface (solid or liquid) between a region at the pressure to be measured and a region at a certain known reference pressure, sometimes the reference is atmospheric pressure.

A large category of mechanical vacuum gauges are diaphragm gauges, the best known of these being the barometer. This contains a hermetically sealed evacuated thin-walled capsule made from a copper–beryllium alloy. This capsule is evacuated to the reference pressure. The external side of the capsule is connected to the vacuum vessel, and as the pressure is reduced, the diaphragm moves outwards. This movement is transfered by a lever system to a pointer which indicates the pressure on a linear scale. Such a reading, due to the sealed reference pressure is independent of atmospheric pressure. A typical type of modern diaphragm gauge is shown in Figure A1.11. Such

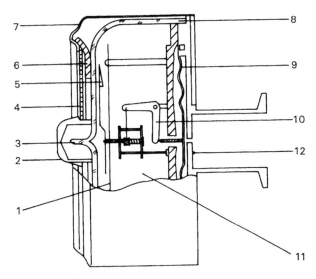

Figure A1.11 A section through a modern diaphragm vacuum gauge: 1, reflecting cover plate; 2, protective cap; 3, seal-off point; 4, Plexiglass disc; 5, pointer; 6, scale; 7, metal cover; 8, glass chamber; 9, diaphragm; 10, lever system; 11, reference vacuum; 12, connecting flange

a pressure gauge will read from atmospheric pressure down to a few mbar with an accuracy of about ±10 mbar. If accurate readings are needed in the range below 50 mbar, then a gauge with the reference capsule evacuated to below 10^{-3} mbar is superior. Such a gauge can measure pressures in the range 100–1 mbar with an accuracy of 0.3 mbar. Such gauges are quite sensitive to vibrational interference. In a modern gauge the movement of the diaphragm will often be linked to an electrical transducer and then the pressure can be displayed digitally on a panel meter.

One of the simplest but most exact ways of measuring the pressure in the rough vacuum region is the mercury manometer. Here the evacuated limb of the U-tube is maintain at a constant pressure equal to the vapour pressure of mercury at room temperature (10^{-3} mbar). The other limb of the tube is connected to the vacuum vessel. From the difference in levels of the two columns of mercury, the pressure can be directly determined in mbar. The major drawback of such gauges for frequent application is their size and vulnerability to breakage.

A compression type of manometer developed by McLeod in 1874 still has important applications today. The measurement of pressure in such gauges results from the fact that a quantity of gas which occupies a large volume at the pressure which is to be determined is then compressed by trapping it behind a column of mercury. This increased pressure can then be determined in the same way as with the conventional mercury manometer. The absolute pressure in the vessel can then be determined knowing the volume of the enclosed gas and the total volume of the gauge (for more detail of such gauges see reference [1]). When operating these gauges it must be noted that the pressure reading is not continuous, but for every pressure measurement the mercury must be raised into the gauge. These gauges can provide an accurate determination of the absolute pressure to within $\pm 2\,\%$ in the rough and medium vacuum regime and can even provide readings extending into the high vacuum region, down to $\approx 10^{-5}$ mbar.

With these gauges however, as with any vacuum gauge involving a compression stage, the presence of condensable vapours can influence the pressure reading obtained. If vapours condense out in the gauge, further compression will not increase the pressure in the entrapped volume and a false reading will be given.

Direct pressure measurement is rarely used in surface analysis instrumentation; however, these types of measurement are used to calibrate other gauges or where higher accuracy is needed in the low vacuum region – for example, when pre-mixing gases prior to leaking them into the vacuum chamber.

A1.4.2 GAUGES USING INDIRECT MEANS OF PRESSURE MEASUREMENT

Gauges which measure the pressure indirectly, invariably make measurements on the gas of an electrical nature and convert this to a pressure reading. The apparatus consists of a gauge head, which is connected to the vacuum system and a control unit which is normally remote from the head.

One of the most common types of gauge used in vacuum science and indeed surface science is the thermal conductivity or Pirani vacuum gauge. These gauges utilize the variation of mean free path (a function of the number density of particles) and the corresponding variation in thermal conductivity of the gas to monitor the pressure. Such gauges are exploited extensively for measurement in the medium pressure region from 1 to 10^{-3} mbar.

The gauge head of a Pirani gauge has a sensing filament which is open to the vacuum chamber. A current is passed through the filament which produces heat. This heat can be transferred away from the filament by radiation or by thermal transfer to the surrounding gas. In the rough vacuum region, the rate of heat transferred away due to convection is almost independent of pressure. However, as the mean free path of the gas molecules becomes of the order of the diameter of the filament, the convection of heat away becomes strongly dependent on pressure. This continues to be so until the pressure reaches $\approx 10^{-3}$ mbar, where the dominant heat transfer process is radiation, which is independent of pressure.

In practice there are two methods used to measure the pressure in this way; those in which the sensing filament is of varying resistance and those in which the resistance of the filament is kept constant. In the first case the sensing filament in the gauge head forms one branch of a Wheatstone's bridge circuit. As the rate of heat transfer changes, so the temperature of the sensing filament changes; for example, if the pressure increases, the rate of heat transfer will increase, the temperature of the filament will decrease and its resistance will decrease, so the bridge becomes out of balance. The bridge current serves as a measure of the pressure which is indicated on a meter. In the second approach to the measurement, the sensing filament is also part of a Wheatstone's bridge; however, in this case the voltage applied to the filament is regulated so as to keep the resistance (and temperature) of the filament constant and the bridge always balanced. As the pressure changes, therefore, the applied voltage must change to compensate for the variation in heat transfer. In this case it is the applied voltage which is a measure of the pressure.

The Pirani gauge of variable resistance can only cope with pressures in the range $10-10^{-3}$ mbar, whereas the constant resistance type can operate in the range 10^3-10^{-3} mbar. Such gauges have an accuracy of typically $\pm 10\%$. Due to their pressure range characteristics and their relatively robust nature, these types of gauge are found extremely frequently in surface science, and are most frequently used for monitoring the pressure above a rotary pump. Common examples of this are the monitoring of the backing pressure for a high vacuum pump or indication of the pressure in a gas inlet system. These gauges are always calibrated for nitrogen or for air, but for other small molecular mass species the reading is within the inherent error of the gauge. For large organic molecules, however, the error is increased and can become significant, especially in the low pressure region below 10^{-2} mbar. Pirani gauges are most frequently used to monitor the low–high vacuum in the backing stages of high vacuum pumps, where they will also activate safety procedures if the pressure in this stage gets too high.

The most common gauges for measuring the pressure in the high and ultra-high vacuum regions are ionization gauges. These gauges measure the pressure in terms of the number density of molecules in the gas. A portion of the atoms or molecules in the gas are ionized by electron impact and the positive ions thus produced are collected by an electrode in the system and the resulting current is measured. There are two basic types of these gauges distinguishable by the method of generation of the ionizing electrons.

In the gauge head of the so-called cold cathode or Penning or inverted Magnetron gauge there are two unheated electrodes – the cathode and the anode – between which a self-sustaining discharge is excited by applying a DC voltage of about 2 kV across the electrodes. The discharge is maintained by the application of a strong magnetic field perpendicular to the lines of electric field such that the electrons in the discharge have long spiralling paths and subsequently a high probability for collision with gas particles. The positive ions generated in the discharge will travel to the cathode and the pressure is measured by monitoring the generated discharge current which is indicated on a meter. Since the ionization cross-section is a function of the gas species being ionized, the pressure reading will be gas-dependent. There is an upper limit on the measuring region of $\approx 10^{-2}$ mbar; this is due to the fact that above this pressure a glow discharge will be generated in the gauge head and in this region the discharge current will be far less dependent on pressure. Although these gauges have a reading limit at 10^{-2} mbar, it is generally safe to operate them at pressures of up to atmospheric pressure, this being one of the great advantages of such gauges, especially where the application requires that the system is frequently let up to atmospheric pressure and then pumped to the high vacuum region again. Penning ionization gauges have a lower pressure measurement limit of the order of 10^{-8} mbar. Problems can arise from stray magnetic fields from such gauges in techniques which are very sensitive to magnetic fields such as low energy electron spectroscopies.

It can be seen that such gauges work on very similar principles to sputter ion pumps and they therefore have a self-pumping effect with a pumping speed of order of $10^{-2} \, l \, s^{-1}$. This effect leads to quite high inaccuracies in the readings from such gauges, up to $\approx \pm 50\%$. Despite this, Penning gauges are frequently found in surface science.

Hot cathode ionization gauges are one of the most common gauges found in surface science, since they are the only commercially available gauges for measuring pressures in the ultra-high vacuum region. Such gauges use a hot cathode (or filament) as the source of the ionizing electrons. In the gauge head of these gauges there are three electrodes, the cathode or filament, the anode and an ion collector (a schematic representation of such an assembly is shown in Figure A1.12). When the filament is heated by passing an electric current through it, it will emit electrons by thermal emission; such a cathode is a very abundant source of electrons. These electrons are then accelerated in an electric field between the cathode and the anode. The anode is in the form of a grid such that a high fraction of the emitted electrons pass through it. The electrical potentials applied to the anode and the cathode are such that the electrons have sufficient energy to ionize gas particles in the system on collision. Gas particles which are ionized on the far side of the anode from the filament will be attracted towards the ion collector which is at a negative potential with respect to the anode. It is this ion current collected that is proportional to the number density of particles in the system and is therefore expressed as a pressure. The abundance of electrons from the filament source means that no magnetic fields are required in these types of gauges and therefore the electrode assembly of the gauge head can be placed directly in the vacuum system, generally without causing interference with other components in the vacuum system. Only where the technique being used is sensitive to stray electrons or light is this not the case.

Except for the case of specially designed gauge heads, the upper limit on pressure measurement from hot cathode ionization gauges is $\approx 10^{-2}$ mbar;

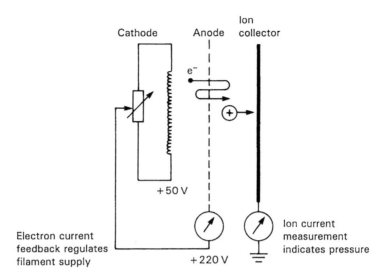

Figure A1.12 Schematic of a hot cathode ionization gauge showing typical operating voltages

above this pressure a glow discharge will form in the region of the electrodes which prevents operation of the gauge. Operation above this pressure will also ultimately result in filament burn out. The limit on the low pressure range of measurement of $\approx 10^{-12}$ mbar is due to two effects, the so-called X-ray and ion desorption effects.

The X-ray effect is caused by the electrons which impinge on the anode. This will emit soft X-ray photons, which may then strike the ion collector causing the emission of further electrons from it. This electron current from the ion collector will be indistinguishable from an ion current flowing to the collector and will simulate a high pressure reading. Also the emitted photons can collide with the walls of the vacuum chamber surrounding the gauge head and so produce electrons. If the electrical potentials in the system are such that these electrons can travel to the ion collector, then they will cause an electron current to flow which will simulate a lower pressure reading. The scale of these effects will be dependent on the anode voltage and ion collector voltages with respect to their surroundings and the surface area of the ion collector.

When electrons impinge on the anode they can cause the desorption of gas species from the surface of the anode, often as positive ions. These emitted ions will travel to the ion collector leading to a falsely high pressure reading – this is the so-called ion desorption effect. The magnitude of this effect will generally be independent of pressure but will to a point increase with increasing emission current. At small emission currents the effect will increase proportionally with current but as the current increases further the process will have the effect of cleaning up the anode and therefore further increases of current will result in a decrease in the effect.

A schematic representation of an ion gauge head is shown in Figure A1.12. The cathode (or filament) is generally made of tungsten. The electrons oscillate through the anode grid giving them long flight paths so increasing the probability for ions being created.

To ensure a linear relation between ion current and pressure the X-ray effect must be minimized. To this end, Bayard and Alpert designed a gauge head where the electrode assembly is such that the hot tungsten cathode lies outside the cylindrical anode grid, with the ion collector being a thin wire, with therefore minimum surface area, on the axis of the electrode system. With this design the X-ray effect is reduced by two or three orders of magnitude over that for the early types of gauge due to the great reduction in the surface area of the collector. These gauges can read pressures down into the ultra-high vacuum region of order of 10^{-10} mbar.

Other ion gauges have been designed for specialist areas of application such as the 'Bayard–Alpert gauge with modulator' and the 'Extractor' type gauge. The coverage of these, however, is beyond the scope of this book but a more thorough description can be found in reference [1].

The pressure measurements from all types of ionization gauge will again be a function of the gas being measured because the ionization cross-section will change from species to species. These gauges are again calibrated for air or nitrogen and relative sensitivity factors are needed for accurate pressure determination for other gases. Although these correction factors are to a certain extent dependent on the specific gauge type used, a table of typical correction factors or different gases is presented in Table A1.2. If a gas other than nitrogen or air is predominant in the vessel, then the pressure reading has to be multiplied by the corresponding factor to give the correct pressure reading.

Hot cathode ion gauges exhibit a pumping action, but this is a considerably smaller effect than that for cold cathode gauges.

A1.4.3 PARTIAL PRESSURE MEASURING INSTRUMENTS

In various vacuum processes it is important to know the composition gas or vapour mixture, i.e. the partial pressures of the gases in the system. The different gases can essentially be distinguished from each other by their molecular masses. A partial pressure measuring device is therefore a sensitive mass analyser in which the measuring system has dimensions which are small enough for it to fit easily into a vacuum system. Also in the case of the measurement of partial pressures in the high and ultra-high vacuum regions they can be baked with the vacuum system they are installed within.

Table A1.2 Correction factors for ionization gauge head readings for different gases

Predominantly present	Factors for gauge reading related to	
	N_2	Air
He	6.9	6.04
Ne	4.35	3.73
Ar	0.83	0.713
Kr	0.59	0.504
Xe	0.33	0.326
Hg	0.303	0.27
H_2	2.4	1.83
CO	0.92	0.85
CO_2	0.69	0.59
CH_4	0.8	0.7
Higher hydrocarbons	0.1–0.4	0.1–0.4

A typical partial pressure measuring instrument (as represented schematically in Figure A1.13) generally consists of three components:

(i)   An ion source where the gas particles in the system are ionized, so that they can then be mass analysed and detected.

(ii) An ion separation system so that ions of different masses can be distinguished easily.

(iii) An ion collector to measure the ion current at each mass.

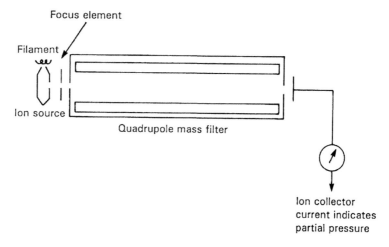

Figure A1.13   Schematic of a simple mass spectrometer for partial pressure measurement

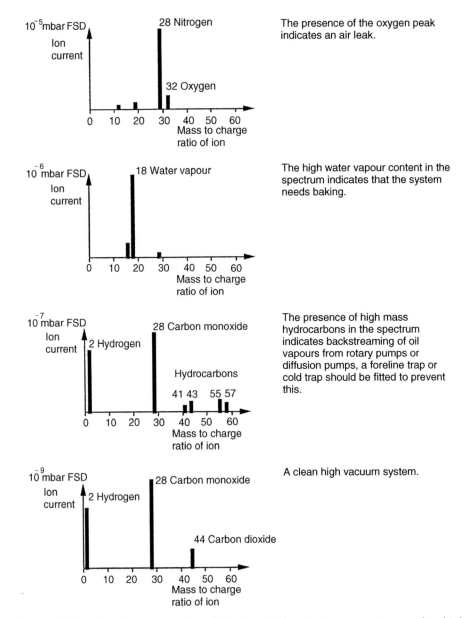

The presence of the oxygen peak indicates an air leak.

The high water vapour content in the spectrum indicates that the system needs baking.

The presence of high mass hydrocarbons in the spectrum indicates backstreaming of oil vapours from rotary pumps or diffusion pumps, a foreline trap or cold trap should be fitted to prevent this.

A clean high vacuum system.

Figure A1.14 Typical mass spectra of the four different stages to arrive at ultra-high vacuum conditions

These devices most commonly take the form of a small quadrupole mass filter fitted with a hot cathode ion source and a 'Faraday cup' detector, or when more sensitivity is required, a secondary electron multiplier detector. The mass analyser will typically have an operating range up to 100 amu and certainly no more than 300 amu and will allow the resolution of peaks 1 amu apart over this whole range.

The output of these devices is shown in spectral form representing a direct measure of the collected ion current or an equivalent as the mass analyser scans through the mass range required. In the interpretation of these spectra, it must be taken into account that different species of gas can have different detection probabilities. This is due not only to their different ionization probabilities, but also to variations in the transmission of the mass analyser as a function of mass and, in the case of an electron multiplier, the detectability as a function of mass. To add to this, molecular species, especially the higher mass hydrocarbons, can dissociate in the ionization process giving several peaks from one gas species.

Typical partial pressure spectral for the four different stages passed through, in achieving a clean ultra-high vacuum environment, are shown in Figure A1.14.

In Figure A1.14:

- The presence of the oxygen peak indicates an air leak.

- The high water vapour content in the spectrum indicates that the system needs baking.

- The presence of high mass hydrocarbons in the spectrum indicates backstreaming of oil vapours from rotary pumps or diffusion pumps; a foreline trap or cold trap should be fitted to prevent this.

- A clean high vacuum system.

Acknowledgement

Helpful discussions with Dr J. Gordon, Dr N. Aas and many other colleagues are gratefully acknowledged.

📖 References

1. James M. Lafferty (Ed). *Foundations of Vacuum Science and Technology*, John Wiley & Sons Ltd, 1998.

Appendix 2
Units, Fundamental Physical Constants and Conversions

A2.1 Base Units of the SI

The International System of Units, the SI (Système International (d'Unitès)), is the internationally agreed basis for expressing measurements at all levels and in all areas of science and technology [1]. There are two classes of units in the SI; base units and derived units. The seven base units of the SI and their base quantities provide the reference used to define all the measurement units of the International System. The seven base units are given in the table below.

Quantity	Base unit	Symbol
length	metre	m
mass	kilogram	kg
time, duration	second	s
electric current	ampere	A
thermodynamic temperature	kelvin	K
amount of substance	mole	mol
luminous intensity	candela	cd

Derived units are defined as products of powers of the base units and are used to measure derived quantities. For example area (m^2) and mass density (kg/m^3). For more information see the NPL website [2].

Surface Analysis – The Principal Techniques 2nd Edition Edited by John Vickerman and Ian Gilmore
© 2009 John Wiley & Sons, Ltd

A2.2 Fundamental Physical Constants

The fundamental physical constants, such as the speed of light, the Planck constant and the mass of the electron provide a system of natural units. The constants provide the link between the SI units and theory and also between scientific fields [2]. The values of the fundamental physical constants are taken from the recommended values of the constants which are produced by the CODATA Task Group on Fundamental Constants [3], based on a review of all the available data. The latest review is available at the CODATA fundamental physical constants webpage [4]. Selected fundamental physical constants and their values are given in the table below which is taken from the latest (2006) values of the constants [4]. The figures in parentheses in the 'Value' column represent the best estimates of the standard deviation uncertainties in the last two digits quoted. Where no uncertainty is given the value is exact through definition.

Quantity	Symbol	Value	Unit
speed of light in vacuum	c	299 792 458	m s^{-1}
magnetic constant	μ_0	$4\pi \times 10^{-7}$ $= 12.566\,370\,614\ldots \times 10^{-7}$	N A^{-2}
electric constant $\dfrac{1}{\mu_0 c^2}$	ε_0	$8.854\,187\,817\ldots \times 10^{-12}$	F m^{-1}
planck constant	h	$6.626\,068\,96(33) \times 10^{-34}$	J s
(in eV s)	h	$4.135\,667\,33(10) \times 10^{-15}$	eV s
elementary charge	e	$1.602\,176\,487(40) \times 10^{-19}$	C
electron mass	m_e	$9.109\,382\,15(45) \times 10^{-31}$	kg
(in u)		$5.485\,799\,0943(23) \times 10^{-4}$	u
proton mass	m_p	$1.672\,621\,637(83) \times 10^{-27}$	kg
(in u)		$1.007\,276\,466\,77(10)$	u
neutron mass	m_n	$1.674\,927\,211(84) \times 10^{-27}$	kg
(in u)		$1.008\,664\,915\,97(43)$	u
Avogadro constant	N_A	$6.022\,141\,79(30) \times 10^{23}$	mol^{-1}
atomic mass constant $m_u = \dfrac{1}{12} m(^{12}C) = 1\text{u}$	m_u	$1.660\,538\,782(83) \times 10^{-27}$	kg
molar gas constant	R	$8.314\,472(15)$	$\text{J mol}^{-1}\,\text{K}^{-1}$
Boltzmann constant R/N_A	k	$1.380\,6504(24) \times 10^{-23}$	J K^{-1}

Quantity	Symbol	Value	Unit
Molar volume of ideal gas RT/p $T = 273.15$ K $p = 100$ kPa	V_m	$22.710\,981(40) \times 10^{-3}$	$m^3\,mol^{-1}$

A2.3 Other Units and Conversions to SI

There are some particular issues in mass spectrometry that analysts should note. The unit of mass used is the 'unified atomic mass unit' which is based on the fundamental physical constant 'atomic mass constant' (see above). The atomic mass constant is defined as one twelfth of the mass, in kg, of a carbon-12 atom in its ground state.

The 'atomic mass unit', abbreviated 'amu', is an archaic unit for relative molecular or atomic mass based on the mass of an oxygen-16 atom. Its use is strongly discouraged [5, 6].

Mass spectrometers measure the mass-to-charge ratio, symbolized by m/z, where m is the mass in unified atomic mass units and z is the electrical charge measured in units of the integer number of elementary charges. Normally, the sign of the charge is not used. For mass spectra with multiply charged ions, for example, electrospray ionisation, then m/z should be used.

Other units that are in common usage, but are not SI, and their conversion to equivalent SI units are given in the table below.

Unit (not SI)	Symbol	SI value and unit	Comment
eV	eV	$1.602\,176\,487(40)$ $\times 10^{-19}$ J	
Unified atomic mass unit	u	$1\ m_u =$ $1.660\,538\,782(83)$ $\times 10^{-27}$ kg	See notes above. Accepted by the SI [6]
Dalton	Da	$1\ Da = 1\ m_u =$ $1.660\,538\,782(83)$ $\times 10^{-27}$ kg	Often used in biochemistry and molecular biology [7]. Accepted by the SI [6]
bar	bar	1×10^5 Pa (exactly)	See the NPL website [8]

Unit (not SI)	Symbol	SI value and unit	Comment
millibar	mbar	100 Pa (exactly)	See the NPL website [8]
torr	torr	101325/760 Pa	See the NPL website [8]

A very useful resource of physical and chemical constants may be found on the on-line edition of Kay and Laby from the 16th Edition published in 1995 [9].

References

1. International Bureau of Weights and Measures (BIPM) (http://www.bipm.org/en/SI).

2. NPL website. The SI Base Units (http://www.npl.co.uk/server.php?show=nav.364).

3. CODATA Task Group on fundamental constants (http://www.codata.org/about/index.html).

4. CODATA fundamental constants website (http://physics.nist.gov/cuu/Constants/index.html).

5. *Rapid Communications in Mass Spectrometry*, Guide for Authors (http://www3.interscience.wiley.com/journal/4849/home/ForAuthors.html).

6. ISO 18115-1 'Surface chemical analysis – Vocabulary – Part 1: General terms and terms for the spectroscopies', Term 5.480.

7. *IUPAC Gold Book* (http://goldbook.iupac.org/D01514.html).

8. NPL website. Pressure Units (http://www.npl.co.uk/server.php?show=ConWebDoc.401).

9. Kaye and Laby, *Tables of Physical and Chemical Constants*, 16th Edition (http://www.kayelaby.npl.co.uk).

Index

Note: Page numbers in *italic* refer to figures or tables

Surface Analysis – The Principal Techniques 2nd Edition Edited by John Vickerman and Ian Gilmore
© 2009 John Wiley & Sons, Ltd

Printed and bound by CPI Group (UK) Ltd, Croydon, CR0 4YY

27/10/2024

14580165-0004